# — Handbook of —
# Molecular and
# Cellular Methods
# in
# Biology and
# Medicine

# Handbook of Molecular and Cellular Methods in Biology and Medicine

Peter B. Kaufman, Ph.D.
Department of Biology and Bioengineering Program
University of Michigan
Ann Arbor, Michigan

William Wu, Ph.D.
Medical School
University of Michigan
Ann Arbor, Michigan

Donghern Kim, Ph.D.
Agricultural Biotechnology Institute
Rural Development Administration
Suwon, Republic of South Korea

Leland J. Cseke, M.Sc.
Department of Biology, Cellular and Molecular Biology Group
University of Michigan
Ann Arbor, Michigan

CRC Press
Boca Raton   London   Tokyo

**Library of Congress Cataloging-in-Publication Data**

Kaufman, Peter B.
 Handbook of molecular & cellular methods in biology & medicine / Peter B. Kaufman, William Wu,
Donghern Kim, Leland Cseke.
  p. cm.
 Includes bibliographical references and index.
 ISBN 0-8493-2511-0
 1. Molecular biology—Laboratory manuals. 2. Cytology—Laboratory manuals.
I. Wu, William. II. Kim, Donghern. III. Title. IV. Title: Handbook of molecular and cellular methods in
biology and medicine.
QH506.K38 1995
574.8′8′072—dc20                     94-22515
                                    CIP

No claim to original U.S. Government works
International Standard Book Number 0-8493-2511-0
Library of Congress Card Number 94-22515
Printed in the United States of America  2 3 4 5 6 7 8 9 0
Printed on acid-free paper

# Preface

This book was conceived and written to help researchers, teachers, and students properly understand and use the major techniques and methods employed in cellular and molecular biology today. Some protocols are widely dispersed in various books and scientific journals, which are not always readily accessible. Thus, one of our objectives is to bring these diverse protocols together within a logical framework of 21 chapters so that they are available to anyone interested in cellular and molecular biology.

The book covers a wide range of current techniques and methods that have been developed and well-tested. The methods and protocols described in appropriate chapters include: isolation and purification of DNA, RNA, and proteins; preparation of hybridization probes; blotting techniques of nucleic acids and proteins; construction of cDNA and genomic DNA libraries; DNA sequencing; PCR technology; DNA recombination and mutagenesis; transcription *in vitro* and translation *in vitro*; gene cloning, gene transfer, and analysis of gene expression in transgenic animals and plants; cell culture and tissue culture; and microscopy as well as bioseparation techniques. Each chapter covers the principle(s) behind the various methods and techniques, detailed step-by-step description of each protocol, and applications of protocols to different systems. Many chapters include a troubleshooting guide. We have found that many of the currently available books in cellular and molecular biology contain only protocols to different systems. Unfortunately, many fail to explain the principles and concepts underlying the methods outlined or to inform the reader of possible pitfalls in the methods described. We intend to fill these gaps. It is our hope that this book is as useful as possible to those who are interested in reading about, or using a particular method in cellular and molecular biology.

For the information of the reader, Chapters 1 through 16 and Chapter 21 were written by William Wu. Chapters 17 and 18 were written by Donghern Kim, Chapter 19 by Peter Kaufman, and Chapter 20 by Leland Cseke and Peter Kaufman. Conclusively, Peter Kaufman made important contributions to each of the chapters.

**P.B. Kaufman**
**W. Wu**
**D. Kim**
**L. Cseke**

# Authors

**Peter B. Kaufman, Ph.D.,** is a Professor of Biology in the Plant Cellular and Molecular Biology Program of the Biology Department and a member of the faculty of the Bioengineering Program at the University of Michign, Ann Arbor. Dr. Kaufman received his B.Sc. degree from Cornell University, Ithaca, New York and his Ph.D. degree in 1954 in plant biology at the University of California, Davis.

Dr. Kaufman is a Fellow of the American Association for the Advancement of Science and Secretary-Treasurer of the American Society for Gravitational and Space Biology. He served on the Editoral Board of *Plant Physiology* for ten years, and is the author of more than 190 research papers. Dr. Kaufman has published 7 professional books to date and teaches a popular course in Plant Biotechnology yearly at the University of Michigan. He has received research grants from the National Science Foundation, the National Aeronautics and Space Administration, and Parke-Davis Pharmaceutical Research Laboratories in Ann Arbor, Michigan.

**William Wu, Ph.D.,** is a Research Investigator in the Department of Anatomy and Cell Biology in the Medical School at the University of Michigan, Ann Arbor. Dr. Wu received his B.Sc. and M.Sc. in biology from Hunan Normal University, the People's Republic of China in 1981 and 1984, respectively. In 1992, he received his Ph.D. degree in the Molecular and Cellular Biology Program from Ohio University, Athens, Ohio.

Dr. Wu has presented and published over 28 research papers at scientific meetings and in national and international journals. He was awarded a Lee Foundation Fellowship in the U.S., and is a member of American Association for the Advancement of Science. Dr. Wu has extensive research experience in molecular and cellular biology in plant and animal systems. He is currently working on gene overexpression, gene underexpression by antisense DNA/RNA, and gene targeting by DNA homologous recombination in mammalian cells and mice.

**Donghern Kim, Ph.D.,** is a Research Scientist at the Agricultural Biotechnology Institute, Suwon, South Korea. Dr. Kim received his Ph.D. degree in 1993 in plant cellular and molecular biology in the Department of Biology at the University of Michigan.

Dr. Kim received a five-year Graduate Fellowship from the Rockefeller Foundation. He has presented and published more than 25 research papers in meetings and professional journals. He served as Graduate Laboratory Teaching Assistant in the course, Plant Biotechnology, at the University of Michigan and is a co-author of the *Plant Biotechnology Laboratory Manual* (1993) for this course. He is currently working on signal transduction and gene transfer and expression in plants.

**Leland J. Cseke, M.Sc.,** is a Ph.D. student in the Plant Cellular and Molecular Biology Program in the Biology Department at the University of Michigan, Ann Arbor. Mr. Cseke received his B.Sc. degree, specializing in biochemistry, from Michigan Technological University, Houghton, Michigan. He received his M.Sc. degree in cellular and molecular biology from the University of Michigan, Ann Arbor in 1994.

Mr. Cseke has presented his research findings on the physiological effects of stress on secondary metabolite production at several scientific meetings. He has been a Graduate Teaching Assistant in several biotechnology courses at the University of Michigan, including Plant Biotechnology, and he is a co-author of the *Plant Biotechnology Laboratory Manual*. He is currently working on the isolation and characterization of the genes involved in scent compound production and emission by flowers.

# Acknowledgments

We wish to acknowledge and express our thanks and appreciation to the following individuals who assisted us in the preparation of this book: David Bay and Dr. Najati S. Ghosheh for their photographic work; Arlene O'Sullivan and Debbie Hall for their assistance with word processing and preparation of some drawings used in the figures; and Wilbur C. Bigelow for his text on principles of operation of the scanning (SEM) and transmission (TEM) electron microscopes.

# Contents

Chapter

# 1

# Isolation and Purification of DNA

## Contents

DNA isolation and purification are essential techniques for molecular biology studies. The quality and integrity of DNA isolated directly affect the results of scientific research.[1-4] Recently, a number of methods have been developed for extraction of DNA from different cell lines and tissue types of organisms.[4-9] Each method has its own strengths and weaknesses. The present chapter, based on our experience, describes in detail the well-developed protocols that are used for isolation and purification of DNA from animals, bacteria, lambda phages, and plants.

# I.     Isolation and Purification of Genomic DNA from Animals

Because animal cells lack cell walls and chloroplasts as compared to plant cells, it is relatively easy for the cells to be lysed and for the DNA to be purified without chloroplast DNA contamination. The following protocols describe in detail the purification of animal DNA that is suitable for restriction enzyme digestion, genomic library construction in λDNA or cosmid DNA vectors, and Southern blot analysis. In order to obtain high-molecular weight DNA, certain precautions must be taken. All glassware, plastic pipette tips, centrifuge tubes, cell scrapers or policemen, solutions, and buffers should be autoclaved or sterile-filtered to avoid DNase contamination. Molecular biology grade or ultrapure chemicals or reagents are strongly recommended to be used. Gloves should be worn during isolation procedures. Vigorous shaking should be avoided to prevent DNA from shearing.

## *Protocol A: Extraction of Genomic DNA with Organic Solvents*

1.   Harvest tissue or collect cells and add lysis buffer to the samples.

**Extraction from fresh blood cells:**   Collect cells from fresh blood (20 ml per extraction) by centrifuging at 1000 rpm for 15 min at room temperature and carefully decant the supernatant containing lysed red cells without nuclei. Add 1 volume of phosphate-buffered saline (PBS) to the remainder containing white blood cells and centrifuge at 1000 rpm for 15 min at room temperature. Carefully decant the supernatant and resuspend the cells in 15 ml of acid citrate dextrose solution (ACD). Incubate at 37°C for 1 to 1.5 h prior to step 2.

**Extraction from frozen blood cells:**   Thaw frozen blood sample (20 ml per extraction) in a water bath at 22°C and transfer the sample to a fresh centrifuge tube. Add an equal volume of PBS and centrifuge at 1200 rpm for 15 min at room temperature. Decant the supernatant containing lysed red cells and resuspend the white blood cells in 15 ml of ACD solution. Incubate the sample at 37°C for 1 to 1.5 h. Proceed to step 2.

**Extraction from cells grown in monolayers:**   Rinse the confluent monolayers of cells twice with ice-cold PBS. Add 10 to 15 ml ice-cold PBS and carefully scrape the cells, using a sterile policeman or cell scraper, into a clean tube. Collect the cells by centrifugation at 1000 rpm for 10 min at 4°C. Resuspend the cells in 8 volumes of ice-cold PBS and centrifuge at 1000 rpm for 10 min at 4°C. Resuspend the cells at $4 \times 10^7$ cells/ml in TE buffer. Transfer the cell suspension to a clean tube or a flask and add 9 ml of lysis buffer per milliliter of the cell suspension. Incubate the mixture at 37°C for 1 to 1.5 h. Carry out step 2.

**Extraction from cells grown in suspension:**   Collect the cells by centrifugation at 1000 rpm for 10 min at 4°C and wash them twice with 1 volume of ice-cold PBS without calcium. Re-collect the cells by centrifugation at 1000 rpm for 10 min at 4°C. Resuspend the cells at a concentration of $4 \times 10^7$ cells/ml in TE buffer (pH 8.0). Add 10 volumes of lysis buffer and incubate at 37°C for 1 to 1.5 h. Proceed to step 2.

**Extraction from tissue:**   Harvest fresh and soft tissue using sterile scissors or razor blade and immediately freeze in liquid nitrogen. Store at –80°C until use. One suitable method for genomic DNA extraction is to grind the tissue in a clean mortar with pestle using liquid nitrogen. Keep adding liquid nitrogen and grinding until a fine powder is obtained. Allow the liquid nitrogen to evaporate for a few minutes, but never warm or thaw the powder too long before adding lysis buffer. Transfer the powder, little by little, to a 30-ml Corex centrifuge tube or a sterile beaker containing 8 to 10 volumes of lysis buffer. Let the powder spread over the surface of the lysis buffer and gently shake to submerge the material. Incubate the tube at 37°C for 1 to 1.5 h. Proceed to step 2.

2.  Add proteinase K (20 mg/ml in dd.H$_2$O) to a final concentration of 100 µg/ml of the lysed-cell suspension from step 1 and gently mix well.

3.  Incubate the mixture in a water bath at 50°C for 2 to 4 h. Gently swirl the viscous mixture every 20 min.

4.  Allow the mixture to cool to room temperature and transfer the mixture into a Corex centrifuge tube. Add 1 volume of 0.5 *M* Tris-buffer-saturated (pH 8.0) phenol and mix well by gently inverting the tube for 5 min.

**Caution:** *Phenol is very toxic. Gloves should be worn when working with this chemical. After use, the waste phenol must be collected in a special container.*

5.  Centrifuge at 6000 × *g* at room temperature for 15 min and carefully and slowly transfer the top, viscous, aqueous phase to a fresh tube with a wide-bore pipette.

6.  Repeat phenol extraction two to three times as in steps 4 and 5 in order to maximally remove proteins and cellular debris.

7.  Add 0.5 volume of 7.5 *M* ammonium acetate or 0.15 volume of sodium acetate (pH 5.2) to the aqueous phase. Slowly pour the supernatant into a fresh tube or beaker containing 2 to 2.5 volumes (of the supernatant volume) of 100% chilled ethanol. **Do not vortex or invert the tube or beaker until the DNA precipitate floats up to the surface of the ethanol** (15 to 30 min at room temperature).

8.  Carefully "fish" out the DNA into a fresh tube with a glass hook. If the DNA precipitate becomes fragmented, centrifuge the DNA at 5000 × *g* for 5 min at 4°C. Briefly and gently rinse the DNA with 4 ml of 70% ethanol and dry the sample under vacuum for 15 to 30 min.

9.  Resuspend the DNA in an appropriate amount of TE buffer (1 ml/4 × 10$^7$ cells). Because of its high molecular weight, the DNA is usually not easily resuspended. In that case, the DNA sample can be placed at 45 to 50°C for 15 to 30 min with gentle and occasional shaking until it is dissolved. Keep the tube open to let remaining ethanol evaporate. At this stage, the DNA should be very pure with a size of 100 to 200 Kb.

10. Measure the quantity and quality of the DNA using a UV-visible spectrophotometer at the wavelength of 260 and 280 nm. A pure DNA should have a ratio of 1.85 to 2.0 of $A_{260}/A_{280}$ reading numbers. A ratio of $A_{260}/A_{280}$ less than 1.75 means that a significant amount of proteins remains in the DNA sample. In this case, add SDS to a concentration of 0.5% to denature remaining proteins and extract the proteins by repeating steps 2 to 10. Store the DNA sample at –20°C. The quality and size of the DNA preparation may be also checked by 0.3 to 0.8% of agarose gel electrophoresis, by pulsed-field gel electrophoresis, or by electrophoresis through a 0.3% agarose gel poured on a 1% agarose support.

The procedures of DNA measurement are as follows:

a.  Turn on the UV spectrophotometer and set the wavelengths at 260 and 280 nm according to the instructions. Some spectrophotometers (e.g., UV 160U, Shimadzu) contain a computerized monitoring unit that can automatically and simultaneously measure DNA at wavelengths of 260 and 280 nm, and calculate and display DNA concentration together with the ratio of $A_{260}/A_{280}$ reading numbers on the computer screen. If such a spectrophotometer is not available, DNA concentration can be measured by other, relatively simple UV spectrophotometers such as the Hitachi model 100–10 UV-visible spectrophotometer. The disadvantages are that the DNA first have to be measured and recorded at 260 nm and then at 280 nm, and that the DNA concentration and the ratio of $A_{260}/A_{280}$ reading numbers have to be calculated manually.

b. Set up a reference using blank solution. Depending on the DNA sample, add 1 ml of TE buffer (pH 8.0) or dd.$H_2O$ to a clean cuvette or to each of two cuvettes (one half to three quarters full), depending on whether one or two cells are available in the spectrophotometer. Insert the cuvette(s) into cuvette holder(s) in the sample compartment with the optical (clear) sides of the cuvette(s) facing the light path. Close the sample compartment cover and adjust the number to 0.000 (either manually or by pressing the "Auto Zero" button, depending on the specific spectrophotometer).

*Notes:* *Gloves should be worn when handling cuvettes. Cuvettes should be rinsed with 95% ethanol followed by dd.$H_2O$, and wiped dry with Kimwipe paper prior to reuse. Unclean cuvettes may contain contaminated material that may affect readings between samples.*

c. Check for mismatching of cuvettes by filling cuvette(s) with blank solution and recording the absorbance (positive or negative number). Make appropriate additive (for negative number) or subtractive (for positive number) corrections to future DNA sample readings. In most cases, however, this step is optional.

d. In a clean cuvette appropriately dilute the DNA sample to be measured in blank solution to a total volume of 1 ml. For example, for 2, 5, 10, 15, 20, or 25 μl of DNA sample, the blank solution should be 998, 995, 990, 985, 980, or 975 μl, respectively. Wrap the cuvette in a piece of Parafilm and mix well by inverting the cuvette two to three times.

e. Insert the sample cuvette and read the absorbance of the sample cuvette against the blank cuvette according to the spectrophotometer instructions. If only one cell is available, remove the reference cuvette after adjusting the number to 0.000 and insert the sample cuvette.

f. After measuring DNA sample(s) at 260 and 280 nm, respectively, calculate the ratio of $A_{260}/A_{280}$ reading numbers and concentration for each DNA sample. Some spectrophotometers (e.g., UV 160U, Shimadzu) can automatically calculate, display, and print out the ratio and concentration for each DNA sample. A pure DNA should have a ratio of 1.85 to 2.0 of $A_{260}/A_{280}$ reading numbers. The concentration of a DNA sample is calculated as follows:

$$DNA\ (\mu g/\mu l) = A_{260}\ reading\ number \times 50\ \mu g/ml \times dilution\ factor$$

For example, if 25 μl of DNA is diluted to 1000 μl for measuring and its $A_{260}$ reading number is 0.4000, then

$$
\begin{aligned}
DNA\ (\mu g/\mu l)\ &= A_{260}\ reading\ number \times 50\ \mu g/ml \times dilution\ factor\\
&= 0.4000 \times 50\ \mu g/ml \times 1000/25\\
&= 800\ \mu g/ml\\
&= 0.8\ \mu g/\mu l
\end{aligned}
$$

## Protocol B: Genomic DNA Extraction by Formamide

This is a relatively simple and low-cost method as compared with Protocol A. The DNA isolated by this method has a molecular weight higher than 200 Kb, which is suitable for making a genomic library in cosmic vectors in addition to doing restriction enzyme digestion and Southern blotting. The procedures are described as follows:

1. Conduct steps 1 to 3 as described in Protocol A.
2. Allow the suspension of lysed cells to cool to 15°C, and add 0.7 volume of denaturation buffer to the suspension. Gently mix using a glass rod, and then place the mixture overnight at 15°C.

3.  Pool and dialyze the viscous suspension in a collodion bag (Sartorius SM 13200E or equivalent) three times against 4 l of buffer A and five times against 4 l of buffer B.

4.  Measure the quantity and quality of the DNA with a spectrophotometer at 260 and 280 nm (see Protocol A). A pure DNA should have a ratio of 1.85 to 2.0 of 260/280 nm reading numbers. The sample can be stored at –20°C until used. If the ratio $A_{260}/A_{280}$ is less than 1.75, repeat steps 2 to 4.

5.  The quality and size of the DNA preparation may be checked by 0.3 to 0.8% of agarose gel electrophoresis, by pulsed-field gel electrophoresis, or by electrophoresis through a 0.3% agarose gel poured on a 1% agarose support.

## Reagents Needed

### Lysis Buffer

10 m$M$ Tris-HCl, pH 8.0
100 m$M$ EDTA, pH 8.0
20 µg/ml Pancreatic RNase
0.5% SDS

### PBS (pH 7.4) without Calcium

NaCl (8.07 g)
KCl (0.201 g)
$Na_2HPO_4$ (1.15 g)
$KH_2PO_4$ (0.204 g)
Dissolve well after each addition in 800 ml dd.$H_2O$, adjust the pH to 7.4, and add dd.$H_2O$ to a final volume of 1000 ml. Autoclave.

### ACD Solution

Citric acid (4.8 g)
Sodium citrate (13.2 g)
Glucose (14.7 g)
Adjust to a final volume of 1000 ml with dd.$H_2O$.

### Denaturation Buffer

80% Deionized formamide (v/v)
0.8 $M$ NaCl
20 m$M$ Tris-HCl, pH 8.0

### Buffer A

0.1 $M$ NaCl
20 m$M$ Tris-HCl, pH 8.0
10 m$M$ EDTA, pH 8.0

### Buffer B

10 m$M$ NaCl
10 m$M$ Tris-HCl, pH 8.0
1 m$M$ EDTA, pH 8.0

## II.    Isolation and Purification of λDNA

λDNAs are commercially available and are commonly used vectors in cDNA and genomic DNA cloning. After "fishing" out putative cDNA clone(s) or specific genes by screening cDNA or genomic libraries using specific probes, the next step is to isolate the recombinant λDNAs for subcloning or DNA analysis. That is the technique described in this section. There are, so far, two protocols for phage lysate preparation. One is the plate method; the other is liquid culture. With the former it is easier to monitor the growth of phage, whereas the latter method is easier to scale-up. Both methods work very well in our lab. λDNA can be isolated and purified by either mini-preparation or large-scale preparation procedures, or by the CsCl gradient method. Many modifications and shortcuts have been emerging in recent years. For example, λDNA may also be purified using lambdaSorb phage adsorbent, which is available from Promega Corporation (Wisconsin). The reagent is composed of an optimized conjugate of fixed *Staphylococcus aureus* cells and rabbit polyclonal antibodies raised against lambda bacteriophage particles. This method has a high degree of specificity, and it is simple and rapid.

### A. Phage Lysate Preparation by the Plate Method

1.   Prepare LB plates, in a sterile laminar flow hood, by slowly pouring 35 ml of LB agar medium into each 85- or 100-mm-diameter plastic Petri plate. Remove any bubbles with a pipette and keep the plates open for 10 to 15 min before covering them. Allow the medium in the plates to harden for about 1 to 2 h and dry for 2 days at room temperature before use.

*Notes:*   *Do not use plates that are too wet or too dry because the bacteria and bacteriophage cannot grow well under these conditions. The prepared plates can be stored at room temperature for up to 10 days or at 4°C for up to 1 month. Make sure no contamination from bacteria or fungi occurs on the surface of the plates before use.*

2.   Based on the titering experiment of the plaque-forming units (pfu) of the plaque eluate, place about $1 \times 10^5$ to $1 \times 10^6$ pfu of the eluate into a microcentrifuge tube and bring the volume to 50 to 100 μl with phage buffer for each 100-mm plate. Mix the bacteriophage sample with 100 μl of a fresh, overnight culture of appropriate plating bacterial strains such *E. coli* Y1089, Y1090, or LE 392. Incubate the tubes at 37°C for 20 to 30 min to allow the phage to absorb the bacteria.

3.   Add 2.8 ml of melted top agarose into each glass tube and keep in a 50°C water bath until use. Add the absorbed bacteria (about 0.15 to 0.2 ml) prepared in step 2 to each of the glass tubes containing the top agarose solution. Mix, and **immediately** pour onto the center of the LB plate. **Quickly and evenly** distribute the top agarose mixture on the surface of the bottom agar by gently tilting the plate.

*Notes:*   *This step should be performed properly and rapidly. Done too slowly, the top agarose mixture will harden before being poured onto the LB plate. Uneven distribution of the top agarose mixture will affect the bacteria's and bacteriophage's growth. Bottom agar should be avoided due to the presence of inhibitors in it that interfere with restriction enzyme digestions.*

4.   Allow the top agarose to harden for 10 min, invert the plates, and incubate at 37°C for 10 to 12 h or until the plaques become confluent.

5.   Overlay each plate with 4 ml of SM buffer. Carefully scrape the top agarose with a clean spatula into a 30-ml Corex centrifuge tube. Break up the agarose with the spatula and incubate at room temperature with slow shaking for 30 min.

*Note:*    *Never scrape the bottom agar, which contains potential inhibitors for restriction enzymes.*

6.    Centrifuge at 10,000 × g for 10 min at 4°C to pellet the agarose and bacterial debris.

7.    Carefully transfer the supernatant into a fresh tube. Add chloroform to 0.2% if long-term storage is desired. Store at 4°C until use for isolation of λDNA (see Sections II.C and II.D).

## B. Phage Lysate Preparation by the Liquid Method

1.    For each 100 ml of culture, place 20 to 25 μl of $1 \times 10^5$ pfu of plaque eluate into a microcentrifuge tube and add 480 to 475 μl of phage buffer. Mix well with 0.5 ml of fresh, overnight-cultured bacteria, and allow the bacteria and phage to adhere to each other by incubating at 37°C for 30 min.

2.    Add the above mixture into a 250- or 500-ml sterile flask containing 100 ml of LB medium prewarmed at 37°C that is supplemented with 1 ml of 1 $M$ MgSO$_4$. Incubate at 37°C with shaking at 260 rpm until lysis occurs. It usually takes 9 to 11 h for lysis to occur. The medium should be cloudy after several hours of culture and then be clear upon cell lysis. Cellular debris also becomes visible in the lysed culture.

*Notes:*    *There is a density balance between bacteria and bacteriophage. If the bacteria density is much over that of bacteriophage, it takes longer for lysis to occur, or no lysis takes place at all. In contrast, if bacteriophage concentration is much over that of bacteria, lysis is too quick to be visible at the beginning of the incubation, and later on, no lysis happens. The proper combination of bacteria and bacteriophage used in step 1 will assure success. In addition, careful observations should be made after 9 h of culture because the lysis is usually quite rapid after that time. Incubation of cultures should be stopped after lysis occurs. Otherwise, the bacteria grow continuously and the cultures become cloudy again. If that happens, it will take a long time to see lysis, or no lysis takes place at all.*

3.    Immediately centrifuge at 9000 × g for 10 min at 4°C to spin down the cellular debris. Transfer the supernatant to a fresh tube. Store at 4°C until use for the phage DNA purification (see Sections II.C and II.D).

## C. Purification of λDNA Using LambdaSorb Phage Adsorbent (Available from Promega Corporation)

1.    Add 0.15 ml of the well-resuspended adsorbent per 10 ml of phage lysate prepared in a 30-ml Corex centrifuge tube. Mix for 40 min with slow shaking (60 rpm) at room temperature.

2.    Centrifuge at 11,000 × g for 10 min at 4°C and carefully remove the supernatant. Resuspend the pellet in 1 ml of SM buffer per 10 ml of initial phage lysate.

3.    Centrifuge the suspension at 11,000 × g for 8 min at 4°C and decant the supernatant.

4.    Resuspend the pellet in 1 ml of SM buffer per 10 ml of initial phage lysate and repeat step 3.

5.    Resuspend the pellet in 0.5 ml of release buffer per 10 ml of initial phage lysate and incubate at 67°C for 6 min to release the phage DNA.

6.    Centrifuge at 10,000 × g for 10 min at 4°C to pellet the adsorbent. Carefully transfer the supernatant containing the phage DNA to a fresh tube.

7.    Add 5 μl of 5 $M$ NaCl per 0.5 ml of supernatant and mix well. Extract phage proteins by adding an equal volume of TE-saturated phenol/chloroform to the supernatant. Mix for 1 min by vortexing and centrifuge at 10,000 × g for 10 min at 4°C.

*Caution:*   *Phenol is a toxic compound and should be used only in the fume hood. Gloves should be worn when working with this reagent.*

8.   Slowly and carefully transfer the upper, aqueous phase to a fresh tube and add an equal volume of chloroform:isoamyl alcohol (24:1). Mix well by vortexing for 1 min.

*Caution:*   *The bottom phenol/chloroform should be decanted into a special container. Do not dump into sink.*

9.   Centrifuge at 10,000 × $g$ for 10 min at 4°C and gently transfer the upper, aqueous phase into a fresh tube. **Never remove any white material that lies between the two phases.**

10.  **For the DNA purified from a plate lysate (see Section II.A),** add 0.5 volume of 5 $M$ NaCl, mix and add 0.33 volume of 30% polyethylene glycol (PEG; mol wt 8000) in 2 $M$ NaCl. Mix by vortexing and allow the phage DNA to precipitate on ice for 60 min. **For the DNA purified from a liquid lysate (see Section II.B),** add 0.5 volume of 7.5 $M$ ammonium acetate and 2 volumes of chilled 100% ethanol. Mix well and place on ice for 60 min to precipitate the phage DNA. The reason of using two different methods to precipitate the λDNA is that the DNA isolated by the plate method is usually difficult to digest with restriction enzymes such as *EcoR* I. This problem can be solved by PEG precipitation.

11.  Centrifuge at 12,000 × $g$ for 15 min at 4°C and decant the supernatant. **Gently and briefly** rinse the DNA pellet with 4 ml of cold 70% ethanol to remove ammonium acetate or PEG. Dry the pellet under vacuum for 20 min and dissolve the λDNA in 50 to 100 µl TE buffer. Take 2 to 5 µl to measure the DNA concentration (see Protocol A) and store at –20°C until use.

*Notes:*   *Make sure that ethanol is completely removed during the vacuum drying. Any ethanol remaining will slow down the dissolving of the DNA in TE buffer and may inhibit restriction enzyme digestion.*

12.  The λDNA from step 11 can be suitable for restriction enzyme digestion, subcloning, and analysis. However, for DNA sequencing, the λDNA purified here should be treated with DNase-free *RNase* A to final concentration of 100 µg/ml to avoid occurrence of nonspecific bands. Place the sample at 37°C for 30 min.

13.  Add NaCl to a final concentration of 50 m$M$. Extract *RNase* A with 1 volume of TE-saturated phenol/chloroform. Mix and spin in a microcentrifuge tube at 12,000 × $g$ for 5 min.

14.  Transfer the upper, aqueous phase to a fresh tube and add an equal volume of chloroform:isoamyl alcohol (24:1). Mix by vortexing and spin as in step 13.

15.  Carefully transfer the upper, aqueous phase to a fresh tube and add 2.5 volumes of chilled 100% ethanol. Mix and precipitate at –20°C for 60 min.

16.  Centrifuge at 12,000 × $g$ for 10 min and carefully remove the supernatant. Dry the DNA pellet under vacuum for 15 min and resuspend the DNA in 50 µl TE buffer or sterile water. Take 4 µl of the sample to measure the concentration of the DNA and store at –20°C until use.

## D. Mini-Preparation Purification of λDNA

1.   Add *RNase* A and *DNase* I to the lambda lysate supernatant obtained from Sections II.C, each to a final concentration of 1 µg/ml. Place at 37°C for 30 min.

*Notes:*   *RNase A functions to hydrolyze the RNAs. DNase I will hydrolyze chromosomal DNA but not phage DNA, which is packed.*

2. Precipitate the phage particles by adding 1 volume of phage precipitation buffer and incubate for at least 1 h on ice.

3. Centrifuge at $12,000 \times g$ for 15 min at 4°C and allow the pellet to dry at room temperature for 5 min.

4. Resuspend the phage particles with 1 ml of phage buffer per 10 ml of initial phage lysate and mix by vortexing.

5. Centrifuge at $8000 \times g$ for 4 min at 4°C to remove debris and carefully transfer the supernatant into a fresh tube.

6. Extract the phage particle proteins with an equal volume of TE-saturated phenol/chloroform. Mix for 1 min and centrifuge at $10,000 \times g$ for 5 min.

7. Transfer the top, aqueous phase to a fresh tube and extract the supernatant one more time as in step 6.

8. Transfer the top, aqueous phase to a fresh tube and extract once with 1 volume of chloroform:isoamyl alcohol (24:1). Vortex and centrifuge as in step 6.

9. **Carefully** transfer the upper, aqueous phase to a fresh tube and add 1 volume of isopropanol or 2 volumes of chilled 100% ethanol. Mix well and allow the DNA to precipitate at −70°C for 30 min or −20°C for 2 h.

10. Centrifuge at $12,000 \times g$ for 10 min at 4°C and drain the supernatant. Briefly wash the pellet with 2 ml of 70% ethanol and dry the pellet under vacuum for 15 min. Resuspend the DNA in 50 to 100 μl of TE buffer. Measure the DNA concentration and store at 4 or −20°C until use.

## E. Large-Scale Purification of λDNA by the CsCl Gradient Method

1. Place the lysate at 4°C for 5 min to reduce the possible growth of bacteria; then move the lysate to room temperature. To hydrolyze chromosomal DNA and RNAs, add *RNase* A and pancreatic *DNase* I to the lysate, each to a final concentration of 1 μg/ml. Incubate for 30 min at 37°C.

2. Add solid NaCl to a final concentration of 1 $M$ and dissolve. Place the sample on ice for 1 h. This helps the dissociation of bacteriophage particles from cellular debris and increases the precipitation of bacteriophage particles from PEG.

3. Centrifuge at $11,000 \times g$ for 10 min at 4°C. Transfer the supernatant into a fresh tube and discard the bacterial debris.

4. Add solid PEG (mol wt 8000) to a final concentration of 10% (w/v) and dissolve by stirring on a magnetic stirrer at room temperature.

5. Place on ice for 2 h to allow the bacteriophage particles to precipitate.

6. Centrifuge at $11,000 \times g$ for 10 min at 4°C and decant the supernatant. Allow the pellet (bacteriophage particles) to dry at room temperature to remove the remaining fluid.

7. Resuspend the pellet in SM (0.2 ml/10 ml of the initial supernatant).

8. Remove the PEG and cellular debris with 1 volume of chloroform and vortex for 1 min.

9. Centrifuge at $3000 \times g$ for 15 min at 4°C and carefully transfer the top, aqueous phase containing the bacteriophage particles to a clean tube.

10. Add exactly 0.5 g/ml of solid CsCl to the aqueous phase and **gently** mix to dissolve. **Carefully** layer the mixture onto the top of the prepared CsCl step gradients in Beckman SW41 or SW28 (or equivalent) clear polypropylene or polycarbonate centrifuge tubes.

11. Centrifuge at 22,000 rpm for 2 h at 4°C in a Beckman SW41 or SW28. Slowly remove the centrifuge tubes and localize the bands. A bluish band of the bacteriophage particles is usually visible at the interface between the 1.45 g/ml layer and the 1.50 g/ml layer. Place the tube against a black background and shine a light from above to help examine the band. The cellular debris band should be at the interface between the 1.45 g/ml layer and the sample layer.

12. Wipe the outside of the tube dry with ethanol and attach a piece of Scotch™ tape to the outside of the tube. Remove the particle band by **carefully** puncturing the side of the tube through the tape using a 21-gauge needle and **slowly** collect the band. Alternatively, the band can be collected using a sterile pipette or a Pasteur pipette. Starting from the sample layer, carefully remove the sample layer, cellular debris, and the 1.45 g/ml layer. The particle band can then be easily collected with a fresh pipette tip.

13. Place the suspension of the particle band in an ultracentrifuge tube and fill two thirds of the tube with CsCl solution (1.5 g/ml in SM). Centrifuge at 38,000 rpm for 24 h at 4°C (Beckman rotor Ti50) or at 35,000 rpm for 24 h at 4°C (Beckman rotor SW50.1 or equivalent).

14. Remove the band of bacteriophage particles as in step 12 and store at 4°C until use.

15. Remove CsCl from the bacteriophage particle suspension by dialysis for 2 h at room temperature with a 1000-fold volume of the dialysis buffer. Change the buffer in the middle of the dialysis.

16. Transfer the dialyzed sample to a fresh tube and add proteinase K to a final concentration of 50 µg/ml. Incubate at 56°C for 1 h.

17. Extract proteinase K with 1 volume of Tris-saturated (pH 8.0) phenol/chloroform. Mix and centrifuge at $10,000 \times g$ for 10 min.

18. Transfer the upper, aqueous phase to a fresh tube and add an equal volume of chloroform:isoamyl alcohol (24:1). Mix by vortexing and centrifugation as in step 17.

19. Carefully transfer the upper, aqueous phase to a fresh tube and add 2.5 volumes of chilled 100% ethanol. Mix and precipitate at –20°C for 1 h.

20. Centrifuge at $12,000 \times g$ for 10 min and carefully remove the supernatant. Dry the DNA pellet under vacuum for 15 min and resuspend the DNA in 50 to 100 µl TE buffer or sterile water. Take 4 µl of the sample to measure the concentration of the DNA (see Protocol A) and store at –20°C until use.

## Reagents Needed

### SM Buffer (per liter)

> 50 mM Tris-HCl, pH 7.5
> 100 mM NaCl
> 8 mM $MgSO_4$
> 0.01% Gelatin (from 2% stock)
> Sterilize by autoclaving.

### LB (Luria-Bertaini) Medium (per liter)

> 10 g Bacto-tryptone
> 5 g Bacto-yeast extract
> 5 g NaCl
> Adjust pH to 7.5 with 2 $N$ NaOH and autoclave.

### LB Top Agarose (200 ml)

> 2 g Bacto-tryptone
> 1 g Bacto-yeast extract
> 1 g NaCl
> 0.6 g Agarose
> Adjust pH to 7.5 with 2 $N$ NaOH and autoclave. When the solution has cooled, add 2 ml of 1 $M$ $MgSO_4$.

## LB Plates

Add 15 g of Bacto-agar to 1 liter of LB medium and autoclave. When the medium cools to about 50°C, pour 30 to 35 ml per plate of medium into 100-mm Petri dishes. Allow the agar to harden in a sterile laminar flow hood. Store at room temperature for 10 days or at 4°C for up to 1 month.

## Phage Buffer

20 m$M$ Tris-HCl, pH 7.4
100 m$M$ NaCl
10 m$M$ MgSO$_4$

## Release Buffer

10 m$M$ Tris-HCl, pH 7.8
10 m$M$ EDTA

## Phage Precipitation Solution

30% (w/v) PEG (mol wt 8000) in 2 $M$ NaCl solution

## TE Buffer

10 m$M$ Tris-HCl, pH 8.0
1 m$M$ EDTA

## TE-Saturated Phenol/Chloroform

Thaw crystals of phenol in a 65°C water bath and mix with an equal volume of TE buffer. Allow the phases to separate for about 10 to 20 min. Then mix 1 part of the lower phenol phase with 1 part of chloroform:isoamyl alcohol (24:1). Allow the phases to separate and store at 4°C in a light, tight bottle.

## DNase-Free RNase A

To make DNase-free *RNase* A, prepare a 10 mg/ml solution of *RNase* A in 10 m$M$ Tris-HCl, pH 7.5 and 15 m$M$ NaCl. Boil for 15 min and slowly cool to room temperature. Filter if necessary.

## CsCl Step Gradients

| Density (ρ) (g/ml) | CsCl (g) | SM (ml) | Refractive index (η) |
|---|---|---|---|
| 1.45 | 60 | 85 | 1.3768 |
| 1.50 | 67 | 82 | 1.3815 |
| 1.70 | 95 | 75 | 1.3990 |

The step gradients can be made in Beckman SW41 or SW28 clear polypropylene or polycarbonate centrifuge tubes, either by carefully layering equal volumes of each of the above solutions of decreasing density on top of one another or by layering the solutions of increasing density under one another. Store the gradient at 4°C until use.

### Dialysis Buffer

        10 m$M$ NaCl
        50 m$M$ Tris-HCl, pH 8.0
        10 m$M$ MgCl$_2$

### Others

        DNase I
        Solid CsCl
        Solid NaCl
        Proteinase K

# III.    Isolation and Purification of Plasmid DNA

Plasmids such as pGEM series, pBR322, pBIN series, and pUC series are essential vectors and play an important role in gene subcloning, sequencing, and expression. Therefore, extraction and purification of plasmid DNA is very useful technique in modern molecular biology studies. Plasmids are usually purified from liquid cultures by inoculating an appropriate volume of LB medium with a single bacterial colony picked up from an agar plate. Many of the commercial plasmids are high-copy number vectors. They can replicate up to 100 to 300 copies per cell so that they can be purified from cultures in high yield. However, some vectors such as pBR322 are low-copy number plasmids that need to be selectively amplified by chloramphenicol treatment for several hours. Chloramphenicol can inhibit host protein synthesis, prevent replication of the bacterial chromosome, but increase plasmid DNA copy numbers. There are many methods that are currently applied to isolate plasmid DNA. We describe in detail two protocols that are very commonly used and successful in our and other laboratories.

## A.    Isolation of Plasmid DNA by the Alkaline Method

    This protocol allows for the rapid purification of small to large amounts of plasmid DNA without the need for banding in CsCl gradients. The principle of the procedure is to take advantage of the alkaline denaturation of plasmid and chromosomal DNA and of the selective renaturation of plasmid DNA following neutralization of the solution. The isolated DNA is suitable for use in restriction enzyme digestion, *in vitro* transcription, DNA subcloning, and DNA sequencing by predenaturation of the double-stranded DNA.

### Protocol 1: Mini-Preparation of Plasmid DNA

1. Inoculate a single colony or 10 μl of previously frozen cells containing the plasmid of interest in 5 ml of LB medium and 50 μg/ml of appropriate antibiotics (e.g., ampicillin) depending on the specific antibiotic-resistant gene carried by the specific plasmid. Culture the bacteria at 37°C overnight with shaking at 160 rpm.

2. Add 1.5 ml of the overnight culture into two microcentrifuge tubes and centrifuge at $10,000 \times g$ for 2 min. Remove the liquid and invert the tubes on a paper towel to dry the bacterial pellet for 4 min.

3. Resuspend the pellet by adding 0.1 ml of ice-cold lysis buffer and vortex for 2 min. Incubate the tubes for 5 min at room temperature. This step lyses the bacteria by hyperlytic osmosis and releases the DNA and other contents.

4.  Add 0.2 ml of a freshly prepared alkaline solution and mix by inversion. **Never vortex.** Incubate the tubes on ice for 5 min. The function of this step is to denature the plasmid and chromosomal DNAs and proteins.

5.  Add 0.15 ml of ice-cold potassium acetate solution. Mix by inversion for 20 s and incubate on ice for 5 min. The purpose of this step is to selectively renature the plasmid DNA. Some chromosomal DNA may be partially renatured and bound by proteins, which will be extracted by phenol/chloroform in later steps.

6.  Centrifuge at $12,000 \times g$ for 5 min and gently transfer the supernatant to fresh tubes.

7.  Add *RNase* A (DNase-free) into the supernatant to a final concentration of 20 µg/ml. Incubate the tubes at 37°C for 20 min. *RNase* A degrades total RNAs from the sample.

8.  Add an equal volume of TE-saturated phenol/chloroform to each tube and mix by vortexing for 1 min.

9.  Centrifuge at $12,000 \times g$ for 4 min and transfer the upper, aqueous phase into fresh tubes.

10. Add an equal volume of chloroform:isoamyl alcohol (49:1) and mix by vortexing for 1 min and centrifuge as in step 9.

11. Centrifuge at $12,000 \times g$ for 4 min and transfer the supernatant to a fresh tube.

    Steps 8 to 11 act to remove *RNase* A, proteins, and DNA/proteins.

12. Add 2 to 2.5 volumes of 100% ethanol. Mix by inversion and allow the plasmid DNA to precipitate for 20 min at –80°C or 2 h at –20°C.

13. Centrifuge at $12,000 \times g$ for 10 min. Decant the supernatant and briefly and gently rinse the pellet with 1 ml of prechilled 70% ethanol. Dry the plasmid DNA under vacuum for 20 min.

14. Dissolve the dried plasmid DNA in 25 µl of TE buffer or sterile deionized water. Take 2 to 4 µl to measure the concentration and store the sample at –20°C until use.

## Protocol 2: Large-Scale Preparation of Plasmid DNA

The principle is the same as the mini-preparation method.

1.  Inoculate a single colony or 50 µl of previously frozen cells containing the plasmid of interest in 100 ml of LB medium and 50 µg/ml of appropriate antibiotics (e.g., ampicillin). Incubate the bacteria at 37°C overnight with shaking at 160 rpm.

2.  Add 30 ml of the overnight culture into two 30-ml Corex tubes and centrifuge at $9000 \times g$ for 5 min at 4°C. Remove the liquid and invert the tubes on a paper towel to dry the bacterial pellet for 5 min. In the following steps, the two samples in two tubes are recommended to be always processed separately. In case one sample fails by accident, another one can replace the lost sample.

3.  Resuspend the pellet by adding 2 ml of ice-cold lysis buffer and vortex for 2 min. Incubate the tubes for 5 min at room temperature.

4.  Add 4 ml of a freshly prepared alkaline solution and mix by inversion. **Do not vortex.** Incubate the tubes on ice for 5 min.

5.  Add 3 ml of ice-cold potassium acetate solution, mix by inversion for 20 s, and incubate on ice for 5 min.

6.  Centrifuge at $10,000 \times g$ for 5 min at 4°C and gently transfer the supernatant to fresh tubes.

7.  Add *RNase* A (DNase-free) to the supernatant to a final concentration of 20 µg/ml. Incubate the tubes at 37°C for 30 min.

8.  Add 1 volume of TE-saturated phenol/chloroform to each tube and mix by vortexing for 1 min.

9.  Centrifuge at $10,000 \times g$ for 5 min at 4°C. Now transfer the upper, aqueous phase into fresh tubes.

10. Add 1 volume of chloroform:isoamyl alcohol (49:1) and mix by vortexing for 1 min and centrifuge as in step 9.

11.   Centrifuge at $10,000 \times g$ for 5 min at 4°C and transfer the supernatant to a fresh tube.

12.   Add 2 to 2.5 volumes of 100% ethanol. Mix by inversion and allow the plasmid DNA to precipitate for 30 min at –80°C or 2 h at –20°C.

13.   Centrifuge at $11,000 \times g$ for 15 min at 4°C. Remove the supernatant, and gently rinse the pellet with 6 ml of prechilled 70% ethanol. Dry the plasmid DNA under vacuum for 20 min.

14.   Dissolve the dried plasmid DNA in 200 µl of TE buffer or sterile deionized water. Take 25 µl to measure the concentration and store the sample at –20°C until use.

## Reagents Needed

### LB (Luria-Bertaini) Medium (per liter)

10 g Bacto-tryptone
5 g Bacto-yeast extract
5 g NaCl
Adjust pH to 7.5 with 2 $N$ NaOH and autoclave. Add antibiotics after the autoclaved solution has cooled to less than 50°C.

### TE Buffer

10 m$M$ Tris-HCl, pH 8.0
1 m$M$ EDTA

### Lysis Buffer

25 m$M$ Tris-HCl, pH 8.0
10 m$M$ EDTA
50 m$M$ Glucose

### Denaturing Solution

0.2 $N$ NaOH
1% SDS

### Potassium Acetate Solution (pH 4.8)

Prepare 120 ml of 5 $M$ potassium acetate. Add 23 ml of glacial acetic acid and 57 ml of $H_2O$. Total volume is 200 ml. This solution is 3 $M$ with respect to potassium and 5 $M$ with respect to acetate. Store on ice until use.

### DNase-Free RNase A

Make a 10-mg/ml solution of *RNase* A in 10 m$M$ Tris-HCl, pH 7.5, and 15 m$M$ NaCl. Heat at 100°C for 10 min and cool slowly to room temperature.

### TE-Saturated Phenol/Chloroform

Thaw crystals of phenol in 65°C water bath with occasional shaking. Mix equal parts of TE buffer and thawed phenol. Let the mixture stand until the phases separate at room temperature. Mix an equal part of the lower, phenol phase with an equal part of chloroform:isoamyl alcohol (49:1).

*Chloroform:Isoamyl Alcohol (24:1)*

*Ethanol (100% and 70%)*

## B.    Isolation and Purification of Plasmid DNA by the CsCl Gradient Method

This method, as compared with the protocol described above, is a relatively complicated, time-consuming, and expensive one. It usually needs a large-scale culture of bacteria, and is not suitable for mini-preparation of plasmid DNA. However, many investigators comment that the plasmid DNA purified by CsCl is relatively pure and more suitable for DNA sequencing. The detailed procedure is described below.

### Procedure

1.   Inoculate a single colony or 100 µl of previously frozen cells containing the plasmid of interest in 200 ml of LB medium containing 2 ml of 20% of maltose, 2 ml of 1 $M$ $MgSO_4$, and 50 µg/ml of appropriate antibiotics (e.g., ampicillin). Incubate the bacteria at 37°C overnight with shaking at 160 rpm.

2.   Pellet the bacteria by centrifuging at $9000 \times g$ for 5 min at 4°C. Remove the liquid and invert the tubes on a paper towel to dry the bacterial pellet for 5 min.

3.   Resuspend the pellet by adding 5 ml of ice-cold lysis buffer and vortex for 4 min. Incubate the tubes for 10 min at room temperature.

4.   Add 10 ml of a freshly prepared alkaline solution and mix by inversion. **Do not vortex.** Incubate the tubes on ice for 6 min.

5.   Add 7.5 ml of ice-cold potassium acetate solution, mix by inversion for 30 s, and incubate on ice for 6 min.

6.   Centrifuge at $10,000 \times g$ for 10 min at 4°C and gently transfer the supernatant to fresh tubes.

7.   Add 0.8 volume of isopropanol to the supernatant, mix well, and allow the sample to precipitate for 15 min at room temperature.

8.   Centrifuge at $11,000 \times g$ for 15 min at 4°C, remove the supernatant, and gently rinse the pellet with 6 ml of prechilled 70% ethanol. Dry the plasmid DNA under vacuum for 20 min.

9.   Dissolve the dried plasmid DNA in 4 ml of TE buffer or sterile deionized water. Use immediately or store at 4°C until use for CsCl gradient centrifugation.

10.   Accurately measure the volume of the DNA sample and add 1 g of solid CsCl per milliliter of DNA sample. Warm the mixture at 30°C to dissolve the CsCl and gently mix well.

11.   Add 0.35 ml of 10 mg/ml ethidium bromide (EtBr) for every 5 ml of the DNA/CsCl solution. Immediately and gently mix the solution. The final density of the mixture should be 1.55 to 1.6 g/ml.

*Caution:*   *EtBr is a powerful mutagen. Gloves should always be worn when working with the solutions. The used EtBr solution must by collected in a special container to gradually inactivate the chemical before it is disposed. Stock solutions of EtBr should be stored in lighttight bottles at 4°C.*

12.   Centrifuge at 8000 rpm for 6 min at room temperature with a Sorvall SS34 rotor (or its equivalent).

13.    Gently transfer the clear, red DNA–CsCl mixture under the surface scum into "Quick-seal" ultracentrifuge tubes (Beckman) using a Pasteur pipette or a disposable syringe fitted with a large-gauge needle.

*Notes:*    *Fill the tubes right up to the neck of the tube (about 5.2 ml per tube). If the tube is overfilled, it will cause a problem in sealing. However, if underfilled, large air bubble present after sealing the tube may cause the tube to collapse. Carefully balance the tubes with CsCl balance solution.*

14.    Turn on and warm the sealing machine for 5 min. Place the tubes on a metal sealing plate and cover each tube with an appropriate metal cap. Place each tube, one at a time, right under the sealing rack and push and hold the hot rack against the metal cap of the tube for 30 to 60 s. The neck of the tube gradually melts until a sound is heard from the sealing machine. Quickly remove the tube from the melt rack and place the cap of the tube against a cold metal strip. Press and hold the strip against the cap for about 1 to 2 min to completely seal the tube and then remove the cap.

15.    Attach a cap and a metal adapter to each sealed tube for ultracentrifugation and carefully balance again using Scotch™ tape if necessary. Symmetrically place the tubes into the Beckman vertical rotor and cover the tubes with their own caps and adapters.

16.    Centrifuge the CsCl density gradients at 45,000 rpm for 16 h (VTi65), 45,000 rpm for 48 h (Ti50), 60,000 rpm for 24 h (Ti65), or 60,000 rpm for 24 h (Ti70.1) at 20°C.

*Note:*    *Never centrifuge at 4°C. It will cause the CsCl to precipitate out of solution and reduce the density of CsCl in the gradient.*

17.    Stop centrifugation. Slowly remove the tubes from the rotor and place them in a plastic rack. Use long-wavelength UV light (360 nm) to visualize the relaxed and supercoiled plasmid DNA bands (Figure 1.1).

*Notes:*    *Short-wavelength UV light can cause nicking of the DNA. The DNA band will be too low to be identified. Two bands of DNA are visible in the center of the gradient (see Figure 1.1), which usually contains linear bacterial chromosomal DNA and nicked circular plasmid DNA. The lower band consists of closed circular plasmid DNA. The deep-red pellet on the bottom of the tube consists of EtBr/RNA. The materials on the top surface are usually proteins.*

18.    For removal of the DNA, place the tube in a ringstand rack with a clamp. Locate the DNA bands with 360-nm UV light. With ethanol wipe dry one side of the DNA band area and place a piece of Scotch tape on it and wipe dry. **Slowly and carefully** puncture through the tube with a needle and syringe (16 to 20 gauge) until the needle has penetrated into the closed circular plasmid DNA band. Gently withdraw a minimal volume of DNA sample; then, slowly remove the needle. Transfer the sample into a fresh tube (Figure 1). Discard the tubes containing the rest of the solution in a special container.

19.    Extract EtBr from the DNA sample with one volume of isopropanol. Gently invert the tube several times and allow the phases to separate by gravity. The EtBr usually partitions into the upper isopropanol phase and is removed with a pipette. The lower aqueous phase contains DNA–CsCl with some EtBr in it. It should be reextracted four to five times until no pink color is visible.

20.    Transfer the lower phase into a fresh Corex tube and dilute out the CsCl with 2 volumes of TE buffer. Add 2 volumes of chilled 100% ethanol into the diluted sample and precipitate the DNA at –20°C for 5 h to overnight.

*Note:*    *Never place the tubes at –80°C; otherwise, the CsCl will precipitate out.*

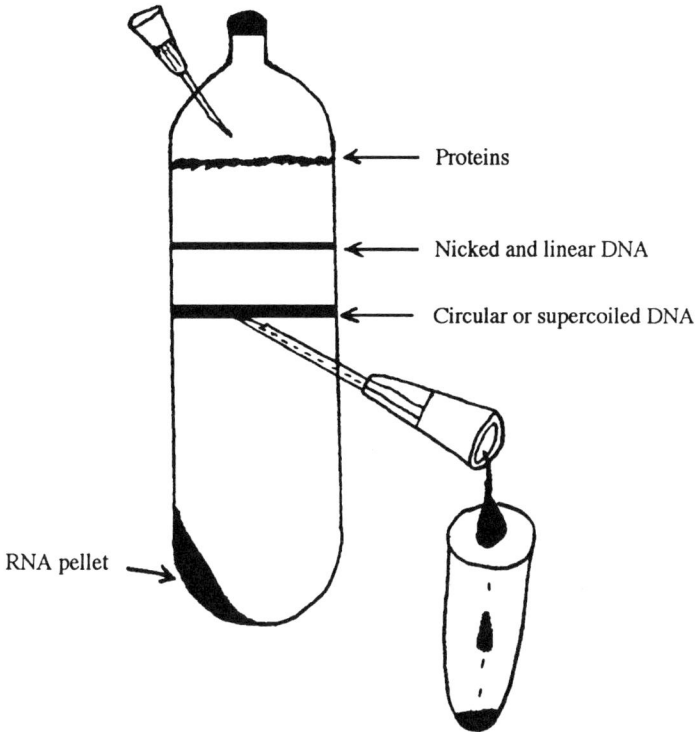

**FIGURE 1.1**
UV-illuminated ultracentrifuge tube following equilibrium sedimentation of plasmid DNA-containing solution. The DNA zones have an orange-red fluorescent appearance and can be individually collected with a syringe needle.

21. Centrifuge at $12,000 \times g$ for 15 min at 4°C. Dry the DNA pellet under vacuum for 15 min, and **gently** resuspend the DNA in 200 µl of TE buffer.

22. Reextract CsCl by repeating step 19.

23. Centrifuge at $12,000 \times g$ for 15 min at 4°C and **gently** wash the DNA pellet with 5 ml of 70% ethanol. Dry the DNA pellet under vacuum for 15 min and **gently** resuspend the DNA in 100 to 200 µl of TE buffer. Measure the DNA concentration and store the sample at −20°C until it is used. The DNA at this stage is pure and suitable for sequencing or cloning.

## Reagents Needed

### LB (Luria-Bertaini) Medium (per liter)

> 10 g Bacto-tryptone
> 5 Bacto-yeast extract
> 5 g NaCl
> Adjust pH to 7.5 with 2 *N* NaOH and autoclave. Add antibiotics after the autoclaved solution has cooled to below 50°C.

### TE Buffer

> 10 m*M* Tris-HCl, pH 8.0
> 1 m*M* EDTA

## Lysis Buffer

25 m$M$ Tris-HCl, pH 8.0
10 m$M$ EDTA
50 m$M$ Glucose

## Alkaline Solution

0.2 $N$ NaOH
1% SDS

## Potassium Acetate Solution (pH 4.8)

Prepare 120 ml of 5 $M$ potassium acetate. Add 23 ml of glacial acetic acid and 57 ml of H$_2$O. This solution is 3 $M$ with respect to potassium and 5 $M$ with respect to acetate. Store on ice until use.

## Isopropanol

## Ethanol (100% and 70%)

# IV.   DNA Isolation and Purification from Plants

## A.    Isolation of Plant Genomic DNA

Plant cells have cell walls and chloroplasts. These features make the procedure to isolate and purify DNA from plants more difficult than from animals. Traditional methods for DNA isolation usually have several drawbacks. These include low yield, shearing of DNA, and being time-consuming and relatively expensive due to use of CsCl gradients. The present protocol is a simple, fast, and low-cost method that has been successfully used in our lab for a wide variety of monocot and dicot plant species. Hexadecyltrimethylammonium bromide (CTAB, Sigma) used in the DNA isolation buffer significantly reduces carbohydrate contamination. The DNA isolated has a higher molecular weight, maximum purity, and a high yield. The tissue used for DNA extraction depends on the specific purpose that the investigator has in mind. For a genomic library construction, PCR, or Southern blot analysis, we strongly recommend that etiolated (dark grown) seedlings be used to avoid chloroplast DNA contamination and to reduce polysaccharide content. This makes the isolation procedure much easier. However, DNA may be isolated from green tissue for protein–DNA interaction experiments. Gloves should be worn during the DNA isolation procedure. All the glassware, pipettes, and solutions should be autoclaved to minimize DNase contamination. Chemicals or reagents of molecular biology grade are preferred. The DNA yield depends on the source of tissue (i.e., leaf, flower, stem, or root) and plant species. Generally, young tissue with meristematic cells has a higher yield of DNA. The DNA isolation protocol is given below.

## Protocol

1.   Harvest the tissue (2 to 5 g per extraction) and quickly freeze in liquid N$_2$. The frozen tissue can be stored in plastic tubes or bags at –20 or –80°C until use.

2.    Prechill a mortar and a pestle with liquid $N_2$. Place frozen tissue in a mortar and add liquid $N_2$. When the liquid $N_2$ evaporates, vigorously grind the tissue. Repeat adding liquid $N_2$ and grinding the tissue until a fine powder is obtained. It takes about 4 to 8 min depending on the tissue.

3.    Briefly warm the powder at room temperature for 2 min, and scrape it with a clean spatula into a conical-bottom plastic chloroform-resistant centrifuge tube containing 5 ml/g tissue of DNA isolation buffer. Gently resuspend the powder in solution by vortexing and then cap the tube loosely. Alternatively, frozen tissue may be ground in DNA isolation buffer preheated at 60°C.

4.    Incubate the samples in a 60°C water bath for at least 40 min with an occasional **gentle** swirling.

5.    Remove the samples to room temperature and add 1 volume of chloroform:isoamyl alcohol (24:1, v/v) in the fume hood. Cap the tubes and gently invert four to five times. Loosen caps to release pressure and let the tubes stand at room temperature for 10 min.

6.    Centrifuge samples at $7000 \times g$ or 5000 rpm in an angle rotor for 10 min at 4°C, or $1600 \times g$ in a swinging bucket rotor for 15 min at room temperature.

7.    **Slowly and gently** transfer the top, aqueous phase to a fresh tube with a wide-bore pipette, avoiding any floating pieces of tissue.

8.    Slowly pour the supernatant into a fresh tube or beaker containing 2 to 2.5 volumes (of the supernatant volume) of 100% chilled ethanol.

*Note:*    *Do not vortex or invert the tube or beaker until the DNA precipitate floats up to the surface of the ethanol (15 to 30 min at room temperature).*

9.    "Fish" out the DNA into a fresh tube with a glass hook and briefly dry the DNA under vacuum for 5 min. Gently resuspend the DNA in 4 ml/g original tissue of TE buffer at 37 to 40°C. At this stage, the DNA is not pure because of the RNAs, proteins, and polysaccharides it contains. It therefore must be extracted and purified using the steps that follow.

10.    Add *RNase* A (DNase-free) to a final concentration of 20 µg/ml DNA sample and incubate at 37°C for 30 min to hydrolyze the RNAs.

11.    Extract *RNase* A, proteins, or polysaccharides once with 1 volume of tris-buffer-saturated phenol/chloroform and gently mix. Centrifuge at $10,000 \times g$ for 10 min at 4°C and slowly transfer the supernatant into a fresh tube.

12.    Add 1 volume of chloroform:isoamyl alcohol (24:1, v/v) to the supernatant and gently mix to avoid DNA shearing. Centrifuge as in step 11.

13.    Add 0.5 volume of 7.5 *M* ammonium acetate or 0.15 volume of sodium acetate (pH 5.2) to the supernatant and gently mix. **Slowly and gently** pour the mixture into 2.5 volumes of 100% ethanol as step 8.

14.    "Fish" out the DNA into a fresh tube with a glass hook or centrifuge the DNA at $5000 \times g$ for 5 min at 4°C. Briefly and **gently** rinse the DNA with 4 ml of 70% ethanol and dry the sample under vacuum for 15 to 30 min. This process removes traces of ethanol and dehydrates the carbohydrates, some of which may become insoluble upon rehydration.

15.    Resuspend the DNA in an appropriate amount of TE buffer (100 µl/g tissue). Because of its high molecular weight, the DNA is usually not easily resuspended. In that case, the DNA sample can be placed at 45 to 50°C for 15 to 30 min with gentle and occasional shaking until it is dissolved. Keep the tube open to let remaining ethanol evaporate. At this stage, the DNA should be very pure with a high molecular weight. It is now suitable for restriction enzyme digestion and genomic library construction.

16.    Measure the quantity and quality of the DNA with a spectrophotometer at 260 and 280 nm. A pure DNA preparation should have a ratio of 1.85 to 2.0 of $A_{260}/A_{280}$ reading numbers. The sample can be stored at –20°C until used. The quality of the DNA preparation may be checked by 0.6 to 0.8% of agarose gel electrophoresis.

*Reagents Needed* —————————————————————————————————

## DNA Isolation Buffer

> 2% (w/v) CTAB (Sigma)
> 1.5 $M$ NaCl
> 25 m$M$ EDTA
> 0.2% (v/v) 2-Mercaptoethanol
> 100 m$M$ Tris-HCl, pH 8.0

## TE Buffer

> 10 m$M$ Tris-HCl, pH 8.0
> 1 m$M$ EDTA

## Tris-Buffer-Saturated Phenol/Chloroform

> Melt crystals of phenol in 65°C water bath and add 1 volume of Tris-HCl buffer, pH 8.0. Mix and let the mixture stand for 30 min. Remove the upper liquid and add an equal volume of Tris buffer and measure the pH using pH paper. Transfer the bottom phase into a fresh brown bottle and add an equal volume of chloroform. Mix completely and store at 4°C until use. The bottom phase is used for DNA extraction.

# B.   Isolation of Chloroplast DNA

Chloroplasts are organelles unique to plant cells. Chloroplast DNA resembles both bacterial and eukaryotic nuclear DNAs in its organization. This chloroplast DNA is a closed circular molecule with a molecular weight of about 85 to 300 Kb in green algae and 120 to 190 Kb in most higher plants. Some chloroplasts are abundant in leaf tissue. This tissue is recommended for the isolation of the chloroplast DNA. The protocol described below represents a modified, easy, and successful procedure.[10]

## Protocol 1: Isolation of Chloroplasts by Sucrose Gradient

1.  Destarch the seedlings in the dark for 24 to 36 h before harvesting this tissue. This avoids starch accumulation in the chloroplasts, which may destroy the chloroplasts during homogenizing of this tissue.

2.  Cut young green leaves from plants (10 g per extraction) with a clean razor blade and place the tissue in ice-cold water. Rinse the materials in 2 volumes of cold 5% commercial bleach such as Clorox for 4 min followed by washing with cold distilled water to completely remove the bleach.

3.  Briefly dry the tissue with a clean paper towel and slice the materials into small pieces (1 to 2 mm) with a razor blade. Transfer the cut pieces to a blender jar containing 6 volumes/g tissue fresh weight of chloroplast isolation buffer. Grind the tissue at low speed using four 4-s bursts in a cold room until a good chloroplasts suspension is obtained.

4.  Immediately filter the homogenate through five layers of cheese cloth and one layer of Miracloth on the top. Collect the filtrate in centrifuge tubes and spin at 3000 × $g$ for 15 to 20 min at 4°C.

5.  Discard the supernatant and **gently** resuspend the green pellet in 5 ml of washing buffer. Layer the resuspended chloroplasts over a 30:45:60% sucrose step gradient. Cover the top of the gradient with 1 ml of washing buffer.

6. Centrifuge in a ultracentrifuge (e.g., Beckman, SW27) at 26,000 rpm for 1 h at 4°C. Remove the intact chloroplast band, which is usually a very dark green at or near the 30:45% interphase using a cut-off pipette tip.

7. Add 3 to 4 volumes of cold washing buffer to the chloroplast sample and centrifuge at $6000 \times g$ for 10 min at 4°C. Gently resuspend the pellet in 5 ml cold washing buffer and store at 4°C until extraction of chloroplast DNA.

## Protocol 2: Isolation of Chloroplast DNA

1. Take 2.5 ml of chloroplast sample and add 0.25 ml of 10 mg/ml proteinase K solution. Gently mix by inversion and incubate at room temperature for 15 to 20 min.

2. Add 0.25 ml of 20% sarcosyl solution and gently mix by inversion to avoid shearing the chloroplast DNA. Incubate the sample at room temperature for 50 min with occasional gentle inversion.

3. Add 9.1 ml of 0.672 g/ml CsCl stock solution and gently mix by inversion 40 times. Allow the mixture to stand for at least 60 min at room temperature.

4. Centrifuge at $12,000 \times g$ for 20 min. Gently transfer the supernatant into a fresh graduated tube and discard the pellet of membranes and starch particles.

5. Adjust the supernatant to 4.8 ml per gradient with TE buffer and add 6.7 ml of 1.24 g/ml CsCl stock solution and 0.3 ml of 10 mg/ml EtBr solution. Slowly and gently mix the sample by inversion and in subdued room light in order to minimize nicking of the DNA.

*Note:* *EtBr is a very toxic chemical; caution must be taken.*

6. Gently transfer the DNA–CsCl mixture into "Quik-Seal" (Beckman) ultracentrifuge tubes and fill the tubes right up to the neck of the tube (about 5.2 ml per tube) with a syringe.

*Notes:* *If the tube is overfilled, it will cause a problem in sealing. However, if underfilled, large air bubbles present after sealing the tube may cause the tube to collapse. Carefully balance the tubes with CsCl balance solution.*

7. Turn on and warm the sealing machine for 5 min. Place the tubes in a metal sealing plate and cover each tube with an appropriate metal cap. Place each tube (one at a time) right under the sealing rack and push and hold the hot rack against the metal cap of the tube for 30 to 60 s. The neck of the tube gradually melts until a sound is heard from the sealing machine. Quickly remove the tube from the melt rack and place the cap of the tube against a cold metal strip. Press and hold the strip against the cap for about 1 to 2 min to completely seal the tube and then remove the cap.

8. Attach a cap and a metal adapter to each sealed tube for ultracentrifugation and carefully balance again using Scotch tape if necessary. Symmetrically place the tubes into the Beckman vertical rotor and cover the tubes with their own caps and adapters.

9. Centrifuge at 50,000 rpm for 9 to 20 h at 20°C.

*Note:* *Never centrifuge at 4°C. It will cause the CsCl to precipitate out of solution and reduce the density of CsCl in the gradient. The DNA band will be too low to be identified.*

10. Stop centrifugation. Slowly remove the tubes from the rotor and place them in a plastic rack. Use long-wavelength UV light (360 nm) to visualize the relaxed and supercoiled chloroplast DNA bands. Short-wavelength UV light can cause nicking of the DNA.

11.    For removal of the DNA, place the tube in a ringstand rack with a clamp. Locate the DNA bands with 360-nm UV light. With ethanol wipe dry one side of the DNA band area and place a piece of Scotch tape on it and wipe dry. **Slowly and carefully** puncture through the tube with a needle and syringe (16 to 20 gauge) until the needle has penetrated into the DNA band area. Gently withdraw a minimal volume (1 to 2 ml) of DNA sample, then slowly remove the needle. Transfer the sample into a fresh tube.

12.    Extract EtBr from the DNA sample with 1 volume of isopropanol. Gently invert the tube several times and allow the phases to separate by gravity. The EtBr usually partitions into the upper isopropanol phase and is removed with a pipette. The lower aqueous phase contains DNA–CsCl with some EtBr in it and should be reextracted four to five times until no pink color is visible.

*Note:*    *EtBr is a mutagenically toxic molecule and should be discarded in a special container to allow it to become chemically inactive before being disposed.*

13.    Transfer the lower phase into a fresh Corex tube and dilute out the CsCl with 2 volumes of TE buffer. Add 2 volumes of chilled 100% ethanol into the diluted sample and precipitate the DNA at –20°C for 5 h to overnight.

*Note:*    *Never place the tubes at –80°C; otherwise, the CsCl will precipitate out.*

14.    Centrifuge at $12,000 \times g$ for 15 min at 4°C. Dry the DNA pellet under vacuum for 15 min and gently resuspend the DNA in 1 ml of TE buffer. It usually takes longer for the large molecular DNA to be dissolved.

15.    Reextract CsCl by repeating step 13.

16.    Centrifuge at $12,000 \times g$ for 15 min at 4°C and gently wash the DNA pellet with 5 ml of 70% ethanol. Dry the DNA pellet under vacuum for 15 min and gently resuspend the DNA in 100 to 200 μl of TE buffer. Measure the DNA concentration and store the sample at –20°C until it is used.

## Solutions Needed

### Stock Solutions

- 1.5 $M$ sorbitol: Dissolve 273.3 g sorbitol in distilled water. Adjust final volume to 1 liter with distilled water and store at –20°C.

- 1 $M$ HEPES/KOH (pH 7.5): Dissolve 119.1 g HEPES in up to 400 ml dd.$H_2O$. Adjust pH with 2 $N$ KOH. Adjust final volume to 500 ml with dd.$H_2O$ and store at 4°C.

- 1 $M$ $MgCl_2$: Dissolve 20.3 g $MgCl_2 \cdot 6H_2O$. Adjust final volume to 100 ml with dd.$H_2O$ and store at 4°C.

- 1 $M$ $MnCl_2$: Dissolve 19.8 g $MnCl_2 \cdot 4H_2O$ in 60 ml dd.$H_2O$. Adjust final volume to 100 ml with dd.$H_2O$ and store at 4°C.

- 500 m$M$ EDTA: Dissolve 93.06 g $Na_2EDTA \cdot 2H_2O$ in 300 ml dd.$H_2O$. Adjust pH to 8.0 with 2 $N$ NaOH solution. Adjust final volume to 500 ml with dd.$H_2O$ and store at room temperature.

- 1 $M$ Tris-HCl (pH 8.0): Dissolve 121.1 g Tris in 800 ml dd.$H_2O$. Adjust pH with 2 $N$ HCl to pH 8.0. Adjust final volume to 1000 ml.

### Chloroplast Isolation Buffer

0.35 $M$ Sorbitol (233.3 ml 1.5 $M$ sorbitol solution)
50 m$M$ HEPES/KOH (50 ml 1 $M$ HEPES/KOH [pH 7.5])
1 m$M$ $MgCl_2$ (1 ml 1 $M$ $MgCl_2$)
2 m$M$ EDTA (4 ml 500 m$M$ EDTA)

1 m$M$ MnCl$_2$ (1 ml 1 $M$ MnCl$_2$)
Adjust final volume to 1 liter with dd.H$_2$O and store at 4°C.
Add 1 g sodium ascorbate prior to use.

## *Washing Buffer*

350 m$M$ Sorbitol (6.4 g sorbitol)
50 m$M$ Tris-HCl (5 ml 1 $M$ Tris-HCl [pH 8.0])
20 m$M$ EDTA (4 ml 500 m$M$ EDTA)
Adjust final volume to 100 ml with dd.H$_2$O.
Autoclave at 15 psi for 20 min.

## *30:45:60% Sucrose Step Gradient*

- Make 30, 45, and 60% sucrose solutions, respectively, by dissolving 30, 45, and 60 g sucrose in a final volume of 100 ml (w/v) dd.H$_2$O, or a final weight of 100 g (w/w) in dd.H$_2$O.

- Add an appropriate volume of 60% sucrose solution into a centrifuge tube and freeze on dry ice or place at −80°C. Add the volume of 45% sucrose solution on the top of frozen 60% sucrose and freeze. Add the volume of 30% sucrose solution on the top of 45% sucrose and freeze.

- Place the centrifuge tube at 4°C overnight prior to being used. The sucrose solutions will gradually thaw and diffuse, forming a gradient.

## *10 mg/ml Proteinase K Solution*

200 mg Proteinase K in 10 ml of wash buffer
Store at −20°C.

## *20% (w/v) Sarcosyl*

Dissolve 40 g sodium lauryl sarcosyl in 100 ml dd.H$_2$O. Adjust to 200 ml with dd.H$_2$O.

## *0.672 g/ml CsCl Stock Solution*

Dissolve 53.76 g CsCl in 40 ml TE buffer. Adjust to a final volume of 80 ml with TE buffer.

## *10 mg/ml Ethidium Bromide*

Dissolve 200 mg ethidium bromide. Adjust to a final volume of 20 ml with dd.H$_2$O. Store in a brown bottle at 4°C.

# V.   Purification of DNA Fragments from an Agarose Gel

Isolation of specific DNA fragment(s) is very useful for preparation of DNA probe(s) for Southern blotting, Northern blotting, or dot blotting. In addition, in cDNA and genomic DNA cloning, DNA inserts of interest may be separated from vectors by appropriate enzymatic digestion and then ligation with new vectors (usually plasmid DNA) for subcloning. One of the most powerful methods is to elute the specific DNA fragment of interest from low-melting temperature agarose gels. DNA purified by this method has a high yield and a high purity, which is desirable for ligation, cloning, labeling, and footprinting.

## A. Elution of DNA Fragments by Gel Slicing

1.  Set up a standard restriction enzyme digestion of DNA containing the sequence of interest as described in the Southern blotting procedure in Chapter 5.

2.  Carry out electrophoresis as described in Southern blotting, but use 0.5 to 1% low-melting temperature agarose instead of normal agarose.

3.  Transfer the gel to a long-wavelength UV transilluminator (305 to 327 nm) to visualize the DNA bands. Excise the band(s) of interest with a clean razor blade and place the slices into a microcentrifuge tube.

*Notes:*   *(1) The gel should not be placed on a short-wavelength (e.g., <270 nm) UV transilluminator because this type of UV light may cause breakage inside the DNA fragments, thus significantly inhibiting the subcloning of the DNA insert(s) of interest. (2) The gel slices containing the DNA bands of interest should be trimmed in order to remove excess unstained gel areas as much as possible.*

*Caution:*   *UV light is harmful to the human body. Protective eyeglasses, gloves, and a lab coat should be worn when using UV light.*

4.  Add 2 volumes of TE buffer to the gel slices and completely melt the gel in a 60 to 70°C water bath.

*Note:*   *The gel slices can be directly melted without adding any TE buffer. The DNA concentration is usually high, but the total yield of the DNA fragments is much lower than that when one adds TE buffer.*

5.  Immediately chill the melted gel solution on dry ice or equivalent and place the tube at –70°C for at least 20 min.

6.  Thaw the gel mixture by tapping the tube vigorously. It takes about 5 to 10 min to thaw the gel into a resuspension state.

7.  Centrifuge at $11,000 \times g$ for 5 min at room temperature.

8.  Carefully transfer the liquid phase containing the eluted DNA fragment into a fresh tube. The DNA solution can be used directly for labeling except at low concentrations.

9.  Extract EtBr with 3 volumes of water-saturated *n*-butanol.

10.  Precipitate the DNA by adding 0.1 volume of 3 *M* sodium acetate buffer (pH 5.2) and 2.5 volumes of chilled 100% ethanol into the DNA solution. Allow to precipitate at –70°C for 1 h and centrifuge at $12,000 \times g$ for 5 min at room temperature.

11.  Discard the supernatant and briefly rinse the pellet with 1 ml of 70% ethanol. Dry the DNA under vacuum for 15 min and dissolve the DNA in an appropriate volume of dd.$H_2O$ or TE buffer. Store at –20°C until use.

## B. Elution of DNA Using NA45 DEAE Membranes

1.  Carry out steps 1 to 2 as in Section V.A, but use 1% normal agarose and insert a membrane in place of the gel during the electrophoresis.

2. Soak a piece of NA45 DEAE membrane (Scheicher and Schuell) in 10 m$M$ EDTA buffer (pH 7.6) for 15 min at room temperature and replace the solution with 0.5 $N$ NaOH solution for 5 min. Wash the membrane with dd.$H_2O$ three to four times and store at 4°C until use.

3. During the electrophoresis, monitor the migration of the DNA bands stained with EtBr in the gel by using a long-wavelength UV lamp. Make an incision below the DNA band of interest and insert a piece of the prepared membrane into the incision in the gel.

4. Continue electrophoresis, while monitoring the migration of the stained band, until the DNA fragments migrate onto the membrane.

5. Remove the membrane strip from the gel and place it in a solution containing 20 m$M$ Tris-HCl (pH 8.0), 0.15 $M$ NaCl, and 0.1 m$M$ EDTA. Shake it for 1 min to remove any agarose.

6. Transfer the strip to 0.2 to 0.3 ml of elution buffer containing 20 m$M$ Tris-HCl (pH 8.0), 1 $M$ NaCl, and 0.1 m$M$ EDTA.

7. Incubate the strip at 55 to 68°C for 30 to 60 min with shaking at 60 rpm.

8. Rinse the strip with 0.1 ml of elution buffer and pool the elution fluid aliquots together.

9. Extract EtBr with 3 volumes of water-saturated $n$-butanol.

10. Precipitate and dissolve the eluted DNA as described in Section V.A.

## C. Elution of DNA Fragments in Wells of Agarose Gel

1. Carry out DNA digestion and gel electrophoresis as described in Sections V.A and V.B, but add the running buffer up to the upper edges of the gel instead of covering the gel.

2. During electrophoresis, monitor the separation of DNA bands stained by EtBr in the gel using a long-wavelength UV lamp. Stop the electrophoresis and use a razor blade or a spatula to make a well in front of the band of interest. Add 20 to 60 μl of running buffer into the well.

3. Continue electrophoresis, while monitoring the band stained by EtBr, until the band migrates into the well.

4. Stop electrophoresis and transfer the solution containing the DNA of interest from the well to a fresh tube.

5. Extract EtBr with 3 volumes of water-saturated $n$-butanol.

6. Precipitate and dissolve the DNA as described in Section V.A.

# References

1. **Taylor, B. H., Manhart, J. R., and Amasino, R. M.,** Isolation and characterization of plant DNAs, in *Methods in Plant Molecular Biology and Biotechnology,* Glick, B. R. and Thompson, J. E., Eds., CRC Press, Inc., Boca Raton, Florida, 1993.

2. **Sambrook, J., Fritsch, E. F., and Maniatis, T.,** *Molecular Cloning: A Laboratory Manual,* 2nd ed., Cold Spring Harbor Press, Cold Spring Harbor, NY, 1989.

3. **Murray, M. G. and Thompson, W. F.,** Rapid isolation of high molecular weight DNA, *Nucleic Acids Res.,* 8, 4321, 1980.

4. **Wu, L.-L., Song, I., Kim, D., and Kaufman, P. B.,** Molecular basis of the increase in invertase activity elicited by gravistimulation of oat-shoot pulvini, *J. Plant Physiol.,* 142, 179, 1993.

5. **Birnboim, H. C. and Doly, J.,** A rapid alkaline extraction procedure for screening recombinant plasmid DNA, *Nucleic Acids Res.,* 7, 1513, 1979.

6. **Lichtenstein, C. P. and Draper, J.,** Genetic engineering of plants, in *DNA Cloning, A Practical Approach,* Vol. 2, Glover, D. M., Ed., IRL Press, Oxford, U.K., 1985.

7. **Kahn, J. D., Yun, E., and Crothers, D. M.,** Detection of localized DNA flexibility, *Nature,* 368, 163, 1994.

8. **Jain, V. K., Gupta, A., and Magrath, I. T.,** Rapid purification of bacteriophage λDNA, *BioTechniques,* 15, 602, 1993.

9. **Chomczynski, P.,** A reagent for the single-step simultaneous isolation of RNA, DNA and proteins from cell and tissue samples, *BioTechniques,* 15, 532, 1993.

10. **Milligan, B. G.,** Purification of chloroplast DNA using hexadecyltrimethylammonium bromide, *Plant Mol. Biol. Rep.,* 7, 144, 1989.

11. **Lunn, G. and Sansone, E. B.,** Ethidium bromide: destruction and decontamination of solutions, *Anal. Biochem.,* 162, 453, 1987.

12. **Quillardet, P. and Hofnung, M.,** Ethidium bromide and safety — Readers suggest alternative solutions. Letter to the editor, *Trends Genet.,* 4, 89, 1988.

# Isolation and Purification of RNA

## Contents

RNAs, as compared with DNA, are very mobile molecules due to their degradation by RNases.[1,2] This makes the procedure of RNA isolation much more complicated. In order to obtain high yield and high quality of RNA, four procedures should be handled carefully and effectively: (1) optimal disruption of cells or tissue; (2) effective denaturation of nucleoprotein complexes; (3) maximum inhibition of exogenous RNase and endogenous RNase released from cells upon cell disruption; (4) effective purification of RNA from DNA and proteins. The purity and integrity of isolated RNA is very critical for Northern blotting, poly(A)+RNA purification, cDNA synthesis, and *in vitro* translation. The most difficult task, however, is to inactivate RNase activity. The procedures given below are modified from previously developed methods[2–7] and are very effective in achieving this objective.

**To inactivate exogenous RNase** — Two common sources of RNase contamination are the user's hands and bacteria and fungi present on airborne dust particles. To prevent this type of contamination:

1.  Gloves should be worn at all times and changed frequently.

2.  Whenever possible, disposable plasticware should be autoclaved and used for RNA isolation. Nondisposable glassware and plasticware should be treated with 0.1% diethyl pyrocarbonate (DEPC) in dd.H$_2$O and be autoclaved prior to use.

3.  After treatment, glassware should be baked at 250°C overnight. Disposable and sterile polypropylene centrifuge tubes are strongly recommended to be used for RNA isolation. The disadvantages of used glass Corex tubes are that the tubes must be deeply cleaned, DEPC treated, and autoclaved followed by being baked at 250°C overnight, thus wasting time and money.

**To inactivate endogenous** RNase — 4 $M$ guanidine thiocyanate and β-mercaptoethanol are strongly recommended for inclusion in the extraction buffer. These are strong inhibitors of RNase. Whenever possible, solutions should be treated with 0.05% DEPC at 37°C for 2 h or overnight at room temperature and then autoclaved for 30 min to remove any trace of DEPC. Tris buffers, which cannot be treated with DEPC, should be made in DEPC-treated dd.H$_2$O.

# I.     Rapid Isolation of Total RNA

## A. Acid Guanidinium Thiocyanate–Phenol–Chloroform Method

1.  **For tissue** — Harvest tissue and immediately drop in liquid nitrogen. Store at –70°C until use. Grind 1 g tissue in a clean blender using liquid nitrogen. Keep adding liquid nitrogen and grinding until a fine powder is obtained. Warm the powder at room temperature for a few minutes and immediately transfer the powder with a sterile spatula into a clean 14- or 50-ml polypropylene tube. Add 10 ml of solution B, mix, and keep the tube on ice. Proceed to step 2.

    **Alternatively,** homogenize the tissue in 10 ml of solution B in a sterile polypropylene tube on ice by using a glass-Teflon homogenizer or equivalent polytron at top speed for two 30-s bursts. Transfer the homogenate to a fresh polypropylene tube and proceed to step 2.

    **For cultured cells grown in suspension culture or in a monolayer** — Collect 1 to 2 × 10$^8$ cells in a sterile 50-ml polypropylene tube by centrifugation at 400 × $g$ for 5 min at 4°C. Wash the cell pellet with 25 ml of ice-cold, sterile 1X PBS or serum-free medium (e.g., DME) and centrifuge at 400 × $g$ for 5 min at 4°C. Repeat washing and centrifugation once more to remove all traces of serum that contains RNase. Remove the supernatant and resuspend in 10 ml solution B. Keep the tube on ice and homogenize the cell suspension using a microtip on a polytron for two 0.5- to 1-min bursts at top speed and carefully transfer homogenate to a fresh tube. Proceed to step 2.

    **Alternatively,** wash and resuspend the cells as described above. Keep the suspension of cells on ice. In order to shear the DNA and lower the viscosity, sonicate the suspension of cells at the maximum power for a clean microtip (60%) using two 0.5- to 1-min bursts. A good sonicated solution should be thin enough to drop freely from the end of a Pasteur pipette. Proceed to step 2.

*Note:*     *As long as the sample is in solution B, RNase released from the tissue may be inhibited by guanidinium thiocyanate.*

*Caution:*     *Guanidinium thiocyanate is a potent chaotropic agent and irritant.*

2.  Add the following to the sample in the order given: 1 ml of 2 $M$ sodium acetate buffer (pH 4.0), 10 ml of water-saturated phenol, and 2 ml of chloroform:isoamyl alcohol. Cap the tube and mix by inversion after each addition. Vigorously shake the tube for 20 s.

***Caution:*** *Phenol is poisonous and can cause severe burns. Proper laboratory clothing including gloves and goggles should be worn when handling these reagents. If phenol contacts your skin, wash the area immediately with large volumes of water, but* **do not rinse with ethanol!**

3. Centrifuge at 11,000 × *g* for 20 min at 4°C using a Corex tube or at 7000 to 8000 rpm for 30 to 40 min at 4°C using a 14-ml polypropylene tube in the swing-out rotor. **Carefully transfer** the top, aqueous phase to a fresh tube. **Decant the phenol phase into a special container.**

4. Add 10 ml of isopropanol, mix, and allow to precipitate at –20°C for 1.5 to 2 h.

5. Centrifuge at 12,000 × *g* for 20 min at 4°C using a Corex tube or at 7000 to 8000 rpm for 30 to 40 min at 4°C using a 14-ml polypropylene tube in the swing-out rotor. Resuspend the RNA pellet in 3 ml of solution B and add 3 ml of isopropanol. Mix and then place at –20°C for 2 h or overnight.

6. Centrifuge at 12,000 × *g* for 20 min at 4°C using a Corex tube or at 7000 to 8000 rpm for 30 to 40 min at 4°C using a 14-ml polypropylene tube in the swing-out rotor. Briefly rinse the total RNA pellet with 4 ml of 75% ethanol. Dry the pellet under vacuum for 15 min.

7. Dissolve the total RNA in 200 μl of 0.5% SDS solution. Take 20 μl to measure the concentration with the spectrophotometer at 260 and 280 nm. The ratio of $A_{260}/A_{280}$ should be 1.8 to 2.0 to be judged as a good RNA preparation. The RNA isolated may be checked by 1% agarose, formaldehyde denaturing gel electrophoresis. Two strong rRNA bands should be visible with some smear through the well (Figure 2.1). At this point, the RNA can be used for Northern blot analysis, or dot blot hybridization. The RNA sample may be stored at –20°C until use for poly(A)+RNA purification.

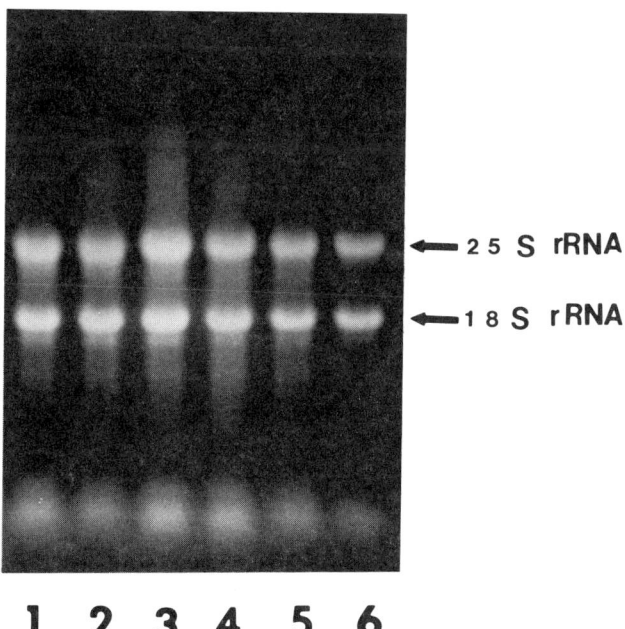

**FIGURE 2.1**

Total RNAs were isolated from different plant tissues and subjected to electrophoresis on a 1% agarose gel containing 17% formaldehyde. Total RNA (15 μg) was loaded to each lane. mRNA appears to be a smear and rRNA has two sharp bands in each lane.

## B. LiCl–Urea and Phenol–Chloroform Method

1. **For tissue** — Harvest tissue and immediately drop in liquid nitrogen. Store at –70°C until use. Grind 1 g tissue in a clean blender using liquid nitrogen. Keep adding liquid nitrogen and grinding until a fine powder is obtained. Warm the powder at room temperature for a few minutes and immediately transfer the powder with a sterile spatula into a clean 14- or 50-ml polypropylene tube. Add 8 ml of RNA extraction solution, mix, and keep the tube on ice. Proceed to step 2.

   **Alternatively,** homogenize the tissue in 8 ml of RNA extraction solution in a sterile polypropylene tube on ice by using a glass-Teflon homogenizer or equivalent polytron at top speed for two 30-s bursts. Transfer the homogenate to a fresh polypropylene tube and proceed to step 2.

   **For cultured cells grown in suspension culture or in a monolayer** — Collect 1 to $2 \times 10^8$ cells in a sterile 50-ml polypropylene tube by centrifugation at $400 \times g$ for 5 min at 4°C. Wash the cell pellet with 25 ml of ice-cold, sterile 1X PBS or serum-free medium (e.g., DME) and centrifuge at $400 \times g$ for 5 min at 4°C. Repeat washing and centrifugation once more to remove all traces of serum that contains RNase. Remove the supernatant and resuspend in 8 ml RNA extraction solution. Keep the tube on ice and homogenize the cell suspension using a microtip on a polytron for two 0.5- to 1-min bursts at top speed and carefully transfer the homogenate to a fresh tube. Proceed to step 2.

   **Alternatively,** wash and resuspend the cells as described above. Keep the suspension of cells on ice. In order to shear the DNA and lower the viscosity, sonicate the suspension of cells at the maximum power for a clean microtip (60%) using two 0.5- to 1-min bursts. A good sonicated solution should be thin enough to drop freely from the end of a Pasteur pipette. Proceed to step 2.

2. Place the homogenate at 0 to 4°C for at least 4 h and then centrifuge at $3000 \times g$ for 25 min at 4°C.

3. Remove the supernatant and add 0.5 volume of cold RNA extraction solution. Mix well by vortexing and centrifuge as in step 2.

4. Carefully decant the supernatant and dissolve the pellet in RNA buffer.

5. Extract proteins, carbohydrates, DNA, and cellular debris with 1 volume of phenol:chloroform: isoamyl alcohol (25:24:1). Shake for 5 min.

*Caution:*   *Phenol is toxic; gloves should be worn when working with it.*

6. Centrifuge at $10,000 \times g$ for 10 min at 4°C and carefully remove the top, aqueous phase to a fresh tube. Never remove any white materials at the interface between two phases. Repeat steps 5 to 6 one more time.

7. Add 0.1 volume of 3 $M$ sodium acetate and 2 volumes of 100% chilled ethanol. Mix well and allow the RNA to precipitate at –20°C for 2 h or on dry ice for 20 min.

8. Centrifuge at $12,000 \times g$ for 10 min at 4°C. Discard the supernatant and wash the RNA pellet with 4 ml of 70% ethanol. Dry the pellet under vacuum for 15 min and dissolve the total RNA in 50 µl of RNA buffer containing 0.2% SDS. Take 5 µl of the sample to measure the concentration of the RNA. Store the sample at –20°C until use for poly(A)+RNA purification.

## C. Isolation of RNA from Oocytes and Embryos by the Phenol–Chloroform and LiCl Method

1. Collect fresh oocytes, fertilized eggs, or embryos from specific organisms of interest and quickly wash the tissue twice with serum-free medium (DEM or PBS buffer). Transfer the tissue in a clean glass homogenizer and decant the fluid.

2. Add 8 volumes of homogenization buffer into the homogenizer and homogenize the tissue on ice until a complete suspension is obtained. Incubate the homogenate for 40 to 60 min at 37°C with occasional mixing.

3. Transfer the homogenate to a polypropylene centrifuge tube or a Corex tube and add 1 volume of water-saturated phenol:chloroform. Shake vigorously for 30 s.

4. Centrifuge at 5000 to 6000 × g for 15 min at 4°C in a swinging-bucket rotor. Carefully transfer the top, aqueous phase to a fresh tube.

5. Repeat the phenol:chloroform extraction once as in steps 3 to 4.

6. Precipitate RNA by adding 0.1 volume of 3 M sodium acetate buffer and 2.5 volumes of chilled 100% ethanol. Place the tube at –20°C for 2 h.

7. Centrifuge at 9000 × g for 10 min at 4°C and discard the supernatant. Briefly rinse the pellet with 1 ml of 70% ethanol and dry the pellet under vacuum for 10 min.

8. Resuspend the RNA in 500 μl of TE buffer. Add 500 μl of 8 M LiCl solution and place at –20°C for 3 to 5 h to precipitate the RNA. LiCl helps remove glycoproteins and other contaminants from the sample.

9. Centrifuge at 11,000 × g for 15 min at 4°C and carefully decant the supernatant. Briefly rinse the pellet with 2 ml of 70% ethanol and dry the pellet under vacuum for 15 min.

10. Dissolve the RNA in 100 μl of water and 0.1 volume of 3 M sodium acetate buffer and 3 volumes of chilled ethanol. Place at –20°C for more than 2 h.

11. Repeat step 9. Dissolve the RNA in 50 μl of TE buffer. Take 5 μl of the sample to measure the concentration and quality. Store the sample at –20°C until use.

## D. Isolation of Total RNA with Guanidinium Thiocyanate and Centrifugation by the Use of a CsCl Gradient

1. Collect and transfer tissue or cells into a homogenizing tube. Add 5 volumes of 4 M guanidinium thiocyanate buffer to 1 g of tissue or to $2 \times 10^7$ cells/ml of cell suspension or confluent monolayer of cells.

2. Homogenize the tissue on ice with a Polytron homogenizer (RNase-free) or equivalent at top speed for two 1-min bursts.

3. Transfer the homogenate to a fresh tube and add sodium lauryl sarcosinate to a final concentration of 0.5% and mix well.

4. Centrifuge at 5000 × g for 10 min at room temperature. Carefully transfer the supernatant to a fresh tube.

5. Carefully layer the supernatant onto a cushion of CsCl/EDTA solution in a clear ultracentrifuge tube. Mark the position of the cushion on the outside of the tube. Add 4 M guanidinium thiocyanate homogenization buffer to fill the tubes and to equalize their weights.

6. Centrifuge at 20°C at the speed and for the time given in Table 2–1 using a swinging-bucket rotor.

### Table 2–1 Swinging-Bucket Rotors Centrifugation Speeds and Times

|  | Rotor types | | |
|---|---|---|---|
|  | SW60 ($^7/_{16}$″ × 23″) | SW40 ($^9/_{16}$″ × 32″) | SW28 (12″ × 31″) |
| Volume of CsCl/EDTA (ml) | 1.2 | 3.5 | 12.0 |
| Volume of homogenate (ml) | 3.1 | 9.7 | 26.5 |
| Time (h) | 12 | 24 | 26 |
| Speed (rpm) | 40,000 | 32,000 | 23,500 |

7.  Turn off the centrifuge brake before decelerating the rotor to prevent the contents of the tube from becoming disturbed. Carefully remove the tubes from the centrifuge. Using a Pasteur pipette, gently remove the fluid, little by little, above the level of the cushion containing the white, viscous cellular DNA. With a fresh pipette and an automatic pipettor, keep removing the gradient fluid until close to the bottom of the tube. Briefly invert the tube to drain away any remaining fluid with pieces of Kimwipe paper and return the tube to an upright position afterwards.

8.  Briefly wash CsCl from the RNA pellet with 2 ml of 70% ethanol and dry the RNA pellet at room temperature. Dissolve the RNA in the 100 µl of water. Wash the ultracentrifuge tube with another 100 µl water and pool the RNA solutions together.

9.  Transfer the RNA sample to a fresh tube with 0.1 volume of 3 $M$ sodium acetate buffer and 2.5 volumes of chilled 100% ethanol. Mix well and allow the RNA to precipitate at –20°C for 2 h or more.

10. Centrifuge at $12,000 \times g$ for 10 min at 4°C. Discard the supernatant. Briefly rinse the pellet with 70% ethanol. Dry under vacuum for 15 or 30 min by air in a laminar flow hood.

11. Dissolve the RNA in 50 µl of TE buffer. Take 5 µl of the sample to measure the concentration and quality of the RNA. Store the sample at –20°C until use.

# II.     Purification of Poly(A)+RNA from Total RNA

## A. Purification of mRNA by Oligo(dT)–Cellulose Column

1.  Add 4 ml of binding buffer to 0.5 g of oligo(dT)–cellulose type 7 (Collaborative Research) in a clean tube and mix well. Transfer the suspension into a 10-ml poly-prep chromatography column (Bio-Rad) or equivalent. Vertically fix the column in a holder with a clamp and place the column at 4°C for 2 h to equilibrate the cellulose resin.

2.  Remove the column to room temperature and wash it by adding 2 ml of 0.1 $N$ NaOH solution through the top of the resin. **The color of the resin changes from white to yellow.** Allow the washing fluid to drain away by gravity and repeat washing the column five times.

3.  Neutralize the column by adding 4 ml of binding buffer to the top of the resin and drain away the binding buffer by gravity. **The color of the resin returns to white.** Repeat five times and store at 4°C until use.

4.  Add 1 volume of loading buffer to the total RNA sample prepared. Heat the mixture for 10 min at 65°C to denature the secondary structure of RNA, and then allow the sample to cool to room temperature.

*Notes:*   *The concentration of SDS in the loading buffer should never be more than 0.5%; otherwise, the SDS will precipitate and block the column. Based on our experience, 0.2% SDS in the binding buffer works well.*

5.  **Slowly load** the total RNA sample to the top of the column capped at the bottom and allow the sample to run through the column by gravity. **Gently loosen** the resin in the column and let the column stand for 5 min. Remove the bottom cap to drain away the fluid from the bottom of the column and collect the fluid containing some unbound mRNA. Reload the fluid onto the column to allow the unbound mRNA to bind to the oligo(dT)–cellulose column and collect the eluate. Repeat this one more time.

6.  Wash the column with 2 ml of washing buffer each time and drain away the buffer. Repeat washing three times.

7.  Cap the bottom of the column and add 1 ml of elution buffer to the column to elute bound poly(A)+RNA. **Gently loosen** the resin using a clean needle and allow the column to stand for

4 min. Remove the bottom cap and collect the eluate in a clean tube. Elute the column twice with 1 ml of elution buffer and collect the eluate.

8.  Pool the eluates together and add 0.15 volume of 3 *M* sodium acetate and 2.5 volumes of 100% chilled ethanol to precipitate the mRNA. Mix and place at –20°C overnight.

9.  Centrifuge at 12,000 × *g* for 15 min at 4°C. Decant the supernatant and briefly wash the mRNA pellet with 2 ml of 70% ethanol. Dry the pellet under vacuum for 15 min to remove ethanol and dissolve the mRNA in 20 to 50 μl of TE buffer. Take 2 or 4 μl of the sample to measure the concentration of the mRNA. At this stage, the sample should have a ratio of $A_{260}/A_{280}$ of 1.9 to 2.0. The quality of the purified mRNA may be checked by 1% agarose, formaldehyde-denaturing gel electrophoresis. A broad-range smear should be visible from the top to the bottom of the well with or without two weak rRNA bands (Figure 2.2). The yield of poly(A)+RNA may be expected to be approximately 6% of total RNA. Store the sample at –70°C until use for Northern blot analysis, *in vitro* transcription, and cDNA synthesis.

## B. Mini-Purification of mRNA by Oligo(dT)–Cellulose Column

1.  Add 1 ml of binding buffer to 0.3 g of oligo(dT)–cellulose type 7 (Collaborative Research) in an Eppendorf tube. Mix and place at 4°C for 30 min.

2.  Wash the resin with 1 ml of 0.1 *N* NaOH solution and mix gently for a few minutes. Centrifuge at 1500 × *g* for 2 min and discard the supernatant and repeat this step eight times.

*Note:*    *Centrifugation greater than 1500 × g may cause damage to the oligo(dT) beads.*

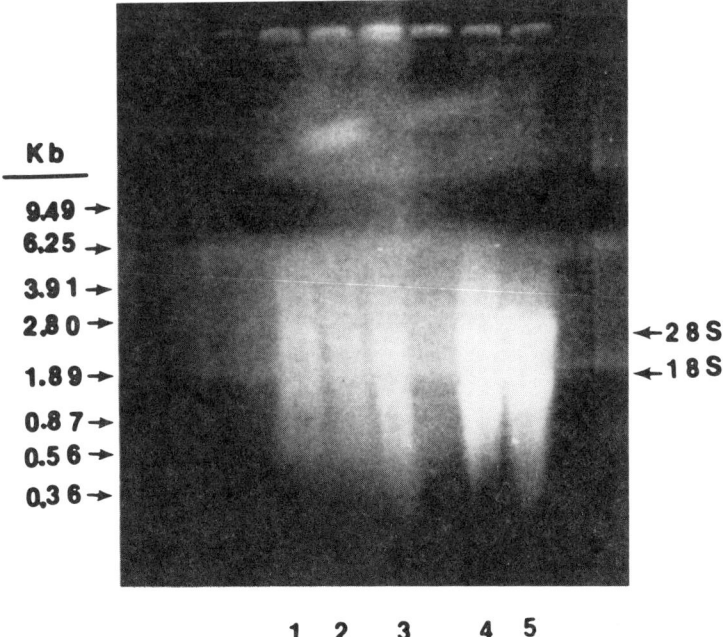

**FIGURE 2.2**
Poly(A)⁺RNA (lanes 1 to 3) was purified from total RNA (lanes 4 to 5) with oligo(dT)–cellulose and subjected to electrophoresis on a 1% agarose gel containing formaldehyde. rRNA bands cannot be seen in purified mRNA (lanes 1 to 3). mRNA (1 μg) or 15 μg of total RNA were loaded to appropriate lanes.

3. Neutralize the resin with 1 ml of binding buffer and gently mix with a sterile needle. Centrifuge at $1500 \times g$ for 2 min and carefully decant the supernatant. Repeat this step eight times. Resuspend the resin with 1 ml of binding buffer.

4. Add 1 volume of loading buffer to 100 μl of the total RNA sample (about 2 μg/μl). Heat the mixture for 10 min to 65°C and allow to cool to room temperature.

5. Load the total RNA sample to the 1-ml slurry of oligo(dT)–cellulose prepared in step 3. **Gently** shake for 2 min and place at room temperature for 15 min to allow the RNA to bind to the resin.

6. Centrifuge at $1500 \times g$ for 5 min and carefully remove the supernatant.

7. Add 0.5 ml of wash buffer to the resin and gently shake for 10 s. Centrifuge at $1500 \times g$ for 2 min. **Carefully** remove the supernatant. Repeat washing the resin three more times.

8. Add 0.2 ml of elution buffer to the resin and gently shake for 4 min. Centrifuge at $1500 \times g$ for 5 min. **Carefully** transfer the supernatant to a fresh tube. Repeat this step two more times.

9. Pool the supernatant and add 0.1 volume 3 $M$ sodium acetate and 2.5 volumes of 100% chilled ethanol to precipitate the mRNA. Mix and place at –20°C overnight.

10. Centrifuge at $12,000 \times g$ for 15 min and briefly wash the mRNA pellet with 1 ml of 70% ethanol. Dry the pellet under vacuum for 15 min and dissolve the mRNA in 10 μl of TE buffer. Take 2 μl of the sample to measure the concentration and quality of the mRNA. Store at –70°C until use.

# III. Fractionation of RNA by Sucrose Gradients Containing Methylmercuric Hydroxide

This is an efficient way to concentrate and find out a specific fraction of RNA of interest. RNA fractionated in this way can be used for cDNA cloning and reduce the large amount of work involved in screening the library to find the cDNA clone of interest. The number of positive cDNA clones can be increased significantly by reducing the background of noninterest clones. The RNA obtained can also be used for *in vitro* translation protein-synthesizing systems.

*Protocol*

1. Prepare sucrose gradients (10 to 30% w/v) containing 10 m$M$ methylmercuric hydroxide in 9/16″ × 3 1/2″ ultracentrifuge tubes (Beckman SW41 or equivalent). Dissolve the sucrose in sterile water, and treat the solutions overnight with 0.1% DEPC at 37°C followed by heating to 100°C for 15 min to remove the DEPC. When the solutions have cooled to room temperature, add 1 $M$ Tris-HCl (pH 7.4) and 0.5 $M$ EDTA (pH 7.4) to final concentrations of 10 m$M$ Tris-HCl and 1 m$M$ EDTA. Finally, add methylmercuric hydroxide to a final concentration of 10 m$M$. The gradients can be made by a gradient maker or by adding the solutions, one by one, in decreasing density of the solutions one over another, or in increasing density of the solution under each other in a tube. The convenient way is to place the tube on dry ice and add a density of sucrose solution. After it is frozen, add another density of solution. Finally, place the tube at 4°C overnight to gradually thaw and diffuse, generating a gradient.

*Caution:* *DEPC is a carcinogen and should be handled with care. Methylmercuric hydroxide is extremely toxic and volatile. Wear gloves when handling such solutions and perform these operations in a fume hood.*

2. Add methylmercuric hydroxide to a final concentration of 20 m$M$ to 100 μl RNA sample (1 μg/ μl) and carefully load the RNA sample onto the gradients.

3. Centrifuge the gradients at 34,000 rpm for 15 to 18 h at 4°C in a Beckman® SW41 rotor (or its equivalent).

4.  Collect 0.2- to 0.3-ml fractions through a hypodermic needle inserted into the bottom of the centrifuge tube. Dilute each of the fractions with 1 volume of sterile, DEPC-treated water containing 5 m$M$ β-mercaptoethanol. Add 60 µl of 3 $M$ sodium acetate buffer and place the tubes at 0°C for 1 h.

5.  Centrifuge at 12,000 × $g$ for 15 min at 0°C and **carefully** decant the supernatants into a fresh tube. Wash the RNA sample with 2 ml of 70% ethanol and allow the pellet to dry for 20 min at room temperature.

6.  Resuspend the RNA in 20 µl of water and add 0.1 volume of 3 $M$ sodium acetate buffer and 3 volumes of chilled 100% ethanol. Mix well and allow to precipitate at –70°C for 30 min.

7.  Centrifuge at 12,000 × $g$ for 10 min at 4°C. Dry the RNA under vacuum for 15 min. Dissolve the RNA in 20 µl of TE buffer. Take 4 µl of the sample to measure the concentration and quality of the RNA. Store the sample at –20°C until use.

## Reagents Needed

### Solution A

4 $M$ Guanidinium thiocyanate
25 m$M$ Sodium citrate, pH 7.0
0.5% Sarcosyl
Sterile-filter and store in a lighttight bottle at room temperature up to 3 months.

### Solution B

100 ml Solution A
0.72 ml β-Mercaptoethanol
Store at room temperature up to 1 month.

### Sodium Acetate Buffer (pH 4.0)

2 $M$ Sodium acetate
Adjust pH to 4.0 with 2 $N$ acetic acid. Autoclave.

### Sodium Acetate Solution

3 $M$ Sodium acetate
Autoclave.

### Water-Saturated Phenol

Thaw crystals of phenol at 65°C in a water bath and mix 1 part of phenol and 1 part of sterile dd.H$_2$O. Mix well and allow two phases to separate. Store at 4°C.

*Caution:*   *Phenol is a dangerous reagent; gloves should be worn and a fume hood should be used.*

### Chloroform:Isoamyl Alcohol (49:1)

98 ml Chloroform
2 ml Isoamyl alcohol
Mix and store at 4°C.

*Ethanol*

> 70% Ethanol: 70 ml of 100% ethanol and 30 ml dd.$H_2O$
> 75% Ethanol: 75 ml of 100% ethanol and 25 ml dd.$H_2O$
> 100% Ethanol: store at –20°C.

*20% SDS*

> 20 g SDS (Sodium dodecyl sulfate) in 100 ml dd.$H_2O$

*0.1 N NaOH*

> 0.4 g NaCl in 100 ml dd.$H_2O$

*RNA Extraction Solution*

> 3 *M* LiCl (126 g)
> 6 *M* Urea (360 g)
> Dissolve LiCl prior to adding urea. Adjust the volume to 1 liter with dd.$H_2O$
> and then sterile-filter.

*RNA Buffer*

> 10 m*M* Tris-HCl, pH 7.5
> 1 m*M* EDTA, pH 8.0
> 0.2% SDS

*Binding Buffer*

> 10 m*M* Tris-HCl, pH 7.5
> 0.5 *M* NaCl
> 1 m*M* EDTA
> 0.5% SDS

*Loading Buffer*

> 20 m*M* Tris-HCl, pH 7.5
> 1 *M* NaCl
> 2 m*M* EDTA
> 0.2% SDS

*Wash Buffer*

> 10 m*M* Tris-HCl, pH 7.5
> 100 m*M* NaCl
> 1 m*M* EDTA
> Autoclave.

*Elution Buffer*

> 10 m*M* Tris-HCl, pH 7.5
> 1 m*M* EDTA

### TE Buffer

> 10 m$M$ Tris-HCl, pH 8.0
> 1 m$M$ EDTA

### Homogenization Buffer

> 50 m$M$ NaCl
> 50 m$M$ Tris-HCl, pH 7.5
> 5 m$M$ EDTA, pH 8.0
> 0.5% SDS
> 200 µg/ml Proteinase K (from a stock solution of 20 mg/ml)

### LiCl Solution

> 8 $M$ LiCl Dissolved in distilled water and autoclaved

### CsCl/EDTA Solution

> 5.7 $M$ CsCl, 96.0 g of solid CsCl
> Dissolve in 90 ml of 0.01 $M$ EDTA, pH 7.5.
> Add DEPC to a final concentration of 0.1%.
> Allow the solution to stand for 1 h.
> Autoclave for 20 min at 15 psi in liquid cycle.
> When it has cooled to room temperature, adjust the volume to 100 ml with DEPC-treated water.

*Caution:* DEPC is a carcinogen and should be handled with care. Gloves should be worn when working with this reagent.

### Others

> 65°C Water bath or oven
> Tissue homogenizer (such as Polytron or Dounce)
> 50-ml Thick-walled polypropylene tubes, DEPC-treated
> Phenol:chloroform:isoamyl alcohol (25:24:1)
> Isopropanol
> RNase-free water

# IV. Measurement of RNA

## Protocol

1. Turn on the ultraviolet (UV) spectrophotometer and set the wavelength to 260 and 280 nm according to the instructions. Some spectrophotometers (e.g., UV 160U, Shimadzu) contain a computer unit that can automatically and simultaneously measure RNA at wavelengths of 260 and 280 nm, and calculate and display RNA concentration together with the ratio of $A_{260}/A_{280}$ reading numbers on the computer screen. If such a spectrophotometer is not available, RNA concentration can be measured by other, relatively simple UV spectrophotometers such as the Hitachi model 100–10 UV-visible spectrophotometer. The disadvantages are that the RNA first have to be

measured and recorded at 260 nm and then at 280 nm, and that the RNA concentration and the ratio of $A_{260}/A_{280}$ reading numbers have to be calculated manually.

2.  Set up a reference using blank solution. Depending on the RNA sample, add 1 ml of TE buffer (pH 8.0) or dd.$H_2O$ to a clean cuvette or to each of two cuvettes (one half to three quarters full), depending on whether one or two cells are available in the spectrophotometer. Insert the cuvette(s) into cuvette holder(s) in the sample compartment with the optical (clear) sides of the cuvette(s) facing the light path. Close the sample compartment cover and adjust the number to 0.000 (either manually or by pressing the "Auto Zero" button, depending on the specific spectrophotometer).

*Caution:*   *Gloves should be worn when handling cuvettes. Cuvettes should be rinsed with 95% ethanol followed by dd.$H_2O$, and wiped dry with Kimwipe paper prior to reuse. Unclean cuvettes may contain contaminated material that may affect readings between samples.*

3.  Check for mismatching of cuvettes by filling cuvette(s) with blank solution and recording the absorbance (positive or negative number). Make appropriate additive (for negative number) or subtractive (for positive number) corrections to future RNA sample readings. In most cases, however, this step is optional.

4.  In a clean cuvette appropriately dilute the RNA sample to be measured in blank solution to a total volume of 1 ml. For example, for 2, 5, 10, 15, 20, or 25 μl of RNA sample, the blank solution should be 998, 995, 990, 985, 980, or 975 μl, respectively. Wrap the cuvette in a piece of Parafilm and mix well by inverting the cuvette two to three times.

5.  Insert the sample cuvette and read the absorbance of the sample cuvette against the blank cuvette according to the spectrophotometer instructions. If only one cell is available, remove the reference cuvette after adjusting the number to 0.000 and insert the sample cuvette.

6.  After measuring RNA sample(s) at 260 and 280 nm, respectively, calculate the ratio of $A_{260}/A_{280}$ reading numbers and concentration for each RNA sample. Some spectrophotometers (e.g., UV 160U, Shimadzu) can automatically calculate, display, and print out the ratio and concentration for each RNA sample. A pure RNA should have a ratio of 1.85 to 2.0 of $A_{260}/A_{280}$ reading numbers. The concentration of a RNA sample can be manually calculated as follows:

$$\text{Total RNA (μg/μl)} = A_{260} \text{ reading number} \times 40 \text{ μg/ml} \times \text{dilution factor}$$

For example, if 25 μl of RNA is diluted to 1000 μl for measuring and its $A_{260}$ reading number is 0.5000, then

$$
\begin{aligned}
\text{Total RNA (μg/μl} &= A_{260} \text{ reading number} \times 40 \text{ μg/ml} \times \text{dilution factor} \\
&= 0.5000 \times 40 \text{ μg/ml} \times 1000/25 \\
&= 800 \text{ μg/ml} \\
&= 0.8 \text{ μg/μl}
\end{aligned}
$$

$$\text{Poly (A) + RNA (μg/μl)} = A_{260} \text{ reading number} \times 42 \text{ μg/ml} \times \text{dilution factor}$$

For example, if 25 μl of mRNA is diluted to 1000 μl for measuring and its $A_{260}$ reading number is 0.4000, then

$$
\begin{aligned}
\text{Poly (A)+RNA (μg/μl)} &= A_{260} \text{ reading number} \times 42 \text{ μg/ml} \times \text{dilution factor} \\
&= 0.4000 \times 42 \text{ μg/ml} \times 1000/25 \\
&= 672 \text{ μg/ml} \\
&= 0.672 \text{ μg/μl}
\end{aligned}
$$

# References

1. **Sambrook, J., Fritsch, E. F., and Maniatis, T.,** *Molecular Cloning, A Laboratory Manual,* 2nd ed., Cold Spring Harbor Press, Cold Spring Harbor, NY, 1989.

2. **Tesniere, C. and Vayda, M. E.,** Method for the isolation of high-quality RNA from grape berry tissues without contaminating tannins or carbohydrates, *Plant Mol. Biol. Rep.,* 9, 242, 1991.

3. **Chomczynski, P. and Sacchi, N.,** Single-step method of RNA isolation by acid guanidinium thiocyanate phenol-chloroform extraction, *Anal. Biochem.,* 162, 156, 1987.

4. **Wu, L.-L., Mitchell, J. P., Cohn, N. S., and Kaufman, P. B.,** Gibberellin (GA$_3$) enhances cell wall invertase activity and mRNA levels in elongating dwarf pea *(Pisum sativum)* shoots, *Int. J. Plant Sci.,* 154, 280, 1993.

5. **Lunn, G. and Sansone, E. B.,** Ethidium bromide: destruction and decontamination of solutions. *Anal. Biochem.,* 162, 453, 1987.

6. **Heikkila, J. J.,** Expression of cloned genes and translation of messenger RNA in microinjected *Xenopus* oocytes, *Int. J. Biochem.,* 22, 1223, 1990.

7. **Coleman, A.,** Translation of eukaryotic messenger RNA in *Xenopus* oocytes, in *Transcription and Translation: A Practical Approach,* Hames, B. D. and Higgins, S. J., Eds., IRL Press, Washington, D.C., 1984.

Chapter 3

# Extraction and Purification of Proteins/Enzymes

## Contents

Proteins and/or enzymes are the final products of gene expression. Although DNA stores the genetic information, it is the proteins that determine the shape and structure of a cell, a tissue, an organ, and the intact organism. Enzymes, which are usually proteins in nature, control the expression of genes and the development of an organism. Overall, proteins constitute half of the dry weight of the cell. In recent years, proteins and enzymes have become one of the core parts of molecular biology studies. This "hot field" primarily comes from the concept of "reverse genetics". A number of molecular biology studies start from specific protein(s) that determine the particular phenotype of an organism, and go back to clone, identify, and characterize the specific gene(s) expressing the protein(s) of interest. In addition, proteins/enzymes are involved in broad molecular biology studies such as the interaction between the *cis*-element and *trans*-factor, DNA/protein interaction in gel retardation and footprinting, screening of cDNA library as probes, immunoblotting or precipitation, and *in vitro* translation.[1–6]

In the study of proteins and enzymes, extraction and purification of these macromolecules are fundamentally important.[1,7,8] The present chapter describes the procedures for extraction and purification in detail.

# I.    General Considerations

Proteins and enzymes are relatively unstable macromolecules as compared with DNA. Their structures or activities are sensitive to a variety of factors or parameters involved in isolation and purification procedures. The following factors should be considered when handling proteins and enzymes.

1.  **Buffer conditions** — A buffer is a solution consisting of a conjugate base and a conjugate acid group that is able to resist pH change to varying degrees. It is strongly recommended that specific buffer conditions be maintained during the extraction and purification since the structures or activities of proteins/enzymes are very sensitive to environmental pH changes. Some investigators use distilled, deionized water to replace the extraction buffer. This may work well for some proteins or enzymes, but other proteins/enzymes may be degraded or lose their enzymatic activities. Therefore, buffer conditions should be optimized for specific cells, tissues, and protein types. In the preparation of buffers, we recommend preparation of a buffer stock solution (100X or 10X), which can be diluted to the desired working buffer concentration prior to use. After being autoclaved or sterile-filtered, the stock buffer solution is quite stable and resists contamination at 4°C or at room temperature.

2.  **Water purity** — Distilled, deionized water (dd.$H_2O$) is highly recommended to be used as the primary ingredient in solutions or buffer systems. Unpurified water contains a lot of microbial organisms or proteases that cause protein degradation.

3.  **Temperature** — Temperature is an important factor that causes protein degradation or loss of enzyme activity. Whenever possible, a cold environment, such as 0 to 4°C in an ice bucket with ice or in a cold room, should be maintained during the extraction and purification of proteins and enzymes in order to achieve the desired quality and integrity of proteins.

4.  **Protease inhibitors** — Proteases that hydrolize proteins or enzymes may exist in solutions and buffers or be released from cells upon their physical disruption. To maximally inhibit those proteases, protease inhibitors are recommended to be added to specific solutions or buffers used for protein/enzyme isolation and purification. Commonly used protease inhibitors include the following:

    a.  Phenylmethanesulfonyl fluoride (PMSF) serves to inhibit serine proteases such as chymotrypsin, trypsin, thrombin, or thiol protease (papain). PMSF is soluble in isopropanol at 10 mg/ml and is quite stable at room temperature as a stock solution.

b. Leupeptin functions to inhibit serine and thiol proteases such as plasmin, cathepsin B, and papain. A stock solution (10 mg/ml in dd.$H_2O$) is stable for 1 week at 4°C or 1 year at –20°C.

c. Ethylenediaminetetraacetic acid (EDTA) is a metalloprotease inhibitor and can also inhibit the activities of other proteases. An EDTA stock solution (0.5 $M$, pH 8.0) is stable for 6 months at 4°C. Use NaOH solution to adjust the pH, or the EDTA will remain insoluble.

d. Pepstatin A acts as an inhibitor of acid proteases such as pepsin, renin, cathepsin D, and chymosin. A stock solution (1 mg/ml in methanol) is stable for 1 year at –20°C.

Others reagents such as aprotinin and benzamidine are also effective protease inhibitors.

5. **Detergents** — The extraction and purification of membrane proteins usually requires detergents. Detergents are amphipathic molecules consisting of a hydrophobic portion of a linear or branched hydrocarbon "tail" and a hydrophilic "head". They can form micelles with the hydrophilic head portions facing outward. Membrane proteins are solubilized by the detergent and form mixed micelles with the detergent. The common detergents used for protein extraction and purification are as follows:

a. Ionic detergents — These detergents contain charged head groups (+/–) and serve to denature proteins in molecular-size separations. For example, sodium dodecyl sulfate (SDS) can denature proteins into their monomeric moieties and make proteins negatively charged. These properties can then be separated, based on their molecular weights (MW), by SDS-PAGE.

b. Nonionic detergents — These compounds have uncharged hydrophilic head groups and are less likely to disrupt protein–protein interactions. They are less denaturing in action than ionic detergents, and may cause proteins to aggregate. Common nonionic detergents include Triton X-100, Triton X-114, Nonidet P-40, octylglucoside, and Tween-20. These are used to block nonspecific protein interactions in solid phase immunochemistry such as enzyme-linked immunosorbent assay (ELISA), radioimmunoassay (RIA), and immunoblotting.

6. **Reducing agents** — Many proteins or enzymes may lose activity when oxidized by $O_2$. This can be avoided by adding a reducing agent to the solution or buffer. 2-Mercaptoethanol can reduce the intramolecular disulfide bonds of proteins. This may result in protein inactivation. Dithiothreitol (DTT) oxidation may cause the formation of stable disulfide bonds in proteins.

7. **Salts, metal ions, and ionic strength** — These should be considered for the extraction and purification of particular proteins and enzymes.

8. **Storage of proteins and enzymes** — The half-life of a protein largely depends on the storage temperature. The best conditions are at 4, –20, or –80°C, or in liquid nitrogen (–196°C) depending on particular uses. Frequently, thawing and freezing may cause protein degradation. Addition of glycerol (20 to 30%, or 50%) is often recommended for the storage of enzymes. Glycerol maintains the protein solution at very low temperature without freezing. Also, addition of 0.02% (3 m$M$) sodium azide is ideal for preventing contamination by bacteria or fungi during long-term storage of proteins or enzymes.

# II.    Protein Extraction Protocols

## A. Extraction of Proteins from Fresh Animal or Plant Tissues

1. Prepare appropriate extraction buffer solutions and place in ice water for at least 1 h prior to use. The type of buffer varies with the particular protein(s) of interest. The commonly used homogenizing buffers are as follows:

### 0.1 M Sodium Phosphate Buffer, pH 6.1

85% (v/v) 0.2 $M$ $NaH_2PO_4$ solution
15% (v/v) 0.2 $M$ $Na_2HPO_4$ solution

1 m$M$ EDTA
1 m$M$ DTT
0.1 m$M$ PMSF
1 m$M$ Benzamidine
Dissolve or mix well after each addition in dd.$H_2O$. Sterile-filtering is optional.

### Tris-Sucrose Buffer

0.1 m$M$ Tris-HCl, pH 7.0
250 m$M$ Sucrose
1 m$M$ MgCl$_2$
50 m$M$ KCl
0.1 m$M$ PMSF
Dissolve well after each addition in dd.$H_2O$. Sterile-filtering is optional.

### 1 M Tris Buffer, pH 7.6

1 $M$ Tris-HCl, pH 7.6
4 m$M$ EDTA
0.2% (v/v) β-Mercaptoethanol (not for enzyme extraction)
2% (w/v) Polyvinylpyrrolidone
1 m$M$ Benzamidine
Dissolve well after each addition in dd.$H_2O$. Sterile-filtering is optional.

### Phosphate-Buffered Saline (PBS), pH 7.4

NaCl (8 g)
KCl (0.2 g)
KH$_2$PO$_4$ (0.24 g)
Na$_2$HPO$_4$ (1.44 g)
Dissolve well after each addition in dd.$H_2O$. Adjust the pH to 7.4 with 3 $N$ HCl. Add dd.$H_2O$ to 1 liter. Autoclave.

2. Rinse the tissue (1 to 5 g per extraction for animal tissue, and 10 to 50 g per extraction for plant tissue) three times with an appropriate homogenizing buffer to wash away traces of blood or other extracellular materials.

3. Chop the washed tissue into small pieces (2 to 5 mm) with a clean razor blade or equivalent knife.

4. Transfer the chopped tissues into a homogenization jar, or a glass mortar or blender. Add 4 to 5 volumes of ice-cold homogenizing buffer to it. Place on ice.

5. Homogenize the tissue on ice to a fine homogenate.

*Notes:* *(1) If using a power-driven Polter-Elvehjem glass-Teflon homogenizer, set the speed at 500 to 1500 rpm and pass the homogenizer through the sample four to six times at 5 to 10 s per stroke. (2) If using a blender or a polytron, homogenize the tissue at top speed for 1.5 to 2 min, pausing for 4 s between each pulse of 20 s. (3) If using a hand homogenizer, pass the sample through 10 to 20 times until a fine homogenate is obtained.*

6. Quickly vacuum-filter the homogenate through four layers of cheesecloth into a beaker on ice.

7. Save the filtrate and directly use it for total proteins (animal tissue) or for total soluble proteins (plant tissue) precipitation in step 8. Residues filtered from plant tissue can be used for cell-wall-bound (insoluble) protein extraction as follows:

    a. Resuspend the residue in 2 volumes of appropriate buffer containing 1 *M* NaCl.

    b. Keep at 4°C for 12 to 24 h with stirring.

    c. Carry out vacuum-filtration as in step 6 and save the filtrate for insoluble protein precipitation. Proceed to step 8.

8. Centrifuge the filtrate (soluble and insoluble) at 4000 to 5000 × *g* for 5 min at 4°C to remove cellular debris. Carefully transfer the supernatant into a fresh beaker or flask and place it on ice.

9. Measure the volume of the supernatant and precipitate proteins as follows:

    a. Precipitation of total proteins:

        i. Add 80% (w/v) solid ammonium sulfate [$(NH_4)_2SO_4$] to the supernatant prepared in step 8.

        ii. Mix well and allow to precipitate at 4°C for 30 to 60 min.

        iii. Centrifuge at 12,000 to 15,000 × *g* at 4°C for 20 min.

        iv. Discard the supernatant and resuspend the protein pellet in 2 ml (for 1 to 2 g of animal tissue) or in 10 ml (for 25 to 50 g plant tissue) of appropriate dialysis buffer such as 50 m*M* sodium acetate buffer (pH 4.5 to 5.5). Keep the sample at 4°C until use.

        v. Soak a dialysis bag in diluted dialysis buffer (1:5 dilution in dd.$H_2O$) for 30 min. Seal the bottom end of the bag by making a tight knot or use clamps.

*Notes:*   *The exclusion size of the dialysis bag should be small enough to allow $(NH_4)_2SO_4$ or other ions to pass through. Total proteins should remain in the bag in order to prevent loss of some proteins.*

        vi. In a cold room, carefully transfer the protein sample in step iv. into a presoaked dialysis bag and seal the other end of the bag. Place the bag and a magnetic stir bar in a large beaker containing 2 l of 1:5 diluted dialysis buffer. Put the beaker on a stir plate and turn on the stir bar. Allow dialysis to occur for 48 h with fresh buffer replacements at 12- or 24-h intervals.

*Notes:*   *After dialysis, the volume of the sample increases due to water moving into the bag during dialysis. Therefore, leave some space (1 to 2 cm³) when sealing the bag for dialysis.*

        vii. Carefully cut the top end of the dialysis bag with a pair of scissors and transfer the dialyzed sample into a fresh tube. Store the sample at 4 or –20°C until use for protein concentration measurement and purification of specific protein(s) or enzyme(s) of interest.

    b. Fractional precipitation of proteins:

    In order to take advantage of different proteins being precipitated at different levels of $(NH_4)_2SO_4$ saturation, total proteins can be divided into several fractions. This will make the purification of specific proteins/enzymes much easier later on. The number of fractions and $(NH_4)_2SO_4$ saturation range of fractions depend on one's particular interest. The procedure below is designed for three $(NH_4)_2SO_4$ precipitation fractions at 0 to 30, 30 to 50, and 50 to 80% saturation.

        i. 0 to 30% $(NH_4)_2SO_4$ precipitation:

          • Add 30% (w/v) solid $(NH_4)_2SO_4$ to the supernatant prepared in step 8.

          • Carry out steps 9.a.ii to 9.a.vii in the procedure for total protein precipitation; however, save the supernatant in step 9.a.iv.

        ii. 30 to 50% $(NH_4)_2SO_4$ precipitation:

          • Add 50% (w/v) solid $(NH_4)_2SO_4$ to the supernatant saved from the 0 to 30% $(NH_4)_2SO_4$ precipitation.

          • Carry out steps 9.a.ii to 9.a.vii in the procedure for total protein precipitation; however save the supernatant in step 9.a.iv.

iii.  50 to 80% $(NH_4)_2SO_4$ precipitation:
- Add 80% (w/v) solid $(NH_4)_2SO_4$ to the supernatant saved from the 30 to 50% $(NH_4)_2SO_4$ precipitation.
- Carry out steps 9.a.ii to 9.a.vii in the procedure for total protein precipitation.

*Note:*    *Different protein samples such as soluble and insoluble proteins should be labeled during the above procedures.*

## B. Extraction of Proteins from Frozen Animal or Plant Tissues

1.   Thaw the tissues on ice.
2.   Carry out steps 1 to 9 described in Section II.A.

*Notes:*    *An alternative method is as follows. (1) Grind the frozen tissues in liquid $N_2$ to a fine powder and briefly warm the powder for 3 min at room temperature. (2) Transfer the powder to a beaker on ice containing 5 volumes of an appropriate extraction buffer. Thoroughly suspend the powder into the buffer and allow it to set for 30 min on ice. (3) Carry out steps 6 to 9 in Section II.A.*

## C. Extraction of Proteins from Cultures of E. coli.

1.   Harvest cells by centrifuging 500 to 1000 ml of *E. coli* liquid culture at $900 \times g$ for 15 min at 4°C.
2.   Discard the supernatant and resuspend the pellet in lysis buffer (3.5 ml/g wet cells).

*Notes:*    *For extraction of secretory proteins, save the supernatant and discard the cell pellet. The proteins can be directly precipitated as described in step 9 in Section II.A.*

3.   To the suspension, add 5 μl of 0.1 *M* PMSF as a protease inhibitor, and then add 0.1 ml of lysozyme (10 mg/ml) per 4 ml of cell suspension.

*Caution:*   *PMSF is extremely toxic to the eyes, skin, and membranes, and it may be fatal if swallowed. Care must be taken when handling this chemical. Gloves should be worn.*

4.   Incubate at 25 to 37°C for 20 min with shaking at 60 rpm.
5.   Add 4 mg of deoxycholic acid per 4 ml of cell suspension.
6.   Incubate at 37°C and stir with a glass rod until the lysate is viscous. Add 2 μl of *DNase* I (10 mg/ml) per 4 ml of cell suspension.
7.   Place the lysate at room temperature until it is no longer viscous (approximately 30 min).
8.   Centrifuge the cell lysate at $12,000 \times g$ for 15 min at 4°C.
9.   Use the supernatant for the crude extraction of part of proteins from *E. coli* by carrying out step 9 in Section II.A. The pellet contains cytoplasmic granules that usually include a high level of expression of proteins in *E. coli*. Therefore, further extraction of the majority of the proteins from these granules or inclusion bodies must be taken.

*Note:*    *In the crude extract, lysozymes and DNase I are included in the protein mixture, and they should be excluded when analyzing the protein in the crude extract.*

10. Resuspend the pellet from step 9 in 5 volumes of extraction buffer.

11. Incubate at room temperature for 10 min and centrifuge at 15,000 × g for 20 min at 4°C.

12. Save the supernatant on ice. Resuspend the pellet in 0.2 ml of lysis buffer containing 8 M urea and 0.1 mM PMSF.

13. Incubate at room temperature for 1 to 1.5 h.

14. Add 9 volumes of potassium buffer.

15. Incubate at room temperature for 30 min and adjust the pH to 8.0 with 3 N HCl.

16. Continue to incubate at room temperature for 30 min.

17. Centrifuge at 15,000 × g for 15 min at room temperature.

18. Pool the supernatant with the supernatant from step 12 and carry out protein precipitation as described in step 9 in Section II.A.

*Notes:*  *The pellet can be resuspended in 2X loading buffer and analyzed by SDS-PAGE. If a lot of the protein of interest is still in the pellet, further lysis and extraction may be needed.*

## Reagents Needed

### Lysis Buffer

50 mM Tris-HCl, pH 8.0
1 mM EDTA, pH 8.0
100 mM NaCl

### Extraction Buffer

10 mM EDTA, pH 8.0
0.5% Triton X-100

### Potassium Buffer

50 mM $KH_2PO_4$, pH 10.7
1 mM EDTA, pH 8.0
50 mM NaCl

## D. Extraction of Cytosolic Proteins from Eukaryotic Cells

1. Centrifuge the cell culture suspension at 500 to 900 × g at 4°C for 10 min.

2. Wash the cells by resuspending the cell pellet in 4 volumes of PBS or TBS and centrifuge as in step 1 and repeat washing twice.

3. Carry out lysis of the cells by either of the following methods.

   a. Nitrogen cavitation:

      i. Resuspend the cells in 10 volumes of cavitation buffer (pH 7.4).

      ii. Subject the cells to a stream of nitrogen gas at 375 psi and 4°C for 20 to 25 min in a cell-disruption chamber (Parr Instrument Co., Moline, IL).

      iii. Collect dropwise the cavitate in a tube and centrifuge at 1000 × g for 5 min at 4°C to remove cellular debris.

      iv. Transfer the supernatant to a fresh tube on ice and prepare to carry out protein precipitation.

    b. Lysis with detergent:

       i.   Resuspend the cell pellet in 10 volumes of lysis buffer.

      ii.   Incubate at 0 to 4°C for 1 to 2 h with occasional shaking.

     iii.   Centrifuge at $1000 \times g$ to remove cellular debris and save the supernatant for protein extraction.

4.   Carry out protein precipitation as described in step 9 in Section II.A.

## Reagents Needed

### TBS Buffer

        10 m$M$ Tris-HCl, pH 7.5
        150 m$M$ NaCl

### Cavitation Buffer

        130 m$M$ KCl
        0.1 m$M$ MgCl$_2$
        1 m$M$ EDTA
        1 m$M$ DTT
        10 m$M$ HEPES
        0.1 m$M$ PMSF
        10 mg/ml Each of leupeptin, soybean trypsin inhibitor, aprotinin, and pepstatin

### Lysis Buffer

        50 m$M$ Tris-HCl, pH 8.0
        150 m$M$ NaCl
        1% Nonidet P-40 or 1% Triton X-100
        0.1 m$M$ PMSF
        0.05% (w/v) SDS (should be omitted for enzyme isolation)

## E. Isolation of Proteins from Cellular Membranes in Cell Suspensions as well as Animal or Plant Tissues

1.   Harvest cells by centrifuging the cell suspension or culture at $900 \times g$ for 10 min at 4°C.

2.   Resuspend the cell pellet in 10 volumes of PBS or TBS buffer and centrifuge as in step 1. Repeat washing twice. For animal and plant tissues, wash the tissues three times with 10 volumes of dd.H$_2$O and slice the tissues into pieces.

3.   Resuspend the washed cells in 10 volumes of HEPES–KOH buffer. Place the animal tissue (2 to 5 g per extraction) or plant tissue (10 to 25 g per extraction) in 5 volumes of HEPES–KOH buffer.

4.   Homogenize the tissues or cell suspension on ice to a fine homogenate using an appropriate cell/tissue homogenizer.

5.   Centrifuge at $9000 \times g$ for 15 min at 4°C.

6.   Transfer the supernatant into fresh ultracentrifuge tubes, and discard the pellet.

7.   Carefully balance the centrifuge tubes and carry out ultracentrifugation at 50,000 to $55,000 \times g$ at 4°C for 60 min.

8.   Discard the supernatant, briefly air dry, and save the membrane pellet (enriched plasma membrane and intracellular membrane fractions).

*Notes:* *The membrane pellet can be directly resuspended in 1X loading buffer for protein analysis by SDS–PAGE (see Western blotting protocol in Chapter 5). If solubilization of membrane proteins is desired, proceed to the next step.*

9.  Resuspend the membrane pellet in 3 volumes of membrane extraction buffer.

10. Incubate at room temperature for 1 to 2 h with occasional shaking.

11. Add 7 volumes of potassium buffer to the lysis mixture and incubate at room temperature for 30 min.

12. Adjust the pH to 8.0 with 3 $N$ HCl and continue to incubate at room temperature for 30 min.

13. Centrifuge at 12,000 × $g$ for 5 min at room temperature.

14. Transfer the supernatant to a fresh tube or beaker, and carry out protein precipitation as described in step 9 in Section II.A.

## Reagents Needed

### HEPES–KOH Buffer

0.25 $M$ Sucrose
30 m$M$ HEPES
3 m$M$ EDTA
0.1 m$M$ MgCl$_2$
14 m$M$ 2-Mercaptoethanol
0.1 m$M$ PMSF
Adjust the pH to 7.5 with 1 $N$ KOH.

### Membrane Extraction Buffer

50 m$M$ Tris-HCl, pH 8.0
1 m$M$ EDTA, pH 8.0
100 m$M$ NaCl
0.1 m$M$ PMSF
8 $M$ Urea

### Potassium Buffer

50 m$M$ KH$_2$PO$_4$, pH 10.7
1 m$M$ EDTA, pH 8.0
50 m$M$ NaCl
Adjust the pH to 10.7 with KOH.

## III.   Protein Concentration Determination

Prior to protein analysis by SDS-PAGE or equivalent gels, or protein purification, the concentration of total or fractional proteins extracted should be determined by one of the following methods that are the most reliable methods used in our laboratory.

### A. Bicinchoninic Acid (BCA) Assay

1.  Prepare the solutions:

*Reagent A (1 l)*

> BCA (10 g)
> $Na_2CO_3 \cdot H_2O$ (20 g)
> $Na_2C_4H_4O_6(2H_2O)$ (1.6 g)
> NaOH (4 g)
> $NaHCO_3$ (9.5 g)
> Dissolve well after each addition in dd.$H_2O$.
> Adjust the pH to 11.25 with NaOH or solid $NaHCO_3$ (optional).
> Add dd.$H_2O$ to 1 liter.
> Store at room temperature for up to 1 year.

*Reagent B (100 ml)*

> $CuSO_4 \cdot 5H_2O$ (4 g)
> Add dd.$H_2O$ to 100 ml.
> Store at room temperature for up to 1 year.

*Working Solution*

> 50 volumes of Reagent A
> 1 volume of Reagent B
> Store at room temperature for up to 1 week.

*0.15% (w/v) Deoxycholate (DOC) Solution*

> Deoxycholate (0.15 g)
> Dissolve in 100 ml dd.$H_2O$.
> Store at 4°C.

*72% (w/v) Trichloroacetic Acid (TCA) Solution*

> TCA (72 g)
> Dissolve in 100 ml dd.$H_2O$.
> Store at 4°C.

2. Precipitate the protein sample using the DOC–TCA method in order to avoid interfering substances.

   a. Dilute protein sample to 1 ml with dd.$H_2O$ and add 0.1 ml of 0.15% DOC solution.

   b. Mix well by vortexing and let stand at room temperature for 10 min.

   c. Add 0.1 ml of 72% TCA solution, vortex, set for 1 min at room temperature, and centrifuge at $12,000 \times g$ for 15 min using a microcentrifuge.

   d. Decant the supernatant and completely air dry at room temperature by inverting the tube on a 3MM Whatman® filter paper.

   e. Dissolve the protein pellet in 70 μl dd.$H_2O$. Proceed to step 4.

3. Prepare the BSA standard curve:

   a. Prepare the BSA solution (1 mg/ml) in dd.$H_2O$. Store at 4°C until use.

   b. Individually add 0, 10, 20, 30, 40, 50, and 60 μl of the BSA solution to microcentrifuge tubes containing 1, 0.99, 0.98, 0.97, 0.96, 0.95, and 0.94 ml dd.$H_2O$, respectively.

   c. Add 0.1 ml of 0.15% DOC solution to each tube.

   d. Carry out steps 2.b to 2.e as described in the protein sample precipitation protocol with DOC–TCA.

4. Add 1.4 ml of working solution to each tube (1 volume of protein sample with 20 volumes of working solution).

5. Mix well and incubate at room temperature for at least 2 h or at 37°C for 30 min.

6. Let cool to room temperature in the case of 37°C incubation.

7. Transfer the liquid into spectrophotometer cuvettes and measure absorbance at 562 nm.

   a. Turn on the spectrophotometer and set the wavelength to 562 nm according to the instructions.

   b. Individually measure the absorbance of each of BSA standards, starting from zero and proceeding to higher concentrations of BSA.

      i. Transfer the liquid from the tube lacking BSA into a cuvette and insert the cuvette into the cell of the spectrophotometer. Use this tube as a blank or a reference and set the reading at 0.0000.

      ii. Replace the blank/reference cuvette with the one containing 10 µg BSA and record the absorbance.

      iii. In the same way, read and record the absorbance for 20, 30, 40, 50, and 60 µg BSA standards, respectively.

   c. Directly measure $OD_{562}$ for protein samples and record the absorbances.

*Note:* *If $OD_{562} > 3.000$, dilute the sample and read again.*

   d. Draw a standard curve and calculate the concentrations of protein samples. If a UV-visible recording spectrophotometer (e.g., UV 160U, Shimadzu) is available, the instrument can automatically record, on the computer monitor screen, each $OD_{562}$ and make a standard curve as well as calculate the concentration of each protein sample. If the spectrophotometer cannot simultaneously record $OD_{562}$, be sure to record each absorbance value and draw a standard curve on a graph paper using BSA concentration as the ordinate and $OD_{562}$ as abscissa. To determine the concentration of each protein sample, find its $OD_{562}$ and draw a horizonal line crossing to the standard curve. Starting from the cross-point on the standard curve, draw a vertical line crossing to the ordinate. This point of crossing on the ordinate is the concentration of BSA equal to the concentration of protein sample.

## B. Bradford Assay

This is a rapid and reliable method.

1. Prepare the solutions and reagents:

### Bradford Stock Solution

   95% Ethanol (100 ml)
   85% Phosphoric acid (200 ml)
   Serva Blue G (350 mg)
   Store at room temperature for up to 1 year.
   (*Note:* Serva Blue G is used for protein measurement and Serva Blue R is used for protein gel staining.)

### Bradford Working Buffer

   dd.$H_2O$ (425 ml)
   95% Ethanol (15 ml)
   85% Phosphoric acid (30 ml)

Bradford stock solution (30 ml)

Filter through Whatman No. 1 paper and store in a brown bottle at room temperature for up to 4 weeks.

2.  Prepare the BSA solution (1 mg/ml) in dd.H$_2$O. Add 0, 10, 20, 30, 40, 50, and 60 μl of BSA solution to each of seven microcentrifuge tubes, respectively. Add dd.H$_2$O to a final volume of 0.1 ml for each tube.

3.  Take 0.1 ml of each of the protein samples and transfer to another set of microcentrifuge tubes.

4.  Add 1 ml of Bradford working buffer to each tube, mix by vortexing, and let stand for 2 min but less than 1 h.

5.  Carry out steps 7.a to d in Section III.A, but set the spectrophotometer's wavelength to 595 nm.

# IV.     Purification of Protein(s) of Interest from Protein Mixture

Protein(s) of interest can be purified from an extracted protein mixture according to size, charge, and binding affinity.[1–3,5,7] Several techniques have been developed to do this. The commonly used methods described in the present chapter are gel filtration, ion-exchange chromatography, affinity chromatography, and gel electrophoresis (Figure 3.1). The high-pressure/performance liquid chromatography (HPLC) method is described in another chapter. However, the specific methods and the order of chromatography protocols used in protein purification depend on the particular protein or enzyme of interest. The detailed protocols described below have been used successfully in our laboratory for the purification of proteins as well as enzymes. One requirement for protein/enzyme purification is that a quick, easy, and inexpensive assay method must be known for enzyme purification, and that a known molecular weight (MW), specific affinity, or immunoaffinity of nonenzymatic protein(s) of interest can be detected using the appropriate methods.

## A.     Gel Filtration

Gel filtration or molecular-exclusion chromatography separates proteins according to molecular sizes. A number of commercial gel matrixes are available, such as Sephadex (G-10, G-25, G-50, and G-75), Sepharose, Sephacryl, Sepharose CL, and Bio-Gel. Sephadex is composed of polysaccharide that is readily contaminated by bacteria. Bio-Gel is made of polyacrylamide and can resist bacterial contamination, but it is very toxic. Different gel resins have different protein size-exclusion ranges. By choosing an appropriate gel resin as a column matrix, suspending the appropriate buffer, and transferring to a chromatography column, proteins can be separated by running them through the column. Because there are tiny holes in billions and billions of gel beads, small-sized proteins will pass through those holes and take a longer time to run out of the column as compared with large-sized proteins that cannot get into those holes, but instead run directly out of the column through void space in the column.

*Procedure*

1.  Suspend the appropriate amount of gel powder (e.g., Sephadex) in a clean beaker containing about 20 volumes of the appropriate buffer and allow to equilibrate for at least 1 to 2 h at 4°C.

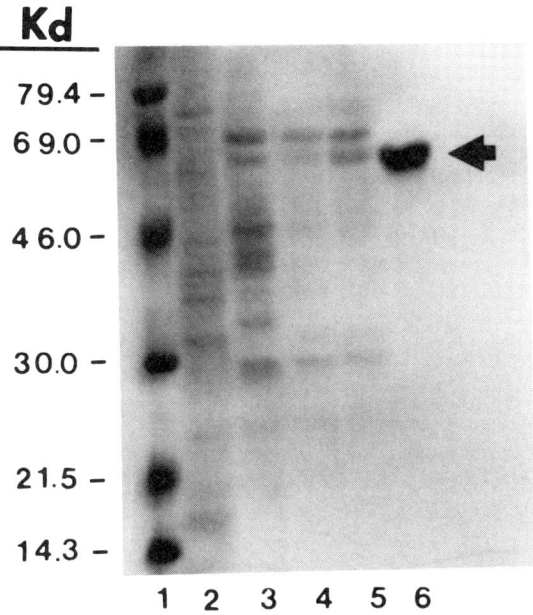

## Kd

79.4 –
6 9.0 –
4 6.0 –
30.0 –
21.5 –
14.3 –

1  2  3  4  5  6

**FIGURE 3.1**

Purification of invertase from a protein mixture subjected to SDS-PAGE. Lane 1: standard protein molecular weight markers. Lane 2: crude extract. Lane 3: CM Sephadex eluate. Lane 4: Sephadex G-75 fraction. Lane 5: ConA-Sepharose affinity column eluate. Lane 6: nondenaturing PAGE eluate.

*Notes:* *(1) All the procedures should be carried out at 4°C for enzyme purification. For other proteins, 4°C is also better than room temperature to prevent bacterial contamination. (2) The selection of the kind of gel-powder to be used as the column matrix depends on particular protein(s) of interest. For example, if the protein of interest is 70 Kd, an appropriate Sephadex or Bio-Gel product that can exclude proteins larger than 65 Kd should be chosen. Therefore, these proteins will run through the void space in the column faster. (3) The equilibrating buffer used for the column varies with the specific protein/enzyme to be purified. Usually, dd.H₂O (pH 7.0) is satisfactory for gel filtration.*

2.  Transfer the matrix suspension into a plastic or glass column (2 × 60 cm to 2 × 100 cm), setting it vertically with the bottom valve closed. The column should be filled with the gel suspension up to a level about 5 cm from the top. Add more buffer or dd.H₂O to the top and let stand for 30 min.

3.  Drain the fluid by opening the bottom valve and close the valve afterwards.

4.  Add 0.5 to 1.0 ml of an appropriate dye solution, such as Blue Dextran, to the top of the column, open the bottom valve, and collect the eluate in a beaker. Continue to add the buffer to the top of the column until the dye runs out of the column. The volume of collected eluate is the estimated void volume of the column.

*Note:* *The MW of the dye should be in the exclusion-size range of the gel matrix in order to accurately estimate the void volume of the column.*

5.  In a cold room, assemble an automatic fraction collector with about 100 collection tubes and place the collector under the column. Set the collection volume per tube at 1 to 2.5 ml. Above the column, connect a beaker or equivalent bottle containing 400 to 500 ml of the appropriate elution

buffer. The bottle or beaker should have a plastic pipe connected to the top of the column. If an automatic protein-peak UV detector is available, connect the detector to the bottom of the column, and then to the collector according to the instructions.

6. Pretest the flow of the assembled components by running the buffer through the column at flow rate of 0.5 to 1.0 ml/min.

7. Stop the addition of elution buffer to the column. Carefully add the extracted protein sample (1 to 10 ml) to the top of the column and start to collect the eluate. When the protein sample solution subsides into the bed-matrix, reconnect the elution buffer to the top of the column and allow the buffer to fill the top headspace of the column. Then adjust the flow rate at the top of the column so that it is the same as the flow rate at the base of the column. Allow the filtration through the column to occur for 10 to 20 h at 4°C.

8. Stop running the column and transfer the tubes in appropriate order to an ice water bath until analysis can be performed.

9. Carry out the appropriate enzyme assay for each of the tubes and pool the active fractions for further purification. For nonenzymatic proteins, appropriate analysis methods should be used to identify the positive fraction(s) containing the protein(s) of interest. These methods include immunoassay, immunoblotting, and MW determination by SDS–PAGE or by elution profile of standard protein markers chromatographed on the same column under the same conditions. The commonly used protein markers are thyroglobulin, bovine (669 Kd), apoferritin, horse spleen (443 Kd), α-amylase, sweet potato (200 Kd), alcohol dehydrogenase, yeast (150 Kd), albumin, bovine serum (66 Kd), carbonic anhydrase, and bovine erythrocytes (29 Kd). After chromatography, make a standard curve using Ve/Vo (Ve: elution volume, Vo: void volume) as the ordinate and log MW as the abscissa. For the protein sample, calculate the Ve/Vo for the particular fraction and determine the MW of proteins in the fraction from the standard curve. Pool the fraction(s) containing the expected proteins for further purification.

10. If necessary, dialyze the pooled samples against diluted elution buffer (1:5 dilution) to reduce the ionic strength prior to the next purification procedure.

11. Concentrate the pooled or dialyzed sample using Amicon centrifuge tubes or equivalent tubes. Add 5 to 7 ml to each tube, assemble the tubes according to the instructions, and centrifuge at 2000 to 3000 × $g$ for 20 min at 4°C. Stop the centrifugation, decant the fluid from the collection tube, and repeat centrifugation twice. Transfer the concentrated protein sample from the inner tube into a fresh tube and proceed to the next step of purification.

## B.    Ion-exchange Chromatography

Unlike gel filtration that separates proteins according to different molecular sizes, ion-exchange chromatography separates proteins based on their charge. The basic principle is that at a given pH most proteins have an overall negative or positive charge depending on their pI value. This makes it possible for them to interact with an oppositely charged chromatographic matrix. Different proteins have different amounts of charge, causing differential retardation in chromatography from which proteins are separated. There are two type of columns commonly used. One is diethylaminoethyl (DEAE) cellulose for binding to net negatively charged proteins. The other is carboxymethyl (CM) Sephadex for binding to net positively charged proteins.

### *Procedure 1: DEAE Cellulose Chromatography*

1. Prepare 2 l of 10 to 20 m$M$ phosphate buffer (pH 6.0) containing 1 m$M$ EDTA, 1 m$M$ benzamidine, and 0.1 m$M$ PMSF. Store at 4°C.

2.  Suspend an appropriate amount of DEAE cellulose in 20 volumes of phosphate buffer and allow it to equilibrate for 1 to 2 h at 4°C.

3.  Carry out steps 2 to 5 in Section IV.A except for the following:

    a.  The column size is $1.5 \times 20$ cm to $2 \times 40$ cm.

    b.  The flow rate is set at 2 ml/min and 3 ml per fraction.

    c.  The elution buffer is a linear gradient of 0 to 0.6 $M$ NaCl in phosphate buffer or dd.H$_2$O or appropriate buffer. This can be made by using a commercial gradient maker consisting of two chambers. First, close the channel between the two chambers. Add a volume (e.g., 250 ml) of phosphate buffer lacking NaCl to the inner chamber. Add an equal volume (e.g., 250 ml) of phosphate buffer containing 0.6 $M$ NaCl to the outer chamber. Drop a stir bar in the inner chamber and place the entire gradient maker on a stir plate that is set above the DEAE cellulose column.

    d.  Add the active-fraction pool purified by gel filtration onto the surface of the gel matrix and start to collect the eluate. Wash the column with two to three bed volumes of phosphate buffer (pH 6.0) and collect the eluate.

    e.  Elute the bound proteins with a 0 to 0.6 $M$ NaCl linear gradient by opening the valve of the gradient maker to the column and then opening the channel between the two chambers with the magnetic stir bar rotating.

*Notes:*  *The flow rate to the column is set at about 3 to 4 ml/min until the top space in the column is filled. Then it is reduced to the same flow rate as that from the bottom of the column. The channel between the two chambers should be open to obtain about the same flow rate as that from the gradient maker to the column. Record the fractions before elution and those fractions obtained during elution.*

4.  Transfer the tubes in the proper order to an ice water bath.

5.  Carry out an appropriate analysis to find the active or positive fractions containing the protein/enzyme of interest (Figure 3.2). Pool those fractions.

6.  Carry out steps 10 (optional) and 11 in Section IV.A.

### Procedure 2: CM Sephadex Chromatography

This chromatography method separates proteins based on their net charge as well as MW. The procedures are almost the same as those used in DEAE cellulose chromatography except for the following:

1.  Use CM Sephadex as the bed matrix for the column.

2.  The buffer pH should be 4.5 to 5.5 or an appropriate value depending on the particular protein of interest.

## C.   Affinity Chromatography

This is a powerful means of purifying proteins. The principle of this technique is based on the fact that some proteins have a very high affinity for specific chemical groups (ligands) covalently attached to a chromatographic bed material, the matrix. After loading and running the protein mixture through the column, only those protein(s) having a high affinity to the column can bind to the column matrix. Other proteins, in contrast, run directly through the column. The bound proteins(s) will be then eluted from the column by a solution containing a high concentration of the soluble form of the ligand or the specific residue that is recognized and bound by the proteins(s). A number of affinity columns are commercially available

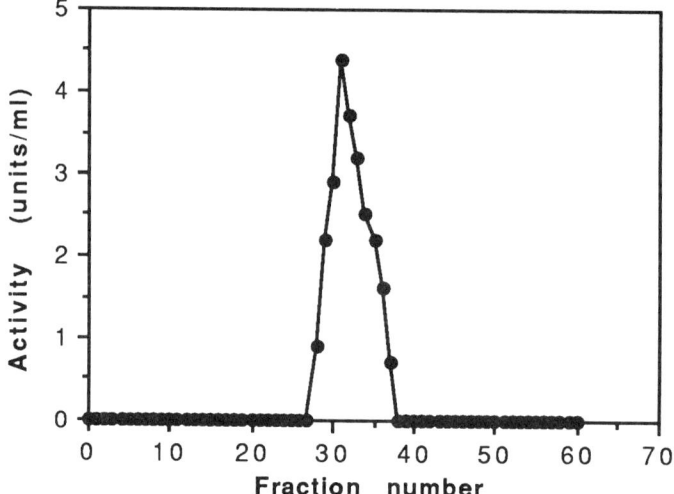

**FIGURE 3.2**
Purification of invertase from a protein mixture by diethylaminoethyl (DEAE) cellulose chromatography. A single peak is evident between fractions 27 and 38.

depending on the particular protein purification desired. One example is ConA-Sepharose affinity chromatography that is specific for glycoprotein purification. The procedures are almost the same as in Section IV.B except for the following:

*Procedure*

1. Use ConA-Sepharose as the bed matrix.

2. The buffer mixture should consist of 20 m$M$ sodium phosphate (pH 6.0), sodium acetate (pH 6.0) which contains 0.5 to 1 $M$ NaCl, 1 m$M$ MgCl$_2$, 1 m$M$ MnCl$_2$, and 1 m$M$ CaCl$_2$.

3. The linear gradient elution solution is 0 to 250 m$M$ α-methyl mannoside.

4. The flow rate is 1 ml/min and 2 ml per fraction.

## D.     Immunoprecipitation of Proteins

This technique is very powerful when a monoclonal antibody is available for a specific antigen of interest.

*Procedure*

1. Mix the protein sample with an excess of monoclonal antibody solution in an appropriate binding buffer that depends on the specific conditions required for the antibody–antigen interaction.

2. Incubate for 3 to 6 h at 4°C.

3. Add the appropriate reagent or resin to help precipitate the antigen–antibody complex and then allow to incubate for another 1 to 2 h at 4°C.

4. Centrifuge at 10,000 × $g$ for 15 min at 4°C to pellet the precipitate. Unprecipitated proteins are still solublized in the buffer.

5. Decant the supernatant and dissolve the pellet in an appropriate volume of buffer. SDS-PAGE analysis should be carried out to check the antigen of interest using a positive antigen as a control.

# E.    Purification of Proteins/Enzymes by Nondenaturing Gel Electrophoresis

The partially purified protein sample obtained from the previous chromatographic separations can be further purified with nondenaturing polyacrylamide gel electrophoresis (PAGE), or native gel electrophoresis. This procedure separates proteins according to their size and charge properties. The acrylamide pore size serves as a molecular sieve to separate different sizes of proteins. Further, proteins that are more highly charged at the pH of the separating gel migrate faster than those with less charged molecules. The major merit of PAGE is that it minimizes the denaturation of proteins in contrast to sodium dodecyl sulfate (SDS)-PAGE, which does denature proteins. Thus, many enzymes still have biological activities after running PAGE. These include, for example, esterase, dehydrogenase alkaline phosphatase, $\alpha$-amylase, transferases, hydrolases, lyases, $\beta$-fructosidase (invertase), and isomerases. The enzyme activity can be assayed either directly within the gel or following protein elution from the gel. Procedures for carrying out discontinuous native gel electrophoresis can be exactly the same as for SDS-PAGE except that there is no SDS component. The following procedure is a modified method that works well in our laboratory.

## *1. Preparation of the Separating Gel*

1.  Thoroughly clean glass plates and spacers with detergent; then wash with tap water, rinse with distilled water several times, and air dry.

2.  Wear gloves and rinse the glass plates and spacer with 100% ethanol, and wipe dry with Kimwipe paper tissues. Assemble a vertical slab gel unit, such as the SE 600 Vertical Slab Gel Unit (Hoeffer Co., San Francisco, CA), in the casting mode according to the instructions. Place one spacer (1.5 mm thick) at each of the two sides between the two glass plates and cam with clamps, forming a sandwich. Repeat for the second sandwich. The size of the separating gel varies with individuals. A standard size is $120 \times 140 \times 1.5$ mm.

*Note:*    *Apply a little stopcock grease to the spacer areas at both the top and bottom ends of the sandwiches to prevent potential leaking.*

3.  Check potential leaking by pipetting some distilled water into the sandwich and drain away the water afterwards by inverting the unit. Set the unit at an even level using a water balancer or its equivalent.

4.  Prepare the separating gel solution in a clean 100-ml beaker or 125-ml flask in the order shown below (for two pieces of standard-sized gels):
    *   12.5 ml Monomer solution for the separating gel
    *   16.25 ml 4X Separating gel buffer
    *   36.85 ml dd.H$_2$O
    *   Mix well after each addition.

5.  Degas the solution to remove any air bubbles by applying a vacuum for 5 min (optional). Add 0.2 ml of freshly prepared 10% ammonium persulfate solution (APS) and 80 µl TEMED. Mix by gently swirling. **Do not generate air bubbles.**

6.  Immediately pipette the mixture with a 50-ml syringe into the assembled sandwich up to a level 4 cm from the top.

7.  Take up 0.8 ml of dd.H$_2$O in a 1-ml syringe equipped with a 2-in. 22-gauge needle or its equivalent and carefully load 0.3 ml of distilled water, starting from one top corner next to the spacer, onto the surface of the acrylamide gel solution. Repeat on the other side of the slab next to the other spacer. The water will layer evenly across the surface of the gel mixture. The purpose of applying

water is to make the surface of the gel very even. A very sharp gel–water interface will be visible after the gel has polymerized.

8.  After polymerization, drain away the water layer by gently tilting the casting unit, and rinse with 2 ml of overlay solution once. Add 1 ml of overlay solution on the top of the gel and allow the gel to sit for 30 min.

## 2. Preparation of the Stacking Gel

1.  Prepare the stacking gel solution in a clean 25-ml beaker or 25-ml flask in the order shown below (for two pieces of standard-sized gels):

    12 ml Monomer solution for the stacking gel

    6 ml 4X Stacking gel buffer

    Degas (optional)

2.  Drain the overlay solution on the separating gel by tilting the casting unit and rinse with 2 ml of the 4X stacking gel mixture. Drain away the stacking gel solution and insert the comb (1.5 mm thick) into the top of the glass sandwich.

3.  Add 98 µl of 10% AP solution and 20 µl of TEMED to the gel mixture. Gently mix well and add the gel mixture by pipette or its equivalent into the sandwich from both upper corners next to the spacers, filling up to the top edge. Allow the gel to polymerize for 30 min at room temperature.

*Notes:*    *The stacking gel solution may be added prior to inserting the comb into the sandwich. However, air bubbles are easily generated and trapped around the teeth of the comb. We recommend that the comb be inserted into the sandwich prior to filling with the stacking gel mixture.*

## Reagents Needed

### Monomer Solution for Separating Gel

[38% (w/v) Acrylamide, 2% bis-acrylamide]
Acrylamide (76 g)
*N, N*-Methylene-bis-acrylamide (4 g)
Dissolve well after each addition in 150 ml dd.H$_2$O.
Add dd.H$_2$O to a final volume of 200 ml.
Wrap the bottle with aluminum foil and store at 4°C in the dark.

*Caution:*    *Acrylamide is neurotoxic. Gloves should be worn when handling this chemical.*

### Monomer Solution for the Stacking Gel

[5% (w/v) Acrylamide, 1.25% bis-acrylamide]
Acrylamide (10 g)
N,N-Methelene-bis-acrylamide (2.5 g)
Dissolve well after each addition in 150 ml dd.H$_2$O.
Add dd.H$_2$O to a final volume of 200 ml.
Wrap the bottle with aluminum foil and store at 4°C in the dark.

### 4X Separating Gel Buffer

[11.47% Tris (w/v), 28.92% 1 *N* HCl (w/v)]
Tris (23 g)

Dissolve well in 80 ml dd.$H_2O$.
1 $N$ HCl (57.84 ml)
Add dd.$H_2O$ to a final volume of 200 ml.
Store at 4°C.

## 4X Stacking Gel Buffer

0.158 m$M$ Tris (1.92 g)
Dissolve well in 50 ml dd.$H_2O$.
1 $N$ Phosphoric acid ($H_3PO_4$, 85%) (25.6 ml)
Add dd.$H_2O$ to a final volume of 100 ml.
Store at 4°C.

## 10% Ammonium Persulfate Solution (APS)

Ammonium persulfate (0.2 g)
Dissolve well in 2.0 ml dd.$H_2O$.
Store at 4°C for up to 10 days.

## Overlay Buffer

Separating gel buffer (25 ml)
Add dd.$H_2O$ to a final volume of 100 ml.
Store at 4°C.

## 3. Loading the Samples and Protein Standard Markers into the Gel

1.  While the stacking gel is polymerizing, add 0.2 to 0.5 volume of sample loading buffer to each of the partially purified protein samples and protein standard markers.

*Notes:* *The amount of proteins loaded into one well should be 10 to 35 μg in a total volume of 5 to 35 μl. The amount of protein standard marker for one well should be 2 to 10 μg.*

2.  Slowly and vertically pull the comb straight up from the gel. Rinse each well with upper tank buffer using a syringe or pipette to add the buffer. Carefully invert the casting stand to drain the wells and repeat twice.

3.  Place the casting stand upright and fill each well with 10 μl of upper tank buffer. Take up sample or markers with a syringe equipped with a 2-in. 22-gauge needle or pipette with an equivalent tip attached. Carefully insert the needle into the buffer in each well and **underlay** the sample or standard marker in the well. Repeat until all the samples are loaded.

*Notes:* *The volume loaded into each well should be less than 40 μl for standard-sized gels or less than 15 μl for mini-gels. Overloading may cause samples to float and cause contamination among wells. It is recommended that markers be loaded into the front left or right well, or both wells, leaving a blank well between markers and samples.*

4.  Carefully overlay the wells with upper tank buffer, using a syringe or pipette, until the buffer reaches the top of the gel.

*Notes:* *The samples and markers containing blue dye should be visible at the bottom of each well. Mark the orientation of samples and record in lab notebook.*

## Reagents Needed

### Sample Loading Solution

> Bromophenol blue (0.05 g)
> 50% Glycerol (25 ml)
> Dissolve well after each addition in 15 ml dd.$H_2O$.
> Add dd.$H_2O$ to a final volume of 50 ml and store at $-20°C$

### Upper-Tank Buffer (pH 8.9)

> 37.6 m$M$ Tris (4.56 g)
> 40 m$M$ Glycine (3 g)
> Add dd.$H_2O$ to a final volume of 1 liter.

### Lower-Tank Buffer (pH 7.5)

> 63 m$M$ Tris (22.7 g)
> 50 m$M$ HCl (1 $N$) (150 ml)
> Add dd.$H_2O$ to a final volume of 3 liters.

## 4. Electrophoresis

1.  Carefully place the upper buffer chamber in its proper position according to the instructions. Remove lower clamps and clamp the sandwiches to the bottom of the upper buffer chamber. **Do not disturb the wells.** Check any potential leaking by filling 10 to 20 ml of upper-tank buffer into the chamber.

*Note:*  *Apply a little stopcock grease on the spacer areas at the top corners of the sandwiches to prevent potential leaking.*

2.  Carefully transfer the assembled unit into the lower buffer chamber according to the instructions. Place the entire tank apparatus on a magnetic stirrer. Slowly fill the lower chamber with 3 liters of lower-tank buffer and add 1 liter of upper-tank buffer to the upper chamber, or use an appropriate volume depending the sizes of the chambers.

*Notes:*  *The bottom-chamber buffer should be of sufficient volume to cover two thirds of the slabs; otherwise, heating will not occur evenly, thus causing distortion of the band patterns of the gel. For the upper chamber, gently fill the running buffer from one corner into the chamber. In order to avoid washing out and/or mixing the samples, do not pour the buffer into the well area.*

3.  Before running the gel, put a stir bar in the lower chamber and stir the buffer for 1 min to remove any air bubbles trapped under the ends of the sandwiches.

4.  Add a few drops of dye solution, such as phenol red, into the upper-chamber buffer if the amount of dye in the samples is not sufficient.

5.  Place the lid on top of the unit and connect the power supply. The cathode (negative pole) should be connected to the upper buffer chamber and the anode (positive pole) should be connected to the lower buffer chamber.

6.  Set the power supply at constant power or current and turn on the power supply. Adjust the current to 25 to 30 mA/1.5-mm thick standard-sized gel. For two pieces of gel, set the current at 50 to

60 mA. The voltage will increase during the time the gel is running. Allow the gel run for several hours at 0 to 4°C until the dye reaches the bottom of the gel.

7.   Turn off the power supply and disconnect the power cables. Loosen and remove the clamps at both sides of the sandwiches. Place the gel sandwiches on the table or bench. Slowly remove the spacers. Use an extra spacer to carefully separate the two glass plates starting from one corner. Remove the top plate; the gel should be on the bottom glass plate. Using a razor blade, make a small cut at the upper left or right corner of the gel to record the orientation.

## 5. Staining and Destaining of the Gel Using Coomassie Blue (CB)

1.   Cut a small strip about 1 cm wide from each of the gels and transfer into a glass or plastic tray containing 100 ml of CB solution. Stain for 4 to 8 h at room temperature with slow shaking at 60 rpm. Alternatively, tightly cover the tray and stain for 20 to 30 min in a 50°C water bath. Wrap the rest of the gels with SaranWrap™ and keep at 4°C.

2.   Remove the staining solution and replace it with destaining solution I. Allow destaining for 1 to 1.5 h.

3.   Gently transfer the stained gel and carefully place it between two sheets of matrix. Gently roll the sandwich and insert it into a cylinder holder and cover the two ends of the holder. An alternative way is to keep the gel in the tray and replace the destaining solution I with destaining solution II. Destain the gel with shaking at 60 rpm until a clear background is obtained.

4.   Place the cylinder into the SE 530 Destainer (Hoeffer Co.), or its equivalent filled with destaining solution II. Put a stir bar into the tank and place the tank on a magnetic stirrer. Turn on the stir bar and allow the gel to destain overnight up to a few days until a clear background is achieved.

Notes:   *A fast destaining method is to destain the gel with 0.5 to 1% (v/v) Clorox® (commercial sodium hypochlorite bleach) in distilled water. Constantly monitor the destaining of the background. When relatively clear protein bands are visible, immediately rinse the gel with a large volume of distilled water several times until the blue color is stable. The disadvantage of this procedure is that if the gel is overdestained, some bands may disappear. Never use >1.5% Clorox.*

## Reagents Needed

### 1% Coomassie Blue (CB) Solution

Coomassie Blue R-250 (2 g)
Dissolve well in 200 ml dd.$H_2O$.
Filter.

### Staining Solution

1% CB solution (125 ml)
Methanol (500 ml)
Acetic acid (100 ml)
Add dd.$H_2O$ to a final volume of 1000 ml.

### Destaining Solution I

Methanol (500 ml)
Acetic acid (100 ml)
Add dd.$H_2O$ to a final volume of 1000 ml.

### *Destaining Solution II*

> Acetic acid (245 ml)
> Methanol (175 ml)
> Add dd.H$_2$O to a final volume of 3.5 l.

## 6. Identifying the Band(s) in the Unstained Gels

1. After staining and destaining, carefully match the stained and unstained portions of the gel on a glass plate.

2. Cut individual gel strips from the unstained portions of the gel, which are equivalent to the same areas as the stained bands.

3. Slice individually the cut gel strips into pieces using a clean razor blade. Use part of the slices (approximately one third) to directly assay the enzyme activity of interest using an appropriate enzymatic method or other detecting method such as an immunoassay. Keep the rest of the slices at 4°C until protein/enzyme elution is performed.

4. Based on the assay data, identify the positive band(s) of interest and proceed to protein/enzyme elution.

## 7. Electroelution of Protein/Enzyme of Interest from Gel Slices (Using the Model 422 Electro-Eluter of Bio-Rad Inc.)

1. Prepare the elution buffer as follows:
   - 25 m*M* Tris base
   - 192 m*M* Glycine
   - 0.1% SDS (omitted for enzyme elution)

2. Soak the membrane caps in the elution buffer at 60°C for 2 h prior to use.

3. Wear gloves and carefully attach the cap, filled with the fresh elution buffer, to the bottom of the glass tube and insert the tube into the rack of the eluter. There is a dialysis lid at the bottom of the tube to hold the gel slices in place. The protein/enzyme can pass through the lid by eletroelution down to the cap containing a membrane at the bottom that only allows the electrons to pass through.

*Note:*   *Avoid any air bubbles in the membrane cap.*

4. Fill the tube with elution buffer and carefully transfer the gel slices containing the protein/enzyme of interest into the elution tube. Avoid any air bubbles inside the tube.

5. Assemble the eluter according to the instructions. Fill the upper and lower chambers with appropriate volumes of elution buffer. Place a stir bar in the lower chamber.

6. Connect to the power supply with the positive pole at the bottom tank and the negative pole at the upper tank. Carry out the elution of protein/enzyme at 9 mA per glass tube (constant current) for 1 to 3 h at 4°C or at room temperature.

7. Reverse the positive and negative electrodes and continue to elute for 1 to 5 min to loosen the protein/enzyme adhering to the membrane of the cap, allowing it to enter the elution buffer.

8. Transfer the rack containing the elution tubes onto the lab bench. Avoid shaking during this process. Gently remove some of the elution buffer from the tube to a level slightly above the bottom of the tube and carefully remove the cap from the bottom of the tube. Suspend the eluted solution in the cap by sucking it up and down with a pipette.

9. Transfer the solution (approximately 0.6 ml) to a fresh tube and add 50 to 100 µl fresh elution buffer to wash the cap and collect the solution. Pool the solutions and measure the concentration of the elution protein(s). Assay enzyme activity if necessary.

# References

1. **Deutcher, M. P., Ed.,** *Guide to Protein Purification. Methods in Enzymology,* Vol. 182, Academic Press, New York, 1990.

2. **Dean, P. D. G., Johnson, W. S., and Middle, F. A., Eds.,** *Affinity Chromatography: A Practical Approach,* IRL Press, Oxford, 1985.

3. **Anonymous,** *Affinity Chromatography. Principles and Methods Handbook,* Pharmacia Fine Chemicals, Ljungforetagen AB Orebro, Sweden, 1983; **Scopes, R. K.,** *Protein Purification. Principles and Practice,* 2nd ed., Springer-Verlag, New York, 1987, chap. 5, 7.

4. **Davis, L. G., Dibner, M. D., and Battey, J. F.,** *Basic Methods in Molecular Biology,* Elsevier, New York, 1986, chap. 19.

5. **Wu, L. -L., Mitchell, J. P., Cohn, N. S., and Kaufman, P. B.,** Gibberellin ($GA_3$) enhances cell wall invertase activity and mRNA levels in elongating dwarf pea *(Pisum sativum)* shoots, *Int. J. Plant Sci.,* 154, 280, 1993.

6. **Smith, B. J.,** SDS polyacrylamide gel electrophoresis of proteins, in *Methods in Molecular Biology, Proteins,* Vol. 1, Walker, J. M, Ed., Humana Press, Clifton, NJ, 1984, chap. 6.

7. **Harrington, M. G.,** Elution of protein from gels, in *Guide to Protein Purification. Methods in Enzymology,* Vol. 182, Deutscher, M. P., Ed., Academic Press, San Diego, 1990, chap. 37.

8. **Merril, C. R.,** Gel-staining techniques, in *Guide to Protein Purification. Methods in Enzymology,* Vol. 182, Deutscher, M. P., Ed., Academic Press, San Diego, 1990, chap. 36.

# Chapter 4

# Preparation of Nucleic Acid Probes

## Contents

Molecular biology studies usually involve molecular cloning, characterization, and analysis of gene expression. These procedures rely heavily on nucleic acid hybridization, such as Southern blots, Northern blots, and dot blots. One of the most important aspects of these protocols is the use of either radioactively or nonradioactively labeled nucleic acids (DNA or RNA) as probes that specifically hybridize with their complementary DNA or RNA strands. The quality of the probe plays an essential role in detecting specific DNA or RNA sequences of interest. Therefore, the preparation of a probe with high specific activity is critical in nucleic acid hybridization.[1] The present chapter describes in detail the most recent methods for the labeling of DNA and RNA.[2–5] These methods are well established and have been routinely used in our laboratory.

# I.     Radioactive Labeling Methods

## A.     Nick-Translation Labeling of dsDNA

The principle of nick translation is that one strand of a double-stranded DNA molecule is nicked with *DNase* I, generating free 3'-hydroxyl ends within the unlabeled DNA. *E. coli* DNA polymerase I, by virtue of its 5' to 3' exonucleolytic activity, removes nucleotides from the 5' side of the nick and simultaneously adds new nucleotides to the 3'-hydroxyl terminus of the nick. During the incorporation of new nucleotides one of four deoxyribonucleotides is radioactively labeled (e.g., $\alpha$-$^{32}$PdATP or $\alpha$-$^{32}$PdCTP, commercially available), and is incorporated into the new strand by a base complementary to the template. In this way, a high specific activity ($10^8$ cpm/µg) of labeled DNA can be obtained using $^{32}$PdATP or $^{32}$PdCTP. The protocol is given below.

*Protocol*

1.  Set up a reaction on ice as follows:
    *   10X Nick-translation buffer (5 µl)
    *   DNA sample (0.4 to 1 µg in <2 µl)
    *   Mixture of three unlabeled dNTPs (10 µl)
    *   [$\alpha$-$^{32}$P]dATP or [$\alpha$-$^{32}$P]dCTP (>3000 Ci/mmol; 4 to 7 µl)
    *   Diluted *DNase* I (10 ng/ml) (5 µl)
    *   *E. coli* DNA polymerase I (2.5 to 5 units)
    *   Add dd.H$_2$O to a final volume of 50 µl.

*Notes:*  *(1) Three unlabeled dNTPs can be made by mixing equal volumes of the three nucleotides (each is 1.5 mM stock solution from a commercial source) and minus [$\alpha$-$^{32}$P]dATP or [$\alpha$-$^{32}$P]dCTP selected as a label. (2) The amount of DNase I used in the reaction mixture should be optimized. Normally, a $10^5$-fold dilution of a stock solution of pancreatic DNase I (1 mg/ml) in ice-cold 1X nick-translation buffer containing 50% of glycerol is ready to use. (3) The amount of sample DNA can be as low as 25 ng in a reaction of 5 to 20 µl.*

2.  Incubate the reaction for 1 h at 15°C.

*Notes:*  *(1) The temperature should not be higher than 18°C, which can generate "snapback" DNA by E. coli DNA polymerase I. "Snapback" DNA can lower the efficiency of hybridization. (2) Relatively longer incubation (1 to 2 h) is acceptable. However, too-long incubation may reduce the overall length of the labeled DNA.*

3.  Stop the reaction by adding 5 µl of 0.2 *M* EDTA (pH 8.0) solution. Place the tube on ice.
4.  Determine the percentage of [$\alpha$-$^{32}$P]dCTP incorporated into the DNA.
    a.  DE-81 filter-binding assay.
        i.  Dilute 1 µl of the labeled mixture in 99 µl (1:100) of 0.2 *M* EDTA solution. Spot 3 µl of the diluted sample, in duplicate, on Whatman DE-81 circular filters (2.3 cm diameter). Dry the filters under a heat lamp.

ii.   Wash one of the two filters in 50 ml of 0.5 $M$ sodium phosphate buffer (pH 6.8) for 5 min to remove unincorporated cpm. Repeat washing once. The other filter will be used directly for total cpm in the sample.

iii.  Add an appropriate volume of scintillation fluid (about 10 ml) to each tube containing one of the filters. Count the cpm in a scintillation counter according to the instructions.

b.  TCA precipitation.

i.    Dilute 1 µl of the labeled reaction in 99 µl (1:100) of 0.2 $M$ EDTA solution. Spot 3 µl of the diluted sample on a glass-fiber filter or a nitrocellulose filter for determinating the total cpm in the sample. Air dry the filter.

ii.   Add 3 µl of the same diluted sample into a tube containing 100 µl of 0.1 mg/ml carrier DNA or acetylated BSA and 20 m$M$ EDTA. Mix well.

iii.  Add 1.3 ml of ice-cold 10% trichloroacetic acid (TCA) and 1% sodium pyrophosphate to the mixture. Mix well and incubate the tube on ice for 20 to 25 min to precipitate the DNA.

iv.   Filter the precipitated DNA on a glass-fiber filter or a nitrocellulose filter under vacuum. Wash the filter with 5 ml of ice-cold 10% TCA four times under vacuum. Rinse the filter with 5 ml acetone (for glass-fiber filters only) or 5 ml of 95% ethanol. Air dry the filter.

v.    Transfer the filters to two cpm counting tubes and add 10 to 15 ml of scintillation fluid to each tube. Count the total cpm and incorporated cpm in a scintillation counter according to the instructions.

5.   Calculate the specific activity of the probe.

a.  Calculate theoretical yield as follows:

$$\text{Theoretical yield (ng)} = \frac{\text{dNTP added} \times 4 \times 330 \text{ ng/nmol}}{\text{specific activity of the labeled dNTP (µCi/nmol)}}$$

b.  Calculate the percentage of incorporation:

$$\text{Percent incorporation} = \frac{\text{cpm incorporated}}{\text{total cpm}} \times 100$$

c.  Calculate the amount of DNA synthesized:

$$\text{DNA synthesized (ng)} = \text{percent incorporation} \times 0.01 \times \text{theoretical yield}$$

d.  Calculate the specific activity of the prepared probe:

$$\text{Specific activity (cpm/µg)} = \frac{\text{total cpm incorporated}}{(\text{DNA synthesized + input DNA) (ng)} \times 0.001 \text{ µg/ng}} \times 100$$

**Notes:**   *The total cpm incorporated is equal to the cpm incorporated $\times 33.3 \times 50$. The factor 33.3 comes from using 3 µl of 1:100 dilution for the filter-binding or TCA precipitation assay. The factor 50 is derived from using 1 µl of the 50-µl reaction for 1:100 dilution.*

**Example:**   Given that 25 ng DNA is to be labeled and that 50 µCi [$\alpha$-$^{32}$P]dCTP (3000 Ci/mmol) is used in 50 µl of a standard reaction, and assuming that $4.92 \times 10^4$ cpm is precipitated by TCA and that $5.28 \times 10^4$ cpm is the total cpm in the sample, the calculations are made as follows:

$$\text{Theoretical yield} \quad = \quad \frac{50\ \mu Ci \times 4 \times 330\ ng/nmol}{3000\ \mu Ci/nmol}$$

$$= \quad 22\ ng$$

$$\text{Percent incorporation} \quad = \quad \frac{4.92 \times 10^4}{5.28 \times 10^4} \times 100$$

$$= \quad 93\%$$

$$\text{DNA synthesized} \quad = \quad 0.93 \times 0.01 \times 22$$

$$= \quad 20.5\ ng$$

$$\text{Specific activity} \quad = \quad \frac{4.92 \times 10^4 \times 33.3 \times 50}{(20.5\ ng + 25\ ng) \times 0.001\ \mu g/ng} \times 100$$

$$= \quad 1.8 \times 10^9\ cpm/\mu g\ DNA$$

*Note:* *The specific activity of $2 \times 10^8$ to $2 \times 10^9$ cpm/μg is considered to be a high specific activity that should be used in Southern, Northern, or equivalent hybridizations.*

6.  Purify the probe by removing unincorporated isotope. This is an optional step but we strongly recommend that it be carried out because the unpurified probe may produce a background (unexpected black spots) on the hybridization filter.

    Chromatography on Sephadex G-50 spin columns Sephadex G-50 or Bio-Gel P-60 spin column is very effective for separating labeled DNA from unincorporated radioactive precursor such as [α-32P]dCTP or [α-32P]dATP and oligomers that are retained in the column. This is very useful when an optimal signal-to-noise ratio probe, 150 to 1500 bases in length, is generated for optimal hybridization.

    a.  Resuspend 2 to 4 g Sephadex G-50 or Bio-Gel P-60 in 50 to 100 ml of TEN buffer and allow to equilibrate for at least 1 h. Store at 4°C until use.

    b.  Insert a small amount of sterile glass wool in the bottom of a 1-ml disposable syringe using the barrel of the syringe to tamp the glass wool in place.

    c.  Fill the syringe completely with the Sephadex G-50 or Bio-Gel P-60 suspension.

    d.  Insert the syringe containing the suspension into a 15-ml disposable plastic tube and place the tube in a swinging-bucket rotor in a bench-top centrifuge. Centrifuge at $1600 \times g$ for 4 min at room temperature.

    e.  Repeat adding the suspended resin to the syringe and centrifuging at $1600 \times g$ for 4 min until the packaged volume reaches 0.9 ml in the syringe and remains unchanged after centrifugation (Figure 4.1).

    f.  Add 100 μl of 1X TEN buffer to the top of the column and recentrifuge as above. Repeat this step two to three times.

*Note:* *The volume of the column should remain unchanged.*

    g.  Transfer the spin column to a fresh 15-ml disposable tube. Add the labeled DNA sample onto the top of the resin dropwise using a pipette.

*Notes:* *If the labeled DNA sample is less than 0.1 ml, dilute it to 0.1 ml in 1X TEN buffer. The spin column can be attached at the bottom to a sterile, decapped microcentrifuge tube. The whole unit will then be inserted into a fresh 15-ml tube. The microcentrifuge tube serves to collect fluid during centrifugation. But this is optional as the fluid can*

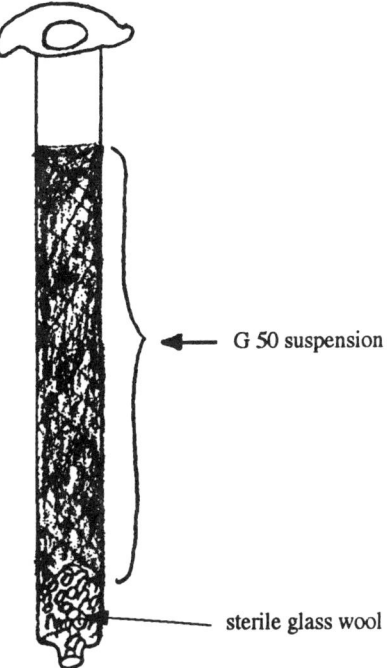

G 50 suspension

sterile glass wool

**FIGURE 4.1**
Preparation of a G-50 column for purification of labeled probe from unincorporated, radioactive nucleotides.

*be collected directly in the bottom of the 15-ml disposable tube without losing any of the fluid.*

h. Centrifuge at 1600 × g for 4 min at room temperature. Remove and discard the column containing unincorporated radioactive label in a radioactive waste container. Carefully transfer the effluent (about 0.1 ml) from the bottom of the decapped microcentrifuge tube or of the 15-ml tube to a fresh microcentrifuge tube. Cap it and store at −20°C until use.

## Reagents Needed

### 10X Nick-Translation Buffer

      500 m$M$ Tris-HCl, pH 7.5
      100 m$M$ MgSO$_4$
      1 m$M$ DTT
      500 µg/ml BSA (Fraction V, Sigma) (optional)
      Aliquot the stock solution and store at −20°C until use.

### Unlabeled dNTP Stock Solutions

      1.5 m$M$ Each dNTP

### Radioactively Labeled dNTP (Commercial Source)

      [α-$^{32}$P]dATP or [α-$^{32}$P]dCTP (3000 Ci/mmol)

### Pancreatic DNase I Solution

*DNase* I (1 mg/ml) in a solution containing 0.15 *M* NaCl and 50% glycerol. Aliquot and store at –20°C.

### E. coli DNA Polymerase I Solution

Commercial suppliers

### Stop Solution

0.2 *M* EDTA, pH 8.0

### 1X TEN Buffer

10 m*M* Tris-HCl, pH 8.0
1 m*M* EDTA, pH 8.0
100 m*M* NaCl

### Sephadex G-50 or Bio-Gel P-60 Powder

Commercial suppliers

## B.     Random-Primer Labeling of dsDNA

A population of random hexanucleotides synthesized by an automatic DNA synthesizer contains all four bases in every position. Such a mixture of random primers is available from commercial companies such as Promega Corporation, Pharmacia, and Boehringer Mannhein Chemicals. The primers are used to prime DNA synthesis *in vitro* from any denatured, closed circular, or linear dsDNA as template using the Klenow fragment of *E. coli* DNA polymerase I. Because this enzyme lacks 5′ to 3′ exonuclease activity, the DNA product is synthesized exclusively by primer extension instead of by nick translation. During the synthesis, one of the four dNTPs is radioactively labeled and incorporated into the new DNA strand being radioactively labeled, which is the probe that is needed. By primer extension, it is possible to generate a probe with extremely high specific activity ($10^8$ to $10^9$ cpm/μg) as more than 70% of the labeled dNTP can be incorporated into the new DNA strand. In addition, the DNA template can be as low as 25 ng as the template remains intact during the reaction, instead of being nicked at some points as seen in nick translation, which usually produces a shorter probe as compared with random-primer labeling.

### Protocol

1.   Add 1 μl of DNA template (0.5 to 1.0 μg/μl in dd.$H_2O$ or TE buffer) to a sterile microcentrifuge tube. Add an appropriate volume (19 to 38 μl dd.$H_2O$ or TE buffer) to the tube to dilute the DNA template to 25 to 50 ng/μl.

*Note:*   *If the concentration of DNA template is about 50 ng/μl, the dilution of the DNA is not necessary.*

2.   Cap the tube and denature the dsDNA into ssDNA by placing the tube in a water bath and boiling for 3 to 6 min. Quickly chill the tube on ice water for 2 to 4 min, and briefly spin down.

3.   Set up the reaction by adding the following components, in the order shown below, into a fresh microcentrifuge tube on ice.

- 5X Labeling buffer (10 μl)
- Denatured DNA template (25 to 50 ng) (1 μl)
- Three unlabeled dNTP mixture (500 μ$M$ each) (2 μl)
- Acetylated BSA (10 mg/ml) (optional) (2 μl)
- [α-$^{32}$P]dCTP or [α-$^{32}$P]dATP (>3000 Ci/mmol) (2.5 to 4 μl)
- Klenow enzyme (5 units)
- Add dd.H$_2$O to a final volume of 50 μl.

*Notes:*   *(1) The standard amount of DNA template used in the reaction should be 25 to 50 ng since the specific activity of the probe depends on the amount of the template. The lower the amount, the higher the specific activity of the probe. (2) Three unlabeled dNTPs can be made by diluting 1 μl of each of the dNTPs (1.5 mM stock solutions) to yield 3 μl of dNTPs mixture at 500 μM for each dNTP. (3) [α-$^{32}$P]dCTP or [α-$^{32}$P]dATP (3000 Ci/mmol) should not be >5 μl used in the reaction, otherwise, the background will be high.*

4.   Mix well and incubate the reaction at room temperature for 1 to 1.5 h.

5.   Stop the reaction mixture by adding 5 μl of 0.2 $M$ EDTA and place on ice. Store at –20°C until use.

*Notes:*   *(1) The labeled probe can be directly used for hybridization after being boiled for 5 to 10 min and chilled on ice for 4 min. However, the unpurified probe usually generates a high background (black spots) on the filter. We recommend that the probe be purified from unincorporated [α-$^{32}$P]dCTP or [α-$^{32}$P]dATP using a Sephadex G-50 or a Bio-Gel P-60 column. This is described under the protocol for nick translations. (2) The specific activity of the probe should be determined by the DE-81 filter-binding assay or the TCA precipitation assay (see the nick-translation protocol). The hybridization signal depends on the specific activity of the probe. The higher the specific activity of the probe, the stronger the signal of the hybridized bands or spots.*

## Reagents Needed

### 5X Labeling Buffer

> 250 m$M$ Tris-HCl, pH 8.0 (from stock solution)
> 25 m$M$ MgCl$_2$
> 10 m$M$ DTT
> 1 m$M$ HEPES, pH 6.6 (from stock solution)
> 26 A$_{260}$ units/ml Random hexadeoxyribonucleotides

### Three Unlabeled dNTPs Solutions

> 1.5 m$M$ Each dNTP

### Klenow Enzyme

> 5 units/μl, Labeling grade

## [α-³²P]dCTP or [α-³²P]dATP

        Commercially available source
        Specific activity of 3000 Ci/mmol

## TE Buffer

        10 m$M$ Tris-HCl, pH 8.0
        1 m$M$ EDTA

## Stop Solution

        0.2 $M$ EDTA

## C.    Labeling of ssDNA

    As compared with labeling of dsDNA, a ssDNA probe only has one complementary strand of a given DNA sequence. The advantage here is the elimination of the potential formation of nonproductive hybrids of the reannealed probe. During the detection of cross-hybridizing sequences in a distantly related species, hybridization between the complementary probes is much more probable and more stable than the hybridization between the probe and its target sequence of interest. This drawback is overcome by the use of ssDNA probe hybridization. ssDNA can be prepared from bacteriophage M13 or phagemid, or first-strand cDNA reversely transcribed from mRNA by reverse transcriptase.

    The labeling procedure is almost the same as for random-primer labeling of dsDNA except that the DNA template should be 0.5 to 1.0 µg in a 25-µl standard reaction mixture.

## D.    3′-End Labeling to Fill Recessed 3′ Ends of dsDNA

### Protocol

1. Set up a restriction enzyme digestion in order to create 3′ ends. In a microcentrifuge tube on ice, add the following components in the order shown:
   - DNA template (1 µg)
   - Appropriate restriction enzyme 10X buffer (2.5 µl)
   - Acetylated BSA (1 mg/ml) (optional) (2.5 µl)
   - Appropriate restriction enzyme (5 units)
   - Add dd.H₂O to a final volume of 25 µl.

*Notes:*   *(1) The DNA template can be cDNA, the genomic DNA insert, or plasmid DNA that is dissolved in dd.H₂O or TE buffer. (2) The restriction enzyme selected should generate recessed 3′ ends in order to be filled with radioactive dNTP. The common restriction enzymes used are EcoR I, Hind III, and BamH I.*

2. Incubate the reaction for 1 to 3 h at the appropriate temperature, depending on particular restriction enzyme.

*Notes:*    It is not necessary to inactivate and remove the enzyme. The digested reaction can be directly used for the labeling reaction.

3.    Add 2 to 4 µl of appropriate [α-³²P]dNTP (400 to 3000 Ci/mmol) to the digested reaction.

*Notes:*    *[α-³²P]dNTP used in the reaction depends on the sequence of the protruding 5′ end of the DNA. For EcoR I-digested DNA, the 3′ end can be labeled with [α-³²P]dATP. However, for BamH I-digested DNA, the 3′ end should be labeled with [α-³²P]dGTP.*

dsDNA <u>*EcoR*</u> I         ...pCpTpTpApAp 5′ <u>[α-³²P]dATP</u>        ...pCpTpTpApAp 5′

...pG_OH 3′                                    ...pGp*Ap*A_OH 3′

dsDNA <u>*BamH*</u> I        ...CpCpTpApGp 5′ <u>[α-³²P]dGTP</u>        ...CpCpTpApGp 5′

...G_OH 3′                                      ...Gp*G_OH 3′

4.    Add 2 to 5 units of the Klenow fragment of *E. coli* DNA polymerase I to the reaction and incubate the reaction at room temperature for 15 min.

*Note:*    *If the dsDNA to be labeled is predigested and purified, a labeling reaction can be set up as follows (on ice):*

- Digested DNA (1 µg)
- Klenow 5X buffer (10 µl)
- Acetylated BSA (1 mg/ml) (optional) (1 µl)
- Three unlabeled dNTPs (2 m*M*) (1 µl)
- Appropriate [α-³²P]dNTP (400 to 3000 Ci/mmol) (2µl)
- Klenow DNA polymerase I (2 to 5 units)
- Add dd.H₂O to a final volume of 50 µl.
- Incubate the reaction at room temperature for 15 min.

5.    Heat the reaction at 70°C for 5 min to stop the reaction. Place on ice until use.

6.    Separate the labeled DNA from unincorporated dNTP as described for nick-translation labeling.

7.    Measure the specific activity of the probe as described in nick-translation labeling (optional).

8.    Add 1 volume of TE-saturated phenol/chloroform to the reaction mixture in step 5. Mix well by vortexing for 1 min and centrifuging at $11,000 \times g$ for 5 min at room temperature.

9.    Carefully transfer the upper, aqueous layer to a fresh tube and add 0.1 volume of 2 *M* NaCl solution. Mix well. Add 2.5 volumes of chilled 100% ethanol. Allow to precipitate at –20°C for 1 h.

10.   Centrifuge at $12,000 \times g$ for 5 min, discard the supernatant, and briefly rinse the pellet with 1 ml of 70% ethanol.

11.   Dry the pellet under vacuum for 15 min and dissolve the pellet (labeled DNA) in 50 µl of TE buffer. Store at –20°C until use.

## *Reagents Required*

### *10X Appropriate Restriction Enzyme buffer*

Appropriate restriction enzyme from commercial source

### *Klenow 5X Buffer*

0.25 *M* Tris-HCl, pH 7.2
50 m*M* MgSO$_4$
0.5 m*M* DTT

### *Klenow DNA Polymerase I*

From commercial source

### *Three Appropriate Unlabeled dNTPs*

Make 2 m*M* stock solution for each dNTP.
Prepare a mixture of dNTPs by adding an equal volume of each stock of dNTP.

### *TE-Saturated Phenol/Chloroform*

Thaw phenol crystals at 65°C and mix equal parts of phenol and TE buffer. Mix well and allow the phases to separate at room temperature for 30 min. Take 1 part of the lower, phenol phase and mix with 1 part of chloroform:isoamyl alcohol (24:1). Mix well and allow the phases to separate. Store at 4°C until use.

### *TE Buffer*

10 m*M* Tris-HCl, pH 8.0
1 m*M* EDTA

## E.    3′-End Labeling of ssDNA with Terminal Transferase

### *Protocol*

1.  3′ Tailing of ssDNA primers
    a.  Set up a reaction as follows:
        - 5X Terminal transferase buffer (5 µl)
        - ssDNA primers (2 pmol)
        - [α-$^{32}$P]dATP (800 Ci/mmol) (1.6 µl)
        - Terminal transferase (10 to 20 units/µl) (1 µl)
        - Add dd.H$_2$O to a final volume of 25 µl.

    b.  Incubate at 37°C for 1 h and stop the reaction by heating at 70°C for 10 min to inactivate the enzyme.

*Note:*  *Multiple [α-$^{32}$P]dATPs were added to the 3′ end of the DNA.*

    c.  Calculate the percentage of incorporation and the specific activity of the probe. The procedures for the TCA precipitation assay and the DE-81 filter-binding assay are described in the section on nick-translation labeling.

$$\text{Percent incorporation} = \frac{\text{cpm incorporated}}{\text{total cpm}} \times 100$$

$$\text{Specific activity} = \frac{\text{percent incorporation} \times \text{total cpm added to the reaction}}{\text{milligrams of DNA in the reaction}} \times 100$$

**Example:**

$$49000/51000 \times 100 = 96\%$$

$$\frac{0.96 \times 3.5 \times 10^6}{0.09 \text{ mg}} \times 100 = 3.7 \times 10^7 \text{ cpm/}\mu\text{g DNA}$$

2.   3′-End labeling of ssDNA with a single [α-32P]cordycepin-5′-triphosphate lacking 3′-OH.

    a.  Set up a reaction as follows:

       •  5X Terminal transferase buffer (10 μl)

       •  ssDNA primers (10 pmol)

       •  [α-32P]Cordycepin-5′-phosphate (3000 Ci/mmol) (7.5 μl)

       •  Terminal transferase (10 to 20 units/μl) (2 μl)

       •  Add dd.$H_2O$ to a final volume of 50 μl.

    b.  Incubate the reaction at 37°C for 1 h.

    c.  Stop the reaction at 70°C for 10 min. The labeled reaction can be directly used in hybridization or purified as described in 3′-end labeling in order to fill the recessed 3′ ends of dsDNA.

    d.  Calculate the percentage of incorporation and the specific activity of the probe (see 3′ tailing of ssDNA).

## Components Needed

### 5X Terminal Transferase Buffer

        0.5 $M$ Cacodylate, pH 6.8
        1 m$M$ $CoCl_2$
        0.5 m$M$ DTT
        500 μg/ml BSA

### ssDNA Primers

        Prepared from bacteriophage M13 or phagemid, sscDNA, or ssDNA isolated
          from DNA.

### Terminal Transferase

        Available from commercial sources

### [α-32P]dATP

        3000 Ci/mmol
        Available from commercial sources.

### [α-32P]Cordycepin-5′-Phosphate Analog

        3000 Ci/mmol

## F.     5′-End Labeling Using Bacteriophage T4 Polynucleotide Kinase

### Protocol

1.  Set up a dephosphorylation reaction to remove phosphate group from both 5′ ends of the linear dsDNA.

    a.  Set up a reaction as follows:
    - 10X Calf intestinal alkaline phosphatase (CIAP) buffer (5 μl)
    - 5′ Ends DNA (2 to 10 pmol)
    - CIAP diluted in 1X CIAP buffer (0.5 units)
    - Add dd.H$_2$O to a final volume of 50 μl.

*Notes:* *(1) For protruding 5′-termini dephosphorylation, incubate at 37°C for 0.5 h and then add 0.5 units of alkaline phosphatase and incubate for 30 min at 37°C. (2) For recessed 5′-termini or blunt ends dephosphorylation, incubate at 37°C for 15 min and at 56°C for another 15 min. Add 0.5 units of alkaline phosphatase and incubate at 37°C and 56°C for 15 min, respectively.*

    b.  Stop and extract the reaction by adding 1 volume of TE-saturated phenol/chloroform. Mix well by vortexing for 1 min.

    c.  Centrifuge at 11,000 × g for 5 min and carefully transfer the top, aqueous phase to a fresh tube.

    d.  Add 1 volume of chloroform:isoamyl alcohol (24:1) to the supernatant. Mix well.

    e.  Centrifuge as in step c. Carefully transfer the supernatant to a fresh tube.

    f.  Add 0.1 volume of 2 *M* NaCl and 2 volumes of chilled 100% ethanol to the supernatant. Allow to precipitate at –70°C for 30 min.

    g.  Centrifuge at 12,000 × g for 5 min. Discard the supernatant and briefly rinse the pellet with 2 ml of 70% ethanol.

    h.  Dry the pellet under vacuum for 15 min and dissolve the pellet in 34 μl of forward exchange 1X buffer.

2.  Carry out the kinase reaction as follows:

    a.  Add the following to the reaction at step 1.h.
    - Reaction above (34 μl)
    - [α-$^{32}$P]dATP (3000 Ci/mmol) (15 μl)
    - T4 Polynucleotide kinase (8 to 10 units/μl) (1 μl)
    - Total volume of 50 μl.

    b.  Incubate at 37°C for 10 min and stop the reaction by adding 2 μl of 0.5 *M* EDTA.

    c.  Extract and precipitate the labeled DNA as in steps 1.b to 1.h.

*Notes:* *Ammonium ions are strong inhibitors of the bacteriophage T4 polynucleotide kinase. Do not use any ammonium acetate buffer prior to running the kinase reaction.*

    d.  Calculate the percentage of incorporation and specific activity as described for nick-translation labeling.

*Reagents Needed* —————————————————————————————————

## 10X CIAP Buffer

> 0.5 *M* Tris-HCl, pH 9.0
> 10 m*M* MgCl$_2$
> 1 m*M* ZnCl$_2$
> 10 m*M* Spermidine

## 10X Forward Exchange Buffer

> 0.5 *M* Tris-HCl, pH 7.5
> 0.1 *M* MgCl$_2$
> 50 m*M* DTT
> 1 m*M* Spermidine

## 0.5 M EDTA Buffer, pH 8.0

## 2 M NaCl Solution

## TE-Saturated Phenol/chloroform

> As described previously

## Chloroform:Isoamyl Alcohol (24:1)

## TE Buffer

> 10 m*M* Tris-HCl, pH 8.0
> 1 m*M* EDTA

## G.    Labeling of RNA by *In Vitro* Transcription

Recently, a number of plasmid vectors have been developed for subcloning of cDNA or genomic DNA inserts of interest. These vectors contain polycloning sites downstream from powerful bacteriophage promoters SP6, T7, or T3 in the vector. The cDNA or genomic DNA insert of interest can be cloned at the polycloning site between promoters SP6 and T7 or T3, forming a recombinant plasmid. The cDNA or genomic DNA inserted can be transcribed *in vitro* into single-strand sense RNA or antisense RNA from a linear plasmid DNA with promoter SP6, T7, or T3. During the process of *in vitro* transcription, one of the rNTPs is radioactively labeled and can be incorporated into the RNA strand, which is the labeled RNA. The labeled RNA probe usually has a high specific activity, and is much "hotter" than a ssDNA probe. As compared with DNA labeling, the yield of the RNA probe is very high because the template can be repeatedly transcribed. RNA probes can be easily purified from a DNA template merely by the use of RNase-free *DNase* I treatment. The greatest advantage of an RNA probe over a DNA probe is that the RNA probe can produce much stronger signals in a variety of different hybridization reactions.

## Protocol

1.  Prepare the linear DNA template for *in vitro* transcription. The plasmid is linearized by an appropriate restriction enzyme in order to produce "run-off" transcripts. To make RNA transcripts from the DNA insert, the recombinant plasmid should be digested by an appropriate restriction enzyme that cuts at one site that is very close to one end of the insert.

    a.  On ice, set up a plasmid linearization reaction as follows:
    *   Recombinant plasmid DNA (μg/μl) (5 μg)
    *   Appropriate restriction enzyme 10 X buffer (5 μl)
    *   Acetylated BSA (1 mg/ml) (optional) (5 μl)
    *   Terminal transferase (10 to 20 units/μl) (20 units)
    *   Add dd.H$_2$O to a final volume of 50 μl.

    b.  Incubate at the appropriate temperature for 2 to 3 h.

    c.  Extract by adding 1 volume of TE-saturated phenol/chloroform and mix well by vortexing. Centrifuge at 11,000 × $g$ for 5 min at room temperature.

    d.  Carefully transfer the top, aqueous phase to a fresh tube and add 1 volume of chloroform:isoamyl alcohol (24:1) to the supernatant. Mix well and centrifuge as in step c.

    e.  Transfer the top, aqueous phase to a fresh tube. Add 0.1 volume of 2 *M* NaCl solution or 0.5 volume of 7.5 *M* ammonium acetate, and 2 volumes of chilled 100% ethanol to the supernatant. Allow precipitation to occur at –70°C for 30 min or at –20°C for 2 h.

    f.  Centrifuge at 12,000 × $g$ for 5 min, decant the supernatant and briefly rinse the pellet with 1 ml of 70% ethanol. Dry the pellet for 15 min under vacuum and dissolve the linearized plasmid in 15 μl dd.H$_2$O.

    g.  Take 2 μl of the sample to measure the concentration of the DNA using UV-absorption spectroscopy at 260 and 280 nm. Store the sample at –20°C until use.

2.  Blunt the 3′ overhang end using the 3′ to 5′ exonuclease activity of Klenow DNA polymerase. Although this is optional, we recommend the 3′ protruding end be converted into a blunt end because some of the RNA sequence is complementary to that of the vector DNA. Enzymes such as *Kpn* I, *Sac* I, *Pst* I, *bgl* I, *Sac* II, *Pvu* I, Sfi, and *Sph* I should not be used to linearize plasmid DNA for *in vitro* transcription.

    a.  Set up, on ice, an *in vitro* transcription reaction as follows:
    *   5X Transcription buffer (8 μl)
    *   100 m*M* DTT (4 μl)
    *   rRNasin ribonuclease inhibitor (40 units)
    *   Linearized template DNA (0.2 to 1.0 g/μl) (2 μl)
    *   Add dd.H$_2$O to a final volume of 15.2 μl.

    b.  Add Klenow DNA polymerase (5 units/μg DNA) to the reaction and incubate at 22°C for 15 min.

    c.  To the reaction, add the following:
    *   Mixture of ATP, GTP, CTP, or UTP (2.5 m*M* each) (8 μl)
    *   120 UTP or CTP (4.8 μl)
    *   [α-³²P]UTP or [α-³²P]CTP (50 μCi at 10 mCi/ml) (10 μl)
    *   SP6 or T7 or T3 RNA polymerase (15 to 20 units/μl) (2 μl).

    d.  Incubate the reaction at 37 to 40°C for 1 h.

3.  Carry out large-scale *in vitro* transcription

   a.  In a microcentrifuge tube on ice, add the following in the order listed below:
       • 5X Transcription buffer (20 µl)
       • 0.1 DTT (8 µl)
       • Ribonuclease inhibitor (100 units)
       • Mixture of ATP, GTP, CTP or UTP (2.5 m*M* each) (20 µl)
       • Linearized DNA template (1 to 2.5 µg/µl) (2 µl)
       • [α-$^{32}$P]UTP or [α-$^{32}$P]CTP (50 µCi at 10 mCi/ml) (25 µl)
       • SP6, or T7 or T3 RNA polymerase (15 to 20 units/µl) (5 µl)
       • Add dd.H$_2$O to a final volume of 100 µl.
   b.  Incubate the reaction at 37 to 40°C for 1 to 2 h.

4. Remove the DNA template using *DNase* I.
   a.  Add RNase-free *DNase* I to a concentration of 1 unit/µg DNA template.
   b.  Incubate for 15 min at 37°C.

5. Purify the RNA probe.
   a.  Extract the enzyme by adding 1 volume of TE-saturated phenol/chloroform. Mix well by vortexing for 1 min and centrifuging at 11,000 × *g* for 5 min at room temperature.
   b.  Transfer the top, aqueous phase to a fresh tube and add 1 volume of chloroform:isoamyl alcohol (24:1). Mix well by vortexing and centrifuge at 11,000 × *g* for 5 min.
   c.  Carefully transfer the upper, aqueous phase into a fresh tube, add 0.5 volume of 7.5 *M* ammonium acetate solution and 2.5 volumes of chilled 100% ethanol. Allow to precipitate at –70°C for 30 min or at –20°C for 2 h.
   d.  Centrifuge at 12,000 × *g* for 5 min. Carefully discard the supernatant and briefly rinse the pellet with 1 ml of 70% ethanol and dry the pellet under vacuum for 15 min.
   e.  Dissolve the RNA probe in 20 to 50 µl of TE buffer and store at –20°C until use.

*Note:*   *The quantity and quality of the labeled RNA can be checked by denaturing agarose gel electrophoresis using 4 to 5 µl of the sample (see Northern blotting protocol).*

6. The percentage of incorporation and the specific activity of RNA probe can be determined right after the *in vitro* transcription reaction.
   a.  Estimate the cpm used in the transcription reaction. For example, if 50 µCi of NTP was used, the cpm is $50 \times 2.2 \times 10^6$ cpm/µCi = $110 \times 10^6$ cpm in 40 µl reaction, or $2.8 \times 10^6$ cpm/µl.
   b.  Carry out a TCA precipitation assay using 1:10 dilution in dd.H$_2$O as described for nick-translation labeling of DNA.
   c.  Calculate the percentage of incorporation.

   percent incorporation = TCA precipitated cpm/total cpm $\times 100$

   d.  Calculate the specific activity of the probe. If 1 µl of a 1:10 dilution was used for TCA precipitation, 10 × cpm precipitated = cpm/µl incorporated. In a 40-µl reaction, 40 × cpm/µl is the total cpm incorporated. If 50 µCi of labeled UTP at 400 µCi/nmol was used, then 50/400 = 0.125 nmol of UTP were added into the reaction. If there was 100% incorporation and UTP represents 25% of the nucleotides in the RNA probe, 4 × 0.125 = 0.5 nmol of nucleotides were incorporated, and 0.5 × 330 ng/nmol = 165 ng of RNA were synthesized. Then, the total ng RNA probe = % incorporation × 165 ng. For example, if 1:10 dilution of the labeled RNA sample has $2.2 \times 10^5$ cpm, the total cpm incorporated were $10 \times 2.2 \times 10^5$ cpm × 40 µl (total reaction) = $88 \times 10^6$ cpm.

$$\text{Percent incorporation} = 88 \times 10^6 \text{ cpm}/110 \times 10^6 \text{ cpm (50 } \mu\text{Ci)} = 80\%$$
$$\text{Total RNA synthesized} = 165 \text{ ng} \times 0.80 = 132 \text{ ng RNA}$$
$$\text{Specific activity of the probe} = 88 \times 10^6 \text{ cpm}/0.132 \text{ } \mu\text{g}$$
$$= 6.7 \times 10^8 \text{ cpm}/\mu\text{g RNA}$$

*Reagents Needed*

*RNase-Free DNase I*

*2 M NaCl Solution*

*7.5 M Ammonium Acetate Solution*

*Ethanol (100%, 70%)*

*TE-Saturated Phenol/Chloroform*

*Chloroform:Isoamyl Alcohol, 24:1 (v/v)*

*TE Buffer*

*5X Transcription Buffer*
>    0.2 $M$ Tris-HCl, pH 7.5
>    30 m$M$ MgCl$_2$
>    10 m$M$ Spermidine
>    50 m$M$ NaCl

*NTPs Stock Solutions*
>    10 m$M$ ATP in dd.H$_2$O, pH 7.0
>    10 m$M$ GTP in dd.H$_2$O, pH 7.0
>    10 m$M$ UTP in dd.H$_2$O, pH 7.0
>    10 m$M$ CTP in dd.H$_2$O, pH 7.0

*Radioactive NTP Solution*
>    [$\alpha$-32P]UTP or [$\alpha$-32P]CTP (400 Ci/nmol)

# II.    Nonradioactive Labeling Methods

Very recently, nonradioactive labeling techniques have been developed for DNA and RNA probe preparations. These methods have major advantages over traditional radioactive labeling in the following aspects.

1.  Safety — Radioactive labels (dNTPs) required for radioactive labeling are totally omitted in nonradioactive labeling methods. It is well known that radioactive dNTPs are absolutely dangerous to human beings.

2.  Simple and cheaper — Traditional radioactive labeling is usually complicated and causes contamination. Since the half-life of the isotope is short, for instance, [$\alpha$-$^{32}$P]dATP or [$\alpha$-$^{32}$P]dCTP only has a 14-day half-life and is very expensive, the labeled DNA or RNA is merely used once and cannot be kept longer. In contrast, nonradioactive labeling has simpler procedures and is relatively inexpensive. The major advantage is that the labeled RNA or DNA can be kept at $-20°C$ for months, and can be reused up to five to six times without any significant decrease in the hybridization signals.

Based on the above major advantages, the nonradioactive labeling methods are of much greater interest to researchers and scientists, and thus have a tendency of totally replacing traditional radioactive labeling in the near future. In the following sections, we describe several techniques for labeling nucleic acids using nonradioactive digoxigenin-dUTP as labeled dNTP.

The basic principle of the digoxigenin labeling and detection method is that, in a random labeling procedure, random hexanucleotides primers anneal to denatured DNA template. The Klenow enzyme catalyzes the synthesis of a new strand DNA complementary to the template DNA. During the incorporation of four nucleotides, one of them is labeled by digoxigenin, i.e., DIG-dUTP. After hybridization of the probe with the target DNA sequence that is immobilized onto a membrane filter, an antidigoxigenin antibody conjugated with an alkaline phosphatase will interact with the digoxigenin-dUTP. The detection of the hybridized signal will then be visualized with the chemiluminescent substrate Lumi-Phos 530 (Boehringer Mannheim Biochemicals). When the substrate is hydrolized by the alkaline phosphatase conjugated to the digoxigenin antibody, photons will be generated by fluorescence and hit the X-ray film. These will be visible as black spots or bands after the film is developed. The detection can also be visualized with the colorimetric substrates NBT and X-Phosphate, which give a purple/blue color.

## A. Random-Primer Labeling of dsDNA

1.  Denature the dsDNA template by boiling for 12 min and quickly chill on ice for 4 min. Briefly spin down and place on ice until use.

2.  In a microcentrifuge tube on ice, set up the following reaction:
    - Denatured DNA template (25 ng to 3 µg) (10 µl)
    - 10X Hexanucleotide primers mixture (4 µl)
    - 10X dNTPs mixture (4 µl)
    - dd·H$_2$O to 38 µl
    - Klenow enzyme (2 units/µl) (2 µl)
    - Total volume of 40 µl.

3.  Incubate at 37°C for 3 to 12 h.

*Notes:*  *The amount of labeled DNA depends on the amount of DNA templates and on the length of the incubation at 37°C. Based on our experience, the longer the incubation within 12 h, the more DNA is synthesized. However, when incubation is longer than 12 h, there is no significant increase in the amount of DNA labeled.*

4.   Add 4 μl of 0.5 $M$ EDTA solution to stop the reaction.

5.   Add 0.15 volumes of 3 $M$ sodium acetate buffer (pH 5.2) and 2.5 volumes of chilled 100% ethanol to the reaction. Allow to precipitate at –70°C for 1 h.

6.   Centrifuge at 12,000 × $g$ for 5 min at room temperature. Carefully discard the supernatant and quickly rinse the pellet with 1 ml of 70% ethanol. Dry the pellet under vacuum for 15 min.

7.   Dissolve the labeled DNA in 50 to 100 μl dd.H$_2$O or TE buffer. Store at –20°C until use.

*Notes:*   *(1) Labeled DNA (dsDNA) must be denatured before being used for hybridization. This can be done by boiling the DNA for 12 min then quickly chilling on ice for 4 min. (2) After hybridization, the probe contained in the hybridization buffer can be stored at –20°C and reused up to five times. Each time, incubate the used probe solution at 68°C for 5 min prior to reuse. (3) The labeled DNA is recommended, but not required, to estimate the yield by dot blotting of serial dilutions of commercially labeled control DNA and labeled sample DNA on nylon membrane or its equivalent. After hybridization and detection, compare the spot intensities of the control and the sample DNA (see dot blotting in Chapter 5).*

## Reagents Needed

### 10X Random Hexanucleotide Mixture

0.5 $M$ Tris-HCl, pH 7.2
0.1 $M$ MgCl$_2$
1 m$M$ Dithioerythritol (DTE)
2 mg/ml BSA
62.5 A$_{260}$ units/ml Random hexanucleotides

### 10X Labeling dNTP Mixture

1 m$M$ dATP
1 m$M$ dCTP
1 m$M$ dGTP
0.65 m$M$ dTTP
0.35 m$M$ DIG-dUTP
pH 6.5

### Klenow DNA Polymerase I

2 units/μl, Labeling grade

### 3 M Sodium Acetate Buffer, pH 5.2

### Ethanol (100%, 70%)

### TE Buffer

10 m$M$ Tris-HCl, pH 8.0
1 m$M$ EDTA

## B. Nick-Translation Labeling of dsDNA with Digoxigenin-11-dUTP

1.  On ice, set up a reaction as follows:
    *   dsDNA template, 1 to 2 μl (2 μg)
    *   10X DIG DNA labeling mixture (4 μl)
    *   10X Reaction buffer (4 μl)
    *   *DNase* I/DNA polymerase I (4 μl)
    *   Add dd.H$_2$O to a final volume of 40 μl.
2.  Incubate at 15°C for 40 min and stop the reaction by adding 4 μl of 0.4 *M* EDTA solution to the reaction and heating to 65°C for 10 to 15 min.
3.  Precipitate and dissolve the DNA as described under the section of random-primer labeling of dsDNA.

## Reagents Needed

### 10X DIG DNA Labeling Mixture

> 1 m*M* dATP
> 1 m*M* dCTP
> 1 m*M* dGTP
> 0.35 m*M* DIG-11-dUTP
> 0.65 m*M* dTTP
> pH 6.5 (+20°C)

### 10X Reaction Buffer

> 0.5 *M* Tris-HCl, pH 7.5
> 0.1 *M* MgCl$_2$
> 10 m*M* DTE

### DNase I/DNA Polymerase I

> 0.08 milliunits/μl *DNase* I
> 0.1 units/μl DNA polymerase I
> 50 m*M* Tris-HCl, pH 7.5
> 10 m*M* MgCl$_2$
> 1 m*M* DTE
> 50% (v/v) Glycerol

### Ethanol (100%, 70%)

## C. 3'-End Labeling with DIG-11-ddUTP

1.  On ice, set up a standard reaction as follows:
    *   DNA fragments or oligonucleotides (20 to 150 pmol)
    *   5X Reaction buffer (8 μl)

- CoCl$_2$ solution (8 μl)
- DIG-11-ddUTP (2 μl)
- Terminal transferase (2 μl)
- Add dd.H$_2$O to a final volume of 40 μl.

2.  Incubate the reaction at 37°C for 30 min and place on ice. Add 2 μl of 0.5 *M* EDTA buffer.

3.  Precipitate and dissolve the labeled DNA in the same manner as for random-primer labeling of dsDNA.

## Components Needed

### 5X Reaction Buffer

125 m*M* Tris-HCl, pH 6.6
1 *M* Potassium cacodylate
1.25 mg/ml BSA

### CoCl$_2$ Solution

25 m*M* Cobalt chloride

### DIG-11-ddUTP Solution

1 m*M* Digoxigenin-11-ddUTP (2′,3′-dideoxyuridine-5′-triphosphate coupled to digoxigenin via an 11-atom spacer) in dd.H$_2$O

### Terminal Transferase

50 units/μl
0.2 *M* Potassium cacodylate
1 m*M* EDTA
0.2 *M* KCl
0.2 mg/ml BSA
50% Glycerol, pH 6.5

## D. 3′ Tailing of DNA with DIG-11-dUTP/dATP

Everything is the same as for 3′-end labeling except that labeled dNTP is DIG-11-dUTP/dATP instead of DIG-11-ddUTP.

## E. 5′-End Labeling with DIG-NHS Ester

1.  Precipitate DNA fragments or oligonucleotides in 50 μl sodium borate solution.

*Note:   A free NH$_2$ group at the 5′ end of the DNA should be generated by a synthesizer.*

2.  Dissolve 1.3 mg DIG-NHS-ester (Boehringer Mannheim Biochemicals) in 50 μl dimethylformamide and add it to the 50 μl DNA sample.

3.  Incubate at room temperature for 12 to 20 h.

4.  Precipitate and dissolve the DNA as described for random labeling of DNA.

## Reagents Needed

### Sodium Borate Solution

>  0.1 $M$ Sodium borate, pH 8.5

### Dimethylformamide (100% ACS-grade)

## F. RNA Labeling

1.  On ice, set up a reaction as follows:
    - Purified and linearized plasmid DNA template containing insert of interest (1 µg)
    - 10X NTP labeling mixture (4 µl)
    - 10X Transcription buffer (4 µl)
    - RNA polymerase (T7, SP6, or T3) (4 µl)
    - Add DEPC-treated dd.$H_2O$ to a final volume of 40 µl.
2.  Incubate at 37°C for 2 to 3 h.
3.  Add 20 units *DNase* I (RNase-free) and incubate at 37°C for 15 min to remove the DNA template. Add 4 µl of 0.5 $M$ EDTA solution.
4.  Precipitate and dissolve RNA as described for random primer labeling of DNA.

## Components Needed

### 10X Transcription Buffer

>  0.4 $M$ Tris-HCl, pH 8.0
>  60 $M$ $MgCl_2$
>  100 m$M$ DTT
>  20 m$M$ Spermidine
>  0.1 $M$ NaCl
>  1 unit RNase inhibitor

### 10X NTP Labeling Mixture

>  100 m$M$ Tris-HCl, pH 7.5
>  10 m$M$ ATP
>  10 m$M$ CTP
>  10 m$M$ GTP
>  6.5 m$M$ UTP
>  3.5 m$M$ DIG-UTP

### DEPC-Treated Water

>  0.1% Diethylpyrocarbonate (DEPC) in dd.$H_2O$
>  Incubate at 37°C overnight followed by autoclaving.

### EDTA Solution

>  0.5 $M$ EDTA in dd.$H_2O$, pH 8.0

### RNA Polymerase

>  20 units/µl T7, SP6, or T3

# References

1. **Sambrook, J., Fritsch, E. F., and Maniatis, T.,** *Molecular Cloning, A Laboratory Manual,* 2nd ed., Cold Spring Harbor Press, Cold Spring Harbor, NY, 1989.

2. **Rigby, P. W. J., Dieckmann, M., Rhodes, C., and Berg, P.,** Labeling deoxyribo-nucleic acid to high specific activity *in vivo* by nick translation with DNA polymerase 1, *J. Mol. Biol.,* 113, 237, 1977.

3. **Feinberg, A. P. and Vogelstein, B.,** A technique for radiolabeling DNA restriction endonuclease fragments to high specific activity, *Anal. Biochem.,* 132, 6, 1983.

4. **Young, W. S.,** Simultaneous use of digoxigenin- and radiolabeled-oligodeoxyribonucleotide probes for hybridization histochemistry, *Neuropeptides,* 13, 271, 1989.

5. **Mitsuhashi, M., Cooper, A., Ogura, M, Shinagawa, T., Yano, K., and Hosokawa, T.,** Oligonucleotide probe design — a new approach, *Nature,* 367, 759, 1994.

# Chapter 5

# Electrophoresis, Blotting, and Hybridization

## Contents

# I.     Southern Blotting

Southern[1] first developed the method of detection of specific nucleotide sequences among a number of DNA fragments after they were separated by gel electrophoresis, so-called Southern blot, which has been subjected to later modification. The basic principle of the technique is that DNA fragments are first separated based on their molecular weights by standard agarose gel electrophoresis, and then they are blotted and immobilized onto a nylon or a nitrocellulose membrane. Specific band(s) of interest can be then detected by being hybridized with a specific DNA probe according to the base complementary rule. This is the most fundamental and useful technology that is used for DNA/DNA hybridization analysis. It is a simple, fast, and very sensitive method and is widely used in gene cloning, screening and isolation, DNA mapping, gene amplification, and general DNA analysis.[2–6]

## A.     Agarose Gel Electrophoresis of DNA

### General Considerations

Agarose gel electrophoresis is a standard method used to separate, identify, and purify DNA fragments. Agarose extracted from seaweed is a linear polymer that is basically composed of D-galactose and L-galactose. When agarose is melted and then allowed to harden, it forms a matrix which serves as a molecular sieve to separate DNA fragments of different sizes. Agarose is commercially available from many companies with a variation in its purity. We recommend that ultrapure grade of agarose that is DNase-free be used for Southern blotting or similar research. The following factors should be considered for the rate at which DNA fragments migrate in an agarose gel.

1.    Agarose concentration should be determined as follows:

| Gel percentage (w/v) | Linear DNA separation range (Kb) |
|:---:|:---:|
| 0.6% | 1–20 |
| 0.7% | 0.8–10 |
| 1% | 0.5–8 |
| 1.2% | 0.4–6 |
| 1.4% | 0.2–4 |

2.    The sizes of DNA fragments — Linear duplex molecules travel through the gel matrix at a rate inversely proportional to the log of their molecular weight. The smaller the fragment, the faster it moves.

3.    DNA conformation — DNA shape also influences the migration rate. For a DNA fragment with a particular size, the migration rate will be closed circular (supercoiled > slightly coiled) > linear > open circular.

4.    Current applied — The migration of DNA fragments is directly proportional to the voltage applied. Too high or too low a voltage is not recommended. The normal voltage should be 5 to 10 V/cm. The distance is referred to the length of the gel between the negative and the positive electrodes.

## Protocol

1. Prepare DNA sample(s) for electrophoresis. DNA isolation and purification are previously described (see Chapter 1 for details). The amount of DNA for one well in an agarose gel, based on our experience, is 20 to 30 µg for genomic DNA Southern blot and 2 to 10 µg for plasmid or λDNA. The minimum amount of specific band to be detected is approximately 15 to 25 ng. If DNA needs to be digested with restriction enzyme, the standard reaction for loading in one well is as follows:

   • DNA in TE buffer or dd.H$_2$O (20 to 30 µg genomic or 2 to 10 µg plasmid or λDNA) (2 to 8 µl)

   • Appropriate restriction 10X buffer (4 µl)

   • Appropriate restriction enzyme (3.4 units/µg DNA)

   • Add dd.H$_2$O to a final volume of 40 µl.

   Incubate at optimal temperature for the enzyme (usually 37°C) for 2 to 3 h and store at 4°C until electrophoresis is performed.

*Notes:* *If the amount of DNA in each well is not enough, the band(s) to be detected will be quite weak. However, if too much DNA is loaded in one well, the enzyme digestion may not be complete and electrophoresis may be not good, thus affecting hybridization and detection. After digestion, the sample without extraction and precipitation can be directly loaded into the gel and electrophoresis performed with no side effects.*

2. Thoroughly clean the appropriate gel apparatus by washing with detergent, completely removing the detergent mixture with tap water and rinsing with distilled water three to five times. Allow the apparatus to dry at room temperature.

3. Prepare an agarose mixture in a clean bottle or a beaker as follows:

| Components | Mini gel | Medi gel | Big gel |
|---|---|---|---|
| 1X TBE or TAE buffer | 30 ml | 75 ml | 120 ml |
| Ultrapure agarose (1% w/v) | 0.3 g | 0.75 g | 1.2 g |

*Notes:* *1X TBE or TAE buffer may be diluted from 5X TBE or 10X TAE stock solution with dd.H$_2$O. Since the agarose is only 1%, its volume can be ignored when calculating the final volume.*

4. Slightly mix and melt the agarose by gently boiling in a microwave oven for 1 to 3 min depending the gel volume. Alternatively, put a magnetic stir bar in the bottle or beaker and heat on a stirring hotplate gently boiling until agarose dissolves. Gently mix and place at room temperature to cool to 50 to 60°C.

5. While the gel mixture is being cooled, seal the air-dried gel tray at the two open ends with a tape or gasket and insert the comb. Add 10 µl of 10 mg/ml ethidium bromide (EtBr) to 100 ml of agarose gel solution (50 to 60°C), gently mix, and slowly pour into the assembled gel tray. Allow the gel to harden for 20 to 30 min at room temperature.

*Caution:* *EtBr is a mutagen and a potential carcinogen. Gloves should be worn when working with this material. The gel running buffer containing EtBr should be collected in a special container.*

*Notes:*   *EtBr is used to stain DNA or RNA molecules, which interlaces between the comple-*
*mentary strands of a double-strand DNA or in the regions of secondary structure in*
*the case of ssDNA or RNA and it fluoresces orange when illuminated with UV light.*
*The merit of adding EtBr to the gel is that DNA bands can be stained and monitored*
*with a UV lamp during electrophoresis. The drawback, however, is that running*
*buffer and gel apparatus are contaminated with EtBr. An alternative way is to carry*
*out electrophoresis without EtBr. The gel is then stained with EtBr for 10 to 30 min*
*following electrophoresis.*

6.   **Carefully** remove the comb and sealing tape or gasket from the gel tray. Place the gel tray in the
electrophoresis tank and add enough 0.5X TBE or TAE buffer to the tank until the gel is covered
to a depth of 1.5 to 2 mm above the gel.

*Notes:*   *The comb should be slowly and vertically removed from the gel, since any cracks*
*inside the wells of the gel will cause leaking when the sample is loaded. The well-side*
*of the gel must be placed at the negative pole end since the negatively charged DNA*
*will migrate toward the positive pole. When the gel is covered with running buffer,*
*each well should be flushed with the buffer using a small pipette tip to flush the buffer*
*up and down inside the well several times. The purpose of doing this is to remove any*
*potential bubbles that will adversely influence the loading of the samples and the*
*electrophoresis.*

7.   Prerun the gel at a constant voltage (5 to 10 V/cm) for 10 min, but this is optional.

8.   Add 5X or 10X loading buffer to the DNA sample and DNA standard marker to a final
concentration of 1X. Mix well. Carefully load the commercial DNA standard markers (0.2 to 2 μg
per well) in the very left or the right well, or in both left and right wells of the gel. Leave one well
blank from the DNA marker well and carefully load the samples, one by one, into the empty wells
in the submerged gel.

*Notes:*   *(1) Do not insert the pipette tip all the way to the bottom of the well; otherwise, it*
*may break the well and cause sample leaking. (2) For genomic DNA, the sample may*
*be heated to 60°C for 5 min prior to loading. (3) The loading buffer must be at least*
*1X to final concentration, or the sample may float out of the well. (4) For nonisotopic*
*detection, we recommend that the DNA markers be prelabeled, precipitated, and*
*directly loaded into the gel without denaturation. The labeled markers are still*
*double-strand DNA fragments. The labeling methods are described in Chapter 4. The*
*prelabeled DNA standard markers allow easy and accurate estimation of the sizes of*
*detected bands in the DNA samples after being exposed onto X-ray film or after being*
*color developed. For isotopic detection, however, we do not recommend prelabeled*
*DNA markers due to potentially massive contamination during various steps. If*
*necessary, the hybridized filter can be stripped off the hybridized probe for the DNA*
*samples and reprobed with labeled DNA standard markers, and the marker positions*
*can be compared with those of the detected bands of interest. For estimation of the*
*size of the bands, a photograph may be taken, with ruler markers and under UV light,*
*of EtBr-stained DNA standard markers and DNA samples and used to compare the*
*sizes of detected bands of interest.*

9.   Measure the length of the gel between the two electrodes and apply power to 5 to 10 V/cm. Allow
the electrophoresis to run for 5 to 6 h or until the first blue dye reaches to 2 cm from the end of
the gel. If overnight running of the gel is desired, the total voltage should be 20 to 25 V.

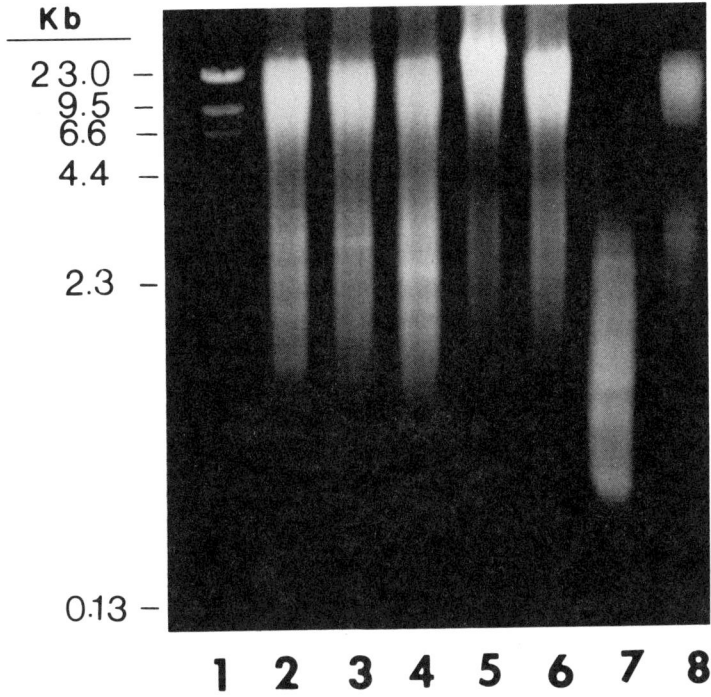

**FIGURE 5.1**

Digestion of 25 µg genomic DNA with different restriction enzymes followed by electrophoresis on a 0.8% agarose gel. Lane 1: DNA standard molecular weight markers. Lane 2: DNA digested by *EcoR* I. Lane 3: *BamH* I. Lane 4: *Hind* III. Lane 5: *Kpn* I. Lane 6: *Pst* I. Lane 7: *Hae* III. Lane 8: *Sac* I. After electrophoresis, the gel was stained by ethidium bromide and photographed.

10.  Stop the electrophoretic running of the gel and remove the gel to observe under UV light. Photograph the gel with a Polaroid™ camera (Figure 5.1) and directly proceed to the blotting procedure.

*Caution:*  *Remember to wear safety glasses and gloves for protection from UV light.*

*Notes:*  *Pictures are recommended to be taken with a ruler at short exposure to obtain sharp band(s) and clear background and at longer exposure to visualize some very weak band(s). For genomic DNA, the picture should be smeared in each lane with some weak visible bands. For plasmid and λDNA containing inserts of interest, the picture should display some sharp bands without a smeared background.*

## Materials Needed

*Ultrapure Agarose*

*Gel Casting Tray*

*Gel Combs*

## Electrophoresis Apparatus

## DC Power Supply

## 5X TBE Buffer

> 600 ml dd.$H_2O$
> 0.45 $M$ Tris base (54 g)
> 0.45 $M$ Boric acid (27.5 g)
> 0.01 $M$ EDTA (20 ml 0.5 $M$ EDTA, pH 8.0)
> Dissolve well after each addition. Add dd.$H_2O$ to 1 liter
>   Autoclave.

## 10X TAE Buffer

> 600 ml dd.$H_2O$
> 0.4 $M$ Tris-acetate (48.4 g tris base, 11.42 ml glacial acetic acid)
> 10 m$M$ EDTA (20 ml 0.5 $M$ EDTA, pH 8.0)
> Dissolve or mix well after each addition. Add dd.$H_2O$ to 1 liter
>   Sterile-filter.

## Ethidium Bromide (EtBr)

> 10 mg/ml in dd.$H_2O$
> Dissolve well and keep in a dark or brown bottle at 4°C.

## 5X Loading Buffer

> 50% Glycerol
> 1 m$M$ EDTA
> 0.25% Bromophenol blue
> 0.25% Xylene cyanol
> Dissolve well and store at 4°C.

# B.     Blotting DNA onto Nylon or Nitrocellulose Membranes

## Capillary Method

1.  Right after photographing the electrophoresed gel, soak the gel for 5 to 8 min in 0.25 $N$ HCl to depurinate DNA. This treatment can increase the efficiency of transferring of DNA fragments that are more than 8 Kb in length.

2.  Place the gel in 500 to 1000 ml of denaturing solution to denature the dsDNA molecules for later hybridization to the probe. Allow the denaturation to occur for 40 to 45 min at room temperature with slow shaking at 60 rpm.

3.  While denaturation of the gel takes place, cut a piece of nylon or nitrocellulose membrane the same size as the gel, a piece of 3MM Whatman™ filter paper the same size as the membrane and a relatively large piece of 3MM Whatman™ filter paper. The membrane filter should be marked at the upper left or the right corner with a pencil.

*Notes:* *(1) Gloves should be worn when cutting the membrane filter. (2) Nylon membrane is stronger and better than nitrocellulose membrane, which is easily broken. Positively charged nylon membrane is better than neutral nylon membrane to interact with negatively charged DNA. However, neutral nylon membrane has a lower background as compared with positively charged nylon filter following detection. (3) For reprobing the hybridized membrane filter, a nylon membrane is strongly recommended. It can be reprobed several times without any significant decrease in the signals. Nitrocellulose membrane, however, cannot endure repeated probing due to its physical condition, and the signal usually decreases significantly during the procedure of stripping. (4) For subsequent reprobing of the hybridized membrane filter, the filter must never become dry (even partially dry) during and after hybridization, washing, and exposure.*

4.  Quickly rinse the denatured gel with 500 to 1000 ml distilled water and place in 500 to 1000 ml of neutralization buffer to partially neutralize the gel but not renature the DNA molecules. Allow the neutralization to occur for 40 to 45 min at room temperature with slow shaking at 60 rpm.

5.  While neutralization of the gel takes place, soak the membrane filters in 10X or 20X SSC or SSPE solution for at least 20 min.

6.  Set up a clean blotting tray, put four test tube caps in the tray to serve as columns, and place an appropriate size of plate on those columns, or equivalent plexiglas blotting plate with standing legs, or use an upside-down gel-forming tray as the platform. Fill the tray with 10X SSC or 10X SSPE up to the edges of the flat plate. **Do not allow fluid to come up over the plate.**

7.  Assemble the blotting apparatus in the order listed below (Figure 5.2):

    a.  Place the soaked large 3MM Whatman filter on the flat plate prepared in step 6. The two ends of the filter should hang into the 10X SSC solution to serve as wicks. Smooth out any bubbles by lifting one end of the filter and slowly laying it down on the plate.

    b.  Carefully place the gel, starting from one side, on the filter with the well-side facing down. Gently press the gel to remove any bubbles underneath the gel.

    c.  Carefully overlay the gel with the membrane filter starting from one side of the gel, then proceeding to the other end. The marked side and left or right orientation should be facing the gel. Wet the filter with some 10X SSC buffer and remove any bubbles between the membrane and the gel by lifting and laying down the membrane. **Do not press the membrane;** otherwise, bubbles that are not visible may be produced underneath the gel.

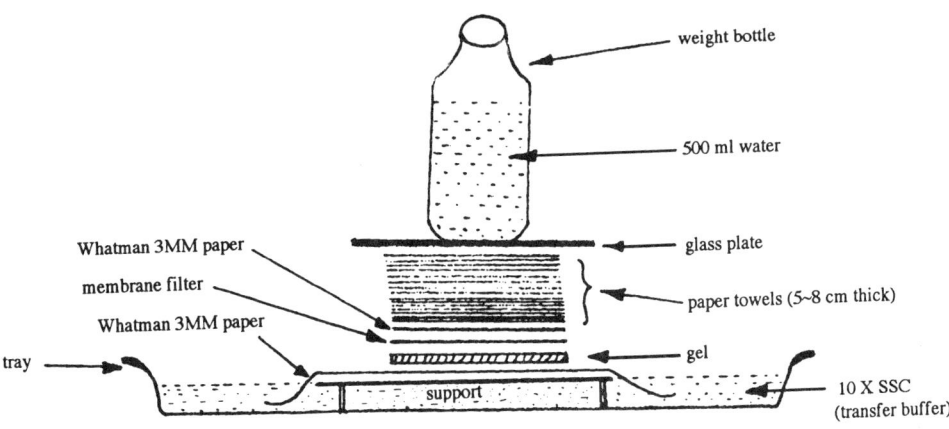

**FIGURE 5.2**
Standard assembly for capillary transfer of nucleic acids (DNA and RNA) from an agarose gel onto nitrocellulose or nylon membrane filters.

d. Gently overlay the membrane with a same-sized piece of 3MM Whatman filter paper with no bubbles.

e. Carefully place a strip of SaranWrap™ of appropriate size against each side of the gel to prevent any potential flow of the 10X SSC buffer to the paper towels.

f. Gently place a precut stack of paper towels (5 to 10 cm thick) of relatively the same size as the membrane filter on top of the 3MM Whatman paper. **Do not distribute the filters lying underneath;** this will produce bubbles.

g. Place a glass plate, or equivalent, on top of the paper towel stack. Wrap the whole apparatus with SaranWrap™ (optional). Put a bottle or beaker containing 500 ml water or equivalent on the glass plate to serve as a weight.

8. Allow the DNA to transfer to the membrane filter by capillary action overnight or for 16 h.

9. Remove the paper towel stack, the Whatman filter, and the membrane filter. Mark the wells on the membrane filter. The filter is then subjected to UV-induced cross-linking for 30 to 60 s at the optimal setting. Check the blotted gel under UV light. An efficient transfer should have no visible DNA staining left in the gel.

10. Air dry the membrane filter for about 20 min. If the membrane is not subjected to UV cross-linking, one should dry the membrane in a oven under vacuum at 80°C for 2 h to immobilize the DNA on the membrane filter.

11. Wrap the membrane filter with aluminum foil and store at 4°C until hybridization is performed.

*Notes:* *Gloves should be worn and changed during the above procedures. Blotting should be handled properly without any bubbles that block local DNA transference onto the membrane. If this happens, it should be visible from the blotted membrane filter and may affect the detection of bands of interest.*

## Vacublot Method

This is a fast but a relatively expensive method of blotting DNA to membrane. The gel to be blotted does not need to be predenatured and neutralized.

1. Cut a nylon membrane to the size of the gel and a plastic mask smaller than the gel.

2. Place a piece of 3MM Whatman filter on the metal grid of a vacublot apparatus and wet it with distilled water.

3. Place the nylon membrane on the 3MM Whatman filter, then the mask, and then the gel (well-side facing up) to form a seal between the gel and nylon membrane.

4. Apply a vacuum (2.5 in. Hg).

*Note:* *You should be able to feel with your finger tips that the vacuum causes the gel to form a seal along the edges of the mask.*

5. Pour 5 to 10 ml of depurination solution to cover the surface of the gel and let it permeate the gel for 5 to 7 min.

6. Remove the depurination solution with a pipette and replace it with 20 ml of denaturing solution. Allow the solution to permeate the gel for 8 to 15 min.

7. Remove the the denaturing solution and replace it with 20 ml of neutralizing solution. Allow the solution to permeate the gel for 8 to 15 min.

8. Remove the neutralizing solution and replace it with 20 ml of 20X SSC or 20X SSPE solution. Allow the solution to permeate the gel for 40 min.

9. After the transfer has finished, mark the wells and remove the membrane. Place the membrane on SaranWrap™ or filter paper and subject it to UV cross-linking at the optimal setting for 20 s.

10. Air dry the filter and place it into a heat-sealable bag and seal the bag about 1 to 2 cm from the edge of the membrane. Store the membrane at this stage at −20°C until hybridization is performed.

## Materials Needed

### Nylon Membranes or Nitrocellulose Membranes

### Heat-Sealable Bags

### 3MM Whatman Paper

### Baking Oven

### Vacublot Apparatus

### Depurination Solution
   0.25 $M$ HCl

### Denaturing Solution
   0.5 $M$ NaOH
   1.5 $M$ NaCl

### Neutralization Solution
   1.5 $M$ NaCl
   1 $M$ Tris-HCl, pH 7.5

### 20X SSC Solution
   175.3 g NaCl
   88.4 Sodium citrate
   Adjust the pH to 7.5 with HCl.

### 20X SSPE Solution
   3 $M$ NaCl
   0.2 $M$ $NaH_2PO_4$
   20 m$M$ EDTA, pH 7.4

## C.   Hybridization Procedure

### Method 1: Hybridization to 32P-Labeled Probe

1. Immerse the filters in 5X SSC for 5 min at room temperature to equilibrate the filters.

*Note:*   *Do not let the filters dry out during subsequent steps. Otherwise, a high background and/or anomalous results will show up.*

2.    Place the filters in the prehybridization solution and carry out prehybridization for 2 to 4 h with slow shaking at 60 rpm.

*Notes:*   *(1) We strongly recommend not using plastic hybridization bags because, in using them, it is usually not easy to get rid of air bubbles nor can they be sealed well. This causes leaking and contamination. An appropriate size of plastic beaker or tray is the best type of hybridization container to use for this purpose. (2) The prehybridization temperature depends on the prehybridization buffer. The temperature should be set at 42°C if the buffer contains 30 to 40% (for low-stringency conditions) or 50% formamide (for high-stringency conditions). If the buffer, on the other hand, does not contain formamide, the temperature is set at 65°C. Low-stringency conditions will help identify cDNAs of a potential multigene family. High-stringency conditions help prevent nonspecific cross-linking hybridization. (3) We strongly recommend the use of a regular culture shaker with a cover and temperature control as the hybridization chamber. Such a chamber is easily handled by placing the hybridization beaker or tray containing the filters and buffer on the shaker in the chamber. A commercial hybridization oven may be difficult to operate because the filters must be covered with a matrix and put inside the hybridization bottle, during which time air bubbles are easily generated. (4) If many filters are to be used for hybridization, one beaker should not contain more than three filter disks. Too many filters in one beaker may cause weak hybridization to occur. (5) The volume of prehybridization solution should be 15 ml per 100-cm² filter disk.*

3.    Denature the labeled double-stranded DNA probe contained in a microcentrifuge tube in boiling water for 10 min and immediately chill on ice for 5 min to denature the probe for hybridization. Briefly spin down prior to use with a microcentrifuge.

*Notes:*   *(1) This is a critical step. If the probes are not completely denatured, a weak or no hybridization signal will occur. Single-strand oligonucleotide probes, however, usually do not require denaturation. (2) The DNA used for labeling can be a specific gene (usually a conserved partial-length fragment), an oligonucleotide (where synthesis is based on the conserved regions of known DNA), or a specific cDNA (partial or full length) from other organisms. The DNA is labeled with [α-³²P]dCTP and is ready for hybridization (see DNA-labeling protocols in Chapter 4). (3) The labeled probe should be separated from the unincorporated nucleotides by use of a G-50 column (see Chapter 4 for details). Otherwise, nonspecific black spots will appear on the filter, causing one to "fish" out false positive plaques. (4) It is recommended, but not required, to calculate the cpm number of the labeled probe prior to hybridization.*

*Caution:*   *[α-³²P]dCTP is a dangerous isotope. A lab coat and gloves should be worn when working with this isotope. Gloves should be changed often and put in a special container. Waste liquid, pipette tips, and papers contaminated with the isotope should be collected in labeled containers. After finishing, a radioactive contamination survey should be performed and recorded.*

4.    Dilute the purified probe with 1 ml of hybridization solution and add the probe at 2 to 10 × 10⁶ cpm/ml to the hybridization buffer. Mix well and carefully transfer the prehybridized filters to the hybridization solution. Allow hybridization to proceed overnight or up to 19 h.

*Notes:* For prehybridization, notes are the same as for hybridization.

5. Wash the hybridized filters according to the following conditions:

    a. High-stringency conditions

        i. Wash the filters in a solution (50 ml per filter) containing 2X SSC and 0.1% SDS (w/v) for 15 min at room temperature with slow shaking. Repeat once.

        ii. Wash the filters in a fresh solution (50 ml per filter) containing 2X SSC and 0.1% SDS (w/v) for 20 min at 65°C with slow shaking. Repeat two to four times.

        iii. Air dry the filters at room temperature for about 40 min and proceed with autoradiography.

    b. Low-stringency conditions

        i. Wash the filters in a solution (50 ml per filter) containing 2X SSC and 0.1% SDS (w/v) for 10 min at room temperature with slow shaking. Repeat once.

        ii. Wash the filters in a fresh solution (50 ml per filter) containing 2X SSC and 0.1% SDS (w/v) for 15 min at 50 to 55°C with slow shaking. Repeat once.

        iii. Air dry the filters at room temperature for 40 min and proceed with autoradiography.

6. Wrap the filters, one by one, with SaranWrap™ and place in an exposure cassette. In a dark room with the safe light on, cover the filters with a piece of X-ray film and place the cassette with an intensifying screen at –80°C for 2 to 24 h prior to their being developed (Figure 5.3).

## Reagents Needed

### 20X SSC Solution (1 l)

    3 *M* NaCl
    0.3 *M* Na$_3$ citrate (trisodium citric acid)
    Autoclave. Adjust the pH to 7.0 with HCl.

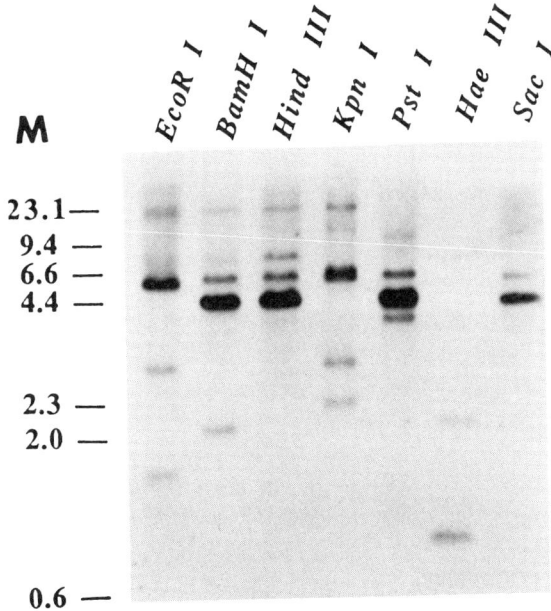

**FIGURE 5.3**
Southern blot analysis of invertase gene(s) in eukaryotic cells. Genomic DNA was digested with different restriction enzymes followed by electrophoresis on a 0.8% agarose gel. DNA was blotted onto a nylon membrane and hybridized with soluble invertase cDNA used as a probe. The weakly hybridized bands may be other members of the invertase gene family. (From Wu, L.-L., Song, I., Kim, D., and Kaufman, P. B., *J. Plant Phys.*, 142, 179, 1993. With permission of © Gustav Fischer Verlag.)

## 5X SSC Solution

Dilute 20X SSC solution four times with sterile water.

## 50X Denhardt's Solution

1% (w/v) BSA (bovine serum albumin)
1% (w/v) Ficoll (Type 400, Pharmacia)
1% (w/v) PVP (polyvinylpyrrolidone)
Dissolve well after each addition, adjust the final volume into 500-ml aliquots with distilled water and sterile-filter. Divide the solution into 50-ml portions and store at −20°C. Dilute tenfold into prehybridization and hybridization buffers.

## Prehybridization Buffer

5X SSC
0.5% SDS
5X Denhardt's reagent
0.2% Denatured and sheared salmon sperm DNA

## Hybridization Buffer

5X SSC
0.5% SDS
5X Denhardt's reagent
0.2% Denatured salmon sperm DNA
[$\alpha$-$^{32}$P]-Labeled DNA probe

## Method 2: Hybridization to Non-Isotope-Labeled Probe

1.  Carry out prehybridization and hybridization as described in Method 1 except that a nonradioactive probe and appropriate buffers are used (see Reagents Needed).

2.  Transfer the hybridized, washed filters into a clean dish containing buffer A (15 ml per filter) for 2 to 4 min.

3.  Transfer the filters in buffer B (15 ml per filter) for 60 min.

4.  Incubate the filters with an antibody solution, which is the anti-DIG-alkaline phosphatase (Boehringer Mannheim Biochemicals) diluted at 1:10,000 in buffer B, at room temperature for 40 to 60 min using 10 ml per filter.

5.  Wash the filters (100 ml per filter) in buffer A for 20 min and repeat once using a fresh washing tray.

*Note:*     *The used antibody solution can be stored at 4°C for up to 2 months and reused for five to six times without any significant decrease in the antibody activity.*

6.  Equilibrate the filters in buffer C for 1 to 4 min.

7.  Detect the hybridized band(s) by one of the following two methods.

    a. Using Lumi-Phos 530 (Boehringer Mannheim Biochemicals) as a substrate:

        i.   Add 0.5 ml of the Lumi-Phos 530 to the center of a clean dish, which should be prewarmed at room temperature for 1 h prior to use.

ii.   Briefly dampen a filter using forceps and completely wet DNA, or the plaque-facing side of the filter and/or both sides of the filter, by slowly laying the filter down on the solution several times.

iii.   Wrap the filter with SaranWrap™ and wipe out the excess Lumi-Phos solution using a paper towel to reduce the black background.

iv.   Place the wrapped filter in an exposure cassette with the side facing up.

v.   Repeat steps ii, iii, and iv until all the filters are done.

vi.   Overlay the wrapped filters with an X-ray film in a darkroom with a safe light on and allow the exposure to occur by placing the closed cassette at room temperature for 2 min to 24 h.

vii.   Develop the film and proceed to positive clone identification.

Notes:   *(1) Exposure for more than 4 h may produce a very black background. Based on our experience, a good hybridization and detection should generate sharp positive spots with 1.5 h of exposure. (2) The film should be slightly overexposed to get a relatively black background, which will help identify the marks made previously.*

b.   Using the NBT and BCIP detection method:

i.   Add 40 ml of NBT solution and 30 µl of BCIP solution in 10 ml buffer C for one filter. NBT and BCIP are available from Boehringer Mannheim Biochemicals.

ii.   Place the filter in the mixture made in step (i) and put in the dark for color development at room temperature for 30 min to 1 day in order to obtain a clean background.

## Reagents Needed

### 20X SSC Solution (1 l)

3 $M$ NaCl
0.3 $M$ Na$_3$ citrate (trisodium citric acid)
Autoclave. Adjust the pH to 7.0 with HCl.

### 5X SSC Solution

Dilute 20X SSC solution four times with sterile dd.H$_2$O.

### Denaturing Solution

1.5 $M$ NaCl
0.5 $M$ NaOH
Autoclave.

### Neutralizing Solution

1.5 $M$ NaCl
0.5 $M$ Tris-HCl, pH 7.4
Autoclave.

### Prehybridization Buffer

5X SSC
0.1% N-Lauroylsarcosine

0.02% Sodium dodecyl sulfate (SDS)

1% Blocking reagent

Dissolve well on a heating and stirring plate at 65°C after each addition. Add sterile water to final volume or add 50% formamide to the mixture if using the formamide method.

## Hybridization Buffer

Add DIG-dUTP-labeled probe to the appropriate volume of fresh prehybridization buffer.

## Buffer A

100 m$M$ Tris-HCl

150 m$M$ NaCl, pH 7.5

## Buffer B

2% (w/v) blocking reagent (Boehringer Mannheim Biochemicals) or equivalent, such as BSA, nonfat milk, or gelatin in buffer A.

Dissolve well by stirring with a magnetic stir bar.

## Buffer C

100 m$M$ Tris-HCl, pH 9.5

100 m$M$ NaCl

50 m$M$ MgCl$_2$

## NBT Solution

75 mg/ml Nitroblue tetrazolium salt in 70% (v/v) dimethylformamide

## BCIP Solution

50 mg/ml 5-Bromo-4-chloro-3-indolyl phosphate (X-phosphate), in 100% dimethylformamide

## Anti-DIG-Alkaline Phosphatase

Anti-digoxigenin conjugated to alkaline phosphatase (Boehringer Mannheim Biochemicals)

## Method 3: Reprobe of the Hybridized Filters

1. Stripping of radioisotopic probe hybridized membranes

   a. For DNA filters, incubate the filters in 0.4 $M$ NaOH at 45°C for 40 min. Transfer the filters in a solution containing 0.1% (w/v) SDS, 0.1X SSC, and 0.2 $M$ Tris-HCl (pH 7.5) and incubate at 45°C for 20 min. Repeat several times. Autoradiograph to check for the removal of the hybridized probe. Carry out prehybridization and hybridization steps with the new probe.

b. For RNA filters, boil 0.1% (w/v) SDS solution and pour onto the filters. Allow to cool to room temperature. Repeat this procedure several times until the hybridized probe has been removed. The filters can then be reprobed.

2. Stripping of nonisotopic Lumi-Phos-detected filters

a. Rinse the filters with sterile distilled water several times to remove the Lumi-Phos from the filters.

b. Incubate the filters in a stripping solution containing 60% formamide, 1% SDS (w/v), and 50 m$M$ Tris-HCl (pH 8.0) at 75°C for 45 min to remove the DIG-labeled probe or its equivalent.

c. Rinse the filters with sterile distilled water several times and carry out reprobing steps.

3. Stripping nonisotopic color-detected filters

a. Incubate the filters in preheated dimethylformamide at 55°C until the blue color is removed.

b. Rinse the filters with sterile distilled water several times.

c. Incubate the filters in a stripping solution containing 60% formamide, 1% (w/v) SDS, and 50 m$M$ Tris-HCl (pH 8.0) to remove the nonisotope.

d. Carry out reprobing steps.

## D.    Troubleshooting Guide

| Symptoms | Solutions |
| --- | --- |
| A smeared signal appears under each lane with no sharp bands | Electrophoresis is too slow or too fast. Try to use 5 to 10 V/cm at a constant level |
| Bands with the same sizes shift but do not occur at the same positions among the different lanes | The thickness of the gel is not even, as a result, the electrophoresis is not uniform |
| Many unexpected bands appear | Nonspecific cross-hybridization problem occurs. Verify the specificity of the probe and try to use high-stringency conditions of hybridization |
| Black background occurs on the filter (for radio-isotopic probe hybridization) | Filter became partially dry or blocking efficiency is low, or the quality of the filter is not good. Try to avoid air drying of the filter, increase the percentage of BSA and denatured salmon sperm DNA, and try to use a fresh neutral nylon membrane filter |
| Purple background occurs on the filter (for nonisotopic probe hybridization) | Color development was too long. Try to stop the color reaction as soon as desired signals appear |
| Unexpected black spots occurs on the filter (for radioisotopic probe hybridization) | Unincorporated $^{32}$PdCTP was not efficiently removed. Try to use a G-50 Sephadex column to purify the labeled DNA |
| Unexpected purple bands occur on the filter (for nonisotopic probe hybridization) | The specificity of the antibody used is not specific. Try to test the quality of the antibody and increase the blocking reagents |
| Signals are weak on the filters | The efficiency of labeling is low or the X-ray film exposure time is not long enough |
| No signal is seen at all in the filter | Sample and/or probe is not denatured |
| No signal occurs at all during subsequent reprobing | The stripping of the hybridized probe was not successful or the filter became dried out during previous probing |

# II.    Northern Blotting

Northern hybridization or RNA blotting is an RNA analysis procedure in which the size and amount of specific mRNA molecules in total RNAs or poly(A)+RNAs can be determined. It basically consists of RNA/DNA or sense RNA and antisense RNA hybridization. RNA molecules are separated according to their sizes by a denaturing agarose gel electrophoresis followed by their being transferred or blotted onto nitrocellulose or nylon membranes. The blotted membrane is hybridized with a specific probe, and the mRNA of interest is then visualized by autoradiography or nonisotopic detecting methods.[3,7,8]

RNA isolation and purification procedures are given in Chapter 2. The probe(s) used in Northern hybridization can be specific genomic DNA, cDNA, oligonucleotides, or antisense RNA. The preparation of probe(s) is described in Chapter 4. The DNA probe is usually stable but needs longer hybridization. The RNA probe, on the other hand, is much hotter but is relatively unstable.

Since RNAs, as compared to DNA, are very mobile molecules due to RNA degradation by RNases, much care should be taken to maintain the purity and integrity of the RNA. This is very critical for Northern blotting. The most difficult task is to inactive RNase activity. Two common sources of RNase contamination are the user's hands and bacteria and fungal molds present on airborne dust particles. To prevent this type of contamination:

1.  Gloves should be worn at all times and changed frequently.
2.  Whenever possible, disposable plasticware should be autoclaved. Nondisposable glass- and plasticware should be treated with 0.1% diethyl pyrocarbonate (DEPC) in dd.$H_2O$ and be autoclaved before use.
3.  After treatment, glassware should be baked at 250°C overnight. Gel apparatus should be deeply cleaned with detergent and thoroughly rinsed with DEPC-treated water. Apparatus used for RNA electrophoresis, if possible, should be separated from DNA or protein electrophoresis apparatus. Chemicals should be of ultrapure grade and RNase-free. Gel mixtures, running buffers, hybridization solutions, and washing solutions should be made with DEPC-treated water.

## A.    Electrophoresis of RNA by the Use of Formaldehyde Agarose Gels

### Protocol

1.  Deeply clean an appropriate gel apparatus by washing with detergent. Completely remove the detergent mixture with tap water and rinse with DEPC-treated distilled water three to five times. Allow the apparatus to dry at room temperature.
2.  Prepare the formaldehyde agarose gels.
    a.  Add 1% (w/v) agarose in a clean bottle or a beaker as follows:

| Components | Mini gel | Medium gel | Large gel |
|---|---|---|---|
| DEPC-treated water | 21.6 ml | 54 ml | 86.4 ml |
| Ultrapure agarose | 0.3 g | 0.75 g | 1.2 g |

    b.  Slightly mix and melt the agarose by bringing it to a gentle boil in a microwave oven for 1 to 3 min depending on the gel volume. Alternatively, put a magnetic stir bar in the bottle or beaker and heat on a stirring hotplate to a gentle boil until the agarose dissolves. Gently mix and place at room temperature to cool to 50 to 60°C.

c. While the gel mixture is being cooled, seal the air-dried gel tray at the two open ends with a tape or gasket and put the comb in place.

d. To the cooled gel mixture, add components as follows:

| Components | Mini gel | Medi gel | Big gel |
|---|---|---|---|
| 10X MOPS buffer | 3 ml | 7.5 ml | 12 ml |
| Ultrapure formaldehyde (18%, v/v) (37%, 12.3 $M$) | 5.4 ml | 13.5 ml | 21.6 ml |
| 10 mg/ml EtBr | 3 µl | 7.5 µl | 12 µl |

e. Gently mix after each addition and slowly pour the mixture into the assembled gel tray placed in a fume hood. Allow the gel to harden for 20 to 30 min at room temperature.

*Caution:* *Formaldehyde vapors are toxic. DEPC and EtBr are carcinogenic. These chemicals should be handled with care. Gloves should be worn when working with these materials. Gel running buffer containing EtBr should be collected in a special container. Formaldehyde serves to denature the secondary structures of RNA. EtBr is used to stain RNA molecules, which interlaces in the regions of secondary structure of RNA and fluoresces orange when illuminated with UV light.*

3. While the gel is hardening, prepare the sample(s) in a sterile tube as follows:

   a. For the sample in one lane, add the components as follows:
   - Total RNA (10 to 35 µg per lane) or poly(A)+RNA (0.2 to 2 µg per lane) (6 µl)
   - 10X MOPS buffer (3.5 µl)
   - Ultrapure formaldehyde (37%, 12.3 $M$) (6.2 µl)
   - Ultrapure formamide (50% v/v) (17.5 µl)
   - Add DEPC-treated dd.H$_2$O to a final volume of 35 µl.

   b. For RNA standard markers, add the components as follows:
   - RNA markers with wide range (0.5 to 1.5 µg) (2 µl)
   - 10X MOPS buffer (1 µl)
   - Ultrapure formaldehyde (37%, 12.3 $M$) (1.8 µl)
   - Ultrapure formamide (50% v/v) (5 µl)
   - Add DEPC-treated dd.H$_2$O to a final volume of 10 µl.

   c. Heat the tubes at 65°C for 15 min and immediately chill on ice to denature the RNA sample. Briefly spin down afterwards.

   d. Add 3.5 and 1 µl DEPC-treated loading buffer to the sample and RNA standard markers, respectively.

4. Electrophoresis

   a. **Carefully** remove the comb and sealing tape or gasket from the gel. Place the gel tray in the electrophoresis tank and add 1X MOPS buffer to the tank right up to the upper edge of the gel or until the gel is covered to a depth of 1.5 to 2 mm above the gel.

*Notes:* *(1) The comb should be slowly and vertically removed from the gel. Any crack inside the wells of the gel will cause leaking when the samples are loaded. (2) The well end of the gel must be placed at the negative pole end because the negatively charged RNA will migrate toward the positive pole. (3) Based on our experience, we recommend that the 1X MOPS running buffer not cover the gel in order to protect the formaldehyde in the gel from diffusing into the buffer, thus affecting the denaturing efficiency of RNA. However, if the gel is covered with running buffer, each well should be flushed with the buffer using a small pipette tip to cause flow of buffer up and down inside the well several times. The purpose of doing this is to remove any potential bubbles that will adversely influence the loading of samples and electrophoresis. (4) For nonisotopic detection, we recommend that the RNA markers be prelabeled, precipitated, and directly loaded into the gel. The labeling methods are described in Chapter 4. The prelabeled RNA standard markers allow easy and accurate identification of the sizes of detected bands in the DNA samples after being exposed onto X-ray film or after being color developed. For isotopic detection, however, we do not recommend prelabeled DNA markers due to potentially massive contamination during various steps. If necessary, the hybridized filter can be stripped off the hybridized probe for the RNA samples and reprobed with labeled RNA standard markers, and the marker positions can be compared with those of the detected bands of interest.*

b. Prerun the gel at a constant voltage (5 to 10 V/cm) for 10 min and immediately load the RNA samples into each well of the gel. Leave one well blank between the RNA standard markers and RNA samples.

*Note:     Do not insert the pipette tip to the bottom of the well; otherwise, it will break the well and cause sample leaking.*

c. Measure the gel length between the two electrodes and apply current to 5 to 10 V/cm (constant), cover the tank, and carry out electrophoresis for 1 h. Cover the gel, which is not submerged with the running buffer, with SaranWrap™ to prevent evaporation of the formaldehyde. Allow the gel to continue to run at a constant voltage for 4 to 5 h or until the first blue dye band reaches a position of 2 cm from the end of the gel.

*Notes:     In the middle of electrophoresis, it is recommended that the gel tray be lifted and the buffer at the two ends of the tank be mixed and the electrophoresis then allowed to continue. Overnight running of the gel is not recommended.*

d. Stop running the gel and remove the gel to observe under UV light. Photograph the gel with a Polaroid™ camera and directly proceed to the blotting procedure.

*Caution:     Remember to wear safety glasses and gloves for protection from UV light.*

*Notes:     Pictures with a ruler are recommended to be taken at relatively longer exposure in order to visualize the smear in each well. A successful electrophoresis of RNA, a long smear with two sharp bands, should be visible from each well for total RNA, or a long smear only for purified poly(A)$^+$RNA (Figure 5.4). The two sharp bands represent rRNAs: one is 28S rRNA in animals or 25S rRNA in plants, the other is 18S rRNA. If the range of the smear is very limited or the rRNA bands are very weak in total RNA, the RNA samples are likely degraded.*

## Materials Needed

*Gel Casting Tray*

*Gel Combs*

*Ultrapure Agarose*

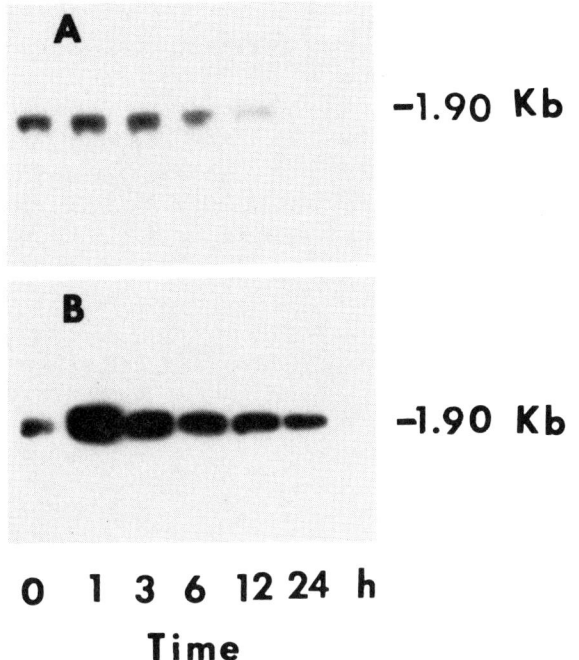

**FIGURE 5.4**

Northern blot analysis of invertase gene expression in oats *(Avena sativa)*. Oats were gravistimulated for different periods of time. Poly(A)+RNA was purified from total RNA that was isolated from tissues A and B, respectively. mRNA (2 μg per lane) was electrophoresed in 1% agarose gel containing formaldehyde and blotted onto a nylon membrane. Invertase cDNA was radioactively labeled and used for hybridization. (From Wu, L.-L. et al., *J. Plant Phys.*, 142, 179, 1993. With permission of © Gustav Fischer Verlag.)

### Electrophoresis Apparatus

### DEPC Water

0.1% DEPC in dd.H$_2$O
Place at 37°C overnight with a stir bar to inactivate any potential RNase.
Autoclave to remove the DEPC.

### 10X MOPS Buffer

0.2 *M* 3-(N-Morpholino)propanesulfonic acid (MOPS)
80 m*M* Sodium acetate
10 m*M* EDTA (pH 8.0)
Dissolve well after each addition in DEPC-treated dd.H$_2$O.
Adjust the pH to 7.0 with 2 *N* NaOH.
Sterile-filter and store at room temperature.

### Ethidium Bromide (EtBr)

10 mg/ml in DEPC-treated dd.H$_2$O
Dissolve well and keep in the dark or in a brown bottle at 4°C.

### 5X Loading Buffer

> 50% Glycerol
> 1 m$M$ EDTA
> 0.25% Bromophenol blue
> 0.25% Xylene cyanol FF
> Dissolve well in DEPC-treated dd.H$_2$O and store at 4°C.

### Nylon Membranes or Nitrocellulose Membranes

### Heat-Sealable Bags

### 20X SSC Solution

> 175.3 g NaCl
> 88.4 Sodium citrate
> Dissolve well in DEPC-treated dd.H$_2$O.
> Adjust the pH to 7.0 and autoclave.

## B.      Blotting RNA onto Nylon or Nitrocellulose Membranes

The general procedures are the same as described under Southern blotting procedures except that all solutions and water should be DEPC-treated.

## C.      Hybridization

The general procedures for prehybridization, hybridization, washing, detection, autoradiography, and reprobing are essentially the same as described for Southern blotting except that all solutions should be DEPC-treated.

## D.      Troubleshooting Guide

| Symptoms | Solutions |
| --- | --- |
| A smeared signal appears under each lane with no sharp bands | Electrophoresis is too slow or too fast. Try to use 5 to 10 V/cm at a constant level |
| Bands with the same sizes shift but do not occur at the same positions among the different lanes | The thickness of the gel is not even, so as a result, the electrophoresis is not uniform |
| Many unexpected bands appear | Nonspecific cross-hybridization problem occurs. Verify the specificity of the probe and try to use high-stringency conditions of hybridization |
| Black background occurs on the filter (for radio-isotopic probe hybridization) | Filter became partially dry or blocking efficiency is low, or the quality of the filter is not good. Try to avoid air drying of the filter, increase the percentage of BSA and denatured salmon sperm DNA, and try to use a fresh neutral nylon membrane filter |

| Symptoms (continued) | Solutions (continued) |
|---|---|
| Purple background occurs on the filter (for nonisotopic probe hybridization) | Color development was too long. Try to stop the color reaction as soon as desired signals appear |
| Unexpected black spots occur on the filter (for probe radioisotopic hybridization) | Unincorporated $^{32}$PdCTP was not efficiently removed. Try to use a G-50 Sephadex column to purify the labeled DNA |
| Unexpected purple bands occur on the filter (for nonisotopic probe hybridization) | The specificity of the antibody used is not specific. Try to test the quality of the antibody and increase the blocking reagents |
| Signals are weak on the filters | The efficiency of labeling is low or the X-ray film exposure time is not long enough |
| No signal is seen at all in the filter | Sample and/or probe is not denatured |
| No signal occurs at all during subsequent reprobing | The stripping of the hybridized probe was not successful or the filter became dried out during previous probing |

# III.    Dot/Slot Blotting

As compared with Southern and Northern blots, dot blotting is a simple, fast, and sensitive method used in DNA/DNA, DNA/RNA, and/or RNA/RNA hybridizations. Denatured DNA or RNA samples can be directly applied to membrane filters without electrophoresis. However, the dot blot method cannot reveal the sizes of specific bands of interest among different DNA or RNA molecules. The basic procedures are given below.

## A. Dot Blotting of DNA

1.  Denature DNA samples at 95°C for 10 to 15 min and immediately chill on ice. Spin down and add one volume of 20X SSC to the sample. An alternative way is the alkaline denaturing method: add 0.2 volume of 2 *M* NaOH solution to the sample, leave at room temperature for 15 min, and add 1 volume of neutralization buffer containing 0.5 *M* Tris-HCl (pH 7.5) and 1.5 *M* NaCl. Leave at room temperature for 15 min.

2.  Cut a piece of nylon or nitrocellulose membrane filter and assemble a vacuum blot apparatus according to the instructions. Turn on vacuum slightly to obtain an appropriate suction that holds the membrane filter onto the apparatus.

*Notes:*   *We recommend the use of an appropriate vacuum for the blotting process. If the vacuum is too low, diffusion of samples loaded is usually a problem. However, the vacuum cannot be too high; otherwise, the vacuum is sucking too fast, such that there is a decrease in the efficiency of blotting.*

3.  Spot samples (about 2 μl per loading without vacuum or 5 μl per loading under vacuum) onto the membrane filter that is prewetted with 10X SSC and air dried. Allow to partially dry between each spot.

4.  After spotting, air dry the filter and place it onto a 3MM Whatman filter paper satured with denaturing solution with the DNA side up. Incubate for 5 to 10 min.

5.  Transfer the membrane filter, DNA-side up, on a 3MM Whatman filter paper presaturated with neutralizing solution for 2 to 5 min.

6. Air dry the membrane filter at room temperature for 20 min.

7. Wrap the filter with SaranWrap™ and place DNA-side down on a transilluminator (312 nm wavelength is recommended) for 4 to 6 min for UV cross-linking.

8. Bake the filter in an oven at 80°C for 2 h under vacuum.

9. Prehybridization, hybridization, washing, and detection are the same as for Southern blotting.

## *Solutions Needed*

### *20X SSC Solution*

> 3 $M$ NaCl
> 0.3 $M$ Sodium citrate
> Dissolve well and autoclave. Adjust the pH to 7.0.

### *Denaturing Solution*

> 1.5 $M$ NaCl
> 0.5 $M$ NaOH
> Dissolve well in sterile distilled water.

### *Neutralizing Buffer*

> 1.5 $M$ NaCl
> 0.5 $M$ Tris-HCl, pH 7.4
> 1 m$M$ EDTA
> Dissolve well in sterile distilled water.

### *B. Dot Blotting of RNA*

1. Denature RNA samples at 65°C for 10 min in 4 volumes of denaturing solution containing 66% (v/v) ultrapure formamide, 21% (v/v) ultrapure formaldehyde (37%), and 13% (v/v) 10X MOPS buffer.

2. After denaturation, quickly chill the sample on ice for 4 min and briefly spin down.

3. Add 1 volume of 20X SSC solution to the sample and spot it on the membrane filter as with DNA dot blotting.

4. Fixing of RNA to the membrane, baking, prehybridization, hybridization, washing, and detection as the same as for Northern blotting.

# IV.    Western Blotting

Western blotting is a protein/protein hybridization technique that is used for immunodetection of specific antigen(s) of interest in a complex mixture of proteins. This is a simple, sensitive, and effective technology used in immunology, molecular and cellular biology, and protein chemistry. The principle of Western blotting is that a protein mixture is first separated according to molecular size using sodium dodecyl sulfate-polyacrylamide gel electrophoresis (SDS-PAGE). The separated protein molecules are then immobilized onto a nitrocellulose membrane or its equivalent. The specific protein band of interest is identified by use of a specific antibody raised by the specific antigen (protein), which can specifically bind to the antigen (protein) of interest in the protein mixture that is immobilized on the nitrocellulose membrane. The antibody–antigen complex is then detected by an enzyme linked to a second antibody and substrate, or by use of [125]I-labeled protein A or its equivalent.[9–12] The antigen can also be directly immunodetected with a fluorescence-labeled antibody, which directly binds to the antigen in a protein mixture. After washing away the nonbound antibody, the antigen-labeled antibody can then be visualized under a microscope, which can be excited by fluorescence using a selected wavelength emitted by a UV lamp. However, the direct method needs a relatively large amount of the labeled antibody to obtain good detection, and the labeling of an antibody of particular interest is very expensive. Therefore, an indirect immunodetection method is a widely used technique today. The following detailed protocols are based on modifications in the method of Towbin et al.[10] who first developed the technique.

## A. Separation of Proteins on the Basis of Molecular Weight by SDS-PAGE

SDS is an anionic detergent that denatures proteins and makes them negatively charged by wrapping around the polypeptides, thus giving equal charge densities per unit length. SDS-PAGE can separate and determine the molecular weights (MW) of proteins using standard protein-size markers. There is a linear relationship between the log of the MW of a polypeptide and its $R_f$, which is the ratio of the distance from the top of the gel to the polypeptide divided by the distance from the top of the gel to the dye front. A standard curve can then be generated by plotting the $R_f$ of each standard polypeptide marker as the abscissa and the $\log_{10}$ of its MW as the ordinate. The MW of an unknown protein can be determined with ease by finding its $R_f$ that vertically crosses on the standard curve and reading the $\log_{10}$ MW that horizontally crosses to the ordinate. The antilog of the $\log_{10}$ MW is the actual MW of the protein.

### 1. Preparation of the Separating Gel

1. Thoroughly clean glass plates and spacers with detergent, wash with tap water, rinse with distilled water several times, and air dry.

2. Wear gloves and wipe dry the glass plates and spacers with 100% ethanol. Assemble the vertical slab gel unit, such as the SE 600 Vertical Slab Gel Unit (Hoeffer), in the casting mode according to the instructions. Place one spacer (1.5 mm thick) at each of the two sides between the two glass plates and fix in place with clamps, forming a sandwich. Repeat for the second sandwich. The size of the separating gel varies with individuals. A standard size is 120 × 140 × 1.5 mm.

*Note:*     *Spray a little grease oil on the spacer areas at both the top and bottom ends of the sandwiches to prevent potential leaking.*

3. Check potential leaking by pipetting some distilled water into the sandwich and then drain away the water by inverting the unit. Set the unit at an even level using a water balancer or its equivalent.

4. Prepare the separating gel solution in a clean 100-ml beaker or a 125-ml flask in the order shown below (for two pieces of standard-sized gels):

   • 20 ml Monomer solution
   • 15 ml Running gel buffer
   • 0.6 ml 10% SDS
   • 24.1 ml dd.$H_2O$
   • Mix well after each addition.

5.  Degas the solution to remove any bubbles by use of vacuum for 5 min (optional). Add 0.25 ml of freshly prepared 10% ammonium persulfate (AP) and 20 μl TEMED and mix by gently swirling. **Do not generate air bubbles.**

6.  Immediately pipette the mixture with a 50-ml syringe into the assembled sandwich up to a level that is 4 cm from the top.

7.  Take up 0.8 ml of dd.H$_2$O in a 1-ml syringe equipped with a 2-in. 22-gauge needle or its equivalent and carefully load 0.3 ml of the water, starting from one top corner next to the spacer, onto the surface of the acrylamide gel solution. Repeat on the other side of the slab next to the other spacer. The water layer will evenly flow across the surface of the gel mixture. The purpose of applying water is to make the surface of the gel very even. A very sharp gel–water interface can be visible after the gel has polymerized.

8.  Drain away the water layer by gently tilting the casting unit and rinse once with 2 ml of overlay solution. Add 1 ml of overlay solution on top of the gel and allow the gel to sit for 2 h or overnight by covering the top of the gel with a piece of Parafilm™ to prevent evaporation.

## 2. Preparation of the Stacking Gel

1.  Prepare the stacking gel solution in a clean 50-ml beaker or a 50-ml flask in the order shown below (for two pieces of standard-sized gels):

    • 2.66 ml Monomer solution

    • 5 ml Stacking gel buffer

    • 0.2 ml 10% SDS

    • 12.2 ml dd.H$_2$O

    • Mix well after each addition.

2.  Drain the overlay solution on the separating gel by tilting the casting unit and rinse with 2 ml of the stacking gel mixture. Drain away the stacking gel mixture and insert the comb (1.5 mm thick) into the glass sandwich according to the instructions for the apparatus.

*Notes:*    *The stacking gel solution may be filled prior to inserting the comb into the sandwich. However, air bubbles are easily generated and trapped around the teeth of the comb. We recommend that the comb be inserted into the sandwich prior to filling the stacking gel mixture.*

3.  Add 100 μl of freshly prepared 10% AP and 10 μl TEMED to the stacking gel solution and mix by gently swirling. **Do not generate air bubbles.**

4.  Immediately and slowly pipette the solution, from both sides next to the spacers, into the sandwich up to the top. Allow the gel to polymerize and sit for 30 min.

## Reagents Needed

### Monomer Solution

> Acrylamide (116.8 g)
> *N,N*-Methylene-*bis*-acrylamide (3.2 g)
> Dissolve well after each addition in 300 ml dd.H$_2$O.
> Add dd.H$_2$O to a final volume of 400 ml.
> Wrap the bottle with aluminum foil and store at 4°C in the dark.

*Caution:*    *Acrylamide is neurotoxic. Gloves should be worn when handling this chemical.*

## Running Gel Buffer

1.5 $M$ Tris (72.6 g)
Dissolve well in 200 ml dd.$H_2O$.
Adjust pH to 8.8 with 2 $N$ HCl.
Add dd.$H_2O$ to a final volume of 400 ml.
Store at 4°C.

## Stacking Gel Buffer

0.5 $M$ Tris (6 g)
Dissolve well in 50 ml dd.$H_2O$.
Adjust pH to 6.8 with 2 $N$ HCl.
Add dd.$H_2O$ to a final volume of 100 ml.
Store at 4°C.

## 10% SDS

SDS (10 g)
Dissolve well in 100 ml dd.$H_2O$.
Store at room temperature.

## 10% Ammonium Persulfate (AP)

AP (0.2 g)
Dissolve well in 2.0 ml dd.$H_2O$.
Store at 4°C for up to 10 days.

## Overlay Buffer

Running gel buffer (25 ml)
10% SDS solution (1 ml)
Add dd.$H_2O$ to 100 ml.
Store at 4°C.

## 3. Loading the Samples and Protein Standard Markers onto the Gel

1.  While the stacking gel is polymerizing, prepare samples and protein standard markers. Add 1 volume of 2X denaturing buffer to each of the samples and to the protein standard markers. Cap the tubes and place them in boiling water for 2 to 4 min. Immediately chill the tubes on ice for 4 min. Briefly spin down, add 10% (v/v) of bromophenol blue solution to the sample, and place on ice until use.

*Notes:   At this stage, the proteins and markers are denatured, which is important for efficient electrophoresis. The isolation and purification of proteins are described in Chapter 3. The amount of proteins loaded into one well should be 10 to 35 μg in a total of 5 to 15 μl for Coomassie Blue staining and Western blotting, and 5 to 15 μg in 5 to 10 μl for highly sensitive silver staining. The amount of protein standard markers for one well should be 2 to 10 μg.*

2.   Slowly and vertically pull the comb straight up from the gel. Rinse each well with running buffer using a syringe or pipette to add the running buffer. Carefully invert the casting stand to drain the wells and repeat twice.

3.   Position the casting stand upright and fill each well with 10 μl of running buffer. Take up sample or markers into a syringe equipped with a 2-in. 22-gauge needle or a pipette with an equivalent tip attached. Carefully insert the needle into the running buffer in each well and **underlay** the sample or standard marker in the well. Repeat until all of the samples are loaded.

*Notes:*   *The volume loaded into each well should be less than 40 μl for a standard-sized gel or less than 15 μl for a mini-gel. Overloading may cause samples to float out and cause contamination among wells. The markers are recommended to be loaded into the front left or right well, or both wells, leaving a blank well between markers and samples. The volume and order of each sample loaded in the two sheets of gels should be identical. One gel will be used for staining in the determination of the MW of the proteins. The other gel will be used in Western blotting.*

4.   Carefully overlay the wells with running buffer, using a syringe or pipette, until the top of the gel is covered.

*Notes:*   *The samples and markers containing blue dye should be visible at the bottom of each well. Mark the orientation of samples and record in lab notebook.*

## Reagents Needed

### 2X Denaturing Buffer

Stacking gel buffer (5 ml)
10% SDS solution (8 ml)
Glycerol (4 ml)
2-Mercaptoethanol (2 ml)
Add dd.$H_2O$ to a final volume of 20 ml.
Divide in aliquots and store at –20°C.

### Bromophenol Blue Solution

Bromophenol blue (0.05 g)
Sucrose (20 g)
Dissolve well after each addition in 15 ml dd.$H_2O$.
Add dd.$H_2O$ to a final volume of 50 ml, aliquot, and store at –20°C.

### Running Buffer

0.25 *M* Tris (12 g)
Glycine (57.6 g)
10% SDS solution (40 ml)
Add dd.$H_2O$ to a final volume of 4 l.

## 4. Electrophoresis

1.   Carefully put the upper buffer chamber in place according to the instructions. Remove lower clamps and clamp the sandwiches to the bottom of the upper buffer chamber. **Do not disturb the wells.** Check for any potential leaking by adding 10 to 20 ml of running buffer into the chamber.

*Note:*    *Apply a little grease oil to the spacer areas at the top corners of the sandwiches to prevent potential leaking.*

2.    Carefully transfer the assembled unit into the lower buffer chamber according to the instructions. Place the entire tank on a magnetic stirrer. Slowly fill the lower chamber with 3 l of running buffer and add 1 l to the upper chamber, or use an appropriate volume depending on the sizes of the chambers.

*Notes:*    *The volume of bottom-chamber buffer should be sufficient to cover two thirds of the slabs; otherwise, the heat generated during electrophoresis will not be distributed evenly, thus causing distortion of the band patterns on the gel. For the upper chamber, gently fill with the running buffer, adding it from one corner into the chamber. Do not pour the buffer into the well areas to avoid washing the samples out.*

3.    Put a stir bar in the lower chamber and stir the buffer for 1 min to remove any air bubbles trapped under the ends of the sandwiches.

4.    Add a few drops of dye solution such as phenol red or equivalent into the upper-chamber buffer if the dye in the samples is not sufficient.

5.    Place the lid, or its equivalent, on the unit and connect the power supply. The cathode (negative pole) should be connected to the upper buffer chamber and the anode (positive pole) should be connected to the bottom buffer chamber. Proteins that become negatively charged by SDS will migrate from the cathode to the anode and will separate according to their MW.

6.    Set the power supply at a constant power or current and turn it on. Adjust the current to 25 to 30 mA/1.5-mm thick standard-sized gel. The voltage will increase during the running process. Allow the gel run electrophoretically for several hours until the dye reaches the bottom of the gel.

7.    Turn off the power supply and disconnect the power cables or their equivalent. Loosen and remove the clamps at both sides of the sandwiches. Place the gel sandwiches on the table or bench. Slowly remove the spacers and use an extra spacer to carefully separate the two glass plates starting from one corner. Remove the top plate; the gel should be on the bottom glass plate. Make a small cut at the upper left or right corner of the gel to record the orientation by using a razor blade. Use one gel for staining to determine the MW of different proteins. The other gel is used for Western blotting.

# B.    Staining and Destaining of the Gel

## 1. Coomassie Blue (CB) Staining and Destaining Method

1.    Carefully transfer one of the gels into a glass or plastic tray containing 100 to 200 ml of CB solution and stain for 4 to 8 h at room temperature with slow shaking at 60 rpm. Alternatively, cover the tray tightly and stain for 20 to 30 min in a 50°C water bath.

2.    Remove the staining solution and replace it with destaining solution I. Allow destaining to take place for 1 to 1.5 h.

3.    Gently transfer the stained gel and carefully place it between two sheets of supporting matrix. Gently roll the sandwich and insert it into a cylinder holder and cover the two ends of the holder. An alternative way is to keep the gel in the tray and replace destaining solution I with destaining solution II. Destain the gel with shaking at 60 rpm until a clear background is obtained.

4.    Place the cylinder into the SE 530 Destainer (Hoeffer) or its equivalent filled up with destaining solution II. Put a stir bar into the tank and place the tank on a magnetic stirrer. Turn on the stir bar and allow the gel to destain overnight or for a few days until a clear background is obtained.

*Notes:*   *(1) Photography of the destained gel to keep as a record is strongly recommended. The photograph may be made after the gel has been dried. However, the gel is sometimes damaged during the drying process due to a vacuum problem. (2) A fast destaining method may be used to destain the gel, using 0.5 to 1% Clorox in distilled water. Constantly monitor the destaining of the background. When clear bands are visible, immediately rinse the gel with a large volume of distilled water several times until the blue color is stable. The disadvantage of this method is that if overdestaining occurs, some bands may disappear.* Never use >1.5% Clorox®.

5.   Carefully place the gel between a water-prewetted thin film (commercially available for gel drying), remove any air bubbles inside the sandwich to prevent any cracks developing in the gel, and dry the gel at 50 to 70°C under vacuum for 1 h. Remove the dried gel after it has cooled to room temperature.

6.   Measure the distance from the top of the gel to each of the bands in the protein standard marker lane, and do the same thing for every single band in the sample lanes. Calculate the $R_f$ for each band by dividing the distance from the top of the gel to the specific band by the distance from the top of the gel to the dye front. Generate a standard curve by plotting the $R_f$ of each standard polypeptide marker as the abscissa and the $log_{10}$ of its MW as the ordinate. The MW of an unknown protein can then be determined with ease by finding its $R_f$ that vertically crosses the standard curve and reading the $log_{10}$ MW horizontally across to the ordinate. The antilog of the $log_{10}$ MW is the actual MW of the protein. Repeat until the MWs of all of the visible bands are determined.

## Reagents Needed

### 1% Coomassie Blue (CB) Solution

>   Coomassie Blue R-250 (2 g)
>   Dissolve well in 200 ml dd.$H_2O$.
>   Filter.

### Staining Solution

>   1% CB solution (125 ml)
>   Methanol (500 ml)
>   Acetic acid (100 ml)
>   Add dd.$H_2O$ to a final volume of 1000 ml.

### Destaining Solution I

>   Methanol (500 ml)
>   Acetic acid (100 ml)
>   Add dd.$H_2O$ to a final volume of 1000 ml.

### Destaining Solution II

>   Acetic acid (245 ml)
>   Methanol (175 ml)
>   Add dd.$H_2O$ to a final volume of 3.5 l.

## 2. Silver Staining Method

Silver staining is a very sensitive technique as compared with Coomassie Blue staining. Some very weak bands that do not become visible using Coomassie Blue staining do become very sharp after silver staining. The disadvantage is that the silver staining procedure is more complicated, and that the background using silver staining may be high. The protocols given below work very well in our experience.

*Notes:* *Silver staining is sensitive enough to stain fingerprinting on the gel; gloves should be worn when handling the gel. Distilled, deionized water should be used in the procedure.*

1.  Carefully transfer the slab gel into a clean container containing 400 ml of fixation solution. Allow fixation to occur for 30 to 60 min with slow shaking at 60 rpm.

2.  Remove the fixation solution and wash the gel with 500 ml of 30% ethanol for 30 min and then 500 ml of water for 30 min with shaking at 60 rpm. Repeat the ethanol/water washing three more times.

*Note:* *The gel should shrink in 30% ethanol and swell in water. This helps in the washing of the gel.*

3.  Stain the gel in 200 ml of staining solution for 20 to 30 min with shaking at 60 rpm.

4.  Immediately and quickly rinse the gel in 400 ml water three times.

5.  Develop the gel in 200 ml of developer solution for 10 min with occasional slow shaking.

6.  Stop the development process in 400 ml of stopping solution for 10 min with shaking at 60 rpm.

7.  Wash the gel in 500 ml water five times at 10-min intervals with shaking at 70 rpm.

8.  Clear the bands in Farmer's reducer solution for just 1 to 2 min and quickly rinse the gel with 500 ml water. Repeat washing at least five times at 10-min intervals in distilled water to completely remove Farmer's reducer solution until a clear background on the gel is obtained.

9.  Repeat staining, rinsing, developing, stopping, and washing as described above.

10. Clear a little bit of the stained background with 200 ml of a 1:5 dilution of Farmer's reducer solution for 2 to 5 min, stop, and wash the gel as above. However, this is optional.

11. Dry the gel and calculate the MWs of the different proteins as described in the Coomassie Blue staining method.

## Reagents Needed

### Fixation Solution

> 30% Ethanol (v/v)
> 10% Acetic acid (v/v)
> Add dd.$H_2O$ to a final volume of 1000 ml.

### 0.1% AgNO₃ Staining Solution

> 0.75 ml 80% $AgNO_3$ (w/v) in 600 ml dd.$H_2O$

### Alternative Staining Solution

> 420 ml 0.36% (w/v) NaOH
> 28 ml 35% (w/v) Ammonia
> Add 80 ml of 20% (w/v) silver nitrate dropwise with stirring.

### Developer or Reducing Solution (Fresh)

> 3% (w/v) Sodium carbonate in 0.02% formaldehyde in dd.$H_2O$ (0.5 ml formaldehyde in 2 l dd.$H_2O$)
> Make up to 500 ml.

### Alternative Developer Solution

> 2.5 ml 1% Citric acid
> 0.26 ml 36% (w/v) Formaldehyde
> Add dd.$H_2O$ to a volume of 500 ml.

### Stopping Solution

> 1% Acetic acid in dd.$H_2O$

### Farmer's Reducer Solution (Fresh)

> 0.5% (w/v) Potassium ferricyanide
> 1% (w/v) Sodium thiosulfate
> Dissolve in and add dd.$H_2O$ to a volume of 400 ml.

## C.    Transfer of Proteins from the Gel to a Nitrocellulose Membrane by Electroblotting

### Protocol

1. Soak the gel, a same size of nitrocellulose membrane or its equivalent, and four pieces of 3MM Whatman filter paper (relatively bigger than the gel) in 500 ml of blotting buffer for 15 to 20 min.

2. Fill a tray that is large enough to hold the cassette with blotting buffer to a depth of 2.5 to 5 cm.

*Notes:* *The cassette should be loaded under the buffer in order to avoid any air bubbles trapped between the different layers. Gloves should be worn when the assembling the electroblotting apparatus.*

3. Place one half of the cassette in the tray with the hook facing up, put one dacron sponge on the cassette half, and press on the sponge several times to force out any air bubbles.

4. Place two pieces of the soaked 3MM Whatman filer paper on the sponge and lay the soaked nitrocellulose membrane or its equivalent on the filter papers. Avoid any air bubbles between the layers.

5. Carefully place the soaked gel on the membrane and lay two soaked pieces of 3MM Whatman filter paper on the gel.

*Note:* *Gently press on the filter papers to force out any trapped air bubbles that will block the local transfer of proteins to the membrane.*

6. Cover the filter papers with another dacron sponge and place the second half of the cassette on the top of the stack so that the hook is down and faces the hole near the edge of the bottom half.

7. Press the two halves together and slide them toward each other so that the hooks are engaged with the opposite half.

8. Insert the assembled cassette into the blotting chamber according to the instructions. Fill the chamber with blotting buffer sufficient to cover the cassette. Place a stir bar at the bottom of the chamber.

*Note:* *Make sure to insert the cassette in the right place so that the membrane is between the gel and the anode.*

9.  Place the whole chamber on a magnetic stirrer plate, turn on the stir bar at low speed, connect to power supply, and turn on the current starting at 0.8 to 1 A.

10. Allow the blotting to take place for 60 to 75 min at 1 to 1.5 A.

*Notes:*   *Because heat will be produced very fast due to the current, the current should be monitored to be not higher than 1.5 A, which will burn the apparatus. The temperature should be controlled so that it is less than 60°C.*

11. When the blotting is complete, quickly mark the orientation of the membrane and immediately rinse it with 200 ml of phosphate-buffered saline (PBS).

*Notes:*   *Do not let the membrane become dry at this stage, as this usually brings about a higher background during immunodetection. If one's time is limited, the membrane at this stage can be wrapped wet with SaranWrap™ and stored at 4°C until use.*

12. Incubate the membrane in 400 ml of blocking solution at room temperature for 45 min with shaking at 60 rpm. This step serves to block the nonbinding sites on the membrane. Proceed to immunodetection.

*Notes:*   *The membrane at this stage may be dried with filter papers, wrapped with filter papers or equivalent, and stored at 4°C until use. To check for the efficiency of blotting, the blotted gel can be stained with Coomassie Blue and then destained. A successful transfer should have no visible bands left.*

## Materials Needed

*Transfer or Blotting Apparatus (such as Hoeffer TE 42 or equivalent)*

*Nitrocellulose Membrane or Equivalent*

*3MM Whatman Filter Papers*

*Blotting Buffer*

> Tris (15.2 g)
> Dissolve in 1 liter dd.$H_2O$
> Glycine (72.1 g)
> Methanol (1 liter)
> Add dd.$H_2O$ to 5 l. (The pH is about 8.3.)

*Phosphate-Buffered Saline (PBS) Solution*

> 10 m$M$ $NaH_2PO_4$
> 150 m$M$ NaCl
> Adjust the pH to 7.2 with 2 $N$ NaOH.
> (Make 4 l.)

*Blocking Solution*

> 2 to 5% (w/v) Bovine serum albumin (BSA) or 5% (w/v) dry nonfat milk or equivalent in PBS solution.

## D.    Immunodetection of Specific Protein(s) of Interest

### *Protocol*

1.   Incubate the unblocked membrane in 400 ml of blocking solution for 45 min with shaking at 60 rpm in order to block the nonbinding sites on the membrane. If the membrane was previously blocked and dried, soak the membrane in 200 ml of PBS solution containing 0.05% (v/v) Tween-20 for 20 min.

*Notes:*   *Gloves should be worn when handling the membrane. The membrane may be cut into a number of strips for repeated loading with the same sample and same amount of proteins. This is useful and the different controls given in Table 5–1 are strongly recommended.*

    The treatments in Table 5–1 are designed to check the activities of the preimmune serum and the first and second antibodies. But it is optional.

2.   Transfer the membrane filter to an appropriately sized tray or place the membrane strips in an assay plate using a pair of forceps. Incubate with first or primary antibody (monoclonal or polyclonal antibodies) diluted in PBS/BSA/T solution [1% (w/v) BSA or equivalent and 0.05% (v/v) Tween-20 in PBS buffer]. Allow incubation to take place at room temperature for 40 to 60 min with slow shaking at 60 rpm.

*Notes:*   *(1) Monoclonal antibodies are more specific than polyclonal antibodies, which usually generate nonspecific band(s). (2) The concentration of antibody used in the incubation varies with different antibodies. It is recommended that different dilutions be tested in order to obtain the optimal concentration for incubation. (3) The volume of antibody solution should be a little in excess in order to cover the membrane. However, too much solution is not preferred. (4) More than two membranes or strips in one tray is not recommended.*

3.   Decant the antibody solution, which can be reused for up to three times, and wash the membrane with 4 volumes of PBS/BSA/T for 5 min. Repeat the washing three times at 5-min intervals with shaking at 60 rpm.

4.   Incubate the membranes in the secondary antibody solution diluted 500 to 3000 times in PBS/BSA/T solution at room temperature for 1 to 1.5 h with shaking at 60 rpm. If the first antibody is a monoclonal one from mice, the secondary antibody should be commercial goat anti-mouse IgA, IgG, or IgM conjugated with alkaline phosphatase. If the first antibody is a polyclonal one from rabbits, the secondary antibody should be goat anti-rabbit IgG conjugated with alkaline phosphatase or alkaline peroxidase.

### Table 5–1 Recommended Controls for the Membrane Strips

| Control groups | Membrane strip number (duplicate for each) | | | | | |
|---|---|---|---|---|---|---|
| | 1 | 2 | 3 | 4 | 5 | 6 |
| Antigen | + | – | – | + | + | + |
| First antibody | – | + | – | – | + | + |
| Second antibody | – | – | + | + | + | + |
| Preimmune serum | – | – | – | + | – | – |
| Color substrate | + | + | + | + | + | + |
| Color developer | + | + | + | + | + | + |

5.  Decant the antibody solution, which can be reused for up to three times, and wash the membrane with 4 volumes of PBS/BSA/T for 5 min. Repeat washing three times with shaking at 60 rpm.

6.  Incubate the membrane in 0.2 $M$ Tris-HCl (pH 9.2) for 2 min.

7.  Detect band(s) by developing the membrane in developer solution for 5 to 40 min until desired band(s) is(are) visible, and quickly stop the development by rinsing the membrane with large volumes of tap water.

8.  Air dry the membranes and photograph for a permanent record (Figure 5.5). The MW of the detected band(s) can be determined by matching the membrane with the Coomassie Blue or silver stained gel, whose MWs have been determined as described previously.

*Note:*  *If different controls have been designed, no bands should be visible from those membrane strips that were only treated with antigen or with primary antibody or secondary antibody, or with preimmune antiserum, or with no antigen.*

## Solutions Needed

### PBS/BSA/T Solution

1% (w/v) BSA
0.05% (v/v) Tween-20 in PBS solution

### Developer Solution

15 ml 0.2 $M$ Tris-HCl, pH 9.2
0.1 ml Nitroblue tetrazolium (NBT) stock solution (1.0 mg NBT/ml Tris/HCl)
0.05 ml 5-Bromo-4-chloro-3-indolylphosphate (BCIP) stock solution (5 mg BCIP/ml dimethyl formamide)
8 µl 2.0 $M$ MgCl$_2$

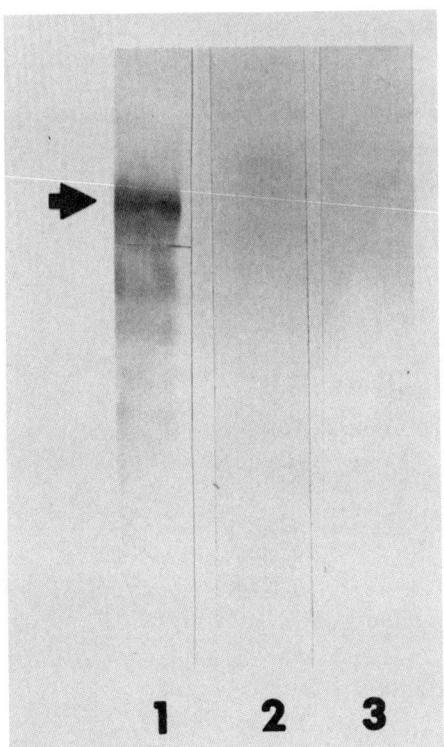

**FIGURE 5.5**
Western blot analysis of invertase in oats. Protein mixture was subjected to SDS-PAGE and blotted onto a nitrocellulose membrane and incubated with polyclonal antibodies against invertase. Lane 1 reveals the invertase band. Lanes 2 and 3 show the negative controls.

## E.     Troubleshooting Guide

| Symptom | Cause and solutions |
| --- | --- |
| No signal at all | (1) Reagents were omitted or added incorrectly. Try to use all reagents in the correct order. (2) Insufficient antigen (protein) was loaded onto the gel. Try to increase the amount of proteins for one well. (3) The primary antibody may not be good. The activity of the antibody may not be higher. (4) The secondary antibody or conjugated alkaline phosphatase loses activity. Try to test the quality of the secondary antibody. (5) Transfer of proteins to the membrane was not successful due to placing the membrane between the gel and negative pole. If that happened, all the proteins were transferred into the blotting buffer in the chamber. Ensure that the blotting is handled properly |
| High background occurs | Reagents may be too concentrated or the amount of blocking reagent is insufficient. Try to use the right concentrations of solutions and increase the amount of blocking reagent; also, incubate the membrane for a longer time in the blocking solution |
| Signal is very weak | The concentration of the antigen or the antibodies or the developer solution was too low, or the incubation time was too short. Try to use the appropriate concentration of each of above and increase the incubation time |
| Nonspecific bands are evident | The primary antibody raised from the antigen is not specific, resulting in cross-reactions. Try to verify the specificity of the antibody before repeating, and increase the blocking reagents to 5% (w/v) in BSA or its equivalent |

# References

1. **Southern, E. M.,** Detection of specific sequences among DNA fragments separated by gel electrophoresis, *J. Mol. Biol.,* 98, 503, 1975.

2. **Fourney, R. M., Aubin, R., Dietrich, K. D., and Paterson, M. C.,** Determination of foreign gene copy number in stably transfected cell lines by Southern transfer analysis, in *Gene Transfer and Expression Protocols,* Murray, E. J., Ed., Humana Press, Clifton, NJ, 1991.

3. **Sambrook, J., Fritsch, E. F., and Maniatis, T.,** *Molecular Cloning: A Laboratory Manual,* 2nd ed., Cold Spring Harbor Press, Cold Spring Harbor, NY, 1989.

4. **Reed, K. C. and Mann, D. A.,** Rapid transfer of DNA from agarose gels to nylon membranes, *Nucleic Acids Res.,* 13, 7207, 1985

5. **Khandjian, E. W.,** Optimized hybridization of DNA blotted and fixed to nitrocellulose and nylon membranes, *Biotechnology,* 5, 165, 1987.

6. **Amasino, R. M.,** Acceleration of nucleic acid hybridization rate by polyethylene glycol, *Anal. Biochem.,* 152, 304 1986.

7. **Thomas, P. S.,** Hybridization of denatured RNA and small DNA fragments transferred to nitrocellulose, *Proc. Natl. Acad. Sci. U.S.A.,* 77, 5201, 1980.

8. **Krumlauf, R.,** Northern blot analysis of gene expression, in *Gene Transfer and Expression Protocols,* Murray, E. J., Ed., Humana Press, Clifton, NJ, 1991.

9. **Knudsen, K. A.,** Proteins transferred to nitrocellulose for use as immunogens, *Anal. Biochem.,* 147, 285, 1985.

10. **Towbin, J., Staehlin, T., and Gordon, J.,** Electrophoretic transfer of proteins from polyacrylamide gels to nitrocellulose sheets: procedure and some applications, *Proc. Natl. Acad. Sci. U.S.A.,* 76, 4350, 1979.

11. **Johnson, D. A., Gautsch, J. W., Sportsman, J. R., and Elder, J. H.,** Improved method for utilizing nonfat dry milk for analysis of proteins and nucleic acids transferred to nitrocellulose, *Gene Anal. Technol.,* 1, 3, 1984.

12. **Kyhse-Anderson, J.,** Electroblotting of multiple gels: a simple apparatus without buffer tank for rapid transfer of proteins from polyacrylamide to nitrocellulose, *J. Biochem. Biophys. Methods,* 10, 203, 1984.

# Chapter

# cDNA Libraries

## Contents

# I.    Construction and Screening of a cDNA Library Using λDNA as a Vector

## A.    Principles and Strategies for Construction of a cDNA Library

Construction of a cDNA library is a highly sophisticated technology that is used in molecular biology studies. The quality and integrity of the cDNA library is directly related to the success or failure of cDNAs that are of interest to investigators.[1,2] In order to obtain a very good library, the person working on the library should have a strong molecular biology background and intensive laboratory experience in molecular biology. Fortunately, there are many methods that have been well developed, and multiple commercial kits are available for cDNA cloning. The present chapter describes the detailed strategies used in cDNA synthesis, construction and screening of cDNA libraries, and in the isolation of putative clones. The step-by-step protocols presented will allow experienced workers as well as beginners to achieve success.

cDNA cloning is a complex series of enzymatic procedures. The general principle is that mRNA is copied into first-stranded DNA that is called complementary DNA or cDNA, based on nucleotide bases complementary to each other. This step is driven by AMV reverse transcriptase using an oligo(dT) primer or random primers. The second-stranded DNA is copied from the first-stranded DNA using DNA polymerase I, generating a double-stranded cDNA. The double-stranded cDNA is subsequently ligated to an adapter for preparing the termini for vector ligation. The recombinant construct will be then packaged *in vitro* and cloned in a specific host, thus constructing a cDNA library. cDNA libraries preserve as much of the original cDNAs as possible and allow one to "fish" out any possible cDNA(s) by screening the cDNA library as long as a specific probe is available.

The classical cDNA cloning method takes advantage of the 3′ hairpins generated by AMV reverse transcriptase during first-strand synthesis. The hairpins are then used to prime second-strand cDNA catalyzed by Klenow DNA polymerase and reverse transcriptase. *S1* nuclease is added to cleave the hairpin loop. However, this digestion is difficult to control, causing low cloning efficiencies and the loss of a significant amount of sequence information corresponding to the 5′ end of the mRNA. An improved strategy utilizes 4 m$M$ sodium pyrophosphate, which greatly suppresses the formation of hairpins during the synthesis of first-stranded cDNA. Second-strand synthesis is then carried out by RNase H to create nicks and gaps in the hybridized mRNA template, generating 3′-OH priming sites for DNA synthesis and repair by DNA polymerase I. After treatment with T4 DNA polymerase to remove any remaining 3′ protruding ends, the blunt-ended, double-stranded cDNAs are ready for adaptor or linker ligation (Figure 6.1). *S1* digestion is avoided in this method, and the cloning efficiency is much higher than in the classic method. In addition, the sequence information can be optimal.

A variety of improved methods have been developed for cloning cDNA molecules. The basic strategies can be grouped into two classes. One is random cloning; the other is orientation-specific cloning. The former is much easier than the latter with respect to the techniques involved. As shown in Figure 6.2, the random or classical cloning of cDNA uses oligo(dT)

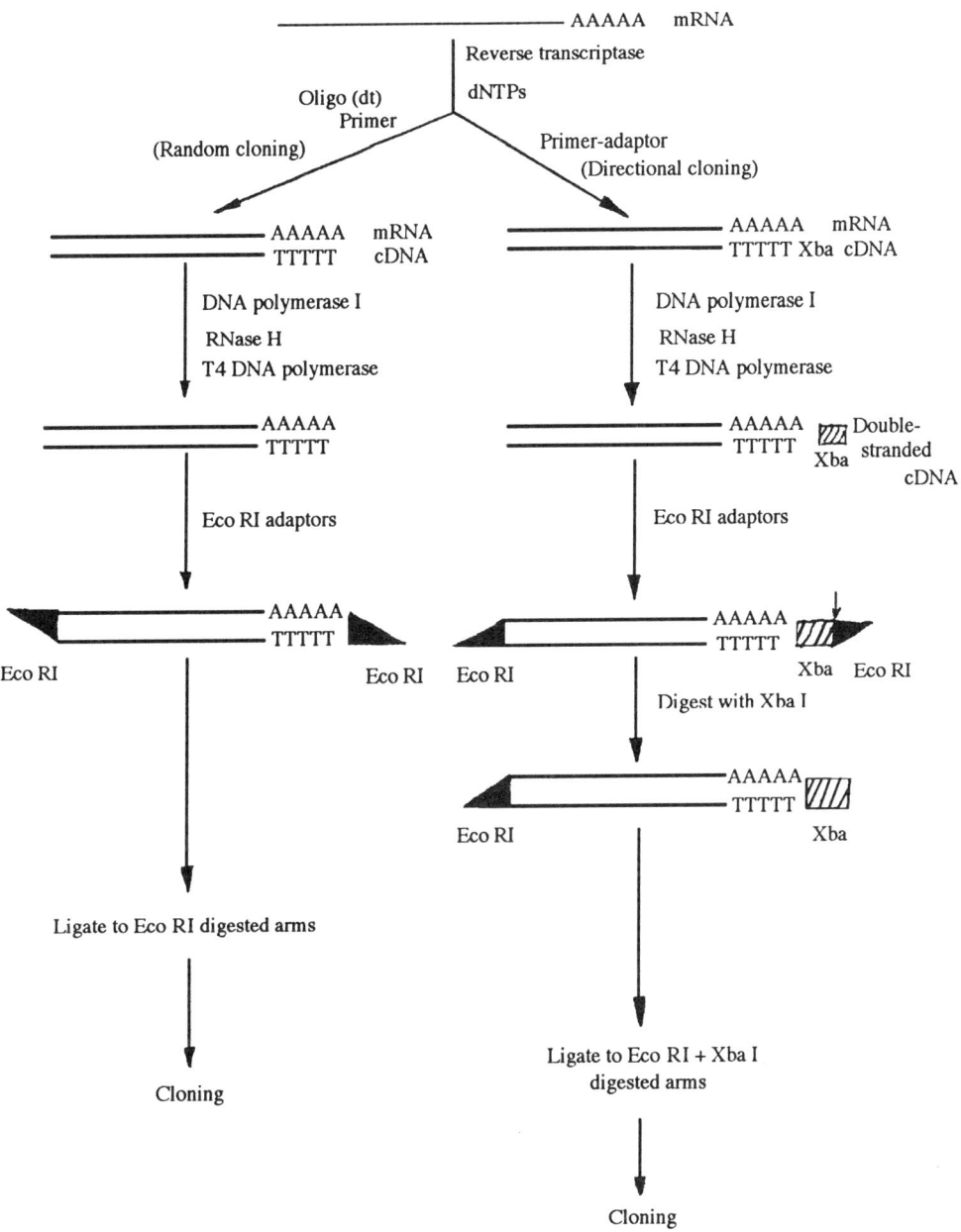

**FIGURE 6.1**

Diagram of cDNA synthesis scheme showing two cloning strategies: random cloning (left) and directional cloning (right).

as a primer and λgt10 or λgt11 as cloning vectors for the cDNA library, where the cDNA are cloned into a single *EcoR* I site via the addition of *EcoR* I adaptors (or linkers) to each end of the cDNA molecules. Because of the single *EcoR* I site cloning, the cDNAs are cloned in a random way, including both sense and antisense orientations. If the cDNA library is constructed using the nonexpressional vector λgt10 (Figure 6.3), it can be screened by using DNA or RNA as a probe. However, if the expression vector λgt11 is used for the cDNA

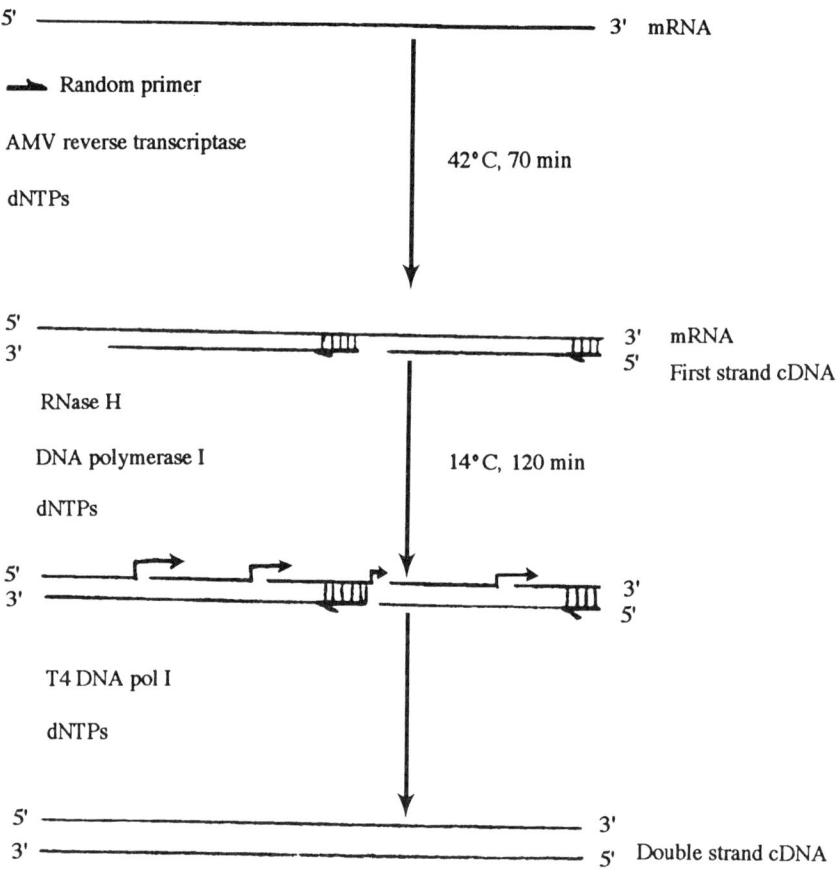

**FIGURE 6.2**
Scheme for cDNA cloning using a random primer.

library, the library can be screened using a specific antibody as well as DNA or RNA as a probe. The disadvantage of using an antibody as a probe to screen the randomly cloned cDNA library is that the possibility of obtaining the positive clones is at least 50% less. This is because approximately 50% of cloned cDNAs are expressed as antisense RNA, which may interfere with sense RNA by inhibiting the translation of the sense RNA into proteins. Because of the shortcomings of random cloning, state-of-the-art strategy is the efficient orientation-specific or directional cloning of cDNA using a primer-adaptor to prime the synthesis of the first-strand cDNA (Figure 6.1). The primer-adaptor consists of oligo(dT) adjacent to a unique restriction site (*Xba* I or *Not* I). The subsequent steps are carried out as in random cloning except that the final double-stranded cDNAs with *EcoR* I adaptors attached are digested with either *Xba* I or *Not* I. The digested cDNA molecule contains one *EcoR* I and one *Xba* I or *Not* I termini, which can be ligated with a vector containing the same restriction enzyme termini. As compared with random cloning protocol, directional cloning is more powerful and valuable. In expression vectors such as λgt11 *Sfi-Not* I, a factor of two is the likelihood of expressing the cDNA insert as the correct polypeptide. This is because of no antisense RNA interference. In transcription vectors, such as the λGEM-2 and λGEM-4 vectors (Figure 6.4), all of the cDNA inserts are cloned in the same direction downstream from the promoters for T7 and SP6 RNA polymerase. Total sense or antisense RNA probes can be obtained from the cDNA library, which represent all of the sequences in the library. These RNAs can be used

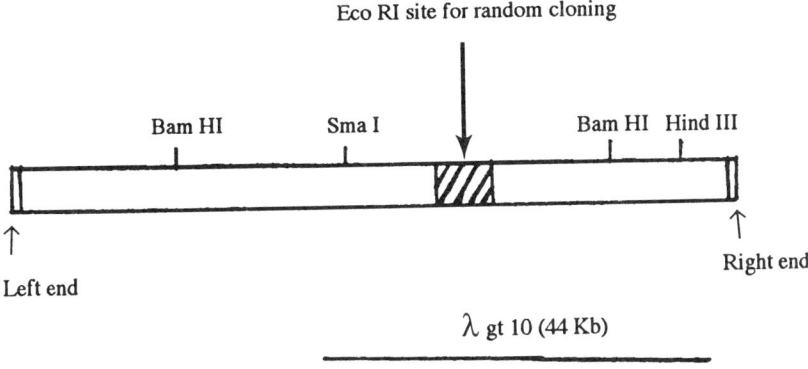

λ gt 10 (44 Kb)

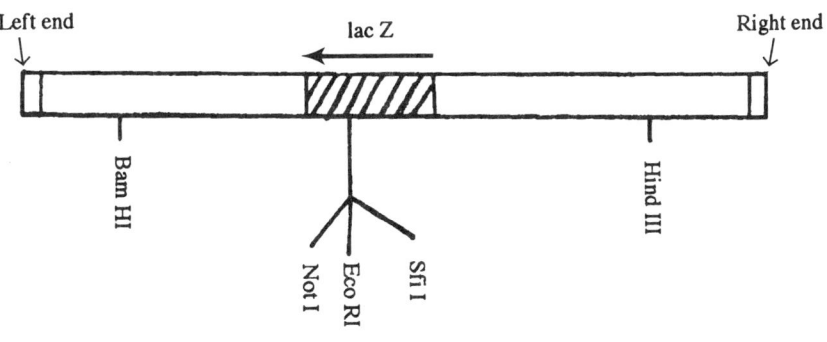

λ gt 11 Sfi-Not (44 Kb)

**FIGURE 6.3**
Structural maps of vectors: λgt10 (top) and λgt11 *Sfi-Not* I (bottom).

for subtraction hybridization to isolate and analyze some rare genes expressed in one organism but not in another, or in different tissues in the same organism. The positive cDNA inserts purified from the directionally cloned library can drive further directional subcloning procedures, using an appropriate plasmid vector such as pGEM-11 (Promega Corporation) that is digested with the same enzymes, *EcoR* I and *Xba* I or *Not* I. The cDNA insert is cloned in the same direction downstream from the promoters for T7 and SP6 RNA polymerase. The

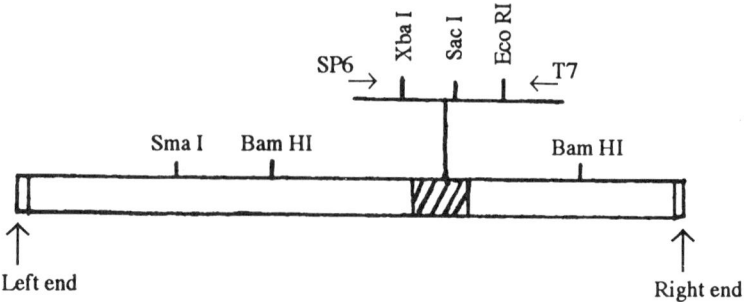

λ GEM 2 (44 Kb)

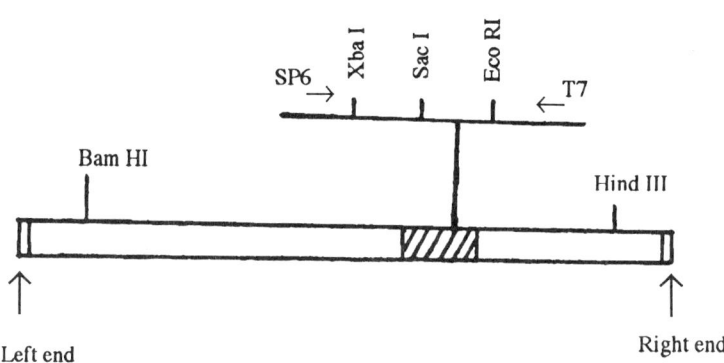

λ GEM 4 (44 Kb)

**FIGURE 6.4**
Structural maps of vectors: λGEM 2 (top) and λGEM 4 (bottom).

poly(A)-tail strand cDNA can be easily identified and sequenced under control by the T7 or SP6 RNA polymerase promoter. The sense RNA or antisense RNA can be utilized as RNA probes to screen cDNA or genomic DNA libraries as well as to perform Northern blot analyses.

## B.     General Preparation and Handling

cDNA cloning is one of the techniques used in state-of-the-art molecular biology. Based on the authors' experience, there are three steps that are critical for success or failure in cDNA library construction. The first is the purity and integrity of the mRNA used for the synthesis of the first-strand cDNA. Any degradation or absence of specific mRNAs will result in partial length cDNA or complete loss of the specific cDNA, especially for some rarely expressed genes. The second important step is to achieve the synthesis of full-length cDNAs. If this procedure is not carried out properly, even if you have a very good mRNA source, the cDNA library to be constructed will be not good, and you may "fish" out only partial-length cDNA or no positive clone at all. As long as double-strand cDNAs are obtained, they will be much more stable than mRNAs. However, there is a third essential step in cDNA cloning. This is the ligation of cDNAs with adaptors to vectors. If the ligation fails or is not good, *in vitro* packaging of recombinant λDNAs cannot be carried out effectively. As a result, the plaque forming units (pfu) number or efficiency is very low. In that case, a poor cDNA library may be not good for screening for specific clones. In order to construct an exceptionally good cDNA library, the following precautions should be taken, in addition to carefully following the protocols given in this chapter. Elimination of the risk of RNase contamination must be carried out whenever possible. Specific strategies for the isolation and purification of high quality and quantity of mRNAs as templates are described in Chapter 2. Procedures for creating an RNase-free laboratory environment for mRNA and cDNA synthesis include the following:

1. Disposable plastic test tubes, micropipette tips and microcentrifuge tubes should be sterilized.

2. Gloves should be worn at all times and changed often to avoid finger-derived RNase contamination.

3. All glassware and electrophoresis apparatus used for cDNA cloning should be separated from other labware. Glassware should be treated with 0.1% diethyl pyrocarbonate (DEPC) solution, autoclaved to remove the DEPC and baked overnight at 250°C before being used.

4. Solutions to be used with mRNA and cDNA synthesis should be treated with 0.1% DEPC (v/v) to inhibit RNase by acylation. Add DEPC to the solutions and stir vigorously for 20 min, followed by autoclaving or heating the solution at 70°C for 1 h to remove the DEPC.

5. Because DEPC reacts with amines and sulfhydryl groups, reagents such as Tris and DTT cannot be treated directly. These solutions can be made by using DEPC-treated and sterile water. If solutions such as DTT cannot be autoclaved, the solution can be sterile-filtered.

*Caution:*    *DEPC is a powerful acylating agent and should be handled in a fume hood. Never add DEPC into solutions containing ammonia because that will cause the formation of ethyl carbamate, a potent carcinogen.*

## C.    Vectors Used for Construction of a cDNA Library

The selection of a vector for cDNA cloning depends on the screening method and on the number of pfu to ensure the presence of the desired cDNA. In order to detect some rare cDNAs from low-abundance mRNAs from mammalian cells, approximately $2 \times 10^5$ pfu of recombinants must be screened using nucleic acid probes. Bacteriophage lambda vectors, as compared with plasmid vectors, are usually recommended when large numbers of recombinants are needed. This is because of the high cloning efficiency of and subsequent infection by the former. The cDNA library constructed can be efficiently screened at different pfu densities with either nucleic acid or antibody probes. However, cloning efficiencies are usually 10- to 50-fold lower using plasmid vectors. The number of colonies screened must be small because, at much higher densities, the colonies will overlap each other, causing difficulty in screening. The screening is much more difficult when antibodies are used as probes. The advantage of using plasmids as vectors is that a commercial plasmid contains multiple cloning sites, and that the plasmids can be used as tailed vectors and for vector-primer cloning strategies (see Section III). In addition, the isolation and manipulation of the resulting clones are relatively easier.

Recently, the advanced features of plasmids have been incorporated into λDNA vectors. For example, the λGEM-2 and λGEM-4 vectors (Promega Corporation) were constructed by inserting a 535bp pGEM-1 vector fragment (λGEM-2) and the entire pGEM-1 plasmid (λGEM-4) into the original cloning site of λgt10. These improved vectors allow directional cloning via *EcoR* I and *Xba* I cloning sites and simplify subcloning procedures. The T7 and SP6 promoters make it possible to synthesize RNA probes, which can be used for subtraction cDNA library construction.

λgt10, λGEM-2, and λGEM-4 are the usually recommended vectors used for cDNA libraries to be screened with DNA or RNA probes. If an antibody is used as a screening probe, the expression vector λgt11 is the better to use. This vector can express the inserted cDNA as part of a β-galactosidase fusion protein. However, since this vector contains a single *EcoR* I cloning site, cDNAs can only be cloned according to a random cloning strategy using oligo(dT) as a primer and *EcoR* I adaptors or linkers. As described in Section I.A, the possibility of detecting the desired cDNA clones is very small because of the antisense RNA interference. Another vector λgt11 *Sfi-Not* I, on the other hand, is used for directional cloning of cDNA. All the cDNAs can be cloned in the same direction, thus doubling the likelihood of expression of cDNA inserts as fusion polypeptides. The cDNA library can be screened using an antibody as a probe as well as nucleic acid probes. This cloning strategy effectively increases the possibility of obtaining the cDNA clones of interest.

## D.    Protocols for Construction of a cDNA Library

### Protocol 1: Synthesis of the First-Strand cDNA

The general scheme is outlined in Figure 6.1. It is recommended that two reactions be set up that are carried out separately all the way to the end. The benefit of doing this is that if one reaction is somehow stopped by accident during the experiments, the other reaction can be used as a backup. Otherwise, one must start over, thus wasting time and money. The standard reaction given below has a total volume of 25 μl using up to 2 μg of mRNA. For each additional microgram of mRNA, increase the reaction volume by 10 μl. A control reaction should be carried out whenever possible under the same conditions.

1.    Anneal 2 μg of mRNA template with 1 μg of primer of oligo(dT), or with 0.7 μg of primer-adaptor of oligo(dT) *Not* I/*Xba* I in a sterile RNase-free microcentrifuge tube. Add nuclease-free dd.H$_2$O to a total volume of 15 μl. Heat the tube at 70°C for 5 min and allow it to slowly cool to room temperature to finish the annealing. Briefly spin down the mixture to the bottom of the microfuge tube.

2.    To the annealed primer/template, add the following in the order shown. To prevent precipitation of sodium pyrophosphate when it is added to the buffer components, the buffer should be preheated at 40 to 42°C for 4 min before the addition of sodium pyrophosphate and AMV reverse transcriptase. Gently mix well after each addition.

   • First-strand 5X buffer (5 μl)

   • rRNasin ribonuclease inhibitor (25 units/μg mRNA) (50 units)

   • 40 m*M* Sodium pyrophosphate (2.5 μl)

   • AMV reverse transcriptase (15 units/μg mRNA) (30 units)

   • Add nuclease-free dd.H$_2$O to a total volume of 25 μl.

**Functions:** The 5X reaction buffer contains components required for cDNA synthesis. Ribonuclease inhibitor inhibits RNase activity and protects the mRNA template. Sodium pyrophosphate suppresses the formation of hairpins that are normally produced in "classical" cloning methods, thus avoiding the S1 digestion step which is difficult to carry out. AMV reverse transcriptase catalyzes the synthesis of the first strand cDNA from mRNA template based on the rule of complementary base pairing.

3. Set up a tracer reaction by transferring 5 µl of the mixture to a fresh tube containing 1 µl of 4 µCi of [α-$^{32}$P]dCTP (>400 Ci/mmol, less than one week old). The first-strand synthesis will be measured by trichloroacetic acid (TCA) precipitation and alkaline agarose gel electrophoresis using the tracer reaction.

*Caution:* *[α-$^{32}$P]dCTP is a dangerous isotope. Gloves should be worn when carrying out the tracer reaction, TCA assay, and gel electrophoresis. Waste materials such as contaminated gloves, pipette tips, solutions, and filter papers should be put in special containers for waste radioactive materials.*

4. Incubate both reactions at 42°C for 1.5 h and place on ice. At this stage, the synthesis of the first-strand cDNA has been completed. To the tracer reaction, add 50 m$M$ EDTA up to a total volume of 100 µl and store on ice for TCA incorporation assays and alkaline agarose gel electrophoresis analysis after extraction. The unlabeled reaction will be directly used for the second strand DNA synthesis.

## Protocol 2: Synthesis of the Second-Strand DNA

1. To the unlabeled first-strand cDNA tube on ice, add the following components in the order given below. Gently mix well after each addition.
   - First-strand reaction mixture (20 µl)
   - Second-strand 10X buffer (10 µl)
   - *E. coli* DNA polymerase I (24 units)
   - *E coli* RNase H (1 unit)
   - Add nuclease-free dd.H$_2$O to a final volume of 100 µl.

   **Functions:** *E. coli* RNase H makes nicks and gaps in the hybridized mRNA template, creating 3′-OH priming sites. *E. coli* polymerase I serves to synthesize DNA and repair the gaps from the priming sites, producing the second-strand cDNA.

2. Set up a tracer reaction for the second-strand cDNA by transferring 10 µl of the mixture to a fresh tube containing 1 µl 4 µCi [α-$^{32}$PJdCTP (>400 Ci/mmol). The tracer reaction will be used for TCA assay and alkaline agarose gel electrophoresis to monitor the quantity and quality of the synthesis of the second-strand DNA.

3. Incubate both reactions at 14°C for 3.5 h. Add 90 µl of 50 m$M$ EDTA to the 10 µl tracer reaction and store on ice.

4. Heat the unlabeled double strand cDNA sample at 70°C for 10 min to stop the reaction. **At this point, the double-strand cDNAs should be generated.** Briefly spin down to collect the contents at the bottom of the tube and place on ice.

5. Add 4 units of T4 DNA polymerase (2 units/µg input mRNA) to the mixture at step 4 and incubate the tube at 37°C for 10 min. This step functions to make blunt-end, double-strand cDNA by T4 polymerase. The blunt ends are required for later adaptor ligations.

6. Stop the T4 polymerase reaction by adding 10 µl of 0.2 $M$ EDTA and place on ice.

7. Extract the cDNAs with 1 volume of TE-saturated phenol/chloroform. Mix well and centrifuge in a microcentrifuge at top speed for 4 min at room temperature.

8.  Carefully transfer the top, aqueous phase to a fresh tube. **Do not take any white materials at the interphase between the two phases.** To the supernatant, add 0.5 volume of 7.5 $M$ ammonium acetate or 0.15 volume of 3 $M$ sodium acetate (pH 5.2), mix well and add 2.5 volumes of chilled (–20°C) 100% ethanol. Gently mix and allow precipitation to occur at –20°C for 2 h.

9.  Centrifuge in a microcentrifuge at the top speed for 5 min. Carefully remove the supernatant, briefly rinse the pellet with 1 ml of cold 70% ethanol, and gently drain away the ethanol.

10. Dry the cDNA pellet under vacuum for 15 min. Dissolve the cDNA pellet in 25 μl of TE buffer. It is important to take 2 to 4 μl of the sample to measure the concentration of cDNAs prior to the next reaction. Store the sample at –20°C until use.

*Note:*     *Before adaptor ligation, it is strongly recommended that the quantity and quality of both first- and second-strand cDNAs be checked by TCA precipitation and gel electrophoresis as described below.*

## Protocol 3: Trichloroacetic Acid (TCA) Assay and Yield Calculations

1.  Spot 4 μl of the first-strand tracer reaction sample and 4 μl of the second-strand reaction sample on glass-fiber filters and air dry. These will represent the total counts per minute (cpm) in the samples.

2.  Add another 4 μl of the same reaction samples to fresh tubes containing 100 μl of carrier DNA solution (1 mg/ml) and mix well. Add 0.5 ml of 5% TCA and mix by vortexing. Allow precipitation to occur on ice for 30 min.

3.  Filter the precipitated samples through glass-fiber filters. Wash the filters with 6 ml cold 5% TCA four times and rinse once with 6 ml of acetone or ethanol. Allow the filters to dry at room temperature. The samples represent incorporated cpm in the reactions.

4.  Place the filters in individual vials and add 10 to 15 ml of scintillation fluid to cover each filter. Count both total and incorporated cpm samples according to the instructions for the scintillation counter. The cpm can also be counted by Cerenkov radiation (without scintillant).

5.  Calculate yield of first-strand cDNA as follows:

$$\text{Percent incorporated} = \frac{\text{incorporated cpm}}{\text{total cpm}} \times 100$$

$$\text{dNTP incorporated (nmol)} = 4 \text{ nmol dNTP/μl} \times \text{reaction volume (μl)} \times \text{percent incorporation/100}$$

$$\text{cDNA synthesized (ng)} = \text{dNTP incorporated (nmol)} \times 330 \text{ ng/nmol}$$

$$\text{Percent mRNA converted to cDNA} = \frac{\text{cDNA synthesied (ng)}}{\text{mRNA in reaction (ng)}}$$

6.  Calculate the yield of second strand cDNA as follows:

$$\text{Percent second-strand DNA incorporated} = \frac{\text{incorporated cpm}}{\text{total cpm}} \times 100$$

$$\text{dNTP incorporated (nmol)} = [0.8 \text{ nmol dNTP/μl} \times \text{reaction volume (μl)} - \text{dNTP (nmol) incorporated in first-strand reaction}] \times \text{percent second-strand incorporation/100}$$

$$\text{Second-strand cDNA synthesized (ng)} = \text{dNTP incorporated (nmol)} \times 330 \text{ ng/nmol}$$

$$\text{Percent converted to double-stranded cDNA} = \frac{\text{second-strand cDNA synthesized (ng)}}{\text{first-strand cDNA synthesized (ng)}} \times 100$$

*Note:*     *Reactions yielding 15 to 30% first-strand conversion and 80 to 200% second-strand conversion values can be used to construct a successful cDNA library.*

### Protocol 4: Alkaline Agarose Gel Electrophoresis of cDNAs

The quality and the size range of cDNAs synthesized in the first and second strand reactions should be checked by 1.4% alkaline agarose gel electrophoresis using tracer reaction samples.

1. Extract the first- and second-strand cDNAs with phenol and precipitate with ethanol. Designate these as the unlabeled reactions.

2. Label λ*Hind* III fragments with $^{32}$P in a fill-in reaction. These are used as DNA markers to estimate the sizes of cDNAs. In a sterile Eppendorf tube, add the following components in the order shown below.
   - *Hind* III 10X buffer (2 μl)
   - dATP (0.2 m$M$)
   - dGTP (0.2 m$M$)
   - [α-$^{32}$P]dCTP (2 μCi)
   - λ*Hind* III markers (1 μg)
   - Klenow DNA polymerase (1 unit)
   - Add dd.H$_2$O to a final volume of 20 μl.

   Mix well after each addition and incubate the reaction for 15 min at room temperature. Add 2 μl of 0.2 $M$ EDTA to stop the reaction. Transfer 6 μl of the sample directly to 6 μl of 2X alkaline buffer and store the remainder at –20°C.

3. Dissolve 1.4% (w/v) agarose in 50 m$M$ NaCl, 1 m$M$ EDTA solution and melt in a microwave for a few minutes. Cool the gel mixture to about 50°C and pour the gel into a mini-electrophoresis apparatus. Allow the gel to harden and equilibrate the gel in alkaline gel running buffer for 1 h prior to electrophoresis.

4. Transfer the same amount of each sample (50,000 cpm) to separate tubes and add an equal volume of TE buffer to each tube. Mix and add 1 volume of 2X alkaline buffer to each tube. Total volume should be approximately 30 μl. It is recommended that the same numbers of incorporated cpm be loaded for each strand to compare the density of signals on the same autoradiograph.

5. Carefully load the samples into wells and immediately run the gel at 7 V/cm until the dye has migrated to about 2 cm from the end.

6. Stop the electrophoresis and soak the gel in 5 volumes of 7% TCA at room temperature until the dye changes from blue to yellow. Dry the gel on a piece of 3MM Whatman paper on a gel dryer or under a weighted stack of paper towels for 6 h. A 7% TCA used here is to neutralize the denatured gel.

7. Wrap the gel with SaranWrap™ and expose to X-ray film at room temperature or at –70°C with an intensifying screen for 1 to 4 h (Figure 6.5).

*Note:*     *For successful synthesis of cDNAs, the signal should range from 0.15 to 8 Kb with a sharp size range of 1.5 to 4 Kb (Figure 6.5).*

a.     Special Troubleshooting Guide

Obtaining good double-strand cDNA is half completed in making a cDNA library. Therefore, the synthesis of first- and second-strand cDNAs is extremely important for successful cDNA library construction. Prior to proceeding further, it is a wise idea to make sure that the double-strand cDNAs synthesized are good enough for the construction of a successful cDNA library. The following items, based on our experience, should be checked before going ahead with any further steps:

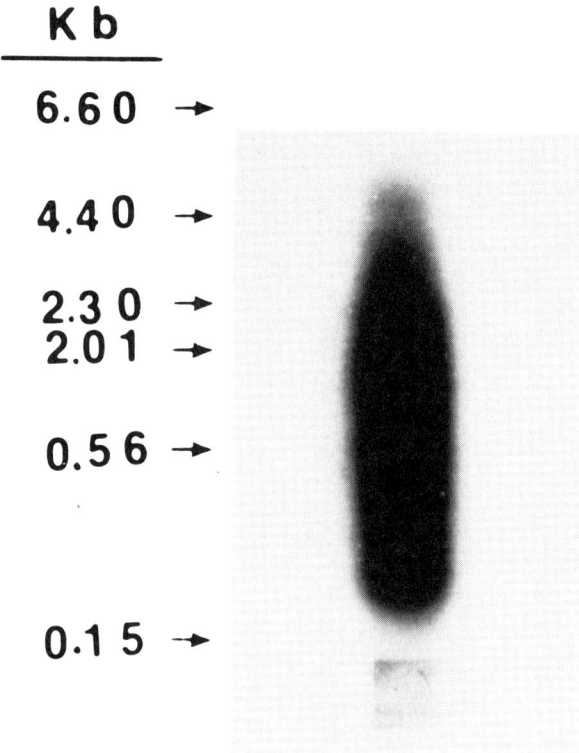

**FIGURE 6.5**
Eukaryotic cDNAs that were synthesized by using an adapter-primer. The cDNAs show a size range from 0.25 to 5.0 Kb.

1.  If the yield of the first-strand cDNA conversion is less than 8%, and if the intensity of the different cDNA sizes shown on the X-ray film is very weak or if the size range is very small and of lower molecular weight, the potential causes for these problems are due to poor mRNA template and/ or the failure of reverse transcriptase to act, or to the presence of some inhibitors.

    The quality of the mRNA template is the most important factor for successful cDNA synthesis. If the poly(A)+RNA is degraded during its isolation and purification procedure, the first strand cDNA synthesized from the template is only partial-length cDNA. To avoid this problem, it is important to check the quality of mRNA template prior to the synthesis of first strand cDNA. This can be done by simply analyzing 2 μg mRNA in 1% agarose–formaldehyde denaturing gel containing ethidium bromide (EtBr). After electrophoresis, take a photograph of the gel under UV light. If the smear range is from 0.65 up to 8.5 Kb, the integrity of the mRNA is very good for cDNA synthesis. Another possible cause is that AMV reverse transcriptase functions at the beginning of the synthesis of first-strand cDNA but fails in the middle of the procedure. In that case, the cDNA produced is likely to be of only partial length. To ensure that this potential problem does not occur, always use positive mRNA template as a control to check the activity of reverse transcriptase. If the mRNA template and reverse transcriptase are very good, then use the control RNA to check for the presence of inhibitors such as SDS, EDTA, and salts. Alternatively, the ratio of the amount of primer:mRNA is not good, which may adversely affect first-strand cDNA synthesis.

2.  Another symptom is that the yield of the second-strand DNA conversion is less than 40%, or that the size range is very small and of abundant lower molecular weight, even though the synthesis of the first-strand cDNA is successful.

    This kind of problem is likely to be due to either RNase H or DNA polymerase I not functioning well. To check the activity of RNase H, treat control RNA with this enzyme and assay by electrophoresis. If the RNase H works well, the RNA is nicked into fragments. This will be shown by gel electrophoresis as compared with untreated RNA. On the other hand, the DNA polymerase I may not function well; as a result, the synthesis of second strand DNA or the repair of gaps will not be completed, thus generating partial-length cDNAs. In that case, set up another reaction using a control experiment, or use fresh DNA polymerase I. If low incorporation is obtained even with the positive control, it is likely that the radioisotope used as a label is not good. Check the specificity and half-life of the isotope and use fresh [$\alpha$-$^{32}$P]dCTP as a control. If you have obtained very good double-strand cDNAs, and if the cDNA of interest is included in the expected size range, you are very successful and may go ahead with adaptor ligation.

*Materials Needed* ───────────────────────────────────────────

*[$\alpha$-$^{32}$P]dCTP (>400 Ci/mmol)*

*50 mM EDTA*

*7.5 M Ammonium Acetate*

*Ethanol (100% and 70%)*

*Chloroform:Isoamyl Alcohol (24:1)*

*2 M NaCl*

*0.2 M EDTA*

*1 mg/ml Carrier DNA (e.g., salmon sperm)*

*Trichloroacetic Acid (5% and 7%)*

*First-Strand 5X Buffer*
        250 m$M$ Tris-HCl, pH 8.3 (at 42°C)
        50 m$M$ MgCl$_2$
        250 m$M$ KCl
        2.5 m$M$ Spermidine
        50 m$M$ DTT
        5 m$M$ Each of dATP, dCTP, dGTP, and dTTP

*Second-Strand 10X Buffer*
        0.5 $M$ Tris-HCl, pH 7.2
        0.9 $M$ KCl
        30 m$M$ MgCl$_2$
        30 m$M$ DTT
        0.5 mg/ml Bovine serum albumin (BSA)

## TE Buffer

> 10 m$M$ Tris-HCl, pH 8.0
> 1 m$M$ EDTA

## TE-Saturated Phenol/Chloroform

> Thaw phenol crystals at 65°C and mix in equal parts of phenol and TE buffer. Mix well and allow the phases to separate at room temperature for 30 min. Take 1 part of the lower, phenol phase and mix with 1 part of chloroform:isoamyl alcohol (24:1). Mix well, allow phases to separate, and store at 4°C until use.

## Alkaline Gel Running Buffer (Fresh)

> 30 m$M$ NaOH
> 1 m$M$ EDTA

## 2X Alkaline Buffer

> 20 m$M$ NaOH
> 20% Glycerol
> 0.025% Bromophenol blue (use fresh each time)

## Protocol 5: Ligation of EcoR I Linkers/Adaptors to the Double-Strand, Blunt-End cDNAs

The *EcoR* I linker or *EcoR* I adaptor ligation system is designed to generate *EcoR* I sites at the termini of blunt-ended cDNA, and is commercially available. If *EcoR* I linkers are used, digestion with *EcoR* I is required after ligation. However, there is no need to digest with *EcoR* I if one uses the *EcoR* I adaptor that has one sticky end. The ligation of *EcoR* I linker or adaptor allows cDNAs to be cloned into unique *EcoR* I sites of vectors λgt10 or λgt11. If cDNAs are synthesized with an *Xba* I or *Not* I primer-adaptor, they can be directionally cloned between the *EcoR* I and *Xba* I or *Not* I sites of λgt11 *Sfi-Not* I vector or vectors such as λGEM-2 and λGEM-4.

1. Size fractionation of double-strand cDNAs

    It is strongly recommended that cDNA samples be size fractionated prior to ligation with linkers to eliminate <200 bp cDNAs. This can be easily done with Sephacryl S-400 spin columns (Promega Corporation) that exclude double-stranded DNA fragments >270 base pairs or with gel filtration.

    a. To a vertical column, slowly add approximately 1 to 2 ml of Sephacryl S-400 slurry and drain the buffer completely. The final height of the gel bed should reach the neck of the column at the lower "ring" marking. Rehydrate the column with buffer, cap the top, and seal the bottom. Store at 4°C until use.

    b. Right before use, drain away the buffer and place the column in the wash tube that is provided or a microcentrifuge tube. Briefly centrifuge in a swinging bucket for 4 min at 800 × $g$ to remove excess buffer.

    c. Place the column in a fresh collection tube and slowly load the cDNA samples (30 to 60 μl), dropwise, to the center of the gel bed. Centrifuge in a swinging bucket at 800 × $g$ for 5 min and collect the eluate. The eluted cDNAs can be directly used or precipitated with ethanol and dissolved in TE buffer.

2.  Protection of the internal *EcoR* I sites of cDNAs by methylation

    a.  Set up the *EcoR* I methylation reaction as follows:

        - 1 m*M* s-Adenosyl-L-methionine (2 µl)
        - *EcoR* I methylase 10X buffer (2 µl)
        - BSA (1 mg/ml) (optional) (2 µl)
        - DNA (1 µg/µl) (2 µl)
        - *EcoR* I methylase (20 units)
        - Add dd.H$_2$O to a final volume of 20 µl.
        - Mix well after each addition.

    b.  Incubate the tube at 37°C for 20 min and inactive the enzyme by heating at 70°C for 10 min.

    c.  Extract the enzyme with 20 µl of TE-saturated phenol/chloroform, mix well by vortexing and centrifuge at 11,000 × *g* for 4 min.

    d.  Carefully transfer the top, aqueous phase to a fresh tube and add an equal volume of chloroform:isoamyl alcohol (24:1). Vortex to mix and centrifuge as in step c.

    e.  Carefully transfer the top, aqueous phase to a fresh tube. **Do not take any white materials.** To the supernatant, add 0.15 volume of 3 *M* sodium acetate buffer, pH 5.2, and 2.5 volumes of chilled 100% ethanol. Allow precipitation to occur at –70°C for 30 min or at –20°C for 2 h.

    f.  Centrifuge at 12,000 × *g* for 10 min and carefully pour off the supernatant. Briefly rinse the DNA pellet with 1 ml of 70% chilled ethanol and dry the pellet under vacuum for 10 min. Dissolve the methylated cDNAs in 15 µl of TE buffer.

3.  Ligation of *EcoR* I linkers to the methylated cDNAs

    a.  Set up the reaction as follows and mix after each addition.

        - cDNA (0.2 µg)(0.1 µg/µl)
        - Ligase 10X buffer (2 µl)
        - BSA (1mg/ml) (optional) (2 µl)
        - Phosphorylated *EcoR* I linkers (60-fold molar excess) (diluted stock linkers, 0.001 A$_{260}$ unit of the linker = 10 pmol)
        - T4 DNA ligase (5 Weiss units)
        - Add dd.H$_2$O to a final volume of 20 µl.
        - **Always add enzyme last.**

    b.  Incubate at 15°C overnight and stop the reaction by heating at 70°C for 10 min.

    c.  Cool the tube on ice for 2 min and carry out *EcoR* I digestion as follows:

        - Ligated cDNAs (20 µl)
        - *EcoR* I 10X buffer (5 µl)
        - *EcoR* I (1 unit/pmol of linker used in step a)
        - Add dd.H$_2$O to a final volume of 50 µl.

    d.  Incubate the reaction at 37°C for 2 h and add 5 µl of 0.2 *M* EDTA to stop the reaction.

    e.  Extract enzymes with 55 µl of TE-saturated phenol/chloroform and mix by vortexing for 1 min.

    f.  Centrifuge at 11,000 × *g* for 4 min and carefully transfer the upper, aqueous phase to a fresh tube.

    Steps g to j are designed for directional cloning of cDNAs that are synthesized with an *Xba* I or *Not* I primer-adaptor. The ligated cDNAs with linkers must be digested with *Xba* I or *Not* I following the *EcoR* I digestion. Alternatively, a double enzyme digestion with *EcoR* I and *Xba* I or *Not* I can be carried out in the same reaction using 10X digestion buffer for *Xba* I or *Not* I in the reaction. However, double digestions are usually of lower efficiency.

g. Set up the *Xba* I or *Not* I digestion reaction as follows:
- cDNAs from step f (40 µl)
- *Xba* I or *Not* I 10X buffer (8 µl)
- *Xba* I or *Not* I (1 unit/pmol of linker used in step a)
- Add dd.H$_2$O to a final volume of 80 µl.

h. Incubate the reaction at 37°C for 2 h and add 8 µl of 0.2 *M* EDTA to stop the reaction.

i. Extract enzymes with 88 µl of TE-saturated phenol/chloroform and mix by vortexing for 1 min.

j. Centrifuge at 11,000 × *g* for 4 min and carefully transfer the upper, aqueous phase to a fresh tube.

k. Remove unligated linkers with a spin column as follows:

Place a spin column in the collection tube and slowly load the restriction enzyme-digested supernatant to the top center of the gel bed. Centrifuge in swinging bucket at 800 × *g* for 5 min and collect the eluate containing cDNA-linkers.

l. Add 0.5 volume of 7.5 *M* ammonium acetate to the eluate, mix and add 2 volumes of chilled 100% ethanol. Allow precipitation at –20°C for 2 h.

m. Centrifuge at 12,000 × *g* for 10 min and briefly rinse the pellet with 1 ml of cold 70% ethanol. Dry the cDNA-linker under vacuum for 10 min and resuspend the pellet in 20 µl of TE buffer. Store the sample at –20°C until use.

n. To check the efficiency of ligation, load 4 µl (0.3 µg) of the ligated sample on a 0.8% agarose gel containing EtBr and an equal amount of the unligated, blunt-ended cDNAs next to the ligated sample. After electrophoresis, take a photograph under UV light. Efficient ligation will cause a shift in the migration rate because of the ligation of linker concatemers, which run more slowly than the unligated sample.

## Reagents Needed

### 7.5 M Ammonium Acetate

### Ethanol (100% and 70%)

### Chloroform:Isoamyl Alcohol (24:1)

### 2 M NaCl

### 0.2 M EDTA

### EcoR I Methylase 10X Buffer
1 *M* Tris-HCl, pH 8.0
0.1 *M* EDTA

### TE Buffer
10 m*M* Tris-HCl, pH 8.0
1 m*M* EDTA

### TE-Saturated Phenol/Chloroform

Thaw phenol crystals at 65°C and mix in equal parts of phenol and TE buffer. Mix well and allow the phases to separate at room temperature for 30 min. Take 1 part of the lower, phenol phase and mix with 1 part of chloroform:isoamyl alcohol (24:1). Mix well and allow phases to separate. Store at 4°C until use.

### Sample Buffer

50% Glycerol
0.2% SDS
0.1% Bromophenol blue
10 m$M$ EDTA

### Ligase 10X Buffer

0.3 $M$ Tris-HCl, pH 7.5
0.1 $M$ MgCl$_2$
0.1 $M$ DTT
10 m$M$ ATP

### EcoR I 10X Buffer

0.9 $M$ Tris-HCl, pH 7.5
0.5 $M$ NaCl
0.1 $M$ MgCl$_2$

### Not I or Xba I 10X Buffer

0.1 $M$ Tris-HCl, pH 7.5
1.5 $M$ NaCl
60 m$M$ MgCl$_2$
10 m$M$ DTT

4.  Ligation of *EcoR* I adaptors to cDNAs

    The *EcoR* I adaptor is a duplex DNA molecule with one *EcoR* I sticky end and one blunt end for ligation to cDNA. Adaptors eliminate the *EcoR* I methylation. The adaptor-cDNA does not need to be digested with *EcoR* I after ligation, which is necessary for *EcoR* I linker ligation methodology. However, a phosphorylation reaction should be carried out if the cloning vector is dephosphorylated.

    a.  Adaptor ligation

       i.  Set up *EcoR* I adaptors to cDNAs in a microcentrifuge tube as follows:
           * cDNA (0.1 µg/µl) (0.2 µg)
           * Ligase 10X buffer (3 µl)
           * BSA (1 mg/ml) (optional) (3 µl)
           * *EcoR* I adaptors (25-fold molar excess)
           * T4 DNA ligase (Weiss units) (8 units)
           * Add dd.H$_2$O to a final volume of 30 µl.
       ii. Incubate the reaction at 15°C overnight and stop the reaction at 70°C for 10 min.

    b.  Kinase reaction and *Xba* I or *Not* I digestion

This reaction is required for the dephosphorylated vector but is not necessary for the phosphorylated vector. For directional cDNA cloning methodology, *Xba* I or *Not* I digestion must be carried out. This can be done simultaneously with the kinase reaction using restriction enzyme 10X buffer instead of the kinase 10X buffer.

i.  From the above step a.ii after ligation, place the tube on ice for 1 min and add the following in the order shown:

**For kinase reaction only**

- Kinase 10X buffer (5 μl)
- 0.1 m*M* ATP (diluted from stock) (2.5 μl)
- T4 polynucleotide kinase (10 units)
- Add dd.H$_2$O to a final volume of 50 μl.

**For both kinase reaction and *Xba* I or *Not* I digestion**

- *Xba* I or *Not* I 10X buffer (5 μl)
- 0.1 m*M* ATP (diluted from stock) (2.5 μl)
- T4 polynucleotide kinase (10 units)
- Either *Xba* I or *Not* I (8 units)
- Add dd.H$_2$O to a final volume of 50 μl.

ii.  Incubate at 37°C for 60 min and extract once by adding 1 volume of TE-saturated phenol/chloroform. Vortex for 1 min.

iii.  Centrifuge at $11,000 \times g$ for 5 min and carefully transfer the top, aqueous phase to a fresh tube.

iv.  Remove unligated adaptors as follows:

Place a spin column in the collection tube and slowly load the supernatant onto the top center of the gel bed. Centrifuge in swinging bucket at $800 \times g$ for 5 min and collect the eluate containing the cDNA-adaptor.

v.  Add 0.5 volume of 7.5 *M* ammonium acetate to the eluate, mix, and add 2 volumes of chilled 100% ethanol. Allow precipitation to occur at –20°C for 2 h.

vi.  Centrifuge at $12,000 \times g$ for 10 min and briefly rinse the pellet with 1 ml of cold 70% ethanol. Dry the cDNA-linker under vacuum for 10 min and resuspend the pellet in 20 μl of TE buffer. Store the sample at –20°C until use.

vii.  To check the efficiency of ligation, load 4 μl (0.3 μg) of the ligated sample onto a 0.8% agarose gel containing EtBr and an equal amount of the unligated, blunt-ended cDNAs next to the ligated sample. After electrophoresis, take a photograph under UV light. Efficient ligation will cause a shift in the migration rate because of the ligation of linker concatemers which run more slowly than the unligated sample.

## Reagents Needed

### Ligase 10X Buffer

300 m*M* Tris-HCl, pH 7.8
100 m*M* MgCl$_2$
100 m*M* DTT
10 m*M* ATP

### Kinase 10X Buffer

700 m*M* Tris-HCl, pH 7.5
100 m*M* MgCl$_2$
50 m*M* DTT

### Xba I or Not I 10X Buffer

>   300 mM Tris-HCl, pH 7.8
>   100 mM MgCl$_2$
>   500 mM NaCl
>   100 mM DTT

### 7.5 M Ammonium Acetate

### Ethanol (100% and 70%)

5. Ligation of cDNA into Lambda Vectors

    **For optimal ligation of vectors and cDNA inserts:** Before performing a large-scale ligation, several ligations of vector arms and cDNA inserts should be carried out to determine their molar ratio and to determine the optimal ligation conditions. The molar concentration of the vectors should remain constant while adjusting the amount of DNA inserts. The molar ratio range of 43 Kb lambda arms to an average 1.4 Kb cDNA inserts is usually from 1:1 to 1:0.5. Two control ligations should be set up. One is the ligation of vectors and positive DNA inserts to check the efficiency, the other is ligation of the vector arms only to determine the background level of religated arms.

    a. Set up the ligation reactions as follows and carry them out separately:

        i. Sample ligations:

| Components | Microcentrifuge tubes | | | |
|---|---|---|---|---|
| | A | B | C | D |
| Ligase 10X buffer | 0.5 | 0.5 | 0.5 | 0.5 (µl) |
| λDNA vectors (0.5 µg/µl; 1 µg = 0.04 pmol) | 1 | 1 | 1 | 1 (µg) |
| Linker/adaptor-cDNA inserts (0.1 µg/µl; 10 ng = 0.01 pmol) | 10 | 20 | 30 | 40 (ng) |
| T4 DNA ligase (Weiss unit) | 2 | 2 | 2 | 2 (units) |
| Add dd.H$_2$O to a final volume of | 5 | 5 | 5 | 5 (µl) |

        ii. Control ligation 1:
            - Ligase 10X buffer (0.5 µl)
            - λDNA vector (0.5 µg/µl) (1.0 µg)
            - Positive control insert DNA (0.1 µg)
            - T4 DNA ligase (2.0 Weiss units)
            - Add dd.H$_2$O to a final volume of 5 µl.
        iii. Control ligation 2:
            - Ligase 10X buffer (0.5 µl)
            - λDNA vector (0.5 µg/µl) (1.0 µg)
            - T4 DNA ligase (2.0 Weiss units)
            - Add dd.H$_2$O to a final volume of 5 µl.

    b. Incubate the ligation reactions at room temperature (22 to 24°C) for 2 h and proceed to packaging.

## Protocol 6: In Vitro Packaging

*In vitro* packaging is performed in order to use a phage-infected *E. coli* cell extract to supply the mixture of proteins and precursors required for encapsulating the recombinant λDNA. The packaging system is commercially available with the specific bacterial strain host and control DNA.

1.    Thaw three Packagene extracts (50 μl per extract) on ice.

*Note:*    *Do not thaw the extract at room temperature or 37°C, and do not freeze the extract once it has thawed.*

2.    When the extracts have thawed, immediately divide each extract into two tubes on ice. Each tube contains 25 μl of the extract. Add each of the 5 μl of the ligation reaction mixture to a 25 μl extract and mix gently.

3.    Incubate at 22 to 24°C for 4 h and add phage buffer to 250 μl and 10 μl of chloroform. Gently mix well and allow the chloroform to settle to the bottom of the tube. Store the packaged phage at 4°C for up to 5 weeks even though the titer may drop.

## Protocol 7: Titration of Packaged Phage

1.    Partially thaw the specific bacterial strain such as *E. coli* LE 392, Y 1090, Y1089, and KW 251 on ice in a sterile laminar flow hood. The bacterial strain is usually kept in 20% glycerol and stored at −70°C until use. Pick up a small amount of the bacteria using a sterile wire transfer loop and immediately inoculate the bacteria by gently drawing several lines on the surface of an LB plate. The LB plate should be freshly prepared and dried at room temperature for a couple of days prior to use. Invert the inoculated LB plate and incubate it in an incubator at 37°C overnight. Multiple bacterial colonies will grow.

2.    Prepare fresh bacterial culture by picking up a single colony from the freshly streaked LB plate using a sterile wire transfer loop and inoculate a culture tube containing 5 ml of LB medium supplemented with 50 μl of 20% maltose and 50 μl of 1 $M$ $MgSO_4$ solution. Shake at 160 rpm at 37°C for 6 to 9 h or until the $OD_{600}$ has reached 0.6. Store the culture at 4°C until use.

3.    Dilute each of the packaged recombinant phage samples at 1000X, 5000X, or 10,000X with phage buffer.

4.    Add 20 μl of 1 $M$ $MgSO_4$ solution and 2.8 ml of melted top agar to 36 sterile glass test tubes in a sterile laminar flow hood. Cap the tubes and immediately place the tubes in a 50°C water bath for at least 30 min.

5.    Mix 0.1 ml of the diluted phage with 0.1 ml of fresh bacterial cells in a microcentrifuge tube. Cap the tube and allow the phage to adsorb the bacteria in an incubator at 37°C for 30 min.

*Notes:*    *There are four package samples and two package controls. Each one should be diluted at 1000X, 5000X, and 10,000X in phage buffer. Each diluted phage should be duplicated at this step. There are a total of 36 plates to be set up.*

6.    Add the incubated phage/bacterial mixture into specified tubes in the water bath. Vortex gently and immediately pour onto the centers of LB plates. Quickly spread the mixture all over the surface of the LB plate by gently tilting them. Cover the plates and allow the top agar to harden for 15 min in the sterile laminar flow hood. Invert the plates and incubate them in an incubator at 37°C overnight or for 15 h.

*Notes:* *The top agar mixture should be evenly distributed over the surface of the LB plate. Otherwise, the growth of bacteriophage will be affected, thus decreasing the pfu number.*

7.  Count the number of plaques for each plate and calculate the titer of the phage (pfu) for each sample and for the control.

Plaque forming units (pfu) per milliliter = number of plaques/plate $\times$ dilution times $\times$ 10

The last 10 of the calculation refers to the 0.1 ml, per milliliter basis, of the packaging extract used for one plate. For example, if there are 100 plaques on a plate made from a 1/5,000 dilution, the pfu per milliliter of the original packaging extract = $100 \times 5000 \times 10 = 5 \times 10^6$.

8.  Compare the titers (pfu/ml) from the different samples; determine the optimal ratio of vector arms and cDNA inserts and prepare a large-scale ligation and packaging reaction using optimum ratio conditions.

## Protocol 8: Large-Scale Ligation and In Vitro Packaging

1.  For "safety" reasons, set up two sample reactions and carry them out separately. In case one sample fails due to accident, another one can serve as a backup. The following is designed for one reaction. It is based on the optimal ratio of vectors and cDNA inserts used in our lab.
    - Ligase 10X buffer (2 µl)
    - λDNA vector (43 Kb; 0.5 µg/µl) (2 µg) (Make sure that the vector is phosphorylated.)
    - Linker/adaptor ligated cDNAs (average 1.5 to 1.8 Kb; 0.2 µg/µl) (0.2 µg)
    - T4 DNA ligase (5 Weiss units)
    - Add dd.$H_2O$ to a final volume of 20 µl.
2.  Incubate the ligation reactions at room temperature (22 to 24°C) for 2 h.
3.  Check the efficiency of ligation by loading 0.1 µg of the ligated sample onto a 0.8% agarose gel containing EtBr and an equal amount of the unligated DNAs and unligated vector arms next to the ligated sample. After electrophoresis, take a photograph under UV light. Efficient ligation will cause a shift in the migration rate due to the ligation of vector concatemers, which run more slowly than the unligated DNAs.

*Notes:* *This is a very important step to check because the ligation is based on optimal conditions. Low efficiency of ligation of the vector will cause trouble in packaging. Thus, the pfu/ml for the library will be less than $1 \times 10^4$, which is considered to be a poor library. If the efficiency of vector ligation is very high, the recombinant DNAs can be directly used for packaging, as described in steps 8 to 10 or for being precipitated from steps 4 to 7.*

4.  For long-term storage of the recombinant DNAs, which actually constitute the cDNA library, extract the enzyme with 1 volume of TE-saturated phenol/chloroform, mix well by vortexing, and centrifuge at $11,000 \times g$ for 4 min.
5.  Carefully transfer the top, aqueous phase to a fresh tube and add an equal volume of chloroform:isoamyl alcohol (24:1). Vortex to mix and centrifuge as in step 4.
6.  Carefully transfer the top, aqueous phase to a fresh tube. **Do not take any white materials.** To the supernatant, add 0.15 volume of 3 *M* sodium acetate buffer, pH 5.2, and 2.5 volumes of chilled 100% ethanol. Allow precipitation to occur at –70°C for 30 min or at –20°C for 2 h.

7. Centrifuge at $12,000 \times g$ for 10 min and carefully pour off the supernatant. Briefly rinse the DNA pellet with 1 ml of 70% chilled ethanol and dry the pellet under vacuum for 10 min. Dissolve the methylated cDNAs in 10 μl TE buffer. Store the recombinant DNAs at –20°C until use.

8. Thaw two packaging extracts (50 μl each) on ice.

*Note:* *Do not thaw the extract at room temperature or 37°C, and do not freeze the extract once it has thawed.*

9. When the extract has thawed, immediately add 9 μl of the ligation mixture without precipitation or 0.5 μg stored vector-cDNA to 50 μl of the extract and mix gently.

*Note:* *Do not add more than 10 μl of the ligation mixture per 50 μl of packaging extract.*

10. Incubate at 22 to 24°C for 4 h and add phage buffer to 500 μl and 25 μl of chloroform. Gently mix well and allow the chloroform to settle to the bottom of the tube. Store the packaged phage at 4°C for up to 5 weeks even though the titer may drop. At this stage, the packaged sample can be used for titering and screening the cDNA library (see the next sections). In order to store the packaged phage for a long time, S buffer but not phage buffer can be added to the mixture and extracted with 1 volume of chloroform (**not phenol**). Add dimethylsulfoxide (DMSO) to final 7% to the top, aqueous phase and store at –70°C.

*Notes:* *The cDNA library stored in this way, based on our experience, may drop in pfu number severalfold. We recommend storing the precipitated recombinant DNAs instead of their being packaged at –70°C. For screening the library, we suggest that a freshly packaged bacteriophage be carried out using the stored recombinant DNAs.*

## Solutions Needed

### Ligase 10X Buffer

> 300 m$M$ Tris-HCl, pH 7 8
> 100 m$M$ MgCl$_2$
> 100 m$M$ DTT
> 10 m$M$ ATP

### Phage Buffer

> 20 m$M$ Tris-HCl, pH 7.4
> 100 m$M$ NaCl
> 10 m$M$ MgSO$_4$

### LB (Luria-Bertaini) Medium (per liter)

> 10 g Bacto-tryptone
> 5 g Bacto-yeast extract
> 5 g NaCl
> Adjust to pH 7.5 with 0.2 $N$ NaOH and autoclave.

### 20% (v/v) Maltose

### 1 M MgSO₄

*1 M MgSO*$_4$

### LB Plates

Add 15 g of Bacto-agar to 1 liter of freshly prepared LB medium and autoclave. Allow the mixture to cool to about 60°C and pour 30 ml of the mixture into 85- or 100-mm Petri dishes in a sterile laminar flow hood with filtered air flowing. Remove any bubbles with a pipette tip and let the plates cool for 5 min prior to their being covered. Allow the agar to harden for 1 h and store the plates at room temperature up for 10 days or at 4°C in a bag for 1 month. The cold plates should be placed at room temperature for 1 to 2 days before use.

### TB Top Agar

Add 3.0 g agar to 500 ml freshly prepared LB medium and autoclave. Store at 4°C until use. For plating, melt the agar in a microwave. When the solution has cooled to 60°C, add 0.1 ml of 1 *M* MgSO$_4$ 10 ml of the mixture. If color selection methodology is used to select recombinants in conjunction with bacterial host strain Y1090, 0.1 ml IPTG (20 mg/ml in water, filter-sterilized) and 0.1 ml X-Gal (50 mg/ml in dimethylformamide) should be added to the cooled (55°C) 10 ml of the top agar mixture.

### S Buffer

Phage buffer + 2% (w/v) gelatin

### *Protocol 9: Immunoscreening of a Lambda Expression cDNA Library*

If the cloning vectors used are expression vector λgt11 *Sfi-Not* I, the cDNA inserts cloned into the *lac* Z gene of the vector can be expressed as a part of a β-galactosidase fusion protein. The expression cDNA library can be screened with a specific antibody used as a probe. The primary positive clones of interest can be "fished" out from the library. After several rounds of retitering and rescreening of the primary positive clones, putative clones will be isolated. The overall schematic procedures are shown in the following:

1. Prepare an LB medium, pH 7.5, by adding 15 g of Bacto-agar to 1 liter of the LB medium and autoclave. Allow the mixture to cool to about 50°C and add ampicillin (100 μg/ml) and tetracycline (15 μg/ml). Mix well and pour 30 to 40 ml of the mixture into each of the 85- or 100-mm Petri dishes in a sterile laminar flow hood with filtered air flowing. Remove any bubbles with a pipette tip and let the plates cool for 5 min prior to being covered. Allow the agar to harden for 1 h and store the plates at room temperature for up 10 days or at 4°C in a bag for 1 month. The cold plates should be placed at room temperature for 1 to 2 days before use.

2. Partially thaw the specific bacterial strain such as *E. coli* Y 1090 on ice. Pick up a small amount of the bacteria using a sterile wire transfer loop and immediately streak out *E. coli* Y1090 on LB plates by gently drawing several lines on the surface of an LB plate. Invert the inoculated LB plate and incubate the plate in an incubator at 37°C overnight. Multiple bacterial colonies will grow.

3. Prepare fresh bacterial cultures by picking up a single colony from the freshly streaked LB plate using a sterile wire loop and inoculate into a culture tube containing 50 ml of LB medium

supplemented with 500 μl of 20% maltose and 500 μl of 1 $M$ $MgSO_4$ solution. Shake at 160 rpm at 37°C overnight or until the $OD_{600}$ has reached 0.6. Store the culture at 4°C until use.

4. Add 20 μl of 1 $M$ $MgSO_4$ solution and 2.8 ml of melted LB top agar to each of the sterile glass test tubes in a sterile laminar flow hood. Cap the tubes and immediately place the tubes in a 50°C water bath for at least 30 min.

5. Based on the titration data, dilute the λcDNA library with phage buffer. Set up 20 to 25 plates for primary screening of the cDNA library. For each 100-mm plate, mix 0.1 ml of the diluted phage containing $2 \times 10^5$ pfu of the library with 0.15 ml of fresh bacterial cells in a microcentrifuge tube. Cap the tube and allow the phage to adsorb the bacteria in an incubator at 37°C for 30 min.

*Notes:*    *The number of clones needed to detect a given probability that a low-abundance mRNA is converted into a cDNA in a library is:*

$$N = \ln (1-P)/\ln (1-1/n)$$

*where N is the number of clones needed, P is the possibility given (0.99), and 1/n is the fractional portion of the total mRNA, which is represented by a single low-abundance mRNA. For example, if P = 0.99, 1/n = 1/37,000, and N = 1.7 × 10⁵.*

6. Add the incubated phage/bacterial mixture to the tubes from the water bath. Vortex gently and immediately pour onto the centers of LB plates. Quickly spread the mixture over the entire surface of the LB plates by gently tilting them. Cover the plates and allow the top agar to harden for 15 min in laminar flow hood. Invert the plates and incubate in an incubator at 42°C for 4 h.

*Note:*    *The top agar mixture should be evenly distributed over the surface of a LB plate. Otherwise, the growth of the bacteriophage will be affected.*

7. Saturate nitrocellulose filter disks in 10 m$M$ IPTG in water for 30 min and air dry the filters at room temperature for 40 min.

8. Carefully overlay each plate with a dried nitrocellulose filter disk from one end and slowly lower it to the other end of the plate, avoiding any air bubbles underneath the filter. Quickly mark the top side and position of each filter in a triangular fashion by punching the filter through the bottom agar using a 20-gauge needle containing India ink. Cover the plates and incubate at 37°C for 4 to 6 h. If a duplicate is needed, a second filter may be overlaid on the plate and incubated for another 5 h at 37°C; however, the signal may be relatively weaker. At this stage, IPTG induces the expression of the cDNA library and the expressed proteins are transferred onto the facing side of the filter.

9. Place the plates at 4°C for 30 min to chill the top agar and to prevent the top agar from sticking to the filter. Move the plates to room temperature and carefully remove the filters using forceps. Label and wrap each plate with Parafilm,™ and store at 4°C until use. The filters may be rinsed briefly in TBST buffer to remove any agar.

*Notes:*    *Due to the diffusion feature of proteins, the filters are strongly recommended to be processed according to the following steps. Storing at 4°C for a couple of days before processing may cause difficulty in the identification of the actual positive plaque in the plate. The filters must not be allowed dry out during any of the subsequent steps, which are carried out at room temperature. Based on our experience, high background and strange results may appear, including even partial drying. If time is limited, the damp filters may be wrapped in SaranWrap™, then in aluminum foil, and stored at 4°C for up to 12 h. The filters should be processed individually during the following steps to obtain an optimal detection signal.*

10. Incubate a filter in 10 ml of TBST buffer containing 2 to 3% BSA or 1% gelatin or 5% nonfatty milk or 20% calf serum to block nonspecific protein-binding sites on the filter. Treat the filter for 40 min with slow shaking at 50 rpm.

*Notes:* *It is recommended that one dish be used for one filter only and that the facing side of the filter be down. Each filter should be incubated with about 10 ml of blocking buffer. Less than 5 ml will cause the filter to dry but more than 15 ml for an 82-mm filter may prevent one from obtaining optimal results.*

11. Carefully transfer the filter to a fresh dish containing 10 ml of TBST buffer with primary antibody.

*Notes:* *The antibody should be diluted 200X to 10,000X depending on the concentration of the antibody prepared. The antibody purified with an IgG fraction or with an affinity column usually produces better signals. The diluted primary antibody can be reused several times if stored at 4 °C.*

12. Wash the filter in 20 ml of TBST buffer containing 0.5% BSA for 10 min with shaking at 50 rpm. Repeat washing two times.
13. Transfer the filter to a fresh dish containing 10 ml of TBST buffer with the second antibody–alkaline phosphatase conjugate at 1:5000 to 10,000 dilution. Incubate for 40 min with shaking at 50 rpm.
14. Wash the filter as in step 12.
15. Briefly damp dry the filter on a filter paper and place the filter into 10 ml of freshly prepared AP color development substrate solution. Allow the color develop for 0.5 to 5 h or overnight with a relatively high background. Positive clones should appear as purple circle spots on the white filter.
16. Stop the color development when the color has developed to the desired intensity using 15 ml of stopping solution. The filter may be stored in the solution or stored dry. The color will fade after drying but can be restored with water.
17. In order to locate the positive plaques, match the filter to the original plate by placing the filter, facing-side down, underneath the plate with the help of the marks previously made. This can be done by placing a glass plate over a lamp and putting the matched filter and plate on the glass plate. Turn on the light; by doing so, the positive clones can easily be identified. Remove individual positive plugs containing phage particles from the plate using a sterile pipette with the tip cut off. Expel the plug into a microcentrifuge tube containing 1 ml of elution buffer. Allow elution to occur for 4 h at room temperature with occasional shaking.
18. Transfer the eluate supernatant into a fresh tube and add 20 μl of chloroform. Store at 4°C for up to 5 weeks.
19. Determine the pfu of the eluate, as previously described. Replate the phage and repeat the screening procedure with the antibody probe several times until 100% of the plaques on the plate are positive. (Figure 6.6)

*Notes:* *During the rescreening process, the plaque number used for one plate should be gradually reduced. In our experience, the plaque density for one 100-mm plate in the rescreening procedures is decreased from 1000 to 500, to 300, and to 100.*

20. Amplify the putative cDNA clones and isolate the recombinant λDNAs by either plate or liquid methods (see the λDNA isolation section in Chapter 1). The purified DNA can be used for subcloning of the cDNA inserts.

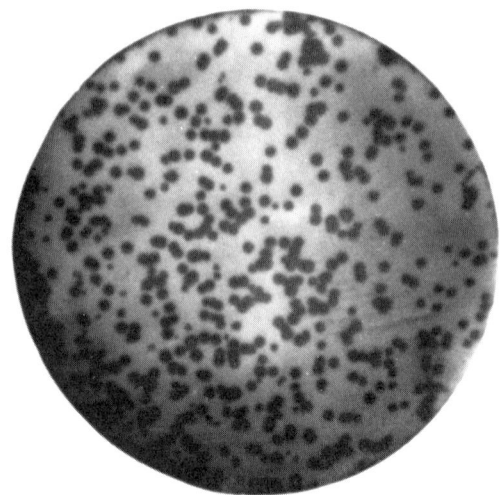

**FIGURE 6.6**
Final step in progressive screening of a cDNA library showing that 100% of plaques are positive.

## a.    Troubleshooting Guide

| Symptoms | Solutions |
| --- | --- |
| Plaques are small in one area and large in other areas | This is due to uneven distribution of the top agar mixture. Make sure that the top agar mixture is spread evenly |
| Too many positive clones in the primary screening | The specificity of the primary antibody is low, or the concentration is too high, or the blocking efficiency is low. Try to use IgG or an affinity-column purified antibody, carrying out different dilutions of the antibody and increasing the percentage of the blocking reagent in TBST buffer |
| Purple background on the filter | Color development is too long. Try to stop the color reaction as soon as desired signals appear |
| Unexpected larger purple spots | Air bubbles may be produced when overlaying the filter on the LB plate. Make sure no air bubbles appear underneath the filter |
| No signal appears at all in any of the primary screening plates | The pfu numbers used for each plate are too low. For rare proteins, $2 \times 10^6$ pfu should be used in primary screening of the library |
| Signals are weak on the filters of subsequent screening | The antibody has low activity, or color development was stopped too early. Try to use an immunoassay to check the quality of the antibody, or increase the color developing time, or try to use high-quality nitrocellulose filters |

## Reagents Needed

### LB (Luria-Bertaini) Medium (per liter)

> 10 g Bacto-tryptone
> 5 g Bacto-yeast extract
> 5 g NaCl
> *Adjust pH to 7.5 with 0.2 N NaOH.*
> Autoclave.

### LB Plates Containing Ampicillin and Tetracycline

> Add 15 g of Bacto-agar to 1 liter LB medium. Autoclave. When the medium cools to 50°C, add ampicillin (0.1 g/ml) and tetracycline (15 µg/ml). Pour 30 to 40 ml into each of the 100-mm Petri dishes in a laminar flow hood with filtered air flowing. Remove any bubbles with a pipette tip and let the plates cool for 5 min prior to being covered. Allow the agar harden for 1 h and store the plates at room temperature up for 10 days or at 4°C in a bag for 1 month. The cold plates should be placed at room temperature for 1 to 2 days before use.

### LB Top Agar (500 ml)

> Add 4 g agar to 500 ml of LB medium and autoclave.

### Elution Buffer

> 10 mM Tris-HCl, pH 7.5
> 10 mM MgCl$_2$

### IPTG Solution

> 10 mM Isopropyl β-D-thiogalactopyranoside in dd.H$_2$O.

### Phage Buffer

> 20 mM Tris-HCl, pH 7.4
> 100 mM NaCl
> 10 mM MgSO$_4$

### TBST Buffer

> 10 mM Tris-HCl pH 8.0
> 150 mM NaCl
> 0.05% Tween 20

### Blocking Solution

> 3% BSA, or 1% gelatin, or 5% nonfatty milk, or 20% calf serum in TBST buffer

*AP Buffer*

        100 m$M$ Tris-HCl, pH 9.5
        100 m$M$ NaCl
        5 m$M$ MgCl$_2$

*Color Development Substrate Solution*

        50 ml AP buffer
        0.33 ml Nitroblue tetrazolium (NBT) stock solution
        0.165 ml 5-Bromo-4-chloro-3-indolyl phosphate (BCIP) stock solution
        Mix well after each addition and protect the solution from strong light. Warm
           the substrate solution to room temperature to prevent precipitation.

*Stop Solution*

        20 m$M$ Tris-HCl, pH 8.0
        5 m$M$ EDTA

## *Protocol 10: Screening a cDNA Library Using Labeled DNA as a Probe*

In addition to the immunoscreening methodology described above, a cDNA library constructed in nonexpression or expression vectors such as $\lambda$gt10 and $\lambda$gt11 *Sfi-Not* I can be screened with labeled DNA as a probe, which can be labeled by isotopic or nonisotopic methods (see Chapter 4 for detailed procedures). The library is screened by *in situ* plaque hybridization to the probe. Identification of cDNAs of a potential multigene family can be obtained at low stringency of hybridization. This will be achieved, based on our experience, by reducing the percentage of formamide (30 to 40%) in hybridization solution and washing temperature (room temperature followed by 50 to 56°C), keeping the temperature of hybridization constant and the salt in the wash solution constant at 2X SSC in 0.1% SDS. To prevent nonspecific cross-linking hybridization, however, a high-stringency condition should be established. Positive plaques will then be isolated to homogeneity by successive rounds of phage titering and rescreening (Figure 6.6). The resulting phage lysates will be used in the large-scale preparation of phage DNAs (see procedures for isolation and purification of phage DNA in Chapter 1).

## *Method A: Screening of a cDNA Library by a 32P-Labeled Probe*

The general procedures are the same as described under the screening of a genomic library in Chapter 7.

## *Method B: Screening of a cDNA Library using a Non-Isotope-Labeled Probe*

The general procedures are the same as described under the section on screening of a genomic library in Chapter 7.

    After obtaining positive clones that contain the cDNA of interest, the cDNA should be subcloned for further characterization (Figure 6.7). The detailed procedures are described in Chapter 7.

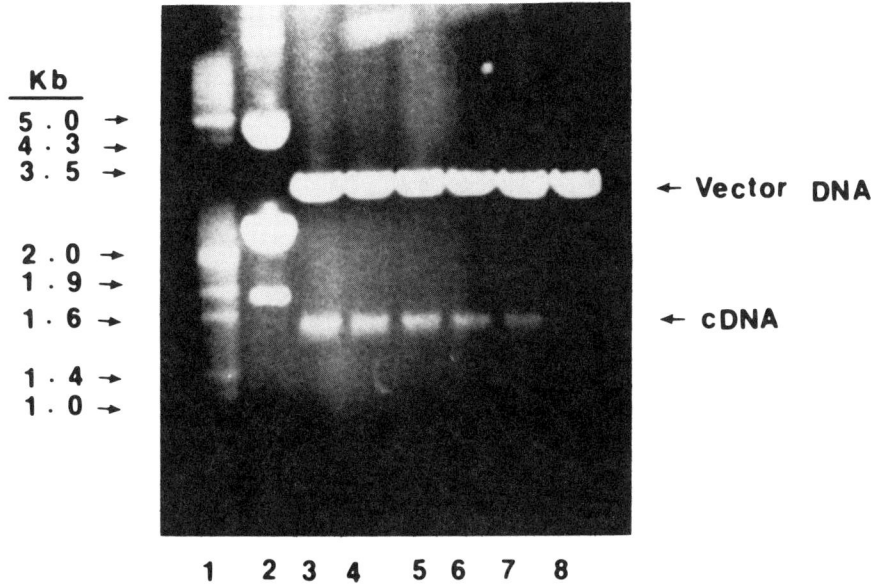

**FIGURE 6.7**

A successful subcloning of a cDNA insert in a plasmid vector. Plasmid DNA was isolated from different transformant colonies by use of a mini-prep. The DNA was digested with *EcoR* I and *Hind* III followed by electrophoresis on a 1% agarose gel containing ethidium bromide. Lane 1: DNA standard molecular weight markers. Lane 2: undigested plasmid DNA showing three bands because of different levels of supercoiling. Lanes 3–8: DNA from different transformant bacterial colonies.

## II.   Construction of a cDNA Library by Subtractive Hybridization Techniques

Subtractive hybridization is a powerful technology used to identify specific cDNA(s) derived from specific mRNAs expressed by the right gene(s) in one tissue or cell type/line but not in another. The first step is to make the first-stranded cDNAs representing all of the mRNA sequences expressed in each of two tissues or cell types/lines. Traditionally, the single-strand cDNAs from two different tissues or cell types/lines can be labeled and used, without subtractive procedures, to screen a conventional cDNA library to identify the sequences differentially expressed in the one tissue or cell type/line but not in the other. The drawback of this "differential screening" method without use of the subtractive technique is the significant hybridization background. In that case, only abundantly expressed sequences can be identified. The rarely expressed low-abundance sequences are missed. In order to solve this problem, subtractive technology was developed to hybridize the single-stranded cDNAs from one tissue or cell type/line to an excess of mRNAs from a second tissue or cell type/line. Any cDNAs from tissue or cell type/line A, which represent sequences expressed in both lines, should form DNA/RNA hybrids with the mRNAs from tissue or cell type/line B. The unhybridized cDNAs are specific for tissue or cell type/line A, but not B, and they remain single stranded. Both single-stranded and hybridized cDNAs can be separated by chromatography on hydroxyapatite columns. The unhybridized cDNAs can then be used to construct a subtracted cDNA library in which the sequences specific to one tissue or cell type/line are greatly enriched. Or, the cDNAs can be used to make subtractive probes to screen conventional cDNA libraries to isolate the specific clones. The subtracted cDNA libraries include an enriched population of the cDNA clones and simplify the screening procedures using much smaller numbers of pfu in the primary screening and speed up the purification process of the cDNA of interest.

## A. Synthesis of the First-Strand cDNA

The detailed protocol is described in Section I.

## B. Removal of the mRNA Template

1.  Add 1 volume (40 µl) of alkaline denaturing buffer to the above 40 µl of the unlabeled reaction mixture.

2.  Incubate at 37°C for 30 min to separate the first-strand cDNA and mRNA template.

3.  Add 1 volume of neutralization buffer to the mixture in step 2.

4.  Add 0.1 volume of 3 *M* sodium acetate buffer, pH 5.2, and 2.5 volume of chilled 100% ethanol. Place at –80°C for 30 min to precipitate the nucleic acids.

5.  Centrifuge at $12,000 \times g$ for 5 min and carefully decant the supernatant. Briefly rinse the pellet with 1 ml of 70% cold ethanol and dry the pellet under vacuum for 15 min.

6.  Dissolve the pellet in 50 µl of TE buffer.

7.  Add RNase A (DNase free) to a final concentration of 20 µg/ml and incubate at 37°C for 30 min.

*Note:*   *The function of this step is to hydrolyze the mRNA templates.*

8.  Extract the enzyme with 1 volume of TE-saturated phenol/chloroform, mix well by vortexing, and centrifuge at $11,000 \times g$ for 4 min.

9.  Carefully transfer the top, aqueous phase to a fresh tube and add an equal volume of chloroform:isoamyl alcohol (24:1). Vortex to mix and centrifuge as in step 8.

10. Carefully transfer the top, aqueous phase to a fresh tube. **Do not take any white material.** To the supernatant, add 0.15 volume of 3 *M* sodium acetate buffer, pH 5.2, and 2.5 volumes of chilled 100% ethanol. Allow precipitation to occur at –70°C for 30 min or at –20°C for 2 h.

11. Centrifuge at $12,000 \times g$ for 10 min and carefully pour off the supernatant. Briefly rinse the cDNA pellet with 1 ml of 70% chilled ethanol and dry the pellet under vacuum for 10 min. Dissolve the cDNAs in 15 µl dd.H$_2$O. At this stage, the cDNA should be single-strand DNA.

*Notes:*   *The separation of the first-strand cDNAs from hybridized mRNA is extremely important for subtraction hybridization. In order to check that the cDNA is of single-strand or double-strand form with mRNA, the purified cDNAs can be subjected to SI-nuclease assay. The resulting mixture is then subjected to 1.4% agarose gel electrophoresis using undigested cDNA as a control. If the cDNA is single strand, and if the SI digestion is complete, no band should be visible as compared with control cDNA.*

## C. Hybridization of cDNA to mRNA

1.  Carry out a hybridization reaction in a microcentrifuge tube as follows:

    •  2 to 5 µg First-strand cDNA from tissue or cell type/line A

    •  15 µl 2X Hybridization buffer

- 25 μg Poly(A)+RNA from tissue or cell type/line B
- Add dd.H$_2$O to a final volume of 30 μl.

or

- 2 to 5 μg First-strand cDNA from tissue or cell type/line B
- 15 μl 2X Hybridization buffer
- 25 μg Poly(A)+RNA from tissue or cell type/line A
- Add dd.H$_2$O to a final volume of 30 μl.

2. Tightly cover the tubes and wrap the top of the tube with parafilm to prevent evaporation. Incubate the tubes at 65 to 68°C overnight to 16 h.

*Note:    An RNase-free environment should be maintained for the hybridization.*

## D. Separation of cDNA/mRNA Hybrids from Single-strand cDNA by Hydroxyapatite (HAP) Chromatography

1. Prepare the HAP column: Add 1g of HAP (DNA Grade Bio-Gel, BioRad, Cat. no. 130–0520) in 0.1 *M* phosphate buffer (PB), pH 6.8, for 5 ml of slurry and mix well. Heat in boiling water for 5 min. Place a 5-ml plastic syringe closed at the bottom in a water bath equilibrated to 60°C. Place a cellulose acetate filter cut to size with a cork borer in the bottom of the syringe and wet filter with 0.1 *M* PB. Alternatively, place some sterile glasswool at the bottom of the syringe. Slowly add 0.5 to 1 ml of HAP slurry to the bottom-closed column and allow to set for 5 min prior to opening the column. Wash the column with 5 volumes of 0.1 *M* PB (60°C) two times.

2. Load the hybridized sample onto the prepared HAP column: Dilute the sample tenfold in prewarmed (60°C) 0.1 *M* PB containing 0.15 *M* NaCl. Gently load the sample onto the column with the bottom closed. Gently loose the mixture in the column with a needle (avoid bubbles) and let it set for 10 min.

3. Open the column from the bottom and collect the effluent containing the single-strand cDNAs. Wash the column twice with 0.5 ml of 0.1 *M* PB containing 0.15 *M* NaCl and collect the effluent. The cDNA/mRNA hybrids in the column can be eluted with 0.5 *M* PB.

4. Pool the effluent together and load onto a fresh HAP column as in steps 2 to 3. The function of this step is to maximally remove any cDNA/mRNA hybrids that potentially remain in the effluent.

5. Pool the effluent together and dialyze against 500 ml dd.H$_2$O overnight to remove the PB salts that are required for ethanol precipitation.

6. Add 0.15 volume of 3 *M* sodium acetate buffer, pH 5.2, and 2.5 volumes of chilled 100% ethanol to the dialyzed sample and place at –80°C for 30 min.

7. Centrifuge at 12,000 × *g* for 5 min and briefly rinse the cDNA pellet with 2 ml of 70% cold ethanol. Dry the pellet under vacuum for 15 min and dissolve the cDNAs in 50 μl TE buffer. Take 4 μl of the sample to measure the concentration of cDNA. Store the sample at –20°C until use. At this stage, the single-strand cDNA should represent the sequences specifically expressed in tissue or cell type/line A or B but not in the other. These cDNAs can be labeled as probes to screen a conventional library or used to construct a subtractive cDNA library.

*Note:    In order to ensure that the cDNA is single-strand DNA, take an appropriate amount of the cDNA sample and carry out an SI digestion assay.*

## E. Making a cDNA Library from the Subtracted First-Strand cDNA

1. Synthesize the second-strand cDNA as follows, using poly(A) primer.

a. Set up the following reaction:
   - Subtractive first-strand cDNA (1 µg)
   - Second-strand 10X buffer (50 µl)
   - Poly(A)12–20 primer (0.2 µg)
   - *E. coli* DNA polymerase I (20 units)
   - Add nuclease-free dd.$H_2O$ to a final volume of 50 µl.

b. Set up a tracer reaction for the second-strand DNA by removing 5 µl of the mixture to a fresh tube containing 1 µl 4 µCi [α-$^{32}$P]dCTP (>400 Ci/mmol). The tracer reaction will be used for TCA assay and alkaline agarose gel electrophoresis to monitor the quantity and quality the synthesis of the second-strand DNA.

c. Incubate both reactions at 14°C for 3.5 h. Add 95 µl of 50 m*M* EDTA to the 5 µl tracer reaction and store on ice.

d. Heat the unlabeled double-strand DNA sample at 70°C for 10 min to stop the reaction. **At this point, the double-strand DNAs should be generated.** Briefly spin down to collect the contents at the bottom of the tube and place on ice.

e. Add 4 units of T4 DNA polymerase (2 units/µg input cDNA) to the mixture and incubate the tube at 37°C for 10 min. This step functions to make the blunt-end, double-strand cDNA by T4 polymerase. The blunt ends are required for later adaptor ligations.

f. Stop the T4 polymerase reaction by adding 10 µl of 0.2 *M* EDTA and place on ice.

g. Extract the cDNAs with 1 volume of TE-saturated phenol/chloroform. Mix well and centrifuge in a microcentrifuge at top speed for 4 min at room temperature.

h. Carefully transfer the top, aqueous phase to a fresh tube. To the supernatant, add 0.5 volume of 7.5 *M* ammonium acetate or 0.15 volume of 3 *M* sodium acetate (pH 5.2), mix well, and add 2.5 volumes of chilled (–20°C) 100% ethanol. Gently mix and allow precipitation to occur at –20°C for 2 h.

i. Centrifuge in a microcentrifuge at the top speed for 5 min. Carefully remove the supernatant, briefly rinse the pellet with 1 ml of cold 70% ethanol, and gently drain away the ethanol.

j. Dry the cDNA pellet under vacuum for 15 min. Dissolve the cDNA pellet in 25 µl of TE buffer. It is important to take 2 to 4 µl of the sample to measure the concentration of cDNAs prior to the next reaction. Store the sample at –20°C until use.

*Note:*   *Before adaptor ligation, it is strongly recommended that the quantity and quality of the double-strand (ds) cDNAs be checked by TCA precipitation and gel electrophoresis as described previously.*

2. Carry out ligation of linkers or adaptors, *Xba* I or *Not* I digestion for directional cloning, ligation to the lambda DNA vectors, *in vitro* packaging, and cloning in *E. coli* host, generating a subtractive cDNA library. The detailed protocols for these procedures are described in the Section I of this chapter.

3. Screen the library to identify clones of interest. To clone the DNA sequences present in tissue or cell type/line A and not B, a library made from A can be screened with either subtracted or nonsubtracted probe from both A and B. The desired phage clones can be identified by hybridization to the probe from A but not from B. Because it is a subtractive cDNA library with enriching cDNAs of interest, the phage library should not be plated densely so that single plaques can be distinguished. Normally, 50 to 2000 plaques on a 100-mm plate are enough. The detailed screening and purification procedures are given in the Section I of this chapter.

4. The specificity of the cDNA inserts in A but not B or in B but not A can be verified by slot-blot screening. A high-titer phage lysate is recommended for this purpose since a single plaque contains about 10$^6$ phages, but a high-titer phage lysate can have 10$^9$ phages per milliliter.

**For one slot**

- 100 µl Phage lysate
- 20 µl 1 *M* Tris-HCl, pH 8.0
- 4 µl 0.5 *M* EDTA
- 124 µl 100% Formamide

Heat for 10 min at 68°C and chill on ice. Load the sample onto the well of a slot-blotter (Schleicher and Schuell) and turn on the vacuum when all samples are loaded. Wash slots with 200 µl 2X SSC. Dry the membrane at room temperature for 10 min and bake at 80°C under vacuum for 2 h. The procedures for prehybridization, hybridization, washing, and exposure are similar to Southern blotting. After exposure, bands should be specific for lysate A or B.

## *Materials Needed*

*[α-³²P]dCTP (>400 Ci/mmol)*

*50 mM EDTA*

*3 M Sodium Acetate Buffer, pH 5.2*

*Ethanol (100% and 70%)*

*Chloroform:Isoamyl Alcohol (24:1)*

*0.2 M EDTA*

*Trichloroacetic Acid (5% and 7%)*

*First-Strand 5X Buffer*
> 250 m*M* Tris-HCl, pH 8.3 (at 42°C)
> 50 m*M* MgCl$_2$
> 250 m*M* KCl
> 2.5 m*M* Spermidine
> 50 m*M* DTT
> 5 m*M* Each of dATP, dCTP, dGTP, and dTTP

*Alkaline Denaturing Buffer*
> 0.5 *M* NaOH
> 1.5 *M* NaCl

*Neutralization Buffer*
> 1.5 *M* NaCl
> 0.5 *M* Tris-HCl, pH 7.4

*Hybridization Buffer (2X)*

> 40 m*M* Tris-HCl, pH 7.7
> 1.2 *M* NaCl
> 4 m*M* EDTA
> 0.4% SDS
> 1 μg/μl Carrier yeast tRNA (optional)

*Phosphate Buffer (Stock)*

> 0.5 *M* Monobasic sodium phosphate
> Adjust the pH with 0.5 *M* dibasic sodium phosphate to 6.8.

# III.   Construction of Fractional cDNA Libraries Using *Xenopus* Oocytes as an Expression System

In addition to the subtractive technology described above, another powerful and state-of-the-art method of enriching a population of cDNA sequences of interest is to construct a fraction or a portion of the actual whole cDNA library. This technology is especially useful when no nucleic acid or protein probe is available but there is a known physiological function such as enzyme activity. Total mRNA is first fractionated using sucrose gradients to concentrate the particular mRNA in a specific fraction, which is individually microinjected into *Xenopus* oocytes to check the expression of interesting mRNA using a specific assay method. The expressed fraction of mRNA will then be used for cDNA synthesis and cloning, and thus greatly increase the possibility of identifying cDNA clones of interest. This is especially useful for those rare mRNAs that are expressed. The *Xenopus* oocyte is a remarkable giant cell that can transcribe injected DNA, mRNA, and postmodified proteins. It is a very good system that has become a well-established tool for gene expression in eukaryotes.

## *Protocol*

1. Isolate the total RNA and purify the poly(A)+RNA from specific organisms of interest as described previously (see the RNA section in Chapter 2).

2. Fractionate the mRNA by sucrose gradients containing methylmercuric hydroxide.

   a. Prepare sucrose gradients (10 to 30% w/v) containing 10 m*M* methylmercuric hydroxide in ⁹/₁₆ in. × 3¹/₂ in. ultracentrifuge tubes (Beckman SW41 or equivalent). Dissolve the sucrose in sterile water, and treat the solutions overnight with 0.1% DEPC at 37°C followed by heating to 100°C for 15 min to remove the DEPC. When the solutions have cooled to room temperature, add 1 *M* Tris-HCl (pH 7.4) and 0.5 *M* EDTA (pH 7.4) to final concentrations of 10 m*M* Tris-HCl and 1 m*M* EDTA. Finally, add methylmercuric hydroxide to a final concentration of 10 m*M*. The gradients can be made by a gradient maker or by adding the solutions, one by one, in decreasing density of the solutions one over another or in increasing density of the solutions under each other to a tube. The most convenient way is to place the tube on dry ice and to add a given density of sucrose solution. After this solution is frozen, add another density of solution. Finally, place the tube at 4°C overnight to gradually allow to thaw and diffuse, thus generating a gradient.

   b. Add methylmercuric hydroxide to a final concentration of 20 m*M* to 100 μl RNA sample (1 μg/μl) and carefully load the RNA sample onto the gradients.

c.  Centrifuge the gradients at 34,000 rpm for 15 to 18 h at 4°C in a Beckman SW41 rotor or its equivalent.

d.  Collect 0.2 to 0.3-ml fractions through a hypodermic needle inserted into the bottom of the centrifuge tube. Dilute each of the fractions with 1 volume of sterile, DEPC-treated water containing 5 m$M$ β-mercaptoethanol. Add 60 μl of 3 $M$ sodium acetate buffer and place the tubes at 0°C for 1 h.

e.  Centrifuge at $12,000 \times g$ for 15 min at 0°C and carefully decant the supernatants to a fresh tube. Wash the RNA sample with 2 ml of 70% ethanol and place the pellets to dry at room temperature.

f.  Resuspend the RNA in 20 μl of water and add 0.1 volume of 3 $M$ sodium acetate buffer and 3 volumes of chilled 100% ethanol. Mix well and leave the preparation at –70°C for 30 min.

g.  Centrifuge at $12,000 \times g$ for 10 min at 4°C. Dry the RNA under vacuum for 15 min. Dissolve the RNA in 20 μl of TE buffer. Take 4 μl of the sample to measure the concentration and quality of the RNA (see Chapter 2). Store the sample at –20°C until use.

3.  Prepare the *Xenopus* oocytes for microinjection.

The *Xenopus* oocyte is an egg-forming cell located in the ovary. It is surrounded by thousands of follicle cells. An oocyte contains a nucleus, an animal hemisphere (the relatively darker one), and a vegetal hemisphere. The development of oocytes includes several stages with the largest or stage VI arrested at meiotic prophase. Stage VI oocytes are usually used for microinjection. Oocytes are ideal for microinjection studies because of their large size (1 to 1.2 mm in diameter), which makes it easy to microinject either mRNA into the cytoplasm or DNA into the nucleus. In addition, the oocyte contains sufficient components needed for the expression of injected genes or RNAs such as RNA polymerases I, II, and III, histones, and ribonucleotide triphosphates. Each oocyte contains more than 200,000 ribosomes, more than 10,000 tRNA molecules, and all of the enzymes required for translation of injected RNA and post-translational modification of proteins.

*Xenopus laevis* adult females are available from Carolina Biological Supply Co. (2700 York Rd., Burlington, NC), and Nasco (901 Janesville Ave., Ft. Atkinson, WI). The frogs can be kept in large plastic tanks with 5 liters of dechlorinated water per frog at a depth of 10 cm with air holes at the tops of the tanks at room temperature. The frogs may be fed twice per week with chopped beef heart, Nasco frog brittle, or Purina trout chow. Newly purchased frogs may not eat for the first 1 to 2 weeks. A normal adult female *Xenopus* contains approximately 30,000 oocytes.

a.  For isolating the oocytes, a female frog can be anesthetized by immersing it in 0.5% ethyl *m*-aminobenzoate for 25 min. Alternatively, the frog can be anesthetized by immersing it in a bowl of ice for 30 to 45 min to lower its body temperature.

***Caution:*** *Gloves should be worn since ethyl* m-*aminobenzoate is a potential carcinogen in humans.*

b.  Rinse the frog with water once it is immobilized, place it on a dissecting tray, and use a sharp scalpel to dissect the frog and remove the oocytes. If the frog is to be reused, sterile instruments and techniques should be used.

c.  Place the frog with the ventral side up on ice to keep the back but not the head on ice. Gently swab the lower abdomen of the frog with a cotton ball soaked in alcohol. Lift and hold the skin with a pair of forceps and make a 1- to 2-cm cut in the skin using a pair of dissecting scissors. Make a similar cut through the underlying muscle. Pull out a section of ovary using forceps and cut it off with a pair of scissors.

d.  Transfer the ovary to a Petri dish containing modified Barth's medium (MBS). Then put the remaining ovary back into the frog. Sew up the muscle with dissolving suture material and stitch the skin with silk sutures. Place the frog on its ventral side in a container with a small amount of water with the head elevated by wet paper towels. The frog can recover completely within a few hours.

e. Remove individual oocytes from the ovarian clumps by gently pulling the oocyte off at its base using two pairs of fine forceps. Select the largest or stage VI oocytes (1 to 1.2 mm in diameter) and transfer the oocytes to a fresh Petri dish containing MBS with a wide-mouth Pasteur pipette. Alternatively, the oocytes can be released by enzymatic method. Treat the clumps of oocytes in a solution containing 2 mg/ml collagenase (Sigma Type II) in MBS for 2 to 4 h at room temperature with slow shaking (60 rpm). The released oocytes should be washed thoroughly MBS.

f. Repeat the enzymatic treatment and washing once more to remove the remaining follicular cells. The well-prepared oocytes should have no red tissue on the surface. The oocytes can be cultured for up to 1 to 2 weeks with daily changes of medium and removal of dead cells. The vegetal hemisphere of dead cells is very pale as compared to that of live oocytes.

4. Microinject the fractionated mRNA into the healthy oocytes.

a. Select microinjection equipment that is relatively simple. It should include a stereomicroscope (10 to 40×) magnification, a light source, micropipettes, and a microinjection system which is composed of a micromanipulator and microsyringe. The microsyringe should be capable of accurately delivering volumes in the nanoliter range. These systems are commercially available: the Pico-Injector (PLI-100, Medical Systems Corp., Greenvale, NY) and the Eppendorf Microinjector (Model 5247, Fremont, CA). Assemble the system according to the instructions and test for nanoliter microinjection capability.

b. Incubate the mature oocytes overnight in MBM prior to injection. In order to immobilize the oocytes during microinjection, a piece of polyethylene mesh or its equivalent should be placed at the bottom of a Petri dish. The size of the mesh wells should be about the right size to secure the oocyte. Add MBM to cover the mesh.

c. Transfer the oocytes to the support mesh with a sterile Pasteur pipette and orient the oocytes with a hair loop or equivalent.

d. Rinse the needle with sterile water and dry with a piece of clean paper. Slowly inject a small volume of mineral oil into the microsyringe with long, fine, plastic tubing attached to a 1-ml sterile syringe. **Avoid any bubbles.**

e. Briefly spin down the specific fraction of mRNA prepared previously in order to prevent potential particles from blocking the microinjector. Place 2 μl of the mRNA (0.5 to 1 μg/μl) as a droplet onto a sheet of Parafilm™ near the needle for ease of filling the needle. Position the micromanipulator close to the mRNA droplet and slowly suck in the droplet. **Avoid any bubbles.**

f. Transfer the oocytes (10 to 20) onto the microinjection support and adjust the fine focus. Slowly adjust the needle to approach individual oocyte and focus again.

g. Under the microscope, orient the oocyte and adjust the needle so that it just reaches the surface of the oocyte in the vegetal region or near the "equator" of the oocyte[9,12] at about a 20 to 30° angle (Figure 6.8). Under the microscope, gently lower the needle toward the inside of the oocyte using the micromanipulator. The local surface of the oocyte gradually becomes depressed at first and then returns to its original shape when the needle punches inside the oocyte. Immediately stop lowering the needle at this time.

h. Slowly inject 30 to 50 nl of the mRNA sample into one oocyte and carefully withdraw the needle. Repeat the injection of the same volume to each of the other oocytes.

*Notes:* *The mRNA amount for each oocyte should determined. It ranges from 4 to 400 ng. The microinjection should be carried out carefully so that the contents inside the oocyte, such as yolk platelets, cannot float out.*

i. Carefully transfer the injected oocytes to a Petri dish containing MBS and allow the mRNA to be translated for 12 h to 3 days at 20°C prior to analyzing the activity of enzyme or protein product.

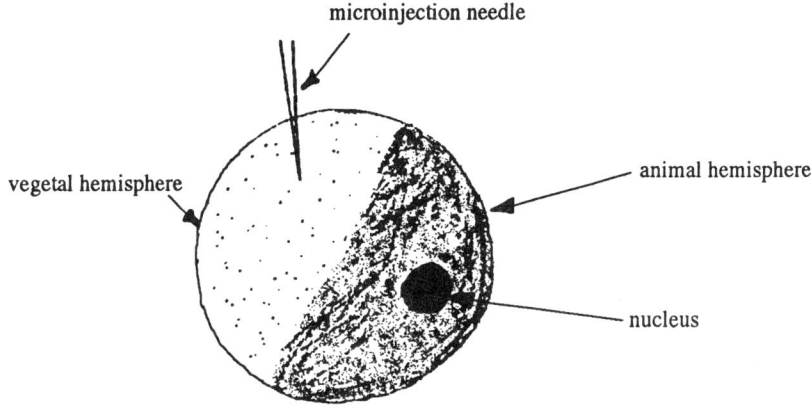

**FIGURE 6.8**

Diagram showing microinjection of RNA into an oocyte of *Xenopus*. The animal hemisphere is shown as a relatively dark region because of the heavy pigmentation.

*Note:*   *Temperatures of 25°C or higher may decrease the oocytes' chances of survival.*

    j.  If labeling of the protein is desired, $^{35}$S-methionine (1 to 5 mCi/ml of medium) can be added to the medium and incubated for 12 h. Alternatively, the radioactive amino acid can be added to the mRNA sample and be coinjected into the oocytes.

5.  Extract the soluble protein to determine the expression of mRNA of interest.

    a.  Homogenize the oocytes on ice in the homogenizing buffer (50 µl per oocyte) and centrifuge the homogenate at $10,000 \times g$ at 4°C for 5 min.

    b.  Carefully transfer the supernatant to a fresh tube. **Avoid any yolk material.** The supernatant can be used for enzyme assay, immunoprecipitation, or electrophysiological response to a specific substrate based on the particular interest and method available.

## *Reagents Needed*

### *Modified Barth's Medium (MBM)*

            88 m$M$ NaCl
            1 m$M$ KCl
            2.4 m$M$ NaHCO$_3$
            15 m$M$ Hepes-NaOH, pH 7.6
            0.82 m$M$ MgSO$_4$·7H$_2$O
            0.33 m$M$ CaNO$_3$·4H$_2$O
            0.41 m$M$ CaCl$_2$·6H$_2$O

### *Homogenization Buffer*

            0.1 $M$ NaCl
            1% Triton-X-100
            1 m$M$ Phenylmethylsulfonyl fluoride
            20 m$M$ Tris-HCl, pH 7.6

6.  Synthesize the cDNA from the fraction of mRNA of interest.

    The procedures for the synthesis of first- and second-strand cDNAs using *Xba* I primer-adaptor with poly(dT), for the ligation of *EcoR* I adaptors to the double-strand cDNAs, and for the digestion of the adaptor-cDNA with *EcoR* I followed by *Xba* I are the same as described in Section I of this chapter. At this stage, the recombinant cDNAs have one *EcoR* I sticky end and one *Xba* I sticky end, both of which are useful for directional cloning given below.

7.  Carry out ligation of cDNAs to the expressional vector, *in vitro* packaging and titration.

    The general procedure of ligation of cDNA inserts to appropriate expression vectors, *in vitro* packaging, and titration have been previously described.

8.  Carry out *in vitro* transcription of the cloned cDNA.

    a.  Isolation of recombinant phage or plasmid DNA ($3 \times 10^4$ clones) containing cDNA inserts has been previously described (see Section I in this chapter).

    b.  Digest the DNA with an appropriate restriction enzyme for making sense RNA. If antisense RNA is desired, the DNA should be digested with another appropriate enzyme depending on the particular expression vector used.

        -   Recombinant DNA (5 µg)
        -   Appropriate enzyme 10X buffer (5 µl)
        -   Appropriate restriction enzyme (20 units)
        -   Add dd.H$_2$O to a final volume of 50 µl.

    c.  Incubate at appropriate temperature for 2 to 3 h and add 2 µl of 500 m*M* EDTA to stop the reaction.

    d.  Extract the enzyme with 50 µl of TE-saturated phenol/chloroform, mix well by vortexing, and centrifuge at $11,000 \times g$ for 4 min.

    e.  Carefully transfer the top, aqueous phase to a fresh tube and add an equal volume of chloroform:isoamyl alcohol (24:1). Vortex to mix and centrifuge as in step d.

    f.  Carefully transfer the top, aqueous phase to a fresh tube. To the supernatant, add 0.15 volume of 3 *M* sodium acetate buffer (pH 5.2) and 2.5 volumes of chilled 100% ethanol. Allow precipitation to occur at –70°C for 30 min or at –20°C for 2 h.

    g.  Centrifuge at $12,000 \times g$ for 10 min and carefully pour off the supernatant. Briefly rinse the DNA pellet with 1 ml of 70% chilled ethanol and dry the pellet under vacuum for 10 min. Dissolve the DNA in 20 µl TE buffer.

    h.  Set up *in vitro* transcription in the presence of the cap analogue GpppG using the linearized DNA templates and T7 RNA polymerase at room temperature, mix well after each addition and incubate the reaction at 37°C for 2 h.

        -   5X Transcription buffer (16 µl)
        -   0.1 *M* DTT (8 µl)
        -   rRNasin ribonuclease inhibitor (80 units)
        -   Nucleotides mixture (2.5 m*M* each of
            ATP, GTP CTP, and 1 m*M* UTP) (16 µl)
        -   5 m*M* m$^7$G(5′)ppp(5′)G (5 µl)
        -   Linearized DNA template (1µg/µl) (5 µl)
        -   T7 RNA polymerase (80 units)
        -   Add dd.H$_2$O to a final volume of 80 µl.

    **Function:**   Cleavage of the recombinant phage DNA is to stop the subsequent *in vitro* transcription of cDNAs cloned in vector immediately after the poly(A) tail, leaving the mRNA of interest intact.

    i.  Add RNase-free *DNase* I to a final concentration of 1 unit/µg template DNA, which will be hydrolyzed by the enzyme after being incubated for 30 min at 37°C.

    j.  Extract the enzyme with 80 µl of TE-saturated phenol/chloroform, mix well by vortexing, and centrifuge at $11,000 \times g$ for 4 min.

    k.  Carefully transfer the top, aqueous phase to a fresh tube and add an equal volume of chloroform:isoamyl alcohol (24:1). Vortex to mix and centrifuge as in step j.

l. Carefully transfer the top, aqueous phase to a fresh tube. To the supernatant, add 0.5 volume of 7.5 *M* ammonium acetate and 2.5 volumes of chilled 100% ethanol. Allow precipitation to occur at –70°C for 30 min or at –20°C for 2 h.

m. Centrifuge at 12,000 × *g* for 10 min and carefully pour off the supernatant. Briefly rinse the DNA pellet with 1 ml of 70% chilled ethanol and dry the pellet under vacuum for 10 min. Dissolve the DNA in 20 μl TE buffer. Measure the concentration of the RNA (see Chapter 2) and store the sample at –70°C until use.

9. Microinject the mRNA transcribed *in vitro* into *Xenopus* oocytes and measure the electrophysiological response with specific substrate as described in step 4.

10. Repeat steps 8–9 several times with the cDNA library with stepwise fractionations from 2000 clones to 1000, 500, 200, and 100 until a single, positive clone is obtained.

## Reagents Needed

### 5X Transcription Buffer

> 200 m*M* Tris-HCl, pH 7.5
> 30 m*M* MgCl$_2$
> 10 m*M* Spermidine
> 50 m*M* NaCl

### 10X Ligation Buffer

> 300 m*M* Tris-HCl, pH 7.8
> 100 m*M* MgCl$_2$
> 100 m*M* DTT
> 10 m*M* ATP

# IV.  cDNA Cloning by the Polymerase Chain Reaction (PCR)

PCR is a powerful technique used for cDNA cloning as long as specific primers are available. It is a relatively fast, simple, and inexpensive procedure as compared with the other strategies described above. Briefly, specific primers are designed according to conserved motifs in two different regions of known proteins or enzymes. The primers are annealed to the first-strand cDNA synthesized from mRNA template of particular organisms of interest. Double-strand cDNAs can be generated and amplified using *Taq* DNA polymerase. There are, however, two major disadvantages of applying the PCR cloning strategy: (1) the cDNA obtained is of partial length ranging from 200 to 800 base pairs; and (2) the annealing of the primer to template, sometimes, is not specific and may amplify nonspecific sequences called "artifacts".

## Protocol

1. Carry out isolation of total RNA and purification of poly(A)+RNA from tissue or cell lines of interest.

2. Design oligonucleic primers based on fully conserved amino acid sequences in two different regions of known proteins or enzymes. Each primer is designed with one specific restriction enzyme site for subcloning of the forthcoming double-strand cDNAs.

For example, two amino acid sequence regions, NDPNG and DPCEW, of a protein are conserved from prokaryotes to eukaryotes, which have been characterized and published in professional journals. The first one is close to the N-terminal, the other is toward the C-terminal. Two oligonucleotides can be designed according to the two conserved amino acid regions. In order to subclone the forthcoming ds cDNAs into a specific vector for sequencing, a *BamH* I restriction site is designed at the N-terminal of the first primer, and a *Hind* III site follows the C-terminal of the second primer. The design is as follows:

Primer 1      5′<u>GGATCC</u>AAC(T)GAT(C)CCIAA(C)TGGI3′      for NDPNG

Primer 2      3′GGTGAGCGTCCCTAG<u>TTCGAA</u>5′      for DPCEW

The sequences underlined are the *BamH* I site at the 5′ end and the *Hind* III site at the 3′ end of the forthcoming ds cDNAs. I stands for the third position of the codon, which can be any of TCAG. The synthesis of the primers can be done with a DNA synthesizer.

Alternatively, primer 1, 5′<u>GGATCC</u>AAC(T)GAT(C)CCIAA(C)TGGI3′, can be designed for NDPNG, which is close to the N-terminal of the protein, and primer 2 can be oligo(dA)<u>TTCGGA</u>. The length of the forthcoming ds cDNA will be longer than that made by the first choice of the two primers.

3.   Synthesize the first-strand cDNA followed by the second-strand cDNA as described previously (see Section I in this chapter). However, if two primers are available, PCR amplification can be immediately carried out directly after the first-strand cDNA.

4.   Carry out PCR amplification of the ds cDNA.

   a.   In a microcentrifuge tube on ice, add the following in the order listed:
   - 10X Amplification buffer (10 µl)
   - dd.$H_2O$ (20 µl)
   - Mixture of four dNTPs (1.25 m$M$ each) (17 µl)
   - Primer 1 (100 to 110 pmol) in dd.$H_2O$ (5 µl)
   - Primer 2 (100 to 110 pmol) in dd.$H_2O$ (5 µl)
   - ss cDNA or ds cDNA in TE buffer (2 µg)
   - Add dd.$H_2O$ to a final volume of 100 µl.

   b.   Cover the tube and heat the reaction at 94°C for 5 min to denature the ds cDNA template.

   c.   Open the tube at 94°C and add 2.5 units of *Taq* DNA polymerase (5 units/µl, Perkin Elmer Cetus). Gently mix well.

   d.   Overlay the mixture with 100 µl of light mineral oil (Sigma or equivalent) to prevent evaporation of the sample.

   e.   Carry out the PCR amplification in a PCR machine programmed as follows:

| Cycle | Denaturation | Annealing | Extension |
|---|---|---|---|
| First | 4 min at 94°C | 2 min at 50°C | 3 min at 72°C |
| Subsequent | 1 min at 94°C | 2 min at 50°C | 3.5 min at 72°C |
| Last | 1 min at 94°C | 2 min at 50°C | 10 min at 72°C |

   Finally, hold at 4°C until the sample is removed.

   f.   Carefully remove the reaction mixture from the mineral oil using a pipette with a relatively long tip attached. Slowly insert the tip into the bottom of the tube and then carefully take the sample until the oil phase. Withdraw the tip from the tube, wipe the outside of the tip with clean tissue, and transfer the sample to a fresh tube.

5. Purify the amplified cDNA by agarose gel elution.
   The detailed protocol is described under the section on DNA elution in Chapter 1.

6. Carry out digestion of cDNAs and vectors, and perform ligation of the cDNA to the vector
   The detailed protocols have been described previously.

7. Carry out sequencing to verify the cDNA insert that is cloned.

## *Reagents Needed*

### *10X Amplification Buffer*

> 100 m*M* Tris-HCl, pH 8.3
> 500 m*M* KCl
> 15 m*M* MgCl$_2$
> 0.1% BSA

### *Ligase 10X Buffer*

> 300 m*M* Tris-HCl, pH 7.8
> 100 m*M* MgCl$_2$
> 100 m*M* DTT
> 10 m*M* ATP

### *Hind III 10X Buffer*

> 300 m*M* Tris-HCl, pH 7.8
> 100 m*M* MgCl$_2$
> 500 m*M* NaCl
> 100 m*M* DTT

# References

1. **Okayama, H. and Berg, P.,** High-efficiency cloning of full-length cDNA, *Mol. Cell Biol.,* 2, 161, 1982.

2. **Wu, L. -L., Song, I., Karuppiah, N., and Kaufman, P. B.,** Kinetic induction of oat shoot pulvinus invertase mRNA by gravistimulation and partial cDNA cloning by the polymerase chain reaction, *Plant Mol. Biol.,* 21, 1175, 1993.

3. **Parimoo, S., Patanjali, S. R., Shukla, H., Chaplin, D. D., and Weissman, S. M.,** cDNA selection: efficient PCR approach for the selection of cDNAs encoded in large chromosomal DNA fragments, *Proc. Natl. Acad. Sci. U.S.A.,* 88, 9623, 1991.

4. **Young, R. A. and Davis, R. W.,** Efficient isolation of genes by using antibody probes, *Proc. Natl. Acad. Sci. U.S.A.,* 80, 1194, 1983.

5. **Sambrook, J., Fritsch, E. F., and Maniatis, T.,** *Molecular Cloning: A Laboratory Manual,* 2nd ed., Cold Spring Harbor Press, Cold Spring Harbor, NY, 1989.

6. **Mignery, G. A., Pikaard, C. S., Hannapel, D. J., and Park, W. D.,** Isolation and sequence analysis of cDNAs for the major tuber proteun patatin, *Nucleic Acids Res.,* 12, 7987, 1984.

7. **Wu, L. -L., Mitchell, J. P., Cohn, N. S., and Kaufman, P. B.,** Gibberellin (GA$_3$) enhances cell wall invertase activity and mRNA levels in elongating dwarf pea *(Pisum sativum)* shoots, *Int. J. Plant Sci.,* 154, 280, 1993.

8. **Frohman, M. A., Dush, M. K., and Martin, G. R.,** Rapid production of full-length cDNAs from rare transcripts: amplification using a single gene-specific oligonucleotide primer, *Proc. Natl. Acad. Sci. U.S.A.,* 85, 8998, 1988

9. **Heikkila, J. J.,** Expression of cloned genes and translation of messenger RNA in microinjected *Xenopus* oocytes, *Int. J. Biochem.,* 22, 1223, 1990.

10. **Coleman, A.,** Translation of eukaryotic messenger RNA in *Xenopus* oocytes, in *Transcription and Translation: A Practical Approach,* Hames, B. D. and Higgins, S. J., Eds., IRL Press, Washington, D. C., 1984.

11. **Winer, M. P.,** Directional cloning of blunt-ended PCR products, *BioTechniques,* 15, 502, 1993.

12. **Hitchcock, M. J. M., Ginns, E. L., and Marcus-Sekura, C. J.,** Microinjection into *Xenopus* oocytes: equipment, in *Methods in Enzymology,* Vol. 152, Berger, S. L. and Kimmel, A. R., Eds., Academic Press, New York, 1987.

# Chapter 7

# Genomic DNA Libraries

## Contents

# I.   General Strategies and Applications

Genomic DNA cloning is a high technology that plays a momentous role in state-of-the-art molecular biology studies. Almost every single gene that has been characterized comes from original genomic DNA cloning. In other words, if one wishes to identify, characterize, and regulate the expression of a full-length unknown genomic gene, he or she may have to start from molecular cloning of genomic DNA, which is described in detail in this chapter.

The quality and integrity of a genomic DNA library are directly correlated with the success or failure of identifying a gene of interest. A very good library is supposed to contain all DNA sequences of the entire genome. The probability of "fishing out" a DNA sequence of interest depends on the size of a library, which in turn relies on the sizes of DNA fragments selected for cloning. The larger the DNA fragments, the smaller the number of clones in the library. The probability of having an interesting DNA sequence in the library can be calculated by the following equation:[1]

$$N = \frac{\ln(1-P)}{\ln(1-f)}$$

where N is the number of recombinants required, P is the desired probability of "fishing out" a DNA sequence, and f is the fractional proportion of the genome in a single recombinant. For example, given a 99% probability of "fishing out" an interesting DNA sequence in a library cloned by 18-Kb fragments of a $4 \times 10^9$-bp genome, the required recombinants are as follows:

$$N = \frac{\ln(1-0.99)}{\ln\left[1-\left(1.8 \times 10^4 / 4 \times 10^9\right)\right]}$$

$$= 1 \times 10^6$$

In general, the average size of DNA fragments selected for cloning is directly related to the cloning vectors that are used to construct genomic DNA libraries. There are three types of vectors that are commonly used for the construction of genomic DNA libraries: (1) bacteriophage lambda vectors; (2) cosmids; and (3) yeast artificial chromosomes (YACs). Lambda vectors can accept 14- to 25-Kb DNA fragments. DNA fragments are ligated with lambda vectors, forming concatemers that are then packaged into bacteriophage lambda ($\lambda$) particles. Recombinants numbering $3 \times 10^6$ to $3 \times 10^7$ are usually necessary to achieve a 99% probability of isolating a specific clone of interest, and the screening procedure of the library is relatively complicated. Cosmids, on the other hand, can accept about 35- to 45-Kb DNA fragments. Only approximately $3 \times 10^5$ recombinant cosmids can achieve a 99% probability of identifying a particular single-copy sequence of interest. However, it is difficult to construct and maintain a genomic DNA library in cosmids as compared with bacteriophage lambda vectors. YAC vectors are very useful to clone 50- to 10,000-Kb DNA fragments. This is powerful when one wishes to clone and isolate extra-large genes (e.g. human factor VIII gene, 180 Kb in length; the dystrophin gene, 1800 Kb in length) in a single recombinant YAC of a smaller size library. This chapter describes and emphasizes the construction of bacteriophage $\lambda$ libraries and YAC libraries.[2-7]

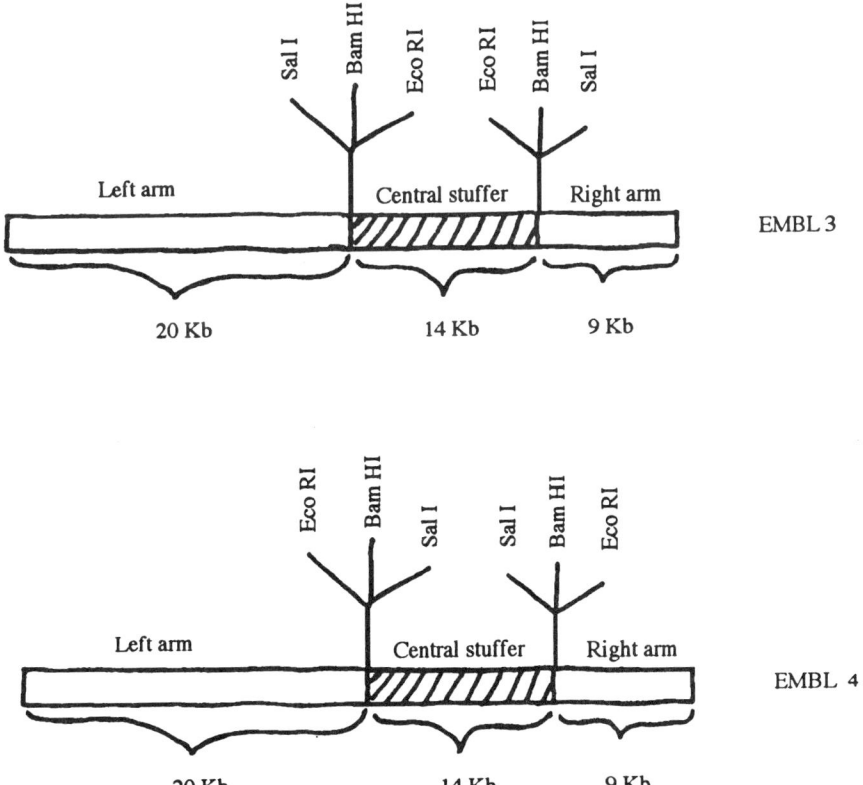

**FIGURE 7.1**
Structural maps of vectors: EMBL 3 (top) and EMBL 4 (bottom).

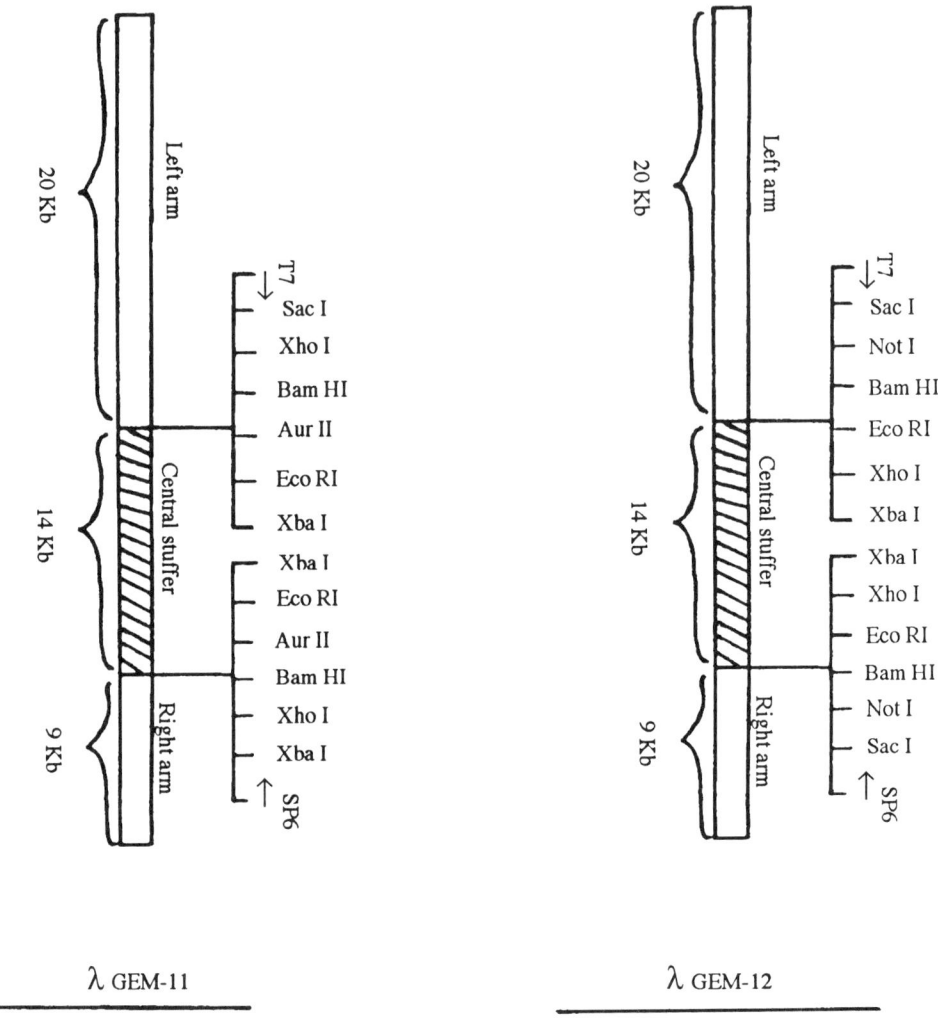

**FIGURE 7.2**
Structural maps of vectors: LambdaGEM-11 (left) and LambdaGEM-12 (right).

## II.    Bacteriophage λ Library

### A.    Construction of a Genomic DNA Library

Several years ago, the commonly used bacteriophage λ DNA vectors were EMBL3 and
EMBL4 (Figure 7.1). Both contain a left arm (20 Kb), a right arm (9 Kb), and a central stuffer
(14 Kb), which can be removed and replaced with a foreign DNA insert (9 to 23 Kb). The only
difference between two λ DNA vectors is the orientations of the polylinker regions.

Recently, new λ DNA vectors have been developed such as LambdaGEM-11 and
LambdaGEM-12 (Figure 7.2). These vectors are modified from EMBL3/EMBL4. The sizes
of arms and capacity of cloning foreign DNA inserts are the same as EMBL3 and EMBL4.
However, there are several advantages over EMBL3 and EMBL4:

1. More restriction enzyme sites are designed in the polylinker region.

2. There are two promoters, T7 and SP6, at the polylinker ends in opposite orientations. These promoters make it possible to directly sequence the DNA insert by using T7 or SP6 primer without subcloning. They also allow one to express the DNA insert into sense RNA or antisense RNA synthesized by T7 or SP6 RNA polymerase.

3. These vectors are designed to be optimized for the highest recombinant efficiencies and lowest nonrecombinant background.

Because of these merits of LambdaGEM-11 and LambdaGEM-12 and based on our successful experience with them, we recommend that these two vectors be used for genomic DNA cloning.

The overall procedure of constructing genomic DNA libraries is that high-molecular weight genomic DNA is partially digested with Sau3A into 14- to 24-Kb fragments. These fragments are ligated with dephosphorylated LambdaGEM-11 or LambdaGEM-12 arms by T4 DNA ligase. The concatemers are then packaged *in vitro* into bacteriophage λ particles, generating a genomic library. The library is screened using specific probe and positive clones will be isolated. After several rounds of rescreening, the putative DNA clones of interest can be purified.

## 1. Protocol: Optimization of Partial Digestion of Genomic DNA with Sau3A I

In order to determine the amount of enzyme that can digest high-molecular weight DNA into 14- to 23-Kb fragments for cloning into λ vectors, small-scale reactions, called the pilot experiments, should be carried out.

1. Prepare 1X *Sau3A* I buffer on ice:
   - 10X *Sau3A* I buffer (0.2 ml)
   - Acetylated BSA (1 mg/ml, optional) (0.2 ml)
   - Add dd.H$_2$O to a final volume of 2.0 ml.

2. Prepare *Sau3A* I dilutions in ten individual microcentrifuge tubes on ice as shown in Table 7–1.

3. Set up 10 individual small-scale digestion reactions on ice in the order shown in Table 7–2.

*Notes:* *(1) Genomic DNA used in the reactions should be high-molecular weight (>150 to 200 Kb) with a ratio of 1.85 to 1.95 of A$_{260}$/A$_{280}$. The purification of high-molecular weight genomic DNA is described in Chapter 1 (Figure 7.3). An unpure DNA sample*

### Table 7–1 Preparation of *Sau3A* I Dilutions

| Tube no. | Preparation of *Sau3A* I (3 units/μl) dilution | Dilution times |
|---|---|---|
| 1 | 2 μl *Sau3A* I + 28 μl 1X *Sau3A* I buffer | 1/15 |
| 2 | 10 μl of 1/15 dilution + 90 μl 1X *Sau3A* I buffer | 1/150 |
| 3 | 10 μl of 1/150 dilution + 10 μl 1X *Sau3A* I buffer | 1/300 |
| 4 | 10 μl of 1/150 dilution + 30 μl 1X *Sau3A* I buffer | 1/600 |
| 5 | 10 μl of 1/150 dilution + 50 μl 1X *Sau3A* I buffer | 1/900 |
| 6 | 10 μl of 1/150 dilution + 70 μl 1X *Sau3A* I buffer | 1/1200 |
| 7 | 10 μl of 1/150 dilution + 90 μl 1X *Sau3A* I buffer | 1/1500 |
| 8 | 10 μl of 1/150 dilution + 110 μl 1X *Sau3A* I buffer | 1/1800 |
| 9 | 10 μl of 1/150 dilution + 190 μl 1X *Sau3A* I buffer | 1/3000 |
| 10 | 10 μl of 1/150 dilution + 290 μl 1X *Sau3A* I buffer | 1/4500 |

## Table 7–2 Small-Scale Digestion Reactions

| Components | Tube number | | | | | | | | | |
|---|---|---|---|---|---|---|---|---|---|---|
| | 1 | 2 | 3 | 4 | 5 | 6 | 7 | 8 | 9 | 10 |
| Genomic DNA (1 µg/µl) | 1 | 1 | 1 | 1 | 1 | 1 | 1 | 1 | 1 | 1 (µl) |
| 10X *Sau3A* I buffer | 5 | 5 | 5 | 5 | 5 | 5 | 5 | 5 | 5 | 5 (µl) |
| Acetylated BSA (1 mg/ml, optional) | 5 | 5 | 5 | 5 | 5 | 5 | 5 | 5 | 5 | 5 (µl) |
| dd.H$_2$O | 34 | 34 | 34 | 34 | 34 | 34 | 34 | 34 | 34 | 34 (µl) |
| *Sau3A* I dilution in the same order as in Table 7–1 | 5 | 5 | 5 | 5 | 5 | 5 | 5 | 5 | 5 | 5 (µl) |
| Final volume | 50 | 50 | 50 | 50 | 50 | 50 | 50 | 50 | 50 | 50 (µl) |

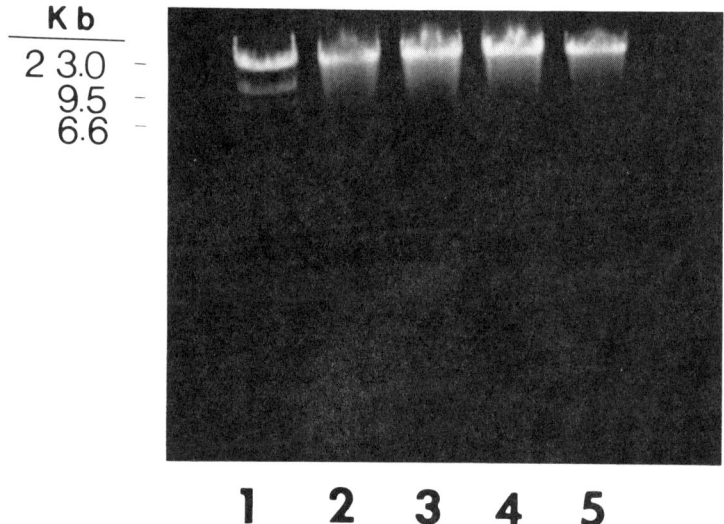

**K b**

2 3.0 –
9.5 –
6.6 –

1   2   3   4   5

**FIGURE 7.3**

Isolation of high-molecular weight genomic DNA from a eukaryotic organism for the construction of a genomic DNA library. Lane 1: DNA standard molecular weight markers. Lanes 2–5: different DNA samples.

*with a ratio <1.75 of A$_{260}$/A$_{280}$ should not be used for construction of a genomic DNA library. (2) Sau3A I dilutions (5 µl) used in the DNA digestion reactions should be always added last from the right dilution tube to the right DNA digestion tube. (3) The final Sau3A I concentration used in tubes 1 to 10 should be 1, 0.1, 0.05, 0.025, 0.015, 0.0125, 0.01, 0.0085, 0.005, and 0.0035 unit/µg DNA, respectively.*

4.  Incubate the ten digestion reactions at the same time at 37°C in a water bath for 30 min. Place the tubes on ice and add 2 µl of 0.2 M EDTA buffer (pH 8.0) to stop the reaction.

5.  While the reactions are being carried out, prepare a large size of 0.4% agarose gel in 1X TBE buffer. Melt the agarose in a microwave oven and allow to cool to about 50°C. Add 10 µl of

**FIGURE 7.4**
Partial digestion of genomic DNA shown in Figure 7.3 with *Sau3A* I at different concentrations. Lanes 1 to 6 show progressively diluted *Sau3A* I. The digested DNA was electrophoresed on a 0.4% agarose gel and stained with ethidium bromide.

10 mg/ml EtBr solution to every 100 ml of gel mixture, mix well, and pour into a prepared gel tray. Allow the gel to harden at room temperature for about 30 min. Place the gel in the apparatus and add 0.5X TBE buffer to cover the gel (1 to 2 mm depth above the gel surface). A mini-gel should not be used for this purpose.

6. Add 10 µl of 5X DNA loading buffer to each of the ten tubes of digested DNA prepared in step 4.

7. Carefully load 30 µl of each sample onto the wells in the order of numbers 1 to 10. Load DNA markers (e.g. λ DNA *Hind* III markers) to the left or the right well of the sample wells to estimate the sizes of digested DNA with ease.

8. Electrophorese the gel at 2 V/cm until the bromphenol blue reaches the bottom of the gel. It usually takes about 10 to 12 h.

9. Photograph the gel under UV light and find the well that shows the maximum intensity of fluorescence in the desired DNA size range of 14 to 23 Kb. The intensity of fluorescence is directly related to the mass distribution of DNA. The amount of *Sau3A* I used to get the maximum intensity of fluorescence from 14 to 23-Kb DNA fragments is the optimal concentration which can guide large-scale digestion of DNA for the construction of genomic DNA libraries (Figure 7.4).

## Table 7–3 Large-Scale Digestion Reactions

| Components | Tube 1 | Tube 2 (duplicate) |
|---|---|---|
| Genomic DNA (1μg/μl) | 50 μl | 50 μl |
| 10X Sau3A I buffer | 250 μl | 250 μl |
| 1 mg/ml Acetylated BSA (optional) | 250 μl | 250 μl |
| dd.H$_2$O | 1.7 ml | 1.7 ml |
| Diluted Sau3A I (0.005 units/μg) prepared as in Table 7–1. tube no. 9 | 250 μl | 250 μl |
| Final volume | 2.5 ml | 2.5 ml |

### 2. Protocol: Large-Scale Preparation of Partially Digested Genomic DNA

1.  Guided by the optimized conditions established in step 1.9, carry out a large-scale digestion of 50 μg of high-molecular-weight genomic DNA using half the number of units of Sau3A I per microgram of DNA that produced the maximum intensity of fluorescence in the DNA size range of 14 to 23 Kb.

*Notes:*  *The DNA concentration, time and temperature should be the same as those used for the small-scale digestion. To ensure success, we recommend that a duplicate large-scale digestion be set up. For example, if tube no. 7 (0.01 unit of* Sau3A *I per microgram DNA) in the small-scale digestion of the DNA displays a maximum intensity of fluorescence in the size range of 14 to 23 Kb, the large-scale digestion of the same DNA can be carried out on ice as shown in Table 7–3.*

*Incubate the reactions at 37°C in a water bath for 30 min. Then stop the reactions by adding 0.1 ml of 0.2 M EDTA buffer (pH 8.0) and place on ice until use.*

2.  To check the size range of digested DNA, take 30 μl of the partially digested DNA from each sample, add 7 μl of 5X loading buffer, and load the mixture onto 0.4% agarose gel in 1X TBE buffer. Electrophorese and photograph as described for the small-scale digestion.

*Notes:*  *(1) If the digestion is adequate, proceed to step 3. (2) If not desired, repeat large-scale digestion with an appropriate amount of enzyme until a size range of 14 to 23 Kb is achieved. If the DNA molecules are under- or overdigested, the undesired DNA fragments will cause failure in packaging* in vitro, *resulting in a bad library.*

3.  Extract with 1 volume of TE-saturated phenol/chloroform. Mix well by inverting for 1 min and centrifuge at 11,000 × g for 5 min at room temperature.

*Caution:*  *Phenol is very toxic. Care should be taken when handling this chemical. Waste phenol should be collected in a special container. Gloves should be worn when dealing with this chemical.*

4.  Carefully transfer the top, aqueous phase to a fresh tube and repeat step 3.

5.  Carefully transfer the upper, aqueous phase to a fresh tube and add 1 volume of chloroform:isoamyl alcohol (24:1). Mix well and centrifuge as in step 3.

6.  Transfer the top, aqueous phase to a fresh tube and add 0.1 volume of 3 *M* sodium acetate buffer (pH 5.2) or 0.5 volume of 7.5 *M* ammonium acetate. Mix and add 2 to 2.5 volumes of chilled 100% ethanol. Allow to precipitate at –70°C for 30 min.

7.  Centrifuge at 12,000 × *g* for 10 min at room temperature. Carefully decant the supernatant and briefly rinse the DNA pellet with 5 ml of 70% ethanol. Dry the pellet under vacuum for 15 min. Dissolve the DNA in 200 to 500 μl of TE buffer. Take 5 to 10 μl of the sample to measure the concentration at 260 nm and store the DNA sample at –20°C until use.

## Reagents Needed

### 10X Sau3A I Buffer

0.1 *M* Tris-HCl, pH 7.5
1 *M* NaCl
70 m*M* MgCl$_2$

### 5X Loading Buffer

38% (w/v) Sucrose
0.1% Bromophenol blue
67 m*M* EDTA

### 5X TBE Buffer

Tris base (54 g)
Boric acid (27.5 g)
20 ml of 0.5 *M* EDTA, pH 8.0

### TE-Saturated Phenol/Chloroform

Completely melt the phenol crystals in a 65°C water bath. Mix equal parts of phenol and TE buffer and allow the phases to separate. Mix 1 part of the lower, phenol phase with 1 part of chloroform:isoamyl alcohol (24:1). Allow the phases to separate and store at 4°C in an aluminum foil-wrapped bottle.

### TE Buffer

10 m*M* Tris-HCl, pH 8.0
1 m*M* EDTA, pH 8.0

### 3 M Sodium Acetate Buffer, pH 5.2

### 7.5 M Ammonium Acetate Solution

### Ethanol (100%, 70%)

## 3. Protocol: Partial Filling of Recessed 3′ Termini of Genomic DNA Fragments

Since partially filled-in *Xho* I sites of λ vectors are commercially available, the partially digested DNA fragments can be partially filled-in in order to be ligated to the vectors. The merit of partially filled-in reactions is to prevent DNA fragments from ligating to each other, thus eliminating the need for size-fractionation of genomic DNA fragments.

1. On ice, set up a standard reaction as follows:
   - Partially digested genomic DNA (20 μg)
   - 10X Fill-in buffer containing dGTP and dATP (10 μl)
   - Klenow fragment of *E. coli* DNA polymerase I (5 units/μl) (4 μl)
   - Add dd.H$_2$O to a final volume of 100 μl.
2. Incubate the reaction at 37°C for 30 to 60 min.
3. Extract and precipitate as described in steps 2.3 to 2.7 except that the DNA pellet should be dissolved in 20 to 30 μl of dd.H$_2$O. Take 2 to 4 μl of the sample to measure the concentration of DNA. Store at –20°C until use.

## Reagents Needed

### 10X Fill-in Buffer

0.5 *M* Tris-HCl, pH 7.2
0.1 *M* MgSO$_4$
1 m*M* DTT
0.5 g/ml Acetylated BSA
10 m*M* dATP
10 m*M* dGTP

## 4. Protocol: Small-Scale Ligation of Partially Filled-in Genomic DNA Fragments and Partially Filled-in LambdaGEM-12 Arms

The purpose of setting up a small-scale ligation is to optimize the conditions for large-scale ligation.

1. Set up, on ice, sample and control reactions as follows:
   a. Sample reactions:

| Components | Tube number | | | | |
|---|---|---|---|---|---|
| | 1 | 2 | 3 | 4 | 5 |
| Insert DNA (0.3 μg/μl, 0.1 μg = 0.01 pmol) | 0 | 4 | 3 | 2 | 0.5 (μl) |
| Vector DNA (0.5 μg/μl, 1 μg = 0.035 pmol) | 2 | 2 | 2 | 2 | 2 (μl) |
| 10X Ligase buffer | 1 | 1 | 1 | 1 | 1 (μl) |
| dd.H$_2$O | 6 | 2 | 3 | 4 | 5.5 (μl) |
| T4 DNA ligase (10 to 15 Weiss units/μl) | 1 | 1 | 1 | 1 | 1 (μl) |
| **Total volume** | **10** | **10** | **10** | **10** | **10 (μl)** |

   b. Positive control reaction:

| | |
|---|---|
| LambdaGEM-12 *Xho* I half-site arms (0.5 μg) | 2 (μl) |
| Positive control insert (0.5 μg) | 2 (μl) |
| 10X Ligase buffer | 1 (μl) |
| dd.H$_2$O | 4 (μl) |
| T4 DNA ligase (10 to 15 Weiss units/μl) | 1 (μl) |
| **Total volume** | **10 (μl)** |

2.   Incubate the ligation reactions for overnight at 4°C for *Xho* I half-site arms.

*Notes:*   *(1) Setting up a positive control is necessary in order to check the efficiency of ligation, packaging and titration. If the efficiency is very low, such that pfu/ml < 10⁴, then either the vectors do not work well or some procedures were carried out incorrectly. Normally, the positive control is pretested by the company, and pfu/ml is usually 10⁶ to 10⁸. (2) Setting up a ligation of vector arms only is in order to check the background induced by the vector arms. Normally, pfu/ml < 10² to 10³. If a higher pfu number such as 10⁴ to 10⁵ is observed, then the vectors' ligation is very high, which lowers the efficiency of ligation between the vector arms and the insert DNA. In those circumstances, we recommend the use of new vector arms that have a very low self-ligation rate.*

## Reagent Needed

### 10X Ligase Buffer

0.3 *M* Tris-HCl, pH 7.8
0.1 *M* MgCl$_2$
0.1 *M* DTT
10 m*M* ATP

## 5. Protocol: In Vitro Packaging of Ligated DNA

1.   Thaw three commercial Packagene extracts (50 μl per extract) on ice. Do not thaw the extract at 37°C.
2.   Quickly divide each extract into two tubes (25 μl per tube) on ice.
3.   Add 4 μl of each of the ligated mixture in step 4.2 into appropriate tubes, each containing 25 μl of Packagene extract. For testing the quality of the Packagene extract, add 0.5 μg of the provided positive packaging DNA into a 25-μl packaging extract.
4.   Incubate at 22 to 24°C (room temperature) for 3 to 4 h.

*Note:*   *Longer packaging (6 to 10 h) is acceptable.*

5.   Add phage buffer to each packaged tube up to 250 μl and add 10 to 12 μl of chloroform. Mix well by inversion and allow the chloroform to settle to the bottom of the tube. Store the packaged phages at 4°C until use or for up to 4 weeks with a consequent severalfold dropping of the titer.

*Note:*   *When the packaged phage solution is used for titration or screening, do not mix the chloroform at the bottom of the tube into the solution; otherwise, the chloroform may kill or inhibit the growth of the bacterial host.*

## Reagent Needed

### Phage Buffer

20 m*M* Tris-HCl, pH 7.5
100 m*M* NaCl
10 m*M* MgSO$_4$

## 6. Protocol: Titration of Packaged Phage on LB Plates

1.  Partially thaw the specific bacterial strain such as *E. coli* LE 392, or KW 251 on ice in a sterile laminar flow hood. The bacterial strain is usually kept in 20% glycerol and stored at –70°C until use. Pick up a tiny bit of the bacteria using a sterile wire transfer loop, and immediately inoculate the bacteria by gently drawing several lines on the surface of an LB plate. The LB plate should be freshly prepared and dried at room temperature for a couple of days prior to use. Invert the inoculated LB plate and incubate the plate in an incubator at 37°C overnight. Multiple bacterial colonies will grow.

2.  Prepare a fresh bacterial culture by picking a single colony from the freshly streaked LB plate using a sterile wire transfer loop and inoculate into a culture tube containing 5 ml of LB medium supplemented with 50 ml of 20% maltose and 50 ml of 1 $M$ MgSO$_4$ solution. Shake at 160 rpm at 37°C for 6 to 9 h or until the OD$_{600}$ has reached 0.6. Store the culture at 4°C until use.

3.  Dilute each of the packaged recombinant phage samples 1000, 5000, or 10,000 times with phage buffer.

4.  Add 20 ml of 1 $M$ MgSO$_4$ solution and 2.8 ml of melted top agar to 36 sterile glass test tubes in a sterile laminar flow hood. Cap the tubes and immediately place the tubes in a 50°C water bath for at least 30 min.

5.  Mix 0.1 ml of the diluted phage with 0.1 ml of fresh bacterial cells in a microcentrifuge tube. Cap the tube and allow the phage to adhere to the bacteria in an incubator at 37°C for 30 min.

6.  Add the incubated phage/bacterial mixture into specific tubes in the water bath. Vortex gently and immediately pour onto the centers of LB plates; then quickly spread the mixture all over the surface of the LB plates by gently tilting them. Cover the plates and allow the top agar to harden for 15 min in the sterile laminar flow hood. Invert the plates and incubate them in an incubator at 37°C overnight or for 15 h.

*Note:*    *The top agar mixture should be evenly distributed over the surface of an LB plate. Otherwise, the growth of bacteriophage will be affected, thus decreasing the pfu number.*

7.  Count the number of plaques for each plate and calculate the titer of the phage (pfu) for each sample and the control.

    Plaque-forming units (pfu) per milliliter = number of plaques per plate × dilution × 10

    The last denoted 10 of the calculation refers to the 0.1 ml, on a per-milliliter basis, of the packaging extract used for one plate. For example, if there are 100 plaques on a plate made from a 1/5000 dilution, the pfu per milliliter of the original packaging extract = 100 × 5000 × 10 = 5 × 10$^6$.

8.  Compare the titers (pfu/ml) from different samples, determine the optimal ratio of vector arms and genomic DNA inserts, and prepare a large-scale ligation and packaging reaction using the optimum ratio of conditions.

## Solutions Needed

### Phage Buffer

20 m$M$ Tris-HCl, pH 7.4
100 m$M$ NaCl
10 m$M$ MgSO$_4$

## LB (Luria-Bertaini) Medium (per liter)

   10 g Bacto-tryptone
   5 g Bacto-yeast extract
   5 g NaCl
   Adjust to pH 7.5 with 0.2 $N$ NaOH and autoclave.

## 20% (v/v) Maltose

## 1 M MgSO$_4$

## LB Plates

   Add 15 g of Bacto-agar to 1 liter of freshly prepared LB medium and autoclave. Allow the mixture to cool to about 60°C and pour 30 ml of the mixture into each 85- or 100-mm diameter Petri dish in a sterile laminar flow hood. Remove any air bubbles with a pipette tip and let the plates cool for 5 min prior to their being covered. Allow the agar to harden for 1 h and store the plates at room temperature up to 10 days or at 4°C in a bag for 1 month. The cold plates should be placed at room temperature for 1 to 2 days before use.

## TB Top Agar

   Add 3.0 g agar to 500 ml freshly prepared LB medium and autoclave. Store at 4°C until use. For plating, melt the agar in a microwave oven. When the solution has cooled to 60°C, add 0.1 ml of 1 $M$ MgSO$_4$ per 10 ml of the mixture.

## S Buffer

   Phage buffer + 2% (w/v) gelatin

## 7. Protocol: Large-Scale Ligation of Partially Filled-in Vector Arms and Partially Filled-in Genomic DNA Fragments

1. Based on the pfu/ml of the small-scale ligations, choose the optimal conditions for large-scale ligation. For example, if tube no. 4 in the small-scale ligation shows the maximum pfu/ml, then a large-scale ligation can be set up as follows:

| | |
|---|---:|
| Partially filled-in DNA inserts (0.3 µg/µl, 0.1 µg = 0.01 pmol) | 10 (µl) |
| Partially filled-in vector arms (*Xho* I half-site arms, 0.5 µg/µl, 1 µg = 0.035 pmol) | 10 (µl) |
| 10X Ligase buffer | 5 (µl) |
| dd.H$_2$O | 20 (µl) |
| T4 DNA ligase (10 to 15 Weiss units/µl) | 5 (µl) |
| **Total volume** | **50 (µl)** |

2. Incubate at 4°C for 12 to 24 h.
3. Carry out *in vitro* packaging of ligated DNA using Packagene extracts thawed on ice. Add 9 µl of the ligated mixture to each of three extracts (50 µl per extract).

4.  Incubate at 22 to 24°C for 5 to 8 h.

5.  Add phage buffer to each tube up to 0.5 ml. Add 25 µl of chloroform to each tube, mix well, and store at 4°C until use or for up to 4 weeks. At this stage, a genomic DNA library has been established.

6.  Perform a titration as in steps 6.1 to 6.8.

*Note:* *A good genomic library should have a pfu/ml up to $10^7$ to $10^8$. If the pfu < $10^4$, do not use for screening.*


## B.     Screening of the Genomic DNA Library Using Appropriate Probe(s)


### Method 1: Screening of the Genomic Library with a 32P-Labeled Probe

1.  Prepare LB plates by adding 15 g of Bacto-agar to 1 liter of LB medium and autoclave. Allow the mixture to cool to about 50°C and pour 30 to 40 ml of the mixture into each 85- or 100-mm petri dish in a sterile laminar flow hood with filtered air flowing. Remove any air bubbles with a pipette tip and let the plates cool for 5 min prior to their being covered. Allow the agar to harden for 1 h and store the plates at room temperature up to 10 days or at 4°C in a bag for 1 month. The cold plates should be placed at room temperature for 1 to 2 days before use.

2.  Partially thaw the specific bacterial strain, *E. coli* LE392, on ice in a sterile laminar flow hood. Pick up a tiny amount of the bacteria using a sterile wire transfer loop and immediately streak out the *E. coli* on LB plates by gently drawing several lines on the surface of each LB plate. Invert the inoculated LB plate and incubate the plate in an incubator at 37°C overnight. Multiple bacterial colonies will grow.

3.  Prepare a fresh bacterial culture by picking up a single colony from the freshly streaked LB plate using a sterile wire transfer loop and inoculate into a culture tube containing 50 ml of LB medium supplemented with 500 ml of 20% maltose and 500 ml of 1 $M$ MgSO$_4$ solution. Shake at 160 rpm at 37°C overnight or until the OD$_{600}$ has reached 0.6. Store the culture at 4°C until use.

4.  Add 20 µl of 1 $M$ MgSO$_4$ solution and 2.8 ml of melted LB top agar to each of the sterile glass test tubes in a sterile laminar flow hood. Cap the tubes and immediately place them in a 50°C water bath for at least 30 min.

5.  Based on the titration data, dilute the λ DNA library with phage buffer. Set up 20 to 25 plates for primary screening of the genomic DNA library. For each of the 100-mm diameter plates, mix 0.1 ml of the diluted phage containing $2 \times 10^5$ pfu of the genomic library with 0.2 ml of fresh bacterial cells in a microcentrifuge tube. Cap the tube and allow the phage to adhere to the bacteria in an incubator at 37°C for 30 min.

6.  Add the incubated phage/bacterial mixture into the tubes from the water bath. Vortex gently and immediately pour onto the center of each LB plate. Quickly spread the mixture over the entire surface of each LB plate by gently tilting the plate. Cover the plates and allow the top agar to harden for 15 min in a laminar flow hood. Invert the plates and incubate in an incubator at 37°C overnight.

*Note:* *The top agar mixture should be evenly distributed over the surface of the LB plate. Otherwise, the growth of the bacteriophage will be uneven.*

7. Chill the plates at 4°C for 1.5 h.

8. Move the plates to room temperature. Carefully overlay each plate with a dry nitrocellulose filter disk or a nylon membrane disk (prewetting treatment is not necessary) from one side of the plate slowly to the other side, carefully preventing any air bubbles from developing underneath the filter. Quickly mark the top side and position of each filter in a triangular pattern by punching the filter through the bottom agar layer with a 20-gauge needle containing some India ink. Allow the phage DNA to mobilize onto the facing side of the membrane filters for 1 to 2 min. If a duplicate is needed, a second filter may be overlaid on the plate for 2 to 3 min.

*Notes:* *We recommend the use of positively charged nylon membrane disks because they tightly bind the negative phosphate groups of the DNA. Nylon membranes are not easily broken, which is usually the case with nitrocellulose membranes.*

9. Carefully remove the filters using forceps and individually place the filters, plaque-side up, on a piece of wet 3MM Whatman filter paper saturated with denaturing solution for 4 min at room temperature. Label and wrap each plate with Parafilm™, and store the plates at 4°C until use. This step serves to denature the double-stranded DNA for hybridization with a probe.

10. Transfer the filters, plaque-side up, on another piece of 3MM Whatman paper saturated with neutralization solution at room temperature for 4 min. This step functions to neutralize the filters for hybridization.

11. Transfer the filters, plaque-side up, on another piece of 3MM Whatman filter paper saturated with 5X SSC at room temperature for 2 min.

12. Air dry the filters at room temperature for 15 min. Then wrap the filters with dry 3MM Whatman paper and bake the filters in a vacuum oven at 80°C for 2 h. Wrap the filters with aluminum foil and store at 4°C until prehybridization is carried out. This step is to fix the DNA to the filters.

13. Immerse the filters in 5X SSC for 5 min at room temperature to equilibrate the filters.

*Note:* *Do not let the filters dry during subsequent steps. Otherwise, a high background and/or anomalous results will show up.*

14. Place the filters in the prehybridization solution and carry out prehybridization for 2 to 4 h with slow shaking at 60 rpm.

*Notes:* *(1) We strongly recommend not using plastic hybridization bags because, in using them, it is usually not easy to get rid of air bubbles nor can they be sealed well. This causes leaking and contamination. An appropriate size of plastic beaker or tray is the best type of hybridization container to use for this purpose. (2) The prehybridization temperature depends on the prehybridization buffer. The temperature should be set at 42°C if the buffer contains 30 to 40% (for low-stringency conditions) or 50% formamide (for high-stringency conditions). If the buffer, on the other hand, does not contain formamide, the temperature is set at 65°C. Low-stringency conditions will help identify cDNAs of a potential multigene family. High-stringency conditions help prevent nonspecific cross-linking hybridization. (3) We strongly recommend the use of a regular culture shaker with a cover and temperature control as the hybridization chamber. Such a chamber is easily handled by placing the hybridization beaker or tray containing the filters and buffer on the shaker in the chamber. A commercial hybridization oven may be difficult to operate because the filters must be covered with a matrix and put inside the hybridization bottle, during which time air bubbles are easily generated. (4) If many filters are to be used for hybridization, one beaker should not contain more than three filter disks. Too many filters in one beaker may cause weak hybridization to occur. (5) The volume of prehybridization solution should be 15 ml per 100-cm² filter disk.*

15.  Denature the labeled double-stranded DNA probe contained in a microcentrifuge tube in boiling water for 10 min and immediately chill on ice for 5 min to denature the probe for hybridization. Briefly spin down prior to use with a microcentrifuge.

*Notes:*    *(1) This is a critical step. If the probes are not completely denatured, a weak or no hybridization signal will occur. Single-strand oligonucleotide probes, however, usually do not require denaturation. (2) The DNA used for labeling can be a specific gene (usually a conserved partial-length fragment), an oligonucleotide (where synthesis is based on the conserved regions of known DNA) or a specific cDNA (partial or full length) from other organisms. The DNA is labeled with [$\alpha$-$^{32}P$]dCTP and is ready for hybridization (see DNA-labeling protocols in Chapter 4). (3) The labeled probe should be separated from the unincorporated nucleotides by use of a G-50 column (see Chapter 4 for details). Otherwise, nonspecific black spots will appear on the filter, causing one to "fish" out false positive plaques. (4) It is recommended, but not required, to calculate the cpm of the labeled probe prior to hybridization.*

*Caution:*  *[$\alpha$-$^{32}P$]dCTP is a dangerous isotope. A lab coat and gloves should be worn when working with this isotope. Gloves should be changed often and put in a special container. Waste liquid, pipette tips, and papers contaminated with the isotope should be collected in labeled containers. After finishing, a radioactive contamination survey should be performed and recorded.*

16.  Dilute the purified probe with 1 ml of hybridization solution and add the probe at 2 to $10 \times 10^6$ cpm/ml to the hybridization buffer. Mix well and carefully transfer the prehybridized filters to the hybridization solution. Allow hybridization to proceed overnight or up to 19 h.

*Notes:*   *For prehybridization, notes are the same as for hybridization.*

17.  Wash the hybridized filters according to the following conditions:

a.  High-stringency conditions

i.   Wash the filters in a solution (50 ml per filter) containing 2X SSC and 0.1% SDS (w/v) for 15 min at room temperature with slow shaking. Repeat once.

ii.  Wash the filters in a fresh solution (50 ml per filter) containing 2X SSC and 0.1% SDS (w/v) for 20 min at 65°C with slow shaking. Repeat two to four times.

iii. Air dry the filters at room temperature for about 40 min and proceed with autoradiography.

b.  Low-stringency conditions

i.   Wash the filters in a solution (50 ml per filter) containing 2X SSC and 0.1% SDS (w/v) for 10 min at room temperature with slow shaking. Repeat once.

ii.  Wash the filters in a fresh solution (50 ml per filter) containing 2X SSC and 0.1% SDS (w/v) for 15 min at 50 to 55°C with slow shaking. Repeat once.

iii. Air dry the filters at room temperature for 40 min and proceed with autoradiography.

18.  Wrap the filters, one by one, with SaranWrap™ and place in an exposure cassette. In a dark room with the safe light on, cover the filters with a piece of X-ray film and place the cassette with an intensifying screen at –80°C for 2 to 24 h prior to their being developed.

*Notes:*   *The film should be slightly overexposed in order to obtain a relatively even background. This will help to identify the marks made previously.*

19.  In order to locate the positive plaques, match the developed film with the original plate by placing the film underneath the plate with the help of previously made markers. This can be done by

placing a glass plate over a lamp and putting the matched film and plate on the glass plate. Turn on the light so that the positive clones can be easily identified. Make sure that the plaque-facing side exposed on the film faces down to identify the actual positive plaques. Any mismatch will cause failure in identification. Remove individual positive plugs containing phage particles from the plate using a sterile pipette with the tip cut off. Expel the plug into a microcentrifuge tube containing 1 ml of elution buffer. Allow elution to occur for 4 h at room temperature with occasional shaking.

20. Transfer the supernatant eluate into a fresh tube and add 20 ml of chloroform. Store at 4°C for up to 5 weeks.

21. Determine the pfu of the eluate as described previously. Replate the phage and repeat the screening procedure with the same isotopic probe several times until 100% of the plaques on the plate are positive.

*Notes:* *During the rescreening process, the plaque number used for one plate should be gradually reduced. In our hands, the plaque density for one 100-mm plate in the rescreening procedures is decreased from 1000 to 500, to 300, and to 100.*

22. Amplify the putative DNA clones and isolate the recombinant λ DNAs by either plating or liquid methods (see the DNA isolation section in Chapter 1). The purified DNA can be used for subcloning of DNA inserts.

a.    Troubleshooting Guide

Various symptoms and solutions are given in Table 7–4.

## Reagents Needed

### LB Medium

Prepare as described previously.

### LB Plates

Prepare as described previously.

### LB Top Agar (500 ml)

Add 4 g agar to 500 ml of LB medium and autoclave.

### Elution Buffer

10 m$M$ Tris-HCl, pH 7.5
10 m$M$ MgCl$_2$

### Phage Buffer

20 m$M$ Tris-HCl, pH 7.4
100 m$M$ NaCl
10 m$M$ MgSO$_4$

### 20X SSC Solution (1 l)

3 $M$ NaCl
0.3 $M$ Na$_3$ citrate (trisodium citric acid)
Autoclave.

## Table 7–4 Troubleshooting Guide

| Symptoms | Solutions |
| --- | --- |
| Plaques are small on one side and large on other side | This is due to uneven distribution of the top agar mixture. Make sure that the top agar mixture is spread evenly |
| Too many positive clones occur in the primary screening | A nonspecific cross-linking problem occurs. Try to use high-stringency conditions of hybridization |
| Black background on the filter | Filter was partially dried or blocking efficiency is low, or the quality of the filter is not good. Try to avoid air drying of the filter; increase the percentage of BSA and denatured salmon sperm DNA, and try to use a fresh neutral nylon membrane filter |
| Unexpected black spots on the filter | Unincorporated $^{32}$PdCTP filter was not efficiently removed. Try to use a G-50 gel column to purify the labeled DNA |
| No signal occurs at all in any of the plates used for the primary screening procedure | The pfu numbers used for each plate are too low. $2 \times 10^5$ pfu should be used in the primary screening of the library |
| Signals are weak on the filters | The efficiency of labeling is low or the X-ray film exposure time is not long enough |
| No signals occur at all on the latter rescreening of the plaque picked up from the primary screening | False positive clones were most likely picked up due to mismatch of the exposed film with the original LB plate |

### 5X SSC Solution
Dilute 20X SSC solution four times with sterile water.

### Denaturing Solution
1.5 $M$ NaCl
0.5 $M$ NaOH
Autoclave.

### Neutralizing Solution
1.5 $M$ NaCl
0.5 $M$ Tris-HCl, pH 7.4
Autoclave.

## 50X Denhardt's Solution

1% (w/v) BSA (bovine serum albumin)
1% (w/v) Ficoll (Type 400, Pharmacia)
1% (w/v) PVP (polyvinylpyrrolidone)
Dissolve well after each addition. Adjust to the final volume to 500 ml with distilled water and sterile filter. Divide the solution into 50 ml aliquots and store at –20°C. Dilute tenfold into prehybridization and hybridization buffers.

## Prehybridization Buffer

5X SSC
0.5% SDS
5X Denhardt's reagent
0.2% Denatured salmon sperm DNA

## Hybridization Buffer

5X SSC
0.5% SDS
5X Denhardt's reagent
0.2% denatured salmon sperm DNA
$[\alpha\text{-}^{32}P]$-Labeled DNA probe

## Method 2: Screening of a Genomic Library Using a Non-Isotope-Labeled Probe

1. Carry out steps 1 to 13 as described in the screening by using a $^{32}P$-labeled probe in Method 1.

2. Carry out prehybridization and hybridization as described in Method 1 except that a nonradioactive probe and appropriate buffers are used (see Reagents Needed).

3. Transfer the hybridized, washed filters into a clean dish containing buffer A (15 ml per filter) for 2 to 4 min.

4. Transfer the filters in buffer B (15 ml per filter) for 60 min.

5. Incubate the filters with an antibody solution, which is the anti-DIG-alkaline phosphatase (Boehringer Mannheim Biochemicals) diluted at 1:10,000 in buffer B, at room temperature for 40 to 60 min using 10 ml per filter.

6. Wash the filters (100 ml per filter) in buffer A for 20 min and repeat once using a fresh washing tray.

*Note:* *The used antibody solution can be stored at 4°C for up to 2 months and reused five to six times without any significant decrease in the antibody activity.*

7. Equilibrate the filters in buffer C for 1 to 4 min.

8. Detect the hybridized band(s) by one of the following two methods.

   a. Using Lumi-Phos 530 (Boehringer Mannheim Biochemicals) as a substrate:

      i. Add 0.5 ml of the Lumi-Phos 530 to the center of a clean dish, which should be prewarmed at room temperature for 1 h prior to use.

      ii. Briefly damp a filter using a forceps and completely wet the plaque-facing side of the filter and/or both sides of the filter by slowly laying the filter down on the solution several times.

    iii.  Wrap the filter with SaranWrap™ and wipe out the excess Lumi-Phos solution using a paper towel to reduce the black background.

    iv.  Place the wrapped filter in an exposure cassette with the side facing up.

    v.  Repeat steps ii, iii, and iv until all the filters are done.

    vi.  Overlay the wrapped filters with an X-ray film in a darkroom with a safe light on and allow the exposure to occur by placing the closed cassette at room temperature for 2 min to 24 h.

    vii.  Develop the film and proceed to positive clone identification.

*Notes:*   *1. Exposure for more than 4 h may produce a very black background. Based on our experience, a good hybridization and detection should generate sharp positive spots with 1.5 h of exposure. 2. The film should be slightly exposed to get a relative black background, which will help identify the marks made previously.*

   b.  Using the NBT and BCIP detection method:

    i.  Add 40 μl of NBT solution and 30 μl of BCIP solution in 10 ml buffer C for one filter. NBT and BCIP are available commercially.

    ii.  Place the filter in the mixture made in step i and put in the dark for color development at room temperature for 30 min to 1 day in order to obtain a clean background.

    iii.  Air dry the filters and proceed to identify the positive clones in original LB plates as described in steps 19 to 22 in Method 1.

## *Reagents Needed*

### *LB Medium*

        Prepare as described previously.

### *LB Plates*

        Prepare as described previously.

### *LB Top Agar (500 ml)*

        Add 4 g agar to 500 ml of LB medium and autoclave.

### *Elution Buffer*

        10 m$M$ Tris-HCl, pH 7.5
        10 m$M$ $MgCl_2$

### *Phage Buffer*

        20 m$M$ Tris-HCl, pH 7.4
        100 m$M$ NaCl
        10 m$M$ $MgSO_4$

### *20X SSC Solution (1 l)*

        3 $M$ NaCl
        0.3 $M$ $Na_3$ citrate (trisodium citric acid), pH 7.0.
        Autoclave.

## 5X SSC Solution

Dilute 20X SSC solution four times with sterile dd.H$_2$O.

## Denaturing Solution

1.5 *M* NaCl
0.5 *M* NaOH
Autoclave.

## Neutralizing Solution

1.5 *M* NaCl
0.5 *M* Tris-HCl, pH 7.4
Autoclave.

## Prehybridization Buffer

5X SSC
0.1% N-Lauroylsarcosine
0.02% Sodium dodecyl sulfate (SDS)
1% Blocking reagent
Dissolve well on a heating and stirring plate at 65°C after each addition. Add sterile water to final volume or add 50% formamide to the mixture if using the formamide method.

## Hybridization Buffer

Add DIG-dUTP-labeled probe to the appropriate volume of fresh prehybridization buffer.

## Buffer A

100 m*M* Tris-HCl
150 m*M* NaCl, pH 7.5

## Buffer B

2% (w/v) Blocking reagent (Boehringer Mannheim Biochemicals) or equivalent such as BSA, nonfat milk, or gelatin in buffer A
Dissolve well by stirring with a magnetic stir bar.

## Buffer C

100 m*M* Tris-HCl, pH 9.5
100 m*M* NaCl
50 m*M* MgCl$_2$

## NBT Solution

75 mg/ml Nitroblue tetrazolium salt in 70% (v/v) dimethylformamide

## Table 7–5  Single- and Double-Enzyme Digestions

| Components | \multicolumn{6}{c}{Tube number} |
|---|---|---|---|---|---|---|
| | 1 | 2 | 3 | 4 | 5 | 6 |
| Recombinant positive phage DNA | 20 | 20 | 20 | 20 | 20 | 20 (µg) |
| Appropriate 10X restriction enzyme buffer | 4 | 4 | 4 | 4 | 4 | 4 (µl) |
| 1 mg/ml Acetylate BSA (optional) | 4 | 4 | 4 | 4 | 4 | 4 (µl) |
| *EcoR* I (10 units/l) | 6.7 | 0 | 0 | 6.7 | 6.7 | 0 (µl) |
| *BamH* I (10 units/l) | 0 | 6.7 | 0 | 6.7 | 0 | 6.7 (µl) |
| *Xho* I (10 units/l) | 0 | 0 | 6.7 | 0 | 6.7 | 6.7 (µl) |
| Add dd.H$_2$O to a final volume of | 40 | 40 | 40 | 40 | 40 | 40 (µl) |
| Enzyme digestion | *EcoR* I | *BamH* I | *Xho* I | *EcoR* I +*BamH* I | *EcoR* I +*Xho* I | *BamH* I +*Xho* I |

### BCIP Solution

50 mg/ml 5-Bromo-4-chloro-3-indolyl phosphate (X-phosphate), in 100% dimethylformamide

### Anti-DIG-Alkaline Phosphatase

Anti-digoxigenin conjugated to alkaline phosphatase (Boehringer Mannheim Biochemicals)

## C.  Restriction Mapping of Recombinant Positive Bacteriophage DNA Containing Genomic DNA Insert of Interest

After putative clones are isolated, the next logical step is to locate the gene of interest within the insert. This will tell you how big the gene is and where it is located in the insert. It is necessary to have this information in order to subclone or directly sequence the gene. For this purpose, restriction mapping should be carried out using three to four restriction enzymes. The selection of enzymes depends on the restriction enzyme sites contained in the vectors. When LambdaGEM-11 or LambdaGEM-12 (Promega Incorporation) is used as the cloning vector, the mapping procedure is as follows:

### Procedure

1.  Set up, on ice, a series of single- and double-enzyme digestions as shown in Table 7–5.
2.  Incubate the tubes at an appropriate temperature for 2 to 3 h.
3.  After restriction-enzyme digestion, carry out electrophoresis on a 0.9% agarose gel (Figure 7.5), blot onto a nylon membrane, and perform prehybridization and hybridization using the same probe for screening of the genomic library, washing and detection as described in detail under Southern blotting in Chapter 5.

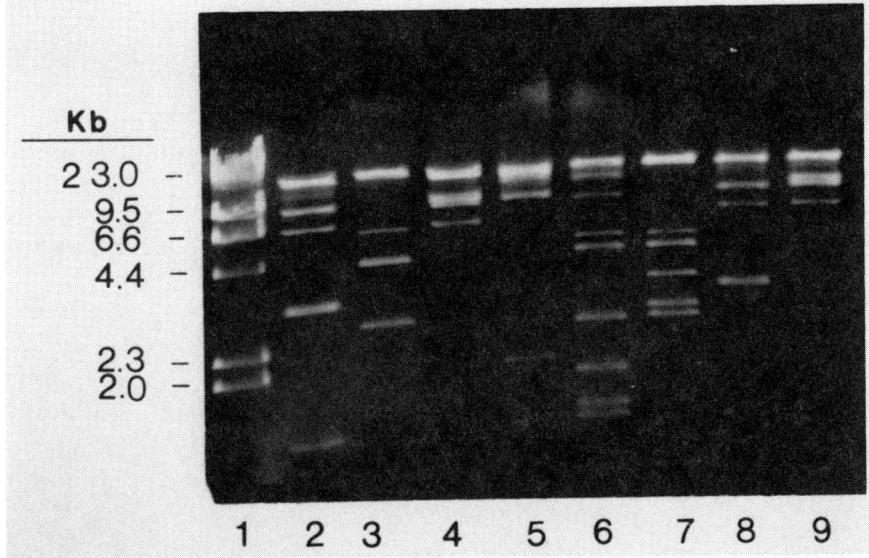

**FIGURE 7.5**
Digestion of a DNA fragment containing the gene of interest with different restriction enzymes for restriction mapping. The digested DNA was subjected to 0.9% agarose gel electrophoresis, stained with ethidium bromide, and photographed prior to being blotted onto a nylon membrane filter. Lane 1: DNA markers. Lanes 2–9: DNA digested with different restriction enzymes, including single- and double-enzyme digestions.

4.  To ensure that the correct results are obtained, we recommend that steps 1 to 3 be repeated. Both primary and repeated Southern blots should have identical hybridization patterns. An example of such results is given below:

| Restriction enzyme digestion | Bands observed on gel (Kb)* |
|---|---|
| EcoR I | 20, 9, <u>10</u>, 6, 4 |
| BamH I | 20, 9, <u>10</u>, 6, 4 |
| Xho I | 20, 9, <u>12</u>, 6, 2 |
| EcoR I + BamH I | 20, 9, <u>6</u>, <u>4</u>, 2 |
| EcorR I + Xho I | 20, 9, <u>6</u>, <u>4</u>, 2 |
| BamH I + Xho I | 20, 9, <u>6</u>, 4, <u>2</u> |

*  Hybridized bands are indicated by underlined numbers.  Doublets or triplets are possible.

5.  Identify and locate the gene in the insert according to the Southern blot hybridization results and draw a restriction map (Figure 7.6). Check out the fragments based on the map as follows:

| Restriction enzyme digestion | Bands observed on gel (Kb)* |
|---|---|
| EcoR I | 20, 9, <u>10</u>, 6, 4 |
| BamH I | 20, 9, <u>10</u>, <u>6</u>, 4 |
| Xho I | 20, 9, <u>12</u>, <u>6</u>, 2 |
| EcoR I + BamH I | 20, 9, <u>6</u>, <u>4</u>, 4, 4, 2 |
| EcorR I + Xho I | 20, 9, <u>6</u>, 6, <u>4</u>, 2, 2 |
| BamH I + Xho I | 20, 9, <u>6</u>, <u>6</u>, 4, <u>2</u>, 2 |

*  See note above.

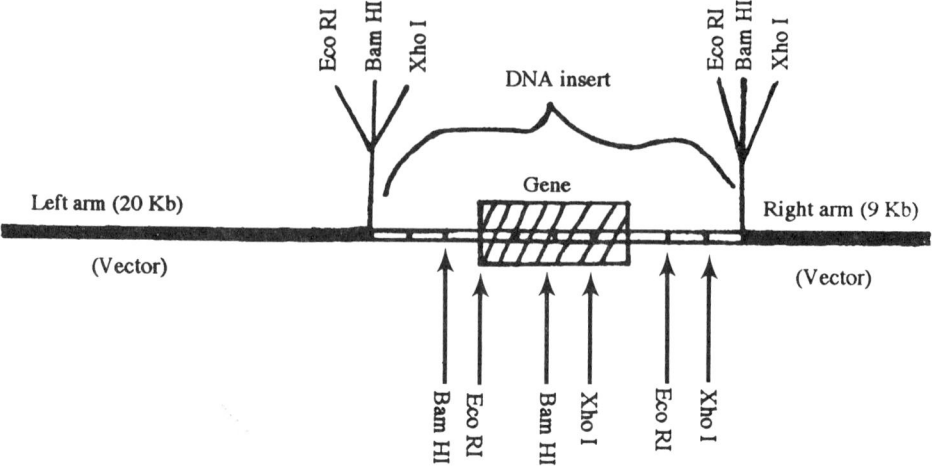

**FIGURE 7.6**

Diagram showing a restriction map and the location of a given gene of interest. The gene is 8 Kb in size and is located in the middle of the genomic DNA fragment that has been cloned in an EMBL 4 vector.

In conclusion, the gene identified is 8 Kb in size and is located in the middle of the insert (Figure 7.6).

6.  After identifying the location of the gene in the insert, carry out the purification of the gene from recombinant LambdaGEM-11 or LambdaGEM-12. Based on the map in step 5, the gene can be cut out with *EcoR I* digestion of the recombinant LambdaGEM-11 or LambdaGEM-12 followed by electrophoresis on a 0.9% low-melting-point agarose gel. When the electrophoresis is completed, a 10-Kb band containing the 8-Kb gene can be eluted out of the agarose gel (see Chapter 1). The eluted fragment will then be used for subcloning using an expressional plasmid vector for sequencing, or it can be religated with LambdaGEM-11 or LambdaGEM-12 arms for direct sequencing by the use of SP6 and/or T7 promoters.

*Notes:    The gene can be directly sequenced together with its other insert fragment using SP6 or T7 primer or both. However, it takes longer and is more expensive to sequence the entire 20-Kb insert. Even more difficult is the analysis of the sequences, which is both tedious and easily subject to errors. Therefore, we recommend that the gene be separated from any other insert sequence as much as possible.*

## D.    Subcloning of the DNA Fragment of Interest

There are at least four advantages of subcloning a DNA fragment of interest. First, the DNA fragment can be amplified up to 300-fold using a high copy number of plasmids that replicate in short-life cycle *E. coli*. Second, plasmids used for subcloning are designed to contain SP6, T7, or T3 promoters upstream from the polycloning sites, thus allowing one to prepare sense RNA or antisense RNA of the insert for analysis. Third, the merit of designing appropriate primer corresponding to SP6, T7, or T3 allows one to sequence the insert of interest on both strands in opposite directions. In addition, a known DNA sequence such as cDNA or a genomic gene, can be ligated to appropriate vectors for gene transfer and expression analysis. For these purposes, the recombinant vectors (usually plasmids) may need to be subcloned. Therefore, subcloning is currently an essential technique used in molecular biology studies. The present section describes the detailed protocol for subcloning the DNA fragment, gene, or cDNA of interest, and selection of transformants. All procedures have been tested successfully in our laboratory.

## 1.     Restriction Enzyme Digestion of Vector and DNA Insert for Subcloning

Commercial plasmids such as pGEM series and pBluescript-SK II are available for cloning. Selection of a particular plasmid vector depends on the individual investigator. Generally speaking, a standard plasmid for cloning should have the following necessary characteristics: (1) a polycloning site for the insertion of the foreign DNA of interest; (2) SP6, T7, or T3, or equivalent promoters upstream from the polylinker site located in opposite directions in order to express the DNA insert for sense RNA, or antisense RNA, or protein analysis; (3) the origin of replication for the duplication of the recombinant plasmid in the host cell; (4) a selectable marker gene such as $Amp^r$ for antibiotic selections of transformants; and (5) a selectable marker gene such as the *lac* Z gene containing the polycloning site for color (e.g., blue/white) screening of interesting bacterial colonies that contain the recombinant plasmids.

### Protocol a: Preparation of Vectors

1.    Set up, on ice, a standard single-restriction enzyme digestion as follows:
   • Plasmid DNA (10 μg)
   • 10X Appropriate restriction enzyme buffer (10 μl)
   • 1 mg/ml Acetylated BSA (optional) (10 μl)
   • Appropriate restriction enzyme (3.3 units/μg DNA)
   • Add dd.H$_2$O to a final volume of 100 μl.

*Notes:*   *(1) The restriction enzyme used for vector and insert DNA digestions should be the same in order to ensure optimal ligation. (2) For directional cloning, the plasmid and insert DNA should be digested using two different restriction enzymes. The double-enzyme digestion of DNA may be set up as a single reaction at the same time or be carried out as two single-enzyme digestions at different times. (3) For double-restriction enzyme digestions, the appropriate 10X buffer containing a higher NaCl concentration than the other buffer may be chosen for the double enzyme digestion buffer.*

Set up, on ice, the double-restriction enzyme reaction as follows:
   • Plasmid DNA (10 μg)
   • 10X Appropriate restriction enzyme buffer (10 μl)
   • 1 mg/ml Acetylated BSA (optional) (10 μl)
   • Appropriate restriction enzyme A (3.3 units/μg DNA)
   • Appropriate restriction enzyme B (3.3 units/μg DNA)
   • Add dd.H$_2$O to a final volume of 100 μl.

2.    Incubate at an appropriate temperature (e.g., 37°C) for 2 to 3 h. For single-enzyme-digested DNA, proceed to step 3. For double-enzyme-digested DNA, proceed to step 5.

*Notes:*  *To ensure that an optimal ligation to insert DNA occurs, the vector plasmid should be completely digested. The digestion efficiency can be checked by loading 1 μg of the digested DNA (10 μl) with loading buffer to a 1% agarose mini-gel. In the meantime, undigested plasmid DNA (1 μg) and standard DNA markers should be loaded in the adjacent wells. After electrophoresis, the undigested plasmid DNA may reveal multiple bands because of different levels of supercoiled plasmids. However, one band will be visible for a complete single-enzyme digestion; one major band and one tiny band (<70 bp) may be visible after digestion with two different restriction enzymes. (2) Double-restriction enzyme digestion is usually more difficult than a single-enzyme digestion since DNA may be cut by one of two enzymes due to salt concentration in the digestion buffer or other factor(s). To ensure that the vector is digested by two restriction enzymes, the digested vector should be purified and subject to religation by using T4 DNA ligase. The efficiency of digestion is then checked by loading the religated vector DNA onto a 1% agarose gel followed by electrophoresis. When the vector DNA is digested with two restriction enzymes and purified, it cannot be religated. Therefore, only one band appears, which is the linearized DNA. On the other hand, when the DNA is digested with one of two restriction enzymes and purified, it will be religated during the ligation reaction. When that occurs, multiple bands may be visible due to different levels of supercoiled plasmids. (3) After completion of the restriction enzyme digestion, calf intestinal alkaline phosphatase (CIAP) treatment should be carried out for the above single-restriction enzyme digestion. This treatment removes 5′-phosphate groups, thus preventing recircularization of the vector during ligation. Otherwise, the efficiency of ligation between vector and insert DNA is very low. For double-restriction enzyme-digested vectors, the CIAP treatment is not necessary.*

3.  Carry out the CIAP treatment by adding the following directly to the single-enzyme-digested DNA sample (90 μl).
    - 10X CIAP buffer (15 μl)
    - CIAP diluted in 10X CIAP buffer (0.01 unit/pmol ends)
    - Add dd.H$_2$O to a final volume of 150 μl.

*Notes:*  *(1) CIAP and 10X CIAP should be kept at 4°C. CIAP treatment should be set up at 0°C. (2) Calculation of the amount of ends is as follows: There is 9 μg digested DNA left after taking 1 μg of 10 μg digested DNA for checking on agarose gel. If the vector is 3.2 Kb, the amount of ends can be calculated by the formula below:*

$$\text{pmol ends} = \frac{\text{amount of DNA}}{\left(\text{base pairs} \times 660/\text{base pair}\right)} \times 2$$

$$= \frac{9}{3.2 \times 1000 \times 660} \times 2$$

$$= 4.2 \times 10^{-6} \times 2$$

$$= 8.4 \times 10^{-6} \ \mu M$$

$$8.4 \times 10^{-6} \times 10^{-6} = 8.4 \text{ pmol ends}$$

4.  Incubate at 37°C for 1 h and add 2 μl of 0.5 *M* EDTA buffer (pH 8.0) to stop the reaction.

5.  Extract with 1 volume of TE-saturated phenol/chloroform. Mix well by vortexing for 1 min and centrifuge at $11,000 \times g$ for 5 min at room temperature.

6.  Carefully transfer the top, aqueous phase to a fresh tube and add 1 volume of chloroform:isoamyl alcohol (24:1) to the supernatant. Mix well and centrifuge as in step 5.

7.  Carefully transfer the upper, aqueous phase to a fresh tube and add 0.1 volume of 3 $M$ sodium acetate buffer (pH 5.2) or 0.5 volume of 7.5 $M$ ammonium acetate to the supernatant. Briefly mix and add 2 to 2.5 volumes of chilled 100% ethanol to the supernatant. Allow to precipitate at $-70°C$ for 1 h or at $-20°C$ for 2 h.

8.  Centrifuge at $12,000 \times g$ for 10 min and carefully decant the supernatant. Briefly rinse the DNA pellet with 1 ml of 70% ethanol and dry the pellet under vacuum for 20 min. Dissolve the DNA pellet in 20 to 40 $\mu$l dd.$H_2O$. Take 4 $\mu$l of the sample to measure the concentration of the DNA at 260 nm. Store the sample at $-20°C$ until use.

*Note:*   *Adding 0.5 volume of 7.5* M *ammonium acetate to the supernatant at step 7 yields a higher amount of DNA precipitation than by adding 0.1 volume of 3* M *sodium acetate buffer (pH 5.2).*

### Protocol b: Preparation of Insert DNA

1.  Purify insert DNA from an agarose gel as described under the DNA elution section in Chapter 1.

*Note:*   *Insert DNA whose size is <4 Kb is easier for successful subcloning than DNA whose size is 4 to 12 Kb.*

2.  Carry out restriction enzyme digestion, purification, and precipitation the same as for vector DNA [see steps 1.1 to 1.8].

## Reagents Needed

### Appropriate Enzymes

10X Appropriate Restriction Enzyme Buffer
1% Agarose Mini-Gel
TE-Saturated Phenol/Chloroform
Chloroform:Isoamyl Alcohol (24:1)
3 $M$ Sodium Acetate Buffer, pH 5.2
7.5 $M$ Ammonium Acetate
Ethanol (100%, 70%)
0.5 $M$ EDTA, pH 8.0
Calf Intestinal Alkaline Phosphatase (CIAP)
TE Buffer

### 10X CIAP Buffer

0.5 $M$ Tris-HCl, pH 9.0
10 m$M$ $MgCl_2$
1 m$M$ $ZnCl_2$
10 m$M$ Spermidine

## 2.     Ligation of Plasmid Vector and Insert DNA

To achieve optimal ligation, the ratio of vector to insert DNA (1:1, 1:2, 1:3, and 3:1 molar ratios) should be optimized by using a small-scale reaction. The following reaction is standard for the ligation of a 3.2 Kb plasmid vector and a 3.0 Kb insert DNA.

### *Protocol*

1. Calculate the molar weights of vector and insert DNA:

$$1\,M \text{ Plasmid vector} = 3.2 \times 1000 \times 660 = 2.112 \times 10^6$$

$$1\,M \text{ Insert DNA} = 3 \times 1000 \times 660 = 1.98 \times 10^6$$

2. Calculate the molar ratio of vector to insert DNA using Table 7–6.
3. Set up the ligations in Table 7–7 on ice.

*Note:*     *The restriction enzyme-digested plasmid (vector) and insert DNA should be dissolved in dd.H₂O (nuclease free) at 0.5 to 1.0 µg/µl. If the DNA is less than 0.4 µg/µl, the DNA should be precipitated so as to dissolve at about 1 µg/µl.*

4. Incubate the reactions at 4°C for 12 to 24 h, or 16°C for 4 to 6 h, or at room temperature (22 to 25°C) for 1 to 2 h.

*Note:*     *After the ligations are completed at the above temperatures, the mixture can be stored at 4°C until use.*

### Table 7–6 Calculation of Molar Ratios

| Vector DNA:insert DNA | Amount of DNA (µg) | |
| :---: | :---: | :---: |
| | Vector | Insert |
| 1:1 | 1 | 0.792 |
| 1:2 | 1 | 1.584 |
| 1:3 | 1 | 2.376 |
| 3:1 | 1 | 0.264 |

### Table 7–7 Components of Ligation Reactions

| Components | Ligation reactions | | | |
| :--- | :---: | :---: | :---: | :---: |
| | 1 (1:1) | 2 (1:2) | 3 (1:3) | 4 (3:1) |
| Plasmid DNA as vector (µg) | 1 | 1 | 1 | 1 |
| Insert DNA (µg) | 0.792 | 1.584 | 2.376 | 0.244 |
| 10X Ligase buffer (µl) | 1 | 1 | 1 | 1 |
| T4 DNA ligase (Weiss units) | 4 | 4 | 4 | 4 |
| Add dd.H₂O to (µl) | 10 | 10 | 10 | 10 |

5. Check the efficiency of the ligations by 1% agarose electrophoresis. When the electrophoresis is complete, photograph the gel stained with EtBr under UV light. As compared with unligated vector or insert DNA, high-efficiency ligation should make it possible to visualize less than approximately 10% unligated vector and insert DNA by estimation of the intensity of fluorescence. Approximately 90% of the vector and insert DNA are ligated to each other and show strong band(s) with molecular weight shifts compared to the vector and insert DNA sizes. By comparing the efficiency of ligations using different molar ratios, the optimal conditions can be determined with ease. These can be used as a guide for large-scale ligation.

*Note:* *The small-scale ligation above is optional, but it is strongly recommended that it be carried out.*

6. The large-scale ligation of vector and insert DNA is based on the optimal conditions determined by small-scale ligations. For example, if one uses a 1:2 molar ratio of plasmid DNA:insert DNA as the optimal ligation condition, a large-scale ligation can be carried out as follows:
   - Plasmid DNA as vector (3 µg)
   - Insert DNA (4.75 µg)
   - 10X Ligase buffer (3 µl)
   - T4 DNA ligase (Weiss units) (15 to 50 units)
   - Add dd.$H_2O$ to a final volume of 30 µl.

   Incubate the ligation mixture at 4°C for 12 to 24 h, or at 16°C for 4 to 6 h, or at room temperature (22 to 25°C) for 1 to 2 h. Store at 4°C until use. Proceed to carrying out transformation.

## 3. Carry Out Transformation

1. Prepare the LB medium and LB plates as described previously. This should be done before ligation.
2. Prepare competent cells as follows. This should be completed before ligation.

### Protocol a: Preparation of Competent Cells for CaCl$_2$ Transformation

1. Streak the appropriate *E. coli* strain (DH5αF′ or JM109 for color screening) directly from small amount of frozen stock stored at –70°C onto the surface of an LB plate using a sterile platinum wire loop. Invert the plate and incubate in a 37°C incubator for 12 to 16 h. Bacterial colonies will become visible.

*Notes:* *(1) It is not necessary to thaw the frozen bacteria at room temperature or at 0°C. A small amount of bacteria adhering to the wire loop is sufficient for inoculation. (2) LB plates should contain thiamine-HCl for the selection of F′ in the cell, which is necessary for color screening.*

2. Inoculate a well-isolated colony from the plate at step 1 into 50 ml of LB medium supplemented with 0.5 ml of 20% maltose and 0.5 ml of 1 *M* MgSO$_4$ solution. Incubate at 37°C overnight with shaking at 160 rpm.
3. Add 0.5 ml of cells from step 2 to 100 ml of LB medium containing 1 ml of 20% maltose and 1 ml of 1 *M* MgSO$_4$ solution. Prepare four 100-ml cell cultures. Incubate at 37°C with shaking at 160 rpm. Measure the OD$_{600}$ or A$_{600}$ every 20 to 30 min until the A$_{600}$ reaches 0.45 to 0.6. It usually takes 2 to 5 h.
4. Chill the cells in ice water for 2 h at centrifuge at 3000 × *g* for 15 min at 4°C.
5. Resuspend the cells in 20 ml of ice-cold trituration buffer and dilute to 400 ml with the same buffer.

6. Incubate the cells on ice for 45 to 60 min.

7. Centrifuge at $2000 \times g$ for 10 min at 4°C and gently resuspend the cells in 40 ml of ice-cold trituration buffer.

8. Add glycerol dropwise with gentle swirling to the cell solution to final concentration of 15% (v/v). Aliquot the cells at 0.2 ml per tube, freeze on dry ice, and then store at –70°C until use.

### Protocol b: Preparation of Competent Cells for Electroporation

1. Carry out steps 1 to 4 in Protocol a.

2. Extensively wash the cells with 100 ml distilled water or low-salt buffer to reduce the ionic strength of the cell suspension.

3. Centrifuge at $2000 \times g$ for 10 min at 4°C and carefully decant the supernatant.

4. Repeat steps 2 and 3 twice.

5. Resuspend the cells in 200 ml of low-salt buffer or distilled water. Add glycerol dropwise with gentle swirling to 10% (v/v). Dispense the cell suspension into 20 µl per tube aliquots at approximately $3 \times 10^9$ cells/ml. Freeze on dry ice and then store at –70°C until use.

## 4.     Transformation Using the CaCl$_2$ Method

### Protocol

1. Thaw three aliquots of 0.2 ml of frozen CaCl$_2$-treated competent cells on ice.

2. Add 3 µl of frozen-thawed DMSO to every 0.2-ml aliquot, mix, and add recombinant plasmid DNA as shown in Table 7–8.

3. Incubate on ice for 30 min.

4. Heat shock at 42°C for 2 min and place on ice for 1 min (optional).

5. Transfer the cell suspension to sterile culture tubes and add 2 ml of LB medium containing 20 µl of 20% maltose and 20 µl of 1 $M$ MgSO$_4$ solution. Incubate at 37°C for 1 to 2 h with shaking at 140 rpm to recover the cells.

6. Add 50 to 150 µl of the culture per plate to the centers of LB plates containing 50 µg/ml ampicillin, 0.5 m$M$ IPTG, and 40 µg/ml X-Gal. Quickly spread the cells over the entire surface of the LB plates using a sterile, bent glass rod.

*Note:*   *Transformations using different amounts of DNA at step 2 should be plated at the same volume in order to determine the optimal conditions.*

### Table 7–8 Aliquot Components

| Components | Aliquot number | | |
|---|---|---|---|
|  | 1 | 2 | 3 |
| Cells (µl) | 20 | 20 | 20 |
| DMSO (µl) | 3 | 3 | 3 |
| DNA (µl) | 1 (15 ng) | 1 (100 ng) | 1 (200 ng) |

## 5.    Transformation by Electroporation

### *Protocol*

1.  Thaw three aliquots of 20 µl of frozen, non-CaCl$_2$-treated competent cells on ice. Place recombinant DNA sample on ice.

2.  Chill three disposable microelectroporation chambers (BRL) on ice.

3.  Connect the power cable to a BRL-Porator pulse control + power supply apparatus, to a BRL-Porator voltage booster, and between these two units.

4.  Set up the pulse control as follows:

    Power: charge

    Capacitance (µF): 330

    High ohm/low ohm: low ohm

    Charge rate: fast

    Set the voltage booster at 4 kV for *E. coli*.

5.  Add ice water to the chamber safe up to four fifths of the volume, and place the chamber rack in the chamber safe.

6.  Add 1, 2, and 3 µl of recombinant plasmid DNA (e.g., pGEM, pBKS II, 0.5 to 1 µl/µl) to three aliquots of 20 µl of ice-thawed cells, respectively. Gently mix and immediately place on ice until use.

*Note:*    *Avoid any air bubbles during mixing. The total volume of the transformation mixture should be <25 µl.*

7.  Use a pipette to transfer one aliquot and carefully place the mixture drop between the electrode poles in the microelectroporation chamber. Gently cover the chamber.

*Note:*    *The liquid should not be allowed to drop into the bottom of the chamber.*

8.  Gently place the chamber into the cell in the safe-rack, cover the chamber safe and turn the electric-shock pointer toward the cell containing the bacterial cell chamber.

9.  Connect the power from the voltage booster to the chamber safe, and turn on power for both the pulse control and the voltage booster units.

10. Press the charge button on the pulse control unit up to 365. When the DC voltage goes down to 345 to 350 V, turn the button from "charge" to "arm". Quickly push the tigger button for 1 s. The voltage goes down to <10 and the voltage booster should read 1.9 to 2.0 kV.

11. Turn off the power and carefully remove the chamber from the cell. The transformed cell suspension should still be between the positive and negative electrode poles.

12. Quickly transfer the transformed cell suspension into a test tube containing 2 ml LB medium without ampicillin as the cells are quite weak. Mix well and place at room temperature for not longer than 15 min.

13. Repeat steps 7 to 12 until all the samples are transformed.

14. Incubate at 37°C for 1 to 2 h with shaking at 150 rpm to recover the cells.

15. Use a sterile, bent glass rod to spread 20, 50, 100, and 200 μl of each of the three recovered transformant cells over the entire surface of LB plates containing 50 μg/ml ampicillin, 0.5 m$M$ IPTG, and 40 μg/ml X-Gal.

## 6.   Selection of Transformants Containing Recombinant Plasmids

### Protocol

1. Invert all the plates prepared at steps 3.1 to 3.6 and 4.1 to 4.15 and incubate in a 37°C incubator for 12 to 16 h until colonies are visible.

2. Chill the plates at 4°C for 1 h to maximally expose the blue colonies that may be not obvious when they are first taken from the incubator.

*Notes:*   *Blue colonies contain non-recombinant plasmid. β-Galactosidase expressed by the* lac Z *gene hydrolyzes X-Gal, forming a blue color. White colonies are supposed to bear recombinant plasmids in which foreign DNA was inserted at the polycloning site in the* lac Z *gene. The interrupted* lac Z *gene cannot express β-galactosidase activity. Therefore, the colonies are white.*

3. Inoculate individual white colonies into 5 ml of LB medium. Incubate at 37°C overnight with shaking at 160 rpm.

*Note:*   *To verify white colonies, at least 20 individual colonies should be analyzed.*

4. Isolate plasmids as described in Chapter 1.

5. Digest the plasmids with the same restriction enzyme(s) digestion as used for subcloning of the DNA insert of interest.

6. Carry out electrophoresis as previously described.

7. Photograph and verify the sizes of the vector and insert DNA as compared with unligated vector and insert DNA.

8. Elute the insert from the agarose gel (see Chapter 1) and carry out dot blotting (see Chapter 5) using unligated insert DNA as a probe. Hybridized bands indicate that the insert DNA of interest has been successfully subcloned in the plasmids.

9. Inoculate 0.5 ml of the verified white colony cells into 100 ml of LB medium containing 50 μg/ml ampicillin and 1 ml of 1 $M$ MgSO$_4$ solution. Incubate at 37°C overnight while shaking at 160 rpm.

10. Aliquot 1 ml of the culture to Eppendorf tubes and add glycerol dropwise to 15% (w/v). Freeze on dry ice and store at –70°C for further use.

11. Carry out isolation and purification of the recombinant plasmids as described in Chapter 1. Use the plasmid DNA for sequencing or RNA analysis of the insert of interest.

### Reagents Needed

#### LB (Luria-Bertaini) Medium

        Bacto-tryptone (10 g)
        Bacto-yeast extract (5 g)
        NaCl (5 g)
        Adjust the pH to 7.5 with 2 $N$ NaOH solution and autoclave. When it has cooled, store at 4°C until use. Make 4 l.

## LB Plates

Add 15 g/l unautoclaved LB medium. Adjust the pH to 7.5 with 2 $N$ NaOH solution. Autoclave. When it has cooled to 50 to 55°C, add 50 µg/ml ampicillin, 0.5 m$M$ IPTG, and 40 µg/ml X-Gal. Mix well and pour into LB plates (30 to 35 ml per plate) in a sterile laminar flow hood. Cover the plates and allow to harden for 1 h. Let the plates set at room temperature for 2 days before use. The plates can be placed at room temperature for up to 10 days, or wrapped and stored at 4°C for up to 1 month. The plates should be placed at room temperature prior to being used.

## Thiamine-HCl Plates

$Na_2HPO_4$ (6 g)
$KH_2PO_4$ (3 g)
NaCl (0.5 g)
$NH_4Cl$ (1 g)
Agarose (15 g)
Add dd.$H_2O$ to 1 liter and autoclave.
Cool to 50°C and add:
　1 $M$ $MgSO_4$ (2 ml)
　1 $M$ $CaCl_2$ (0.1 ml)
　20% Glucose (10 ml)
　1 $M$ Thiamine-HCl (1 ml)

## Trituration Buffer

0.1 $M$ $CaCl_2$
70 m$M$ $MgCl_2$
40 m$M$ Sodium acetate, pH 5.5
Freshly prepare and sterile-filter.

## 0.1 M IPTG Solution

1.2 g IPTG in 50 ml dd.$H_2O$
Filter-sterilize and store at 4°C.

## X-Gal Stock Solution

50 mg/ml Stock in N,N′-dimethylformamide

# III.　YAC Libraries

In bacteriophage λ libraries, genomic DNA is digested by a four-base "cutter", *Sau3A* I, and fragments of 14 to 23 Kb are ligated to λ DNA vectors. The flaw in this strategy is that large genes (>23 Kb) are usually divided into multiple fragments contained in different clones. The gene sequence is located in multiple overlapping fragments. Therefore, it is impossible to identify an entire gene (>23 Kb) in a single clone. To overcome the above disadvantage, the technique of constructing YAC libraries has been developed. In a YAC library, extra-large genomic DNA molecules with average size of 800 to 1000 Kb can be cloned with ease using YAC vectors. Theoretically, any size of gene can be readily isolated from a single positive clone in a standard YAC library. This is a breakthrough in gene cloning and makes it possible to completely accomplish the Human Genome Project, Plant Genome Project, or other genome projects by genome mapping, physical mapping, and chromosomal walking.

YAC vectors used in cloning are artificially designed to carry necessary sequences such as the centromere (CEN4), an autonomous replicating sequence (ARS), the telomeres (TEL), selectable marker genes, and the cloning site for insertion of the genomic DNA of interest. The YAC vectors, such as pYAC4, are propagated as bacterial plasmids, linearized by an appropriate restriction enzyme (e.g., *EcoR* I) and ligated to genomic insert DNA. The recombinant, linear YAC vectors are then transformed into an appropriate yeast host strain, such as AB1380 (*MAta ade 2–1 ura3 can1–100 lys2–1 trp1 his 5*).

## A.     Preparations of Cells or Tissues for Isolation and Purification of High-Molecular Weight DNA

Construction of a YAC library requires that the MW of genomic DNA be as high as possible. The purity and integrity of the isolated DNA is crucial for pulsed-field gel electrophoresis (PFGE) and cloning. To obtain intact genomic DNA, traditional shearing and solvent (e.g., phenol/chloroform) extraction should be avoided. Instead, cells or protoplasts are lysed *in situ* in an agarose plug, digested with appropriate restriction enzyme and checked by PFGE or field inversion gel electrophoresis (FIGE).

### *Protocol*

1. For cultured cells

   a. Wash the cells with 5 volumes of ice-cold phosphate-buffered saline (PBS) and centrifuge at $100 \times g$ for 5 min at room temperature. Carefully decant the supernatant and repeat the washing twice.

   b. Resuspend the cells in ice-cold cell suspension buffer at approximately $5 \times 10^7$ cells/ml.

2. For fresh animal tissues

   a. Slice the tissues into 1- to 2-mm pieces using a clean razor blade or equivalent and transfer to an ice-cold glass homogenizer with a tight-fitting pestle.

   b. Add 4 volumes of ice-cold PBS to the homogenizer, briefly suspend the tissue slices and homogenize for 2 to 4 min on ice.

   c. Filter the homogenate through two layers of cheesecloth to remove the cell fragments.

   d. Centrifuge the cells at $1000 \times g$ for 5 min at room temperature. Carefully decant the supernatant and wash the cells three times with 5 volumes of ice-cold PBS.

   e. Resuspend the cells in ice-cold cell suspension buffer at approximately $5 \times 10^7$ cells/ml.

3. For frozen animal tissues

   a. Grind frozen tissues in liquid nitrogen to a fine powder using a chilled mortar and pestle.

   b. Transfer the powder to a centrifuge tube containing 5 volumes of ice-cold PBS, suspend the powder into PBS and centrifuge at $1000 \times g$ for 5 min at room temperature.

   c. Decant the supernatant and wash the cells three times in 5 volumes of ice-cold PBS.

   d. Resuspend the cells at $5 \times 10^7$ cells/ml in ice-cold cell suspension buffer.

4. For fresh plant tissues

For plant tissues, we recommend that one prepare protoplasts instead of the entire cells whose cell walls are relatively difficult to digest.

a. Remove six to eight of the youngest, fully expanded leaves from plants grown in the greenhouse or under sterile conditions. Peel off the lower side of the epidermis using a pair of jeweler's forceps and place the leaf tissue in a Petri dish.

b. Surface-sterilize the leaves from greenhouse-grown plants by immersing the leaves in 5 to 10% Clorox solution (sodium hypochlorite) for 5 to 10 min followed by thorough rinsing with 40 ml sterile distilled water four to five times to remove the Clorox.

c. Add 5 to 10 ml of sterile enzyme medium, mix, and incubate in the dark at room temperature (24 to 25°C) for 18 to 20 h without shaking.

*Note:*    *At this stage, the cell walls are hydrolyzed by enzymes such as cellulase. This can be monitored by looking at the cells under a microscope. Cells with hydrolyzed cell walls only have protoplasts.*

d. Add 15 ml of washing medium and gently shake to loosen the protoplasts from undigested leaf materials.

e. Filter through a nylon mesh (50 μm pore diameter) to remove undigested materials. The protoplasts are in the filtrate solution.

f. Centrifuge the protoplasts at $1000 \times g$ for 5 min at room temperature and carefully decant the supernatant.

g. Resuspend the protoplasts in 4 ml of washing medium and centrifuge as in step f.

h. Resuspend the protoplasts in 1 ml of washing medium, add 1 ml of 18% sucrose, which will become an underlayer in the protoplast suspension and centrifuge at $120 \times g$ for 5 min at room temperature.

i. Carefully transfer the protoplasts from the interface using a wide-bore Pasteur pipette to a clean centrifuge tube and add 1 ml protoplast suspension buffer.

j. Count the protoplasts using a microscope and a haemacytometer.

k. Centrifuge at $1000 \times g$ for 5 min at room temperature and resuspend the protoplasts at approximately $5 \times 10^7$ protoplasts/ml in cell suspension buffer.

## *Reagents Needed*

### *Phosphate-Buffered Saline (PBS)*

NaCl (8 g)
KCl (0.2 g)
$Na_2HPO_4$ (1.44 g)
$KH_2PO_4$ (0.24 g)
Dissolve well after each addition in 800 ml dd.$H_2O$.
Adjust the pH to 7.4 with 2 *N* HCl and add dd.$H_2O$ to 1 liter.
Autoclave and store at room temperature.

### *Clorox® Solution*

5 to 10% (v/v) Clorox® (sodium hypochlorite solution or commercial bleach) in dd.$H_2O$

### *CPW-Salt Solution (1 l)*

$KH_2PO_4$ (27.2 g)
KI (0.16 mg)

$CuSO_4 \cdot 5H_2O$ (0.025 mg)
$KNO_3$ (0.101 g)
$MgSO_4 \cdot 7H_2O$ (0.246 g)

## Enzyme Medium

9% (w/v) Mannitol
3 m$M$ 2-(N-Morpholino)-ethane-sulphonic acid (MES)-KOH, pH 5.8
1% (w/v) Cellulase
0.2% (w/v) Macerozyme
Make up in CPW-salt solution.

## Washing Medium

3 m$M$ MES-KOH, pH 5.8
2% (w/v) KCl
Make up in CPW-salt solution. Autoclave.

## Sucrose Solution

18% (w/v) Sucrose
3 m$M$ MES-KOH, pH 5.8
Make up in CPW-salt solution. Autoclave.

## Cell Suspension Buffer

10 m$M$ Tris-HCl, pH 7.6
100 m$M$ EDTA, pH 8.0
20 m$M$ NaCl.

# B.    Isolation of High-Molecular Weight DNA

## Protocol

1.  Prepare an equal volume of 1% (w/v) low-melting temperature agarose in cell suspension buffer. Melt the agarose in a microwave and allow to cool to 42°C.

2.  Warm an equal volume of cell suspension or protoplast suspension ($5 \times 10^7$ cells/ml) to 42°C and add to the agarose gel mixture (42°C). Mix well to ensure that the cells or protoplasts are evenly dispersed throughout the agarose.

3.  Add the melted agarose-cell/protoplast mixture to an ice-cold plug former or to preformed Plexiglas molds (50 to 100 µl) or equivalent tubes using a 1-ml pipette that has the tip cut off.

4.  Allow the plugs to harden for 30 min on ice and carefully remove the plugs by pushing them out of the mold. Cut the cylindrical plugs into smaller blocks, if necessary.

5.  Place the plugs or blocks in 50 volumes of lysis buffer and incubate for 24 h at 50°C with shaking at 60 rpm. Replace the old lysis buffer with fresh lysis buffer and continue to incubate at 50°C for 24 h with shaking at 60 rpm.

*Notes:*    *Cells or protoplasts are lysed and large DNA molecules are released, which remain trapped within the agarose matrix and are protected from mechanical shearing. The degraded cell materials diffuse out of the agarose matrix. After lysis, the plugs or blocks can be stored at 4°C in fresh lysis buffer for years.*

6.  Rinse the plugs four times in $TE_{10.5}$ buffer and store at 4°C until use.

## Reagents Needed ———————————————

### *Lysis Buffer*

0.5 *M* EDTA, pH 8.0
1% (w/v) N-Laurylsarcosine
1 mg/ml Proteinase K

### *TE₁₀.₅ Buffer*

10 m*M* Tris-HCl, pH 7.8
5 m*M* EDTA, pH 8.0

## C.  Isolation of Intact Yeast DNA

### *Protocol* ——————————————————————

1.  Harvest yeast cells from a liquid culture by centrifugation at $1000 \times g$ for 5 min at 4°C and decant the supernatant.
2.  Wash the cell pellet by resuspending the cells in 5 volumes of dd.$H_2O$ and centrifuge as in step 1. Repeat washing once.
3.  Resuspend the cells in 50 m*M* EDTA buffer (pH 8.0) at approximately $4 \times 10^9$ cells/ml on ice.
4.  Prepare an equal volume of 1%(w/v) low-melting temperature agarose gel in dd.$H_2O$. Melt in a microwave and allow to cool to 42°C.
5.  Warm an equal volume of yeast cell suspension to 42°C and add to an equal volume of agarose gel mixture (42°C). Mix well and pour the mixture into a plug mold on ice. Allow to harden at 0°C for 30 min.
6.  Carefully transfer the plugs into 10 volumes of SCEM buffer containing 1 unit/ml of Zymolyase 20-T, and incubate at 37°C for 5 h.
7.  Replace SCEM buffer with 10 volumes of DLS buffer and incubate at 50°C for 3 h. Replace the old DLS buffer with fresh DLS buffer and incubate for another 3 h.
8.  Rinse the plugs four times with 4 volumes of $TE_{10.5}$ buffer and store at 4°C until use.

## Reagents Needed ———————————————

### *SCEM Buffer*

1 *M* Sorbitol
10 m*M* EDTA, pH 8.0
100 m*M* Sodium acetate, pH 5.8
Autoclave, cool to 30°C, and add 30 m*M* 2-mercaptoethanol.

### *DLS Buffer*

1% (w/v) dodecyl lithium sulphate (DLS)
50 m*M* NaCl
10 m*M* Tris-HCl, pH 7.8

## $TE_{10.5}$ Buffer

> 10 m$M$ Tris-HCl, pH 7.8
> 5 m$M$ EDTA, pH 8.0

# D.    Isolation of Yeast DNA for PCR Screening

## Protocol

1.  Carry out steps 1 to 8 in Section C.
2.  Dilute the plugs with 10 volumes of dd.H$_2$O and boil for 5 min and quickly centrifuge at 10,000 × $g$ for 4 min at room temperature.
3.  Transfer the supernatant to a fresh tube, measure the DNA concentration (about 2 to 3 ng/μl) and have it ready for PCR screening.

# E.    Restriction Enzyme Digestion of DNA in Agarose

The DNA purified in agarose plugs or blocks is almost intact and should be partially digested with a rare cutting enzyme used for YAC cloning or be completely digested with an appropriate enzyme for PFGE analysis.

## Protocol

1.  Complete the restriction digestion for PFGE:
    a.  Incubate agarose plugs in 50 volumes of TE$_{10.1}$ buffer (pH 7.6) at room temperature for 30 min.
    b.  Transfer the plugs to individual microcentrifuge tubes and add 10 volumes of appropriate 1X restriction enzyme buffer to each tube. Incubate the tubes for 30 min at 4°C.
    c.  Remove the buffer and add 2 volumes of the same fresh 1X restriction enzyme buffer. Add 40 to 50 units of the appropriate restriction enzyme (e.g., *EcoR* I) to each tube and incubate at the optimal temperature for the enzyme overnight.
    d.  Soak the plugs in 50 volumes of cold TE$_{10.1}$ buffer (pH 7.6) at 4°C for 1 h in order to diffuse out any salt in the restriction buffer from the plugs. The plugs can be individually loaded onto PFGE.
2.  Partial *EcoR* I restriction enzyme digestion of genomic DNA for YAC cloning:
    a.  Rinse plugs containing DNA three times in 50 volumes of 1X restriction enzyme buffer lacking Mg$^{2+}$.
    b.  Remove the buffer and add one volume of 1X restriction buffer lacking Mg$^{2+}$ with 4 units/μg of *EcoR* I at 4°C for 1 h.
    c.  Add Mg$^{2+}$ from a stock solution of 100 m$M$ MgCl$_2$ to the desired concentration to initiate *EcoR* I digestion.
    d.  Immediately incubate at 37°C for 1 h.
    e.  Stop the reaction by removing the restriction buffer and adding 10 volumes of cold TE$_{10.5}$ buffer. Store at 4°C until use.

## Reagents Needed

### 10X EcoR I Buffer

> 100 m$M$ NaCl
> 10 m$M$ Tris-HCl, pH 7.9

### TE$_{10.5}$ Buffer

> 10 m$M$ Tris-HCl, pH 7.9
> 5 m$M$ EDTA, pH 7.8

## F.  Preparation of YAC Vectors for Cloning

YAC4 is the most widely used yeast artificial chromosome vector that is propagated as bacterial plasmids pYAC4 in *E. coli.*

### Protocol

1.  Prepare an *E. coli* culture containing pYAC4. The procedure for incubation of *E. coli* is described in Chapter 1.

2.  Linearize pYAC4 with the restriction enzyme *BamH* I in order to release a HIS3 spacer fragment between the telomeres. The *BamH* I site is then dephosphorylated to prevent ligation between the telomeres and the HIS3 spacer. The procedures for *BamH* I digestion and dephosphorylation with CIAP are described in the section on subcloning of insert DNA in a plasmid vector.

3.  Extract the linearized and dephosphorylated pYAC4 with phenol/chloroform, chloroform:isoamyl alcohol (24:1), precipitate in ethanol, and dissolve the DNA in dd.H$_2$O as described for plasmid DNA isolation in Chapter 1.

4.  Carry out restriction enzyme digestion with *EcoR* I to open the cloning site in the intron of the *SUP*4tRNA gene as described previously.

5.  Extract, precipitate, and resuspend the DNA in dd.H$_2$O as in step 3. Store at –20°C until use.

## G.  Ligation of Partially Digested Genomic DNA Insert to pYAC4 Vector

### Protocol

1.  Briefly rinse the agarose plug containing partially digested genomic DNA twice with 20 volumes of dd.H$_2$O followed by 10 volumes of 1X ligation buffer.

2.  Discard the buffer and add the prepared pYAC4 vector to the plug at a vector:insert DNA molar ratio of 40:1. The mass of vector is approximately equal to the mass of insert.

3.  Melt the plugs in a 68°C water bath for 5 min and transfer to 37°C.

4.  Preheat 2X ligation buffer containing 4000 units/ml DNA T4 to 37°C for 2 min and add 1 volume of the buffer to the gel mixture prepared in step 3. Gently mix and allow the ligation reaction to incubate at 37°C for 2 to 3 h.

5.  Transfer the reaction mixture to room temperature and continue to incubate the reaction overnight.

*Note:*   *Longer incubation at room temperature is acceptable.*

*Reagents Needed* ———————————————————————————————————

*10X Ligase Buffer*

> 300 m$M$ Tris-HCl, pH 7.8
> 100 m$M$ $MgCl_2$
> 100 m$M$ DTT
> 10 m$M$ ATP

## H.  Size Fractionation of DNA by CHEF Gel or Other PFGE

It is important that the ligated reaction be size-fractionated prior to transformation. In 1% (w/v) low-melting point agarose CHEF gel, apply switching conditions to retain fragment above a particular size in a compression zone. By use of switching times of 15 s on the CHEF apparatus, DNA fragments <300 Kb are allowed to migrate as a function of their sizes. However, fragments >300 Kb migrate more slowly in a compression zone without resolution. Electrophoresis is carried out using 0.5X TBE at 10°C.

*Notes:*   *After electrophoresis and staining, multiple DNA bands should be visible. They are, from the bottom to the top, HIS3 spacer fragment, unligated left or right YAC4 arms, ligated right to right arms, ligated left to right arms, ligated left to left arms, <300 Kb zone, and 300 to 1500 Kb compression zone from which the ligated DNA can be recovered.*

## I.  Agarase Hydrolysis of Agarose

Agarase treatment can be directly carried out inside the agarose matrix to hydrolyze agarose and to release the recombinant YAC/DNA insert. The solution containing the ligated YAC/DNA and oligosaccharide can be directly transformed into yeast spheroplasts without purification.

*Protocol* —————————————————————————————————————————————

1.  Equilibrate the agarose slice or block in 30 m$M$ NaCl in dd.H$_2$O.
2.  Discard the solution and place in a 68°C water bath until the agarose has melted.
3.  Transfer to 37°C and add agarase (40 to 80 units per gram of agarose) to the melted agarose mixture.
4.  Incubate the reaction mixture at 37°C for 2 to 3 h. Store at room temperature or 4°C until use for transformation.

## J.  Preparation of Spheroplasts for Transformation

*Protocol* —————————————————————————————————————————————

1.  Inoculate a single AB1380 colony (yeast host) into 200 ml of YPD medium and incubate at 30°C until the culture has reached mid-log growth phase.
2.  Harvest the cells by centrifuging at 900 × $g$ for 5 min at room temperature.

3.  Discard the supernatant and resuspend the cells in 20 ml dd.H$_2$O. Centrifuge as in step 2. Repeat rinsing once.

4.  Resuspend the cells in 10 ml of 1 *M* sorbitol solution, count the cells under the microscope in the haemocytometer, and centrifuge as in step 2.

5.  Resuspend the cells in 5 ml of SCEM and add Zymolyase 20-T (5 to 20 units/1.5 × 10$^9$ cells). Incubate at 30°C for 15 min with gentle shaking at 60 rpm.

6.  Centrifuge at 500 × *g* and wash the pellet once in 10 ml of 1 *M* sorbitol solution.

7.  Centrifuge as in step 6 and resuspend the cells in YPD medium containing 1 *M* sorbitol. Allow the cells to recover for 30 min at room temperature.

8.  Add 5 ml of STC and centrifuge as in step 6. Wash the cells in 10 ml of STC buffer, centrifuge and resuspend the cells in 5 ml of STC buffer.

9.  Check the spheroplasts with a phase-contrast microscope. A good preparation should have <5% of lysed cells in the STC buffer, but 100% lysed cells in water added to the slide. The cells at this point are stable and ready for transformation.

## *Reagents Needed*

### *YPD Medium*

> 1% (w/v) Yeast extract (Difco)
> 2% (w/v) Bactopeptone (Difco)
> 2% (w/v) D-Glucose
> 1.5 to 2% (w/v) Bactoagar (Difco)
> Adjust the pH to 5.8. Autoclave.

### *SCEM Buffer*

> 1 *M* Sorbitol
> 10 m*M* EDTA, pH 8.0
> 100 m*M* Sodium citrate, pH 5.8
> Autoclave. Cool to 35°C and add 30 m*M* 2-mercaptoethanol.

### *1 M Sorbitol Solution*

> 1 *M* sorbitol in dd.H$_2$O
> Autoclave or sterile-filter.

### *STC Buffer*

> 1 *M* Sorbitol
> 10 m*M* Tris-HCl, pH 8.0
> 10 m*M* CaCl$_2$
> Autoclave.

### *Zymolyase 20-Ton Stock Solution*

> 1000 units/ml in dd.H$_2$O
> Filter-sterilize and store at 4°C.

### *Agarase Stock Solution*

> 2000 units/ml in dd.H$_2$O with 50% glycerol
> Store at −20°C.

## K.     Transformation of Spheroplasts with Recombinant YAC/DNA Insert (Prepared at the end of Section I.)

### *Protocol*

1. Add 0.5 to 1.0 volume of 2 $M$ sorbitol solution to the size-fractionated, agarase-treated liquid mixture (Section I) and aliquot into 10-ml tubes (15 μl per aliquot).

2. Add 0.1 ml of the cell suspension (Section J) to each of four aliquots, gently mix, and incubate at room temperature for 15 min.

3. Add 1 ml of PEG solution to each tube and incubate at room temperature for 15 min.

4. Centrifuge at 500 × $g$ for 10 min at room temperature and carefully discard the supernatant.

5. Resuspend the cell pellet in 0.15 ml of SOS per tube and incubate at 30°C for 45 to 60 min.

6. Add 3 ml of TOP lacking uracil, which is prewarmed to 40°C, to each tube and transfer to Petri dishes containing SORB without uracil.

7. Incubate at 30°C for 3 to 4 days until transformants appear.

*Notes:*     *YAC cloning in yeast has a low transformation efficiency, which is usually 3 × 10³ YAC transformants per microgram DNA. This low transformation efficiency is due to the large size of the transforming DNA as well as the particular morphology of yeast.*

### *Reagents Needed*

### *PEG Buffer*

> 20% (w/v) PEG 8000
> 10 m$M$ Tris-HCl, pH 8.0
> 10 m$M$ CaCl$_2$
> Filter-sterilize.

### *SOS Buffer*

> 1 $M$ Sorbitol
> 6.5 m$M$ CaCl$_2$
> 0.25% (w/v) Yeast extract (Difco)
> 0.5% (w/v) Bactopeptone (Difco)
> 20 μg/ml Uracil and tryptophan
> Adjust the pH to 5.8. Filter-sterilize.

### *2 M Sorbitol Solution*

> 2 $M$ Sorbitol in dd.H$_2$O
> Autoclave.

### *SORB Buffer*

> 0.9 $M$ Sorbitol
> 3% (w/v) D-Glucose
> 0.67% (w/v) Yeast Nitrogen Base lacking amino acids
> 1.5 to 2% (w/v) Bactoagar

Adjust the pH to 5.8. Autoclave.
Cool to 37°C and add the amino acids required.

### TOP

1 *M* Sorbitol
2% (w/v) D-Glucose
0.67% (w/v) Yeast Nitrogen Base lacking amino acids
1% (w/v) Bactoagar
Adjust the pH to 5.8. Autoclave.
Cool to 37°C and add the amino acids required.

## L.   Verification of YAC Transformants

### Protocol

1.  Transfer individual, primary transformants onto SD medium without both uracil and tryptophan.

2.  Incubate at 30°C with shaking. Only YAC positive clones can grow on SD medium because of the expression of the tryptophan gene and because they have the red phenotype feature due to the interrupted *SUP*4tRNA gene.

3.  Verify the YAC clones by PFGE and by Southern blot hybridization using genomic DNA insert fragments as probe(s).

## M.   Amplification and Storage of the YAC Library

### Protocol

1.  Triplicate storage of arrayed YAC libraries used for large-scale physical mapping is carried out as follows:
    a.  Incubate single colonies in 0.6 ml of YPD medium in Micronics racks (96 × 1 ml) at 30°C with agitation for 36 to 40 h. Incubate the cultures in SD medium for colony filters and PCR pools.
    b.  Add 0.2 ml of 80% glycerol to each culture and mix well.
    c.  Transfer 0.2 ml of the cell suspension to each of three microtitre plates. One is used as the master library, the other two are used as working libraries. Wrap the plates with SaranWrap™ and quickly store at −80°C.
    d.  Cap the Micronics and store at −80°C.

2.  Nonorganized storage of a pooled YAC library used for identifying one or a few of YAC clones of interest is carried out as follows:
    a.  Scrape off and pool approximately 500 colonies grown on SD medium lacking uracil and tryptophan. Incubate these cells in 500 ml of YPD medium at 30°C for 6 to 10 h with agitation.
    b.  Remove about $7 \times 10^8$ yeast cells in duplicate to make approximately 20 µg total DNA in 2 ml for 500 to 1000 PCR reactions.
    c.  Centrifuge the remaining part of the culture at $900 \times g$ for 10 min at room temperature.
    d.  Carefully discard the supernatant and resuspend the cell pellet in 20% (v/v) glycerol in YPD medium and store the pooled yeast in aliquots at −80°C.
    e.  Incubate approximately 2500 colonies from an aliquot on SD medium lacking uracil at 30°C with agitation for 2 days.

f. Make 5 replicates on nylon filters. Two of the colony filters are stored on 3MM Whatman paper soaked with 20% (v/v) glycerol at –80°C. Three colony filters are to be used for colony screening.

## Reagents Needed

### YPD Medium

### SD Medium

### Amino Acids

| AAs (Sigma) TRP⁻ | URA⁻ (600 mg/l) | URA⁻ (520 mg/l) |
|---|---|---|
| Ade A 9795 | 1 | 1 |
| Arg A 3909 | 4 | — |
| His H 9511 | 2 | 2 |
| Iso I 7383 | 6 | 6 |
| Leu L 1512 | 6 | 6 |
| Llys L 1262 | 5 | 5 |
| Met M 2893 | 2 | 2 |
| Phe P 5030 | 5 | 5 |
| Thr T 1645 | 20 | 20 |
| Trp T 0271 | 4 | — |
| Tyr T 1020 | 5 | 5 |
| **Total** | **60** | **52** |

## N.    Screening of a YAC Library

### Protocol

1. PCR screening of pooled YACs for exclusion of a large part of the library is carried out as follows:

   a. Incubate superpools of 1920 colonies in 20 of 96-well microplates at 30°C for 36 h.

   b. Isolate DNA from yeast cells as described previously.

   c. Carry out PCR reactions using approximately 20 to 40 ng DNA from a yeast pool in a final volume of 20 μl overlayered with a drop of light mineral oil. The PCR procedure is described in detail in Chapter 9. The PCR cycles depend on the primer and the size of the PCR products. They need to be optimized.

   d. Analyze the PCR products on 1.0 to 1.5% (w/v) agarose gels or Seckem gels. Positive pools should have the expected PCR band as compared to the positive control lane and the marker lane.

   e. Repeat steps a to d by screening 96 colonies included in the positive pool.

   f. Store the positive colony pool in 20% (v/v) glycerol at –80°C.

2. Colony screening on a nylon membrane is carried out as follows:

   a. Thaw the glycerol stocks from the positive pool of the PCR screening and incubate in 0.15 ml SD medium without uracil in microtitre plates at 30°C for 3 days.

   b. Inoculate a charged nylon membrane with 96 or 384 YAC clones from cultures at step a.

   c. Place the inoculated membranes on SD medium without uracil and trytophan on 22.5 × 22.5 cm plates and incubate at 30°C for 3 days or until colonies are approximately 2 mm in diameter.

*Note:*   *Avoid any air bubbles under the membranes in these procedures.*

   d. Individually transfer the membranes onto the 22 × 22 cm 3MM Whatman paper plates, saturated in SCEM containing 0.5 unit/ml Zymolyase 20-T for at least 30 min. Seal the plates with parafilm™ and incubate at 30°C overnight.

   e. Incubate the membrane at room temperature on 3MM Whatman paper saturated as follows:

| Saturated with | Incubation time |
|---|---|
| 10% (w/v) SDS | 5 min |
| 0.5 *N* NaOH | 10 min |
| Transfer onto dry 3MM Whatman paper | 5 min |
| 0.2 *M* Tris-HCl, pH 7.5, 2X SSC | $3 \times 5$ min |

   f. Air dry the membranes for 2 h or under vacuum for 1 to 2 h at 80°C. Store the membrane in aluminum foil or in prehybridization buffer until use.

   g. Prehybridize the membranes at 65°C for 3 to 6 h in 20 ml per filter prehybridization solution with 7% (w/v) PEG 8000, 10% (w/v) SDS, and 100 µg/ml of sonicated denatured salmon sperm DNA.

   h. Hybridize the membrane at 65°C using fresh prehybridization buffer with at least $3 \times 10^5$ cpm/ml probe.

   i. Carry out washing and exposing the membranes as described for Southern blotting in Chapter 5. Positive clone(s) should be visualized as black spot(s).

   j. Store individual positive clones in 20% (v/v) glycerol at −80°C.

3. Verification of positive YAC clones is carried out as follows:

   a. Streak out individual positive clones from the glycerol stock on SD medium lacking uracil and incubate at 30°C for 3 days.

   b. Inoculate individual colonies (usually 2 to 4) in 5 ml of YPD medium and incubate at 30°C overnight with agitation.

   c. Harvest the cells and extract the DNA as previously described.

   d. Carry out PFGE, using 2 to 4 of the agarose plugs prepared, followed by Southern blot hybridization.

   e. Check the insert size of the genomic DNA in the positive YAC clones.

   f. Prepare 20% (v/v) glycerol stocks of the putative clones and store at −80°C. In the future, the stocks can be used for genomic and physical mapping.

# References

1. **Sambrook, J., Fritsch, E. F., and Maniatis, T.,** *Molecular Cloning: A Laboratory Manual,* 2nd ed., Cold Spring Harbor Press, Cold Spring Harbor, NY, 1989.

2. **Tonegawa, S., Brock, C., Hozumi, N., and Schuller, R.,** Cloning of an immunoglobin variable region gene from mouse embryo, *Proc. Natl. Acad. Sci. U.S.A.,* 74, 3518, 1977.

3. **Saiki, R. K., Gelfand, D. H., Stoffel, S., Scharf, S., Higuchi, R., Horn, G. T., Mullis, K. B., and Erlich, H. A.,** Primer-directed enzymatic amplification of DNA with a thermostable DNA polymerase, *Science,* 239, 487, 1988.

4. **Wu, L.-L., Song, I., Kim, D., and Kaufman, P. B.,** Molecular basis of the increase in invertase activity elicited by gravistimulation of oat-shoot pulvini, *J. Plant Physiol.,* 142, 179, 1993.

5. **Erlich, H. A.,** *PCR Technology: Principles and Applications for DNA Amplification,* Stockton Press, New York, 1989.

6. **Innis, M. A., Gelfand, D. H., Sninsky, J. J., and White, T. J.,** *PCR Protocols: A Guide to Methods and Applications,* Academic Press, New York, 1989.

7. **Jones, D. H. and Winistorfer, S. C.,** A method for the amplification of unknown flanking DNA: targeted inverted repeat amplification, *BioTechniques,* 15, 894, 1993.

8. **Sukharev, S. I., Blount, P., Martinac, B., Blattner, F. R., and Kung, C.,** A large-conductance mechanosensitive channel in *E. coli* encoded by *mscL* gene, *Nature,* 368, 265, 1994.

# Chapter 8

# DNA Sequencing

## Contents

DNA sequencing is a necessary technique for almost all molecular biology studies including DNA cloning, characterization, mutagenesis, DNA recombination, and regulation of gene expression.[1-3] There are several well-established methods for nucleic acid sequencing. The present chapter describes in detail the protocols for DNA sequencing by dideoxynucleotides chain termination,[2,3] direct sequencing by PCR,[1,4,5] and DNA sequencing following unidirectional deletions[5,6] with modifications. These methods are routinely used in our laboratory.

# I.  DNA Sequencing by Dideoxynucleotides Chain Termination

Since Sanger et al.[2] developed the dideoxynucleotides chain termination method of DNA sequencing, it has been modified and well established by the use of a superior enzyme and superior DNA cloning vectors. The general principles for this method include:

1.  A synthesized oligonucleotide primer aneals to the 3′ end of the DNA template to be sequenced.

2.  A DNA polymerase catalyzes *in vitro* the synthesis (5′ → 3′) of a new DNA strand that is complementary to the template starting from the primer site using deoxynucleoside 5′-triphosphates (dNTPs); one of the dNTPs is $\alpha$-$^{35}$SdATP or $\alpha$-$^{35}$SdCTP.

3.  After the synthesis reaction of the new strand DNA has been carried out for a given time period, it is terminated by the incorporation of a nucleotide analog that is an appropriate 2′,3′-dideoxynucleotide 5′-triphosphate (ddNTP).

All four ddNTPs lack the 3′-OH group, which is required for DNA chain elongation. Based on the nucleotide bases of the DNA template, one of the four ddNTPs in each reaction is used and the enzyme-catalyzed polymerization will then be terminated at each site where the ddNTP is incorporated, generating a population of chains with different sizes. Therefore, by setting up four separate reactions, each with a different ddNTP, complete nucleotide sequence information for the DNA strand will be revealed.

## A.     General Considerations and Strategies

**DNA template** — The purity and integrity of the DNA to be sequenced is essential for obtaining an accurate and complete nucleotide sequence. The DNA template can be single-stranded DNA from M13 cloning vectors, double-stranded plasmid DNA, and double-stranded bacteriophage DNA with appropriate pretreatment.

**DNA polymerase** — We strongly recommend that Sequenase Version 2.0 DNA polymerase (United States Biochemical, or USB) be used for DNA sequencing, which is routinely used in our laboratory with successful results. Sequenase Version 2.0 DNA polymerase is a superior enzyme that is genetically modified from wild-type T7 DNA polymerase. It has no

$3' \rightarrow 5'$ exonuclease activity and the properties of high processivity and high purity. It is fast and simple, and efficiently uses nucleotide analogs for sequencing (e.g, ddNTPs, alpha-thio dNTPs, dITP). Also, it has less radioactivity background as compared with that observed by using AMV reverse transcriptase or the large fragment of *E. coli* DNA polymerase I (Klenow enzyme).

**Primers** — For sequencing the region of the DNA template close to the cloning site of the vector, the primers are complementary to the vector DNA strand(s) and are commercially available from USB, Promega, CLONtech, and other companies. For the sequence beyond 250 to 500 nucleotides from the cloning site, new primers should be designed based on the nucleotide sequence information obtained from the last sequencing of the DNA template. The designed primers, also called extending primers, can then be synthesized with a DNA synthesizer.

**Radioactively labeled dNTP** — $^{32}$P-labeled dNTP (dATP or dCTP) has a high energy level and a short half-life (usually 14 days). A major disadvantage of using $^{32}$P is that it gives diffuse bands on autoradiographic X-ray film, limiting readible information of DNA sequence. In contrast, [$^{35}$S]dATP has a lower energy level and a longer half-life (usually 84 to 90 days). It greatly improves autoradiographic resolution. Therefore, [$^{35}$S]dATP is recommended for DNA sequencing.

**Compressions** — Compressions represent a common problem in DNA sequencing primarily due to dG- and dC-rich regions, which cannot be fully denatured during electrophoresis. This usually causes interruption of the normal pattern of migration of DNA fragments. The bands are usually spaced closer than usual (compressed together) or occur further apart than usual, resulting in a significant loss of sequence information. In order to solve this problem, we use dITP or 7-deaza-dGTP (USB) to replace the nucleotide dGTP, which forms a weaker secondary structure that can be readily denatured during electrophoresis. We found with this modification that bands are sharper and compressions were eliminated by the use of dITP. In some cases, some bands appear to be weak using both dGTP and dITP. This limition can be eliminated by using pyrophosphatase (USB) in the presence of $Mg^{2+}$, $Mn^{2+}$, or both. The manganese can improve band uniformity and sequence information close to the priming site.

## B.    Protocols

The following protocols are adapted from Sanger et al.[2] and Sequenase Version 2.0 (USB) with some modifications.

### Protocol 1: Preparation of DNA Templates to be Sequenced

#### a.    Preparation of Single-Stranded Template DNA

Single-stranded DNA (ssDNA) used as a template usually reveals excellent nucleotide sequencing results. The following protocol works well for purification of ssDNA from plasmid-phage (phagemid) vectors. Phagemid vectors include M13mp9, M13mp12, M13mp13, M13mp18, and M13mp19.

1. Plate out the transformants of an appropriate *E.coli* strain that have the putative phagemids containing the DNA insert of interest on an LB plate. Invert the plate and incubate at 37°C in order to obtain single colonies.

2. Inoculate a single colony in 5 ml of LB medium containing 50 µg/ml ampicillin and 12 µg/ml tetracycline or appropriate antibiotics. Incubate at 37°C overnight with shaking at 150 rpm.

3.  Add 0.3 ml of the overnight culture to 3 ml of superbroth in a 50-ml conical tube. Incubate at 37°C with shaking at 150 rpm for 2 to 3 h.

4.  Add 8 μl of helper phage R408 (pfu = 1 × 10$^{11}$, available from Stratagene) to the culture at step 3 and continue to incubate for 8 to 10 h.

5.  Transfer 1.5 ml of the culture to each of two microcentrifuge tubes and centrifuge at 11,000 × $g$ for 2 min.

6.  Transfer 1.2 ml of the supernatant from each tube into a fresh tube and add 0.3 ml of PEG precipitation buffer containing 3.5 $M$ ammonium acetate buffer (pH 7.5) and 20% (w/v) PEG to the supernatant. Vortex for 1 min and leave at room temperature for 20 min.

7.  Centrifuge at 12,000 × $g$ for 15 min and decant the supernatant completely.

8.  Resuspend the PEG pellet in 0.3 ml of TE buffer (pH 8.0) and extract it with 1 volume of TE-saturated phenol:chloroform:isoamyl alcohol (25:24:1). Mix by vortexing and centrifuge at 11,000 × $g$ for 5 min.

9.  Transfer the top, aqueous phase to a fresh tube and repeat extraction as in step 8 twice.

10. Precipitate the single-stranded DNA by adding 0.5 volume of 7.5 $M$ ammonium acetate (pH 7.5) and 2.5 volumes of chilled 100% ethanol to the supernatant. Place at –70°C for 30 min.

11. Centrifuge at 12,000 × $g$ for 20 min at 4°C, decant the supernatant and briefly rinse the DNA pellet with 1 ml of 70% ethanol. Dry the pellet under vacuum for 30 min and dissolve the DNA in 10 μl of TE buffer (pH 7.6). Combine the two DNA samples into one tube and take 2 to 4 μl to measure the concentration of the single-stranded DNA at 260 and 280 nm. Store the sample at –20°C until use.

## Reagents Needed

### Superbroth Medium

> Bacto-tryptone (12 g)
> Bacto-yeast extract (24 g)
> 0.4% Glycerol (v/v)
> Dissolve in a total volume of 900 ml in dd.H$_2$O and autoclave. Cool to about 50°C and add 100 ml of phosphate buffer containing 170 m$M$ KH$_2$PO$_4$ and 720 m$M$ K$_2$HPO$_4$. Autoclave again.

### PEG Precipitation Buffer

> 3.5 $M$ Ammonium acetate, pH 7.5
> 20% (w/v) PEG, 8000

b.   Preparation of Double-Stranded Plasmid DNA

Double-stranded plasmids containing DNA inserts of interest can be directly sequenced. Both strands of the DNA insert can be simultaneously sequenced at opposite directions using two different primers, which are annealed to the appropriate sites at their 3′ ends, in two separate sequencing reactions. This simultaneous sequencing method greatly speeds up the sequencing procedure; it is especially useful for sequencing large DNA templates. Double-stranded plasmids can be purified using the CsCl gradient method or regular large-scale preparation methods. We have found that the CsCl gradient method reveals excellent results, but it is time-consuming, and relatively expensive. The large-scale preparation method, on the other hand, is simple, cheaper, and works equally well in our laboratory if handled properly. The detailed protocols for isolation and purification of plasmid DNA are described in Chapter 1.

c.     Purification of Double-Stranded Lambda DNA

Double-stranded bacteriophage and other linear double-stranded DNA templates can be directly sequenced without subcloning into plasmid vectors. However, the double-stranded DNA should be treated with T7 Gene 6 Exonuclease (USB) in order to generate a single-stranded DNA template prior to doing the sequencing reaction. The detailed methods for purification of lambda DNA are described in Chapter 1.

## *Protocol 2: Sequencing Reactions*

### Method A: Sequencing of Double-Stranded Plasmid DNA

1. Denature the double-stranded DNA as follows:

   a. Denature plasmid DNA by the alkaline-denaturation method in order to obtain good results. Transfer an appropriate volume of the purified plasmid sample (approximate 1 μg/μl) to a microcentrifuge tube and add 1 volume of freshly prepared denaturation solution containing 0.4 *M* NaOH and 0.4 m*M* EDTA (pH 8.0) to the sample. Incubate at 37°C for 30 to 40 min.

*Notes:*   *The amount of DNA to be denatured should be in excess to the amount of DNA to be sequenced. For example, if using 3 to 5 μg DNA for one primer sequencing reaction, the amount of DNA for two primer sequencing reactions for both strands at opposite directions will be 6 to 7 μg. If the yield of DNA following denaturation, neutralization, and precipitation is 80%, the amount of DNA used for denaturation should be 7.2 to 8.4 μg. We routinely use the double amount of DNA for denaturation and measure the DNA concentration after it is precipitated. This ensures that one has a sufficient amount of DNA for sequencing reactions.*

   b. Add 0.1 volume of 3 *M* sodium acetate buffer (pH 5.2) to the denatured sample to neutralize the mixture.

   c. Add 2 to 4 volumes of 100% ethanol to the mixture and allow precipitation to occur at –70°C for 20 min.

   d. Centrifuge at 12,000 × *g* for 15 min, decant the supernatant, and briefly rinse the DNA pellet with 1 ml of chilled 70% ethanol to wash away the salt. Dry the DNA under vacuum for 30 min to completely evaporate the ethanol.

   e. Dissolve the DNA in 15 μl dd.H$_2$O and place the tube on ice. Quickly measure the concentration of DNA using 2 to 3 μl of the sample at 260 and 280 nm. Immediately proceed to primer annealing.

2. Set up the template-primer annealing reaction as follows:

   a. Transfer 3 to 5 μg freshly denatured plasmid DNA to each of two microcentrifuge tubes on ice for the two opposite primers. Add dd.H$_2$O to a total volume of 7 μl in each tube.

   b. To each tube, add 2 μl of reaction buffer and 1 μl of appropriate primer. This gives approximately 1:1 (template:primer) molar stoichiometry. Each tube contains a total of 10 μl mixture.

   c. Place the tubes in a plastic rack or equivalent and heat at 65°C in a water bath for 2 to 3 min. Quickly transfer the rack together with the tubes to a beaker or tray containing an appropriate volume of 60 to 63°C water, which allows one to slowly cool the sample to <30°C or room temperature over 20 to 30 min. If one uses a heating block, the tubes can be heated at 65°C for 2 min followed by slow cooling to room temperature by turning off the heat. The heating block will slowly cool down. When the temperature drops to <30°C, the annealing is completed. The real annealing temperature is 50 to 52°C. Some laboratories prefer to anneal at 50 to 55°C for 15 to 30 min followed by slow cooling down.

3.   Prepare for labeling and termination as follows:

a.   While the annealing mixture is being cooled, remove the necessary materials from a commercial sequencing kit, which is usually stored at –20°C, and remove [$^{35}$S]dATP from the freezer (–80°C). Thaw the materials on ice.

b.   Label four microcentrifuge tubes for each template-primer mixture, which are A, G, T, and C that, respectively, represent ddATP, ddGTP, ddTTP, and ddCTP.

c.   Add 2.5 µl of termination mixture of ddATP, ddGTP, ddTTP, and ddCTP to the labeled tubes A, G, T, and C, respectively. Cap each tube and keep at room temperature until use.

d.   Dilute the labeling mixture 5-fold as a working concentration and store on ice until use. For example, 2 µl of labeling mixture is diluted to total 10 µl with dd.H$_2$O.

*Notes:*   *(1) Mix the mixture in each stock tube well by pipetting it up and down prior to removal of an appropriate amount of mixture from each tube. (2) There are two sets of labeling and termination mixtures in the sequencing kit (USB). For regular noncompression sequencing, the one labeled as dGTP should be used. However, if compression appears due to G-C-rich sequences, the one labeled as dITP is strongly recommend to be used. dITP replaces the nucleotide analog dGTP. We found that it significantly eliminates the compressed bands.*

e.   Dilute Sequenase Version 2.0 T7 DNA polymerase (1:8 dilution) as follows:
   • Ice-cold enzyme dilution buffer (6.5 µl)
   • Pyrophosphatase (0.5 µl)
   • DNA polymerase stored at –20°C (1 µl)

*Notes:*   *The enzyme should not be diluted in glycerol enzyme dilution buffer if one uses TBE in the gel and in the running buffer. The diluted enzyme should be stored on ice and be used within 50 min.*

4.   Carry out the labeling reaction when the template-primer annealing is completed.

a.   Briefly spin down the annealed mixture and store on ice.

b.   Add the following to the tube containing 10 µl of annealed mixture in the order shown below:
   • DTT (1 µl)
   • Diluted labeling mixture (2 µl)
   • [$^{35}$S]dATP (1000 to 1500 Ci/mmol) (1 µl)
   • Diluted Sequenase DNA polymerase (2 µl)

*Caution:*   *[$^{35}$S]dATP is dangerous. It should be handled carefully using an appropriate Plexiglas protector shell. Gloves should also be worn.*

*Note:*   *Adding 1 µl of Mn buffer may enhance the bands close to the primer.*

c.   Gently mix well to avoid any air bubbles and incubate at room temperature for 3 to 5 min.

d.   While labeling, place the previously labeled tubes containing termination mixture [step 3.c] in a 37°C heating block for at least 2 min.

5.   Carry out the termination reactions as follows:

a.   Carefully and quickly transfer 3.5 µl of the labeled mixture to each termination tube (A, G, T, and C) warmed at 37°C, mix, and quickly return to the 37°C heating block.

b.   Continue to incubate the tube at 37°C for 5 to 6 min.

   c. Add 4 µl of stopping solution to each tube, mix, and cap the tubes. Store at 4°C for immediate use or at –20°C for later use.

*Notes:*    *The samples should be electrophoresed within 4 days even when stored at –20°C. Denature the samples at 75 to 80°C for 2 to 3 min prior to loading them into the sequencing gel.*

## Reagents Needed

### 5X Reaction Buffer

       0.2 *M* Tris-HCl, pH 7.5
       0.1 *M* MgCl$_2$
       0.25 *M* NaCl

### DTT Solution

       100 m*M* Dithiothreitol

### Radioactively Labeled dNTP

       [$^{35}$S]dATP (1000 to 1500 Ci/mmol)

### 5X Labeling Mixture for dGTP

       7.5 µ*M* dGTP
       7.5 µ*M* dTTP
       7.5 µ*M* dCTP

### 5X Labeling Mixture for dITP

       7.5 µ*M* dITP
       7.5 µ*M* dTTP
       7.5 µ*M* dCTP

### ddATP Termination Mixture for dGTP

       80 µ*M* dATP
       80 µ*M* dGTP
       80 µ*M* dCTP
       80 µ*M* dTTP
       8 µ*M* ddATP
       50 m*M* NaCl

### ddGTP Termination Mixture for dGTP

       80 µ*M* dATP
       80 µ*M* dGTP
       80 µ*M* dCTP
       80 µ*M* dTTP
       8 µ*M* ddGTP
       50 m*M* NaCl

## ddCTP Termination Mixture for dGTP

> 80 μ$M$ dATP
> 80 μ$M$ dGTP
> 80 μ$M$ dCTP
> 80 μ$M$ dTTP
> 8 μ$M$ ddCTP
> 50 m$M$ NaCl

## ddTTP Termination Mixture for dGTP

> 80 μ$M$ dATP
> 80 μ$M$ dGTP
> 80 μ$M$ dCTP
> 80 μ$M$ dTTP
> 8 μ$M$ ddTTP
> 50 m$M$ NaCl

## ddATP Termination Mixture for dITP

> 80 μ$M$ dATP
> 80 μ$M$ dITP
> 80 μ$M$ dCTP
> 80 μ$M$ dTTP
> 8 μ$M$ ddATP
> 50 m$M$ NaCl

## ddGTP Termination Mixture for dITP

> 80 μ$M$ dATP
> 160 μ$M$ dITP
> 80 μ$M$ dCTP
> 80 μ$M$ dTTP
> 2 μ$M$ ddGTP
> 50 m$M$ NaCl

## ddCTP Termination Mixture for dITP

> 80 μ$M$ dATP
> 80 μ$M$ dITP
> 80 μ$M$ dCTP
> 80 μ$M$ dTTP
> 8 μ$M$ ddCTP
> 50 m$M$ NaCl

## ddTTP Termination Mixture for dITP

> 80 μ$M$ dATP
> 80 μ$M$ dITP
> 80 μ$M$ dCTP
> 80 μ$M$ dTTP
> 8 μ$M$ ddTTP
> 50 m$M$ NaCl

## Sequence Extending Mixture for dGTP

180 µ$M$ dATP
180 µ$M$ dGTP
180 µ$M$ dCTP
180 µ$M$ dTTP
50 m$M$ NaCl

## Sequence Extending Mixture for dITP

180 µ$M$ dATP
360 µ$M$ dITP
180 µ$M$ dCTP
180 µ$M$ dTTP
50 m$M$ NaCl

## Mn Buffer (only for dGTP)

150 m$M$ sodium isocitrate
100 m$M$ MnCl$_2$

## Enzyme Dilution Buffer

10 m$M$ Tris-HCl, pH 7.5
5 m$M$ DTT
0.5 mg/ml BSA

## Glycerol Enzyme Dilution Buffer

20 m$M$ Tris-HCl, pH 7.5
2 m$M$ DTT
0.1 m$M$ EDTA
50% Glycerol

## Sequenase Version 2.0 T7 DNA Polymerase

13 units/µl in
20 m$M$ KPO$_4$, pH 7.4
1 m$M$ DTT
0.1 m$M$ EDTA, pH 7.4
50% Glycerol

## Pyrophosphatase

5 units/ml in
10 m$M$ Tris-HCl, pH 7.5
0.1 m$M$ EDTA, pH 7.5
50% Glycerol

## Stop Solution

20 m$M$ EDTA, pH 8.0
95% (v/v) Formamide

0.05% Bromophenol blue

0.05% Xylene cyanol FF

*Note:* *All materials should be stored at –20°C except [$^{35}$S]dATP, which should be stored at –70°C and thawed on ice prior to use.*

## Method B: Sequencing of Single-Stranded DNA

1.  Prepare single-stranded M13 DNA.

    a.  Dilute freshly cultured host cells (e.g., JM101) 1:100 in 1.5 ml of 2X YT medium in 4 microcentrifuge tubes.

        • *2X YT medium (1 l)*

        • Bactotryptone (10 g)

        • Yeast extract (10 g)

        • NaCl (5 g)

        • Autoclave. Store at 4°C and warm up to 37°C prior to use.

    b.  Infect each tube with a purified M13 plaque.

    c.  Incubate at 37°C with shaking at 160 rpm for 5 to 9 h.

    d.  Centrifuge at 500 to 1500 × g for 10 min at room temperature.

    e.  Carefully transfer the supernatant containing the phage to fresh tubes and add ¹⁄₉ volume of 40% (w/v) PEG-8000 and ¹⁄₉ volume of 5 *M* sodium acetate (pH 7.0) to each tube.

    f.  Allow the DNA to precipitate at 4°C for 20 to 30 min.

    g.  Centrifuge at 12,000 × g for 15 min and carefully decant the supernatant. Wipe off the inside of the tube to remove PEG as much as possible using a clean Kimwipe.

    h.  Suspend the pellet in each tube in 0.2 ml of TE buffer (pH 7.5) and add 0.2 ml of 50 m*M* Tris-HCl-saturated (pH 7.5) phenol:chloroform (3:1). Mix well by vortexing.

    i.  Centrifuge at 10,000 × g for 5 min and transfer the upper, aqueous phase to fresh tubes.

    j.  Repeat step i once.

    k.  Extract the supernatant with 1 volume of chloroform:isoamyl alcohol (24:1). Centrifuge as in step i. Transfer the supernatant to a fresh tube.

    l.  To the supernatant, add 0.1 volume of 3 *M* sodium acetate (pH 5.2) and 2 to 4 volumes of chilled 100% ethanol. Mix well and place at –70°C for 20 to 30 min.

    m.  Centrifuge at 12,000 × g for 15 min, decant the supernatant and briefly rinse the pellet with 1 ml of 70% ethanol. Dry the DNA pellet under vacuum for 30 min.

    n.  Dissolve the DNA together in 50 to 100 µl of TE buffer. Measure the concentration of the DNA sample and check the quality and quantity by agarose gel electrophoresis. Store the sample at –20°C until sequencing is performed.

2.  Carry out sequencing reactions.

    The procedures are very similar to those described in Method A except for following:

    a.  No double-stranded DNA denaturation is necessary. The appropriate primer and template DNA can be directly annealed in a annealing reaction.

    b.  The temperature for primer-template annealing can be at 50 to 55°C in a heating block for 15 to 25 min followed by slow cooling down by turning off the heating.

## Method C: Sequencing of Bacteriophage Lambda Linear Duplex DNA

For the purposes of sequencing, double-stranded bacteriophage lambda DNA has been denatured by heating or by using alkaline (NaOH) conditions. The drawback is, unlike denatured plasmid DNA, that the denatured lambda DNA readily renatures during the process of primer-template annealing, competing with or preventing the primer from being annealed to the DNA template. This usually causes the failure of complete sequencing. DNA inserts cloned into bacteriophage lambda DNA vectors can now be directly sequenced without being subcloned into appropriate plasmid vectors. This can be achieved by using bacteriophage T7 gene 6 exonuclease (USB). The basic principle is that the enzyme only degrades double-stranded DNA from $5' \rightarrow 3'$. It simultaneously removes deoxynucleotides from both strands of DNA in opposite directions until the middle point of the DNA molecule is reached, generating two half-length, single-stranded DNA fragments. The two half-length DNA fragments represent the sense and antisense orientations of the DNA cloned in the vectors. The DNA insert in the single-stranded products can then be sequenced using appropriate primers annealed to the cloning sites of the vectors in both directions. The sequence of new strand DNA synthesized from the antisense strand fragment is actually the sequence of the sense strand DNA of interest. Therefore, the sequence generated from the sense strand fragment needs to be converted to the complementary sequence in order to be combined with the sequence obtained from the antisense strand fragment at their $3'$ ends.

1.  Carry out a retriction enzyme digestion of the bacteriophage linear duplex DNA.

    This is strongly recommended since the ends of lambda vectors are usually protected and/or modified such that the efficiency of T7 gene 6 exonuclease degradation is not optimal. However, if the DNA is pretreated with an appropriate restriction enzyme, the T7 gene 6 exonuclease works well. The restriction enzyme chosen should be unique and close to the cloning site of the DNA insert of interest.

    a.  Set up a retriction enzyme reaction in a microcentrifuge tube on ice as follows:
        - Purified lambda duplex DNA (2 to 3 µg/µl) (15 µg)
        - 5X Sequencing buffer (3 µl)
        - Appropriate restriction enzyme (45 to 50 units)
        - Add dd.H$_2$O to a final volume of 15 µl.

    b.  Incubate at the appropriate temperature (e.g., 37°C) for 60 min and directly proceed to step 2.

2.  Digest the double-stranded DNA fragments by the use of T7 gene 6 exonuclease.

    a.  Add 75 units of T7 gene 6 exonuclease diluted in TE buffer to 15 µl of the restriction enzyme-digested DNA sample.

    b.  Incubate at 37°C for 20 to 30 min.

    c.  Stop the reaction by heating at 80°C for 10 to 15 min.

3.  Carry out sequencing reactions exactly as for single-stranded M13 DNA sequencing described under Method B.

## Protocol 3: Preparation of Sequencing Gel

Gel preparation is one of the most important steps requisite for successful DNA sequencing. The DNA sequencing gel is a very thin gel (0.2 to 0.4 mm thick). Its quality and integrity directly influence the resolution and maximum reading information of a DNA sequence. There are two common problems in preparing a sequencing gel. One is the trapping of air bubble(s) in the gel, while the other is the leaking of the gel mixture from the bottom or the side-edges of the gel apparatus. Each of these problems can cause failure or a poor quality of the gel to occur. Both can result in false data for the DNA sequence. This, of course, wastes time and money. The detailed protocol given below can solve these problems and is successfully and routinely used in our laboratory.

1.  Prepare a gel mixture as follows:

    a.  Dissolve urea in distilled, deionized water (dd.$H_2O$), in the amounts shown in Table 8–1 or 8–2, in a clean beaker using a stir bar.

### Table 8–1 Recipe if Using the Long Ranger Gel Mixture (AT Biochemicals)

| Components | Small-sized gel (50 ml) | Large-sized gel (100 ml) |
|---|---|---|
| Ultrapure urea (7 *M*) | 21 g | 42 g |
| dd.$H_2O$ | 12 ml | 24 ml |

### Table 8–2 Recipe for Regular Acrylamide Gel

| Components | Small-sized gel (50 ml) | Large-sized gel (100 ml) |
|---|---|---|
| Ultrapure urea (7–8.3 *M*) | 21–25 g | 42–50 g |
| dd.$H_2O$ | 12 ml | 24 ml |

    b.  Warm at 40 to 50°C to dissolve the urea with stirring.

    c.  Add the components shown in Table 8–3, or 8–4, or 8–5 to the mixture with stirring at room temperature.

### Table 8–3 Recipe if Using the Long Ranger Gel Mixture (AT Biochemicals)

| Components | Small-sized gel (50 ml) | Large-sized gel (100 ml) |
|---|---|---|
| 5X TBE | 12 ml | 24 ml |
| Long Ranger mixture | 5 ml | 10 ml |
| Add dd.$H_2O$ up to | 50 ml | 100 ml |

### Table 8–4 Recipe for Regular Acrylamide Gel Using TBE Buffer

| Components | Small-sized gel (50 ml) | Large-sized gel (100 ml) |
|---|---|---|
| 5X TBE | 10 ml | 20 ml |
| 6% Acrylamide/*bis*-acrylamide | 2.8 g/0.15 g | 5.7 g/0.3 g |
| 8% Acrylamide/*bis*-acrylamide | 3.8 g/0.2 g | 7.6 g/0.4 g |
| Add dd.$H_2O$ up to | 50 ml | 100 ml |

### Table 8–5 Recipe for Regular Acrylamide Gel Using Glycerol Gel Buffer (for Sequenase Verison 2.0 DNA Polymerase Diluted with Glycerol Enzyme Dilution Buffer)

| Components | Small-sized gel (50 ml) | Large-sized gel (100 ml) |
|---|---|---|
| 10X TBE | 5 ml | 10 ml |
| 6% Acrylamide/bis-acrylamide | 2.8 g/0.15 g | 5.7 g/0.3 g |
| 8% Acrylamide/bis-acrylamide | 3.8 g/0.2 g | 7.6 g/0.4 g |
| Add dd.$H_2O$ up to | 50 ml | 100 ml |

*Caution:* *The Long Ranger mixture and the acrylamide/bis-acrylamide are very toxic. Care should be taken. Gloves should be worn when handling the gel mixture.*

    d. Filter and degas under vacuum for 2 to 3 min (optional) and transfer the gel mixture to a plastic squeeze bottle or leave it in the beaker if using a syringe to fill the gel apparatus with the liquid.

2. While the gel mixture is being cooled (25 to 30°C) and mixed with a stirring bar, begin to clean the glass plates. Gently place the glass plates, one by one, in a sink. Thoroughly clean the glass plates with detergent using a sponge; thoroughly wash away the soap residue with running tap water followed by distilled water. Spray (95 or 100%) ethanol onto the plates to completely remove the soap and oily residues. Finally, wipe the plates dry using clean paper towels with no dust (e.g., bleached singlefold towels, James River Corporation, Norwalk, CT).

3. Lay one glass plate down on the lab bench and coat its surface with 0.5 to 1.0 ml of either Sigmacoat (Sigma Chemicals) or Gel Slick (AT Biochemicals) to form a thin film. This can prevent the gel from sticking to the glass plates when they are separated from each other after electrophoresis. Mark the coated plate on the back with a marker pen for later identification. The other plate should not be coated and served as the bottom plate when the plates are separated from each other.

4. Assemble a glass plate sandwich by placing a spacer (0.2 or 0.4 mm thick) on each of the two side edges of one glass plate and cover the plate with the other plate. There are different ways to assemble the glass plates depending on manufacturer's instructions and to pour the gel mixture into the sandwich.

## a.    Pouring the Gel Mixture Horizontally

1. Place two styrafoam supporters (approximately 5 to 10 cm thick, one for the top and the other for the bottom of the sandwich) or two plastic pipette-tip boxes with an even surface on a flat bench. Put the glass sandwich on the two supports and check the flatness with a level. It is not necessary to use tape to seal the edges of the glass sandwich especially when two raised edges (about 0.03 to 0.5 cm above the glass surface) on each side of the bottom plate are used to hold the top plate in place. Clamping the side edges of the plates with manufacturer's clamps is optional. The glass sandwich should not be placed directly on the hard surface of the lab bench because tapping the top plate while pouring the gel may cause damage to the glass plate.

2. Immediately prior to pouring the gel, add N,N,N′N′-tetramethylethylenediamine (TEMED) and freshly prepared 10% ammomium persulfate solution (APS) to the gel mixture. (For a 100-ml gel mixture, the volume for TEMED and APS should be 48 μl and 0.49 ml, respectively.) Quickly and gently mix the mixture by swirling the squeeze bottle to avoid air bubbles, and immediately begin to pour the gel mixture into the sandwich. Starting at the middle region at the top of the sandwich, slowly and continuously squeeze the bottle with one hand to cause outward flow of the mixture. Simultaneously use Scotch™ masking tape (1 to 2 cm wide and 10 to 12 cm diameter) to tap the top glass plate along the front of the flowing solution with the other hand until the sandwich is completely filled. The gel mixture distributes into the sandwich by capillary suction and the tapping helps to cause even flow of the gel mixture. If squeezing of the bottle is stopped at any time, make sure to squeeze the bottle to remove any air bubbles before allowing the mixture to flow into the sandwich again. It is strongly recommended that two people work together while pouring the gel. One person can focus on squeezing the bottle to pour the gel, while the other is responsible for tapping the top glass plate to help the flow and prevent bubble formation. The gel mixture can also be loaded into the sandwich by the use of a syringe or a pipette depending on personal perference.

3. Immediately and slowly insert the comb (0.2 or 0.4 mm thick depending the thickness of the spacer), upside down, into the gel mixture to make a straight, even edge along the top of the gel. Avoid any bubbles underneath the comb. To prevent the gel from forming between the comb and

the glass plates, clamp the comb in place together with the glass plates with 2 to 3 appropriately sized clamps. It is recommend that a 24-well comb be used for a small gel and a 48- or 60-well comb for a large gel. The color of the comb should be offwhite or white so that each well can be clearly seen while loading the DNA samples.

4.    Allow the gel to polymerize for 0.5 to 1.5 h. The gel can be subjected to electrophoresis directly by removing the comb and mounting the gel cassette onto the sequencing apparatus according to the manufacturer's instructions. For multiple loadings at approximately 2-h intervals, we recommend that the gel be poured late in the day or at night, and that electrophoresis be carried out early the next morning. If this is the option, the polymerized gel should be covered at the top and bottom edges of the cassette with clean paper towels wetted with distilled water. Wrap the paper towels with SaranWrap™ to keep the gel from drying. Leave it overnight so that the electrophoresis can be performed the next day.

## b.    Pouring a Gel at an Angle

1.    Tightly seal the two side edges of the sandwich using tape. Clamping each side of the sandwich is optional.

2.    Seal the bottom of the sandwich as follows: (1) tightly seal the bottom edge with an appropriate tape; (2) leave the bottom unsealed and attach the special bottom tray (provided by the manufacturer) to the bottom of the sandwich. Some kinds of commercial apparatus have disposable bottom spacers available, which can be inserted into the sandwich at the bottom and may then be sealed by tape.

3.    Pour the gel mixture into the sandwich. One option is to preseal the bottom to ensure no leaking. This can be done by transferring approximately 5 ml of the gel mixture to a tray and adding 5 µl of TEMED and 45 µl of freshly prepared 10% APS. Mix well and immediately immerse the bottom of the sandwich, which is held vertically. The mixture containing a high concentration of APS will quickly polymerize and seal the bottom of the sandwich. After the bottom is sealed, the rest of the gel mixture can be poured.

4.    Immediately prior to pouring the gel, add TEMED and freshly prepared 10% APS to the gel mixture contained in a clean squeeze bottle. For 100 ml of gel mixture, the volume for TEMED and APS should be 48 µl and 0.49 ml, respectively. Quickly and gently mix the mixture by swirling the squeeze bottle to avoid bubbles, and immediately begin to pour the gel mixture into the sandwich. Raise the top of the sandwich to a 45° angle with one hand. Using the other hand, slowly squeeze the bottle to cause flow of the gel mixture at the top corner along one side of the sandwich. The angle of the plate and the rate of flowing should be adjusted to avoid air bubbles. When the sandwich is half-full, place the top of the plate on a small box or support on the lab bench and the bottom of the plate on the bench, forming approximately an angle of 25 to 30°. Continuously add the gel mixture into the sandwich and simultaneously use a roll of Scotch™ masking tape (1 to 2 cm thick and 10 to 12 cm diameter) or its equivalent to tap the top glass plate along the front of flowing gel with the other hand in order to avoid formation of any air bubbles until the sandwich is completely filled. The gel mixture can also be loaded into the sandwich with a syringe or pipette depending on personal preference.

5.    Immediately and slowly insert the comb (0.2 or 0.4 mm thick depending the thickness of the spacer), upside down, into the gel mixture to make a straight, even top front of the gel. Avoid formation of any air bubbles underneath the comb. To prevent gel from forming between the comb and glass plates, clamp the comb in place between the glass plates with 2 to 3 appropriately sized clamps. It is recommend that a 24-well comb be used for a small gel and 48- or 60-well comb for a large gel. The color of the comb should be offwhite or white so that each well can be seen clearly while loading the DNA sample.

6.    Slowly lay the plates on a flat surface and allow the gel to polymerize for 0.5 to 1.5 h. The gel can be directly subjected to electrophoresis by removing the comb and mounting the gel cassette onto the sequencing apparatus according to the manufacturer's instructions. For multiple loadings

(e.g., 4 loadings at 2-h intervals), we recommend that the gel be poured late in the day or at night, and that the electrophoresis be carried out early next morning. If this is the option, the polymerized gel should be covered at the top and bottom edges of the cassette with clean paper towels wetted with distilled water. Wrap the paper towels with SaranWrap™ to keep the gel from drying. Leave it overnight until electrophoresis is carried out.

## Materials Needed

### 5X TBE Buffer

Tris base (54 g)
Boric acid (27.5 g)
0.5 $M$ EDTA buffer (pH 8.0) (20 ml)
Dissolve well after each addition in 700 ml dd.$H_2O$. Then add dd.$H_2O$ to a final volume of 1000 ml. Autoclave. Store at room temperature.

### 10X Glycerol-Tolerant Gel Buffer

Tris base (108 g)
Taurine (36 g)
$Na_2EDTA \cdot 2H_2O$ buffer (pH 8.0) (4.65 g)
Dissolve well after each addition in 700 ml dd.$H_2O$. Then add dd.$H_2O$ to a final volume of 1000 ml. Autoclave. Store at room temperature.

### Urea

Ultrapure grade

### Long Ranger Gel Mixture

AT Biochemicals (Cat. no. 211ISI).

### 10% Ammonium Persulfate Solution (APS)

0.1 g in 1 ml dd.$H_2O$ (fresh)

### High-Power Supply

It should have an upper limit of 2500 to 3000 V.

## Protocol 4: Electrophoresis

1. Remove the comb and carefully insert the cassette into the gel apparatus according to the manufacturer's instructions.
2. Add a sufficient quantity of 0.6X TBE running buffer (diluted from 5X TBE stock buffer) to both top (cathode) and bottom (anode) chambers.
3. Gently wash the top surface of the gel several times by pipetting the running buffer up and down.
4. Vertically hold the left and right sides of the comb using two hands and carefully insert the comb into the top of the sandwich to form the wells. The insertion of the comb should be slow and even until all the teeth just touch the surface of the gel. Remove any air bubbles trapped in the wells by using a pipette.

### Table 8–6 Examples with Constant Power

| Gel volume | Constant Power | Electrophoresis temperature |
|---|---|---|
| 35 cm$^3$ | 35–38.5 W | 45–55°C |
| 70 cm$^3$ | 70–77 W | 45–55°C |

*Notes:*   *It is acceptable if the teeth of the comb protrude a little bit deeply into the gel (<0.5 mm from the surface of the gel). However, if one inserts the teeth too deep into the gel or repeatedly pulls and inserts the comb several times, the surface of the gel will be badly damaged and cause leaking while loading samples, resulting in contamination of the wells and inaccurate sequence data.*

5.   Calculate the volume of the gel (length × width × thickness of the spacer) and set up the power supply. We strongly recommend that the power supply be set at a constant power (watts) using 1.0 to 1.1 W/cm$^3$. If constant power is not available from the power supply, 0.5 to 0.8 mA/cm$^3$ constant current should be set up. Examples are as given in Table 8–6.

*Note:*   *If one sets the power supply at a nonconstant level (e.g., high voltage or high current), the gel may burn or melt during the electrophoresis due to the heat generated.*

6.   Connect the power-supply unit to the gel apparatus with the cathode located at the top of the gel and anode at the bottom of the gel so that the DNA samples bearing a negative charge will migrate downward.

7.   Turn on the power and prerun the gel for 10 to 12 min.

8.   After prerunning for 8 to 10 min, denature the labeled DNA samples at 75°C for 2.5 to 3 min prior to loading them in the wells. The tubes containing the samples should be capped tightly while heating in order to prevent evaporation.

*Caution:*   *$^{35}$SdATP is dangerous. Gloves should be worn during the process.*

*Note:*   *For heating, it is recommended that one place the tubes in a heating block in the order of A, T, G, C, or A, G, T, C in order to prevent a potential mix-up while loading the samples into the gel.*

9.   Turn off the power, wash the wells briefly to remove urea, and quickly and carefully load each of the samples to appropriate wells (3.5 to 4 µl per well).

*Notes:*   *(1) There should be no leaking between wells; otherwise, the DNA samples will be contaminated and result in mix-up of the nucleotide sequence. (2) There are special sample-loading pipette tips available for DNA sequencing. It is not recommended that one tip be used for loading multiple wells in order to prevent any contamination. If one uses a 0.4 mm spacer, we recommend the use of normal 100- to 200-µl pipette tips for loading the wells. This is faster and no air bubbles develop as compared with using the commercial sequencing loading tips which are long and flat. (3) After taking samples from each tube, the tubes should immediately be capped. All tubes should be briefly spun down and stored at 4 or –20°C until the next loading. (4) It is very important that the samples loaded into the respective wells be recorded in a notebook in the order of DNA samples loaded (e.g., A, G, T, C, or A, T, G, C depending on the particular loading sequence). This will help in the reading of the DNA sequence following autoradiography.*

10.  After all the samples are loaded, turn on the power, which is set at an appropriate constant voltage, and allow to electrophorese. The running time depends on the size of DNA. Normally, the run time for 250, 400 to 450, and >500 bp are 2, 4, and 6 to 8 h, respectively. For multiple loadings (e.g., 2 to 5 times loading, depending on the size of the gel and volume of the DNA sample), monitor the migration of the first blue dye (bromophenol blue or BPB), which migrates at 40 bp. When the BPB reaches approximately 2 to 3 cm from the bottom, carry out the next loading so that there are overlapping sequences between the adjacent loads. This is very useful for the correct reading of DNA sequences after autoradiography.

*Notes:*  *The power supply must be turned off and the wells to be loaded should be washed to remove bubbles with the running buffer using a pipette. The DNA samples may be denatured again at 75°C for 2 to 3 min prior to loading. The loading order of the samples should be the same as for the previous loading in order to avoid misreading of DNA sequences.*

*Warning:*  *The electrophoresis is carried out at a high voltage, so care must be taken. Prior to each loading of the samples, make sure that the power is turned off. It is recommended to monitor the running for a few minutes until stable conditions are established, especially when new apparatus is used.*

11.  After the electrophoresis has been completed, disconnect the power supply and decant the upper chamber buffer to a $^{35}$S-waste container. Lay the cassette containing the gel on a table or bench and allow to cool for 15 to 25 min prior to separating the glass plates. Dump the bottom-chamber buffer into a waste container. While the gel is being cooled, cut a piece of 3MM Whatman paper that is relatively larger than the gel. Prepare for gel drying (e.g., dry-ice bucket or freeze-trap, pump and dry apparatus).

*Caution:*  *$^{35}$S is dangerous. Gloves should be worn during the process.*

*Note:*  *If the glass plates are separated immediately following electrophoresis, the gel may be uneven, broken, or it may cause difficulty in separating the glass plates.*

12.  Remove the clamps and/or the heat-cooling plate from the glass plates. Starting at one corner of the spacer site, carefully insert a spatula into the sandwich and slowly lift the top glass plate (the Sigma-coated plate) with one hand, and hold the bottom plate with the other hand until the top glass plate is completely separated from the gel. Remove the top plate; the gel should stick to the bottom plate.

*Notes:*  *As long as the top plate is loosened, continue to lift it until it is completely separated from the other plate. Do not allow the top plate to come back in contact with the gel in the middle of this apparatus or it may damage the gel.*

13.  Carry out fixation. If using the Long Ranger gel mixture, the electrophoresed gel does not need to be fixed. If, instead, regular acrylamide gel mixture is used, fixing of the gel should be performed. Immerse the gel together with the bottom plate in a large tray containing a sufficient volume of fixing solution containing 15 to 20% of ethanol or methanol and 5 to 10% acetic acid for 15 to 20 min. Remove the gel together with the plate from the tray, drain off excess fixing solution, and gently place bleached, clean paper towels on the gel to remove excess solution.

14.  Gently and slowly lay the prepared 3MM Whatman paper on the gel from the bottom side or the upper side of the gel until it covers the entire gel. Gently and firmly press the paper thoroughly onto the gel surface using a styrofoam block (approximately 50 cm long, 5 cm wide, and 5 cm thick). Starting from one corner, slowly peel the 3MM Whatman paper with the gel on it from the

bottom plate. Place it on a flat surface with gel-side up and carefully cover the gel with SaranWrap™ without any air bubbles or wrinkles forming between the SaranWrap™ and the gel.

15. Assemble the gel on the drying apparatus according to the manufacturer's instructions and dry the gel at 70 to 80°C under vacuum for 30 to 55 min depending on the strength of the vacuum pump. Turn off the heat and allow to cool for another 20 to 30 min under vacuum. Proceed to autoradiography.

*Notes:* *(1) The position of the water trap should be lower than the gel dryer for effective suction of the water. If no frozen or refrigerated trap is available, place a water-trap flask or bottle in a bucket containing some dry ice and ethanol. Connect the trap to a vacuum pump and to the gel dryer. This should work fine. (2) If the gel is removed right after the heat is turned off, the gel may crack and/or become uneven. So, it is recommended that it be allowed to cool for a while under vacuum prior to being removed from the gel dryer.*

## Protocol 5: Autoradiography

1. Peel off the SaranWrap™ from the gel and place the gel on an appropriate exposure cassette. In a dark room under a safety light, place an appropriately sized piece of X-ray film (Kodak XAR-film or Amersham X-ray film) on the surface of the gel and close the cassette.

2. Allow exposure to occur at room temperature (for $^{35}$S) by placing the cassette in a dark place for 1 to 4 days depending on the intensity of the signal. It is recommended that one develop the film after exposure for 24 h and decide the length of further exposure (Figure 8.1).

3. Develop the exposed film using an automatic developing machine or its equivalent in a dark room under a safety light. Proceed to DNA sequence reading and analysis (Figure 8.2).

## Protocol 6: Formamide Gels

Compressions are a common phenomenon that usually result in unreadability on a developed sequencing film. The problem is primarily due to intramolecular base pairing in a primer extension that is G-C rich. The local folded structures or hairpin loops migrate faster through the gel matrix than unfolded structures, and persist during the electrophoresis, resulting in bands running very close together with a gap or increased band spacing in the region above. One way to solve this problem is to increase the denaturing condition in the gel matrix using urea and formamide. The procedures for preparing a formamide gel, electrophoresis, and autoradiography are similar to those described above in Protocols 1 to 5 except for the following steps.

1. The gel mixture is made as shown in Table 8–7, 8–8, or 8–9

### Table 8–7 Using the Long Ranger Gel Mixture (AT Biochemicals)

| Components | Small-sized gel (50 ml) | Large-sized gel (100 ml) |
|---|---|---|
| Ultrapure urea | 21 g | 42 g |
| Ultrapure formamide | 20 ml | 40 ml |
| Long Ranger mixture | 8 ml | 16 ml |
| 5X TBE | 10 ml | 20 ml |
| Add dd.H$_2$O up to | 50 ml | 100 ml |

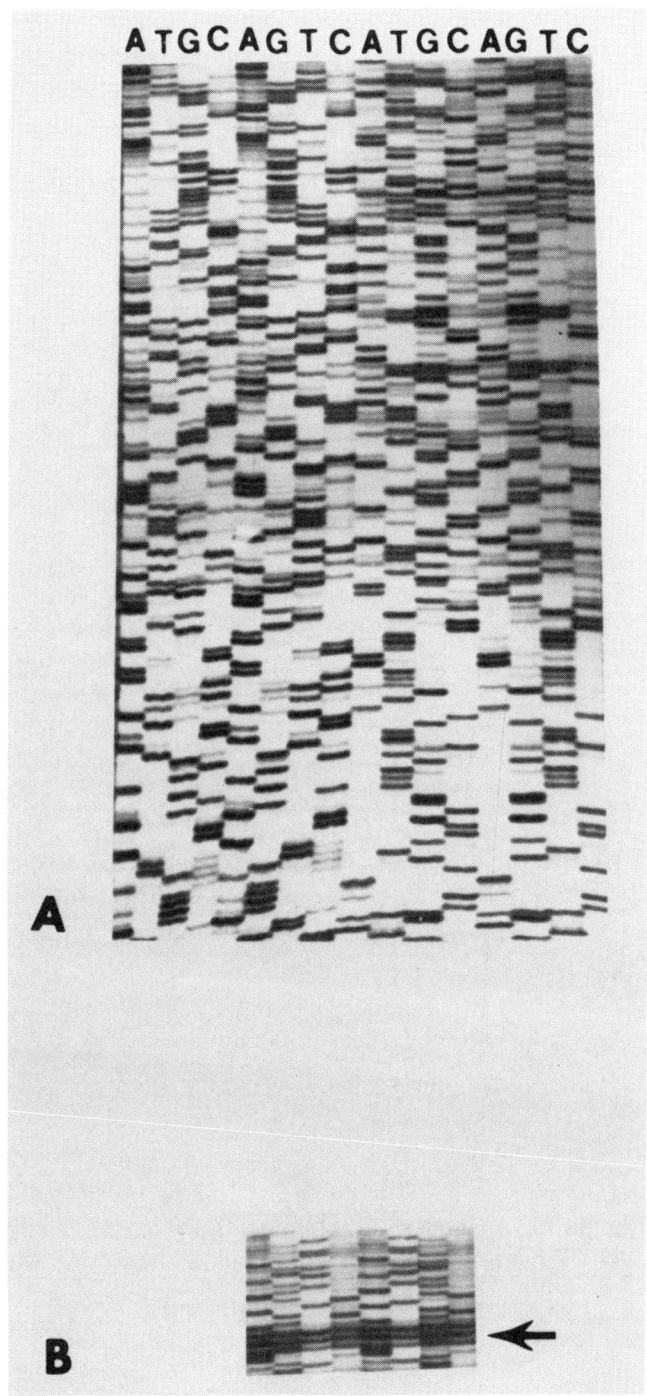

**FIGURE 8.1**

Autoradiograph showing a portion of DNA sequences obtained by the dideoxynucleotide chain termination method. (A) Normal sequencing gel; (B) sequencing gel compressions as denoted by the arrow.

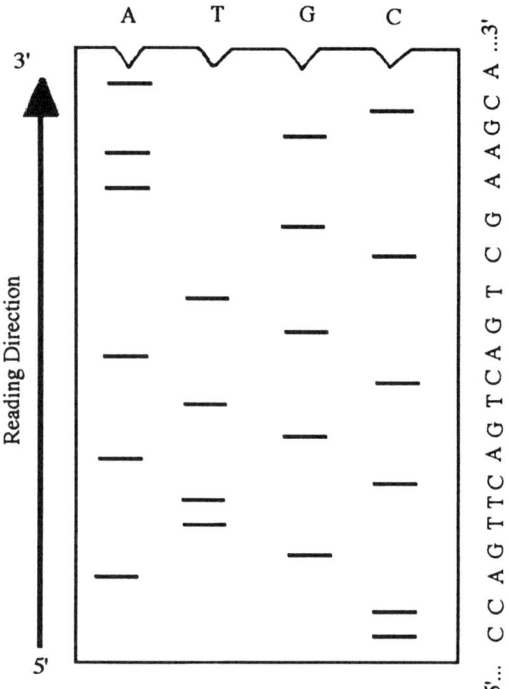

**FIGURE 8.2**

Diagram illustrating the reading of an autoradiograph obtained by dideoxy DNA sequencing.

## Table 8–8 Recipe for a Regular Acrylamide Gel Mixture (if Sequenase Version 2.0 DNA Polymerase is Diluted with a Buffer without Glycerol)

| Components | Small-sized gel (50 ml) | Large-sized gel (100 ml) |
|---|---|---|
| Ultrapure urea (7 $M$) | 21 g | 42 g |
| Ultrapure formamide | 15–20 ml | 30–40 ml |
| 6% Acrylamide/bis-acrylamide | 2.8 g/0.15 g | 5.7 g/0.3 g |
| 8% Acrylamide/bis-acrylamide | 3.8 g/0.2 g | 7.6 g/0.4 g |
| 5X TBE | 5 ml | 10 ml |
| Add dd.H$_2$O up to | 50 ml | 100 ml |

## Table 8–9 Recipe for a Regular Acrylamide Gel Mixture (if the Sequenase Version 2.0 DNA Polymerase is Diluted in Glycerol Enzyme Dilution Buffer)

| Components | Small-sized gel (50 ml) | Large-sized gel (100 ml) |
|---|---|---|
| Ultrapure urea (7 $M$) | 21 g | 42 g |
| Ultrapure formamide | 15–20 ml | 30–40 ml |
| 6% Acrylamide/*bis*-acrylamide | 2.8 g/0.15 g | 5.7 g/0.3 g |
| 8% Acrylamide/*bis*-acrylamide | 3.8 g/0.2 g | 7.6 g/0.4 g |
| 10X Glycerol-tolerant gel buffer | 5 ml | 10 ml |
| Add dd.H$_2$O up to | 50 ml | 100 ml |

Dissolve the urea by placing the container in a 55 to 60°C water bath with shaking. The mixture should be cooled to 25 to 30°C prior to adding TEMED (0.12%, v/v) and 10% APS (0.78%, v/v).

2. Use 1X TBE running buffer instead of 0.6X TBE for both the regular acrylamide gel mixture and the Long Ranger gel mixture (AT Biochemicals). If the DNA polymerase is diluted with the glycerol-enzyme dilution buffer, the running gel buffer should contain 1X glycerol-tolerant gel buffer instead of TBE buffer.

3. Double the electrophoresis time since the formamide slows the migration of DNA by about 50%.

4. After the electrophoresis, fix the gel in 20% ethanol or methanol and 10% acetic acid for 15 min.

### *Protocol 7: Extending Sequencing Far from the Primer*

We routinely use the sequence-extending mixture (USB) for obtaining sequences beyond 600 to 800 bases from the primer with high resolution. Using multiple loadings, more than 1500 to 2000 bases from the primer can be extended. The procedures for extending sequencing are similar to those described in Protocols 1 to 5 except the following:

1. The labeling mixture should be undiluted or diluted for 2- to 3-fold instead of 5-fold dilution.

2. Increase the amount of [$^{35}$S]dATP (1000 to 1500 Ci/mmol) from the regular volume of 1 to 1.5–2 µl.

3. Extend the labeling reaction from the regular time of 2–5 to 4–7 min at room temperature.

4. For the termination reaction, instead of adding 2.5 µl of ddNTPs to the appropriate tubes (A,G,T, and C), add 2.5 µl of a mixture of the appropriate sequence-extending mixture (USB) and appropriate termination mixture (USB) at ratios of 2:0.5, 2:1.0, and 1.5:1 (v/v), depending on the particular DNA sequence, to appropriate tubes.

5. Use a relatively long gel (80 to 100 cm) and extend electrophoresis for 12 to 16 h using a high-quality sequencing apparatus.

## 3. Troubleshooting Guide

1. **Gel pulls away from the comb during electrophoresis.**

   **Cause:**  Too much heat is built up.

   **Solution:**  Be sure to set the power supply at a constant power or a constant current. Do not set the voltage at some level since the current undergoes changes during electrophoresis, generating temperatures high enough to melt the surface of the gel. When that happens, multiple loadings of the samples are impossible.

2. **No bands occur at all on the developed X-ray film.**

   **Possible causes:**

   a. The primer does not work.

   b. Double-stranded DNA is not denatured well, such that the primer fails to anneal to the template.

   c. Some component(s) is(are) missed during the labeling reaction.

   d. Sequenase Version 2.0 T7 DNA polymerase has lost its activity.

   **Solution:** Be sure to denature the DNA template completely and add all components necessary for the labeling reaction. Try to use the control DNA and primer provided.

3. **Bands are fuzzy.**

   **Possible causes:**

   a. Urea is not washed from the wells prior to loading the samples.

   b. Labeled samples are overheated during denaturation.

   c. It takes too long to finish loading all the samples, resulting in some annealing of DNA.

**Solutions:**

a. Be sure to wash the surface of the gel prior to drying the prerun and repeat washing after the prerun prior to loading the samples into the wells. The washing can be done with the running buffer using a pipette tip.

b. Control the time for denaturing of the labeled samples between 2 to 3 min and immediately load the sample into the wells. For many samples, the loading should be carried out quickly so that all the samples should be loaded within 2 min.

4.  **No clear bands appear except a smear in each lane.**

    **Possible causes:**

    a. Preparation of DNA template is bad.

    b. Labeling reaction goes too long and results in very old labeling.

    c. Labeled DNA samples are not denatured well at 75°C prior to their being loaded into the gel.

    d. Gel polymerizes too rapidly (10 to 15 min) due to excess 10% APS added.

    e. The gel is electrophoresed at a too-cold or too-hot temperature.

    f. The gel is dried at a too-warm temperature.

    **Solutions:**

    a. Make sure the DNA template is very pure without any nicks. This can be checked by spectrophotometer measurement with a ratio of $A_{260}/A_{280}$ of 1.9 to 2.0. Furthermore, the purified DNA should be checked by agarose gel electrophoresis.

    b. Use 0.5 ml of freshly prepared APS per 100 ml of gel mixture and make sure the gel mixture is cooled to room temperature prior to its being poured into the glass sandwich.

    c. Keep the labeling reaction time to 2 to 5 min for regular sequencing and 4 to 7 min for extended sequencing.

    d. Be sure to denature the labeled samples at 75 to 80°C for 2 to 3 min prior to their being loaded into the gel.

    e. Dry the gel at 75 to 80°C under vacuum but not above 80°C.

5.  **All the bands are weak.**

    **Possible causes:**

    a. Primer concentration is too low or the annealing of primer and template does not work well.

    b. Double-stranded linear DNA and double-stranded plasmid DNA are too large due to the presence of a large DNA insert, resulting in difficulty in denaturation.

    c. Labeling reaction goes on for too long.

    d. Labeled DNA samples are not be completely denatured before loading into the gel.

    **Solutions:**

    a. Heat the primer and double-stranded DNA template at 65°C for 3 to 4 min and slowly cool down to room temperature over 20 to 35 min.

    b. Use the alkaline-denaturing method to denature large-sized DNA templates. If this still does not work well, try to fragment the DNA insert to be sequenced and subclone for further sequencing.

    c. Make sure the labeling reaction is carried out properly and denature the labeled sample at 75 to 80°C for 3 to 4 min prior to their being loaded into the gel.

6.  **Bands occur across all four lanes in some areas called compressions.**

    **Possible cause:** Occurrence of nucleotide sequences with strong secondary structure.

    **Solution:** Use an appropriate amount of dITP to replace dGTP and an appropriate amount of pyrophosphatase in the labeling reaction.

7. **Bands are faint near the primer.**

   **Possible cause:** Insufficient DNA template or insufficient of primer.

   **Solutions:**

   a. Use 1 to 1.5 µg of single-stranded M13 DNA or 3 to 5 µg plasmid DNA per reaction.

   b. Increase the molar ratio of primer:DNA template from 1:1 to 1:4–5.

   c. Use 1 µl of Mn buffer per regular labeling reaction.

8. **Bands are faint or blank in 1 or 2 lanes.**

   **Possible cause:** Some components may have been misadded or missed.

   **Solution:** Be sure that all the components are added properly as for other lanes revealing bands with good resolution.

# II.    Direct Sequencing by PCR

DNA molecules or fragments (single-stranded, double-stranded, and plasmid DNA) can now be directly sequenced by combining polymerase chain reaction (PCR) technology and the dideoxynucleotide chain termination method. This is a very powerful technique that is a widely used method in biotechnology, medicine, and other molecular biological studies.

## *Performing PCR and Sequencing Reactions*

1. Label four 0.5-ml microcentrifuge tubes (A, G, T, and C) for each set of sequencing reactions for each primer. A, G, T, and C represent ddATP, ddGTP, ddTTP, and ddCTP, respectively.

2. Add 0.5 µl of 2X stock mixture of dNTPs/ddATP, dNTPs/ddGTP, dNTPs/ddTTP, and dNTPs/ddCTP to the labeled tubes A, G, T, and C, respectively. Add 0.5 µl dd.H$_2$O to each tube, generating 1X working mixture solution. Cap the tubes and store on ice until use.

3. Prepare the following mixture for each set of four sequencing reactions for each primer in a microcentrifuge tube (0.5 ml) on ice.

   • Appropriate primer (15mer to 27mer, 10 to 30 ng/µl), 2 to 5 pmol (15 to 27 ng) depending on the size of primer

   • DNA template (0.4 to 7 Kb, 10 to 100 ng/µl), 100 to 1000 ng depending on the size of the template

   • [α-$^{35}$S]dATP (>1000 Ci/mmol), 1 to 1.2 µl

   • **or** [α-$^{32}$P]dATP (800 Ci/mmol), 0.5 µl

   • 5X Sequencing buffer, 4 µl

   • Add dd.H$_2$O to a final volume of 17 µl.

4. Add 1 µl (5 units/µl) of sequencing-grade *Taq* DNA polymerase (Promega Corporation) to the mixture at step 3. Gently mix by pipetting up and down, and store on ice.

5. Remove 4 µl of the primer-template-enzyme mixture at step 4 to the bottom of each tube containing 1 µl of dNTPs and the appropriate ddNTP prepared in step 2.

6. Overlay the mixture in each tube with approximately 20 µl of mineral oil to prevent evaporation during the PCR amplification.

7. Place the tubes in a thermal cycler preheated to 95°C and begin 30 cycles following the cycling profiles given in Table 8–10, depending on the size of primer and the size of DNA template.

## Table 8–10 Thermal Cycling Profiles

| Profile | Predenaturation | Cycling | | | Last |
| --- | --- | --- | --- | --- | --- |
| | | Denaturation | Annealing | Extension | |
| Template (4 Kb); primer is <24 bases or with G-C content <40% | 94°C, 2 min | 94°C, 1 min | 50°C, 1 min | 70°C, 1.5 min | 4°C |
| Template is >4 Kb; primer is >24mer or <24 bases with G-C content >50% | 95°C, 2 min | 95°C, 1 min | 60°C, 1 min | 72°C, 2 min | 4°C |

8.  After the PCR cycling is completed, carefully remove the mineral oil from each tube using pipette tips (optional). Add 3.5 µl of stop solution to inactivate the enzyme activity.

9.  The reaction mixture can be directly subjected to electrophoresis following denaturation at 75°C for 2 min, or stored at 4°C until use. Load 3 µl per well and avoid inclusion of mineral oil. The procedures for electrophoresis and autoradiography are described in Section I of this chapter.

## Reagents Needed

### 2X dNTPs/ddATP Mixture

dATP (80 µ$M$)
dTTP (80 µ$M$)
dCTP (80 µ$M$)
7-Deaza dGTP (80 µ$M$)
ddATP (1.4 m$M$)

### 2X dNTPs/ddTTP Mixture

dATP (80 µ$M$)
dTTP (80 µ$M$)
dCTP (80 µ$M$)
7-Deaza dGTP (80 µ$M$)
ddTTP (2.4 m$M$)

### 2X dNTPs/ddGTP Mixture

dATP (80 µ$M$)
dTTP (80 µ$M$)
dCTP (80 µ$M$)
7-Deaza dGTP (80 µ$M$)
ddGTP (120 µ$M$)

### 2X dNTPs/ddCTP Mixture

dATP (80 µ$M$)
dTTP (80 µ$M$)
dCTP (80 µ$M$)
7-Deaza dGTP (80 µ$M$)
ddCTP (800 µ$M$)

*Amount of Primer per pmole*

15mer or 15 bases (5 ng)
16mer (5.3 ng)
17mer (5.7 ng)
18mer (6.0 ng)
19mer (6.3 ng)
20mer (6.7 ng)
21mer (7.0 ng)
22mer (7.3 ng)
23mer (7.6 ng)
24mer (8.0 ng)
25mer (8.3 ng)
26mer (8.6 ng)
27mer (9.0 ng)
28mer (9.3 ng)
29mer (9.6 ng)
30mer (10.0 ng)

*5X Sequencing Buffer*

0.25 $M$ Tris-HCl, pH 9.0 at room temperature
10 m$M$ MgCl$_2$

*Stop Solution*

10 m$M$ NaOH
95% Formamide
0.05% Bromophenol Blue
0.05% Xylene cyanole

# III.   DNA Sequencing after Unidirectional Deletions of the DNA

In case of large sizes of DNA, it takes a longer time to sequence the entire DNA using regular stepwise sequencing methods. However, using exonuclease III, the gene or large DNA fragment can be unidirectionally deleted to generate a series of shorter fragments with overlapping ends. These progressively deleted fragments can then be simultaneously sequenced in a short time. The general principles and procedures are outlined in Figure 8.3 (p. 239) and are described below.

## A.   Protocols

1.  A recombinant plasmid phagemid, or bacteriophage M13 replicative form DNA, which contains the cloned DNA of interest, is linearized with two appropriate restriction enzymes. Both enzymes should cut the recombinant DNA between one end of the target DNA and the binding site for the universal sequencing primer. One enzyme should cleave near the target DNA and must generate either a recessed 3′ end (or 5′ protruding end) or a blunt end. The other enzyme should cleave near the binding site for the universal sequencing primer and must generate a 4-base protruding 3′ end or be filled-in with α-phosphorothioate dNTPs.

2.  The linearized DNA is progressively deleted with exonuclease III that only digests the DNA from the blunt or 5′ protruding terminus, leaving the 3′ protruding (overhang) or α-phosphorothioate-

filled end intact. The digestion proceeds unidirectionally from the site of cleavage to the target DNA sequence. The digestion is terminated by removing appropriate amounts of the samples at appropriate time intervals, generating a series and progressive deletions of shorter fragments.

3.   The exposed single strands are then cleaved by nuclease S1 or mung-bean nuclease, producing blunt termini at both ends of the DNA fragments.

4.   The shortened DNA is then recircularized by using T4 DNA ligase, which is transformed into an appropriate bacterial host. Transformants can be selected with appropriate antibiotics in the culture medium.

5.   The recombinant plasmids are purified and subjected to nucleotide sequencing.

## Protocol 1: Subcloning of DNA Insert to be Sequenced in Appropriate Vectors

The DNA of interest should be first subcloned in appropriate plasmid vectors such as pGEM-5Zf and pGEM-7Zf (Promega Corporation) whose multiple cloning sites contain two unique restriction sites lying between the end of the DNA insert to be deleted and the binding site for the universal sequencing primer (e.g., SP6 or T7 primer from Promega Corporation). One of the two unique restriction sites should be near the end of the insert and must generate a blunt or 5′ overhang end that is necessary for exonuclease III deletion. The other enzyme should act near the sequencing primer binding site and must produce a 3′ overhang end to protect the end from exonuclease III deletion. Therefore, the exonuclease digestion will be unidirectional and proceed into the insert DNA sequence. The unique restriction enzymes are listed in Tables 8–11 and 8–12. The general procedures for subcloning of the DNA of interest into plasmids are described in detail in Chapter 7.

### Table 8–11 Unique Enzymes for Generating 5′ Protruding or Blunt Ends

| Enzyme | Recognition sequence | Enzyme | Recognition sequence |
|--------|---------------------|--------|---------------------|
| *Not* I | 5′..GC*GGCC GC..3′ | *Sma* I | 5′..CCC*GGG..3′ |
|        | 3′..CG CCGG*CG..5′ |        | 3′..GGG*CCC..5′ |
| *Xba* I | 5′..T*CTAG A..3′ | *Xho* I | 5′..CT*CGA G..3′ |
|        | 3′..A GATC*T..5′ |        | 3′..GA GCT*C..5′ |
| *Sal* I | 5′..G*TCGA C..3′ |  |  |
|        | 3′..C AGCT*G..5′ |  |  |

*Note:* The asterisks indicate the restriction enzyme cutting site.

Although *Hind* III, *EcoR* I, and *BamH* I can also generate 5′ overhangs, they are not usually unique since most of the DNA inserts contain the recognition sites for these enzymes.

### Table 8–12 Unique Enzymes That Produce Exonuclease III-Resistant 3′ Protruding (Overhang) Termini

| Enzyme | Recognition sequence | Enzyme | Recognition sequence |
|--------|---------------------|--------|---------------------|
| *Sph* I | 5′..G CATG*C..3′ | *Pvu* I | 5′..CG AT*CG..3′ |
|        | 3′..C*GTAC G..5′ |        | 3′..GC*TA GC..5′ |
| *Sac* I | 5′..G AGCT*C..3′ | *Kpn* I | 5′..G GTAC*C..3′ |
|        | 3′..C*TCGA G..5′ |        | 3′..C*CATG G..5′ |
| *Aat* II | 5′..G ACGT*C..3′ | *Bgl* I | 5′..GCCN NNN*NGGC..3′ |
|        | 3′..C*TGCA G..5′ |        | 3′..CGGN*NNN NCCG..5′ |

*Note:* The asterisks indicate the restriction enzyme cutting site.

## Protocol 2: Purification of Recombinant Plasmids for Exonuclease Deletions

The quality and integrity of plasmids isolated are critical for successful deletions with exonuclease III. Any nicks in the plasmids may cause exonuclease III digestion to occur from the nick points, resulting in single-stranded gaps. Therefore, uniform, circular double-stranded plasmids should be prepared and used for the enzyme deletion.

a.    Purification of Circular Plasmid DNA by Equilibrium Sedimentation in CsCl-Ethidium Bromide Gradients

The detailed procedures are described under the section on purification of plasmid DNA in Chapter 1.

b.    Purification of Supercoiled Plasmid DNA by the Acid-Phenol Extraction Method

1.  Isolate plasmids using the standard methods as described in Chapter 1, except one should dissolve the DNA in dd.H$_2$O (1 to 2 µg/µl). The plasmid DNA at this point should be free of RNA and proteins, but may contain nicked plasmids.

2.  To every 0.1-ml DNA sample, add 5 µl of 1 $M$ sodium acetate buffer (pH 4.0) and 3.8 µl of 2 $M$ NaCl solution. Mix well by pipetting.

3.  Remove the nicked and linear DNA by adding 1 volume of acid-phenol to the mixture. Mix well by vortexing.

4.  Centrifuge at 11,000 × $g$ for 5 min. Then, carefully transfer the top, aqueous phase (supernatant) to a fresh tube.

5.  Repeat steps 3 to 4 twice.

6.  Add 0.1 volume of 0.5 $M$ Tris-HCl (pH 8.6) and extract once with 1 volume of chloroform:isoamyl alcohol (24:1). Mix well.

7.  Centrifuge at 11,000 × $g$ for 5 min and carefully transfer the upper, aqueous phase to a fresh tube.

8.  Add 0.1 volume of 2 $M$ NaCl solution and 2 to 4 volumes of 100% ethanol. Allow to precipitate at –70°C for 30 min.

9.  Centrifuge at 12,000 × $g$ for 15 min, decant the supernatant, and briefly rinse the pellet with 1 ml of 70% ethanol. Dry the DNA pellet under vacuum for 30 min.

10. Dissolve the DNA in 10 to 50 µl of TE buffer, measure the concentration of the sample, and store at –20°C until use.

## Reagents Needed

*1 M Sodium Acetate Buffer (pH 4.0)*

*2 M Sodium Chloride Solution*

*TE Buffer*

*Chloroform:Isoamyl Alcohol (24:1)*

*Acid-Phenol (pH 4.0)*

- Melt phenol crystals at 55 to 65°C in water bath.
- Add 250 ml of 50 m$M$ sodium acetate buffer (pH 4.0) diluted from 1 $M$ sodium acetate buffer to 250 ml of the melted phenol with stirring for 2 to 3 h at room temperature.

- Stop the stirring and allow the phases to separate for 30 to 45 min.
- Remove the top aqueous phase by aspiration. Remove a small aliquot to determine the pH.
- Repeat the second to fourth steps two or three times until the pH reaches 4.0. Store at 4°C until use.

## Protocol 3: Double-Restriction Enzyme Digestions of Circular Plasmids

This is an essential step for successful deletion of DNA inserts of interest, which will be carried out later. According to the multiple cloning sites of the vectors and restriction enzyme recognition sequence information given in Procedure 2, two unique and different restriction enzymes should be used. One enzyme cleaves the DNA near one end of the target DNA and must produce blunt or 5' overhang ends that allow exonuclease III to progressively and unidirectionally digest the DNA into the target sequence. The other enzyme cuts the DNA near the binding site of universal primer sequence and must generate a 3' overhang end to resist exonuclease III digestion. The detailed protocols for double restriction enzyme digestion are described under the section on subcloning of DNA in Chapter 7.

## Protocol 4: Performing Series Deletion of the Linearized DNA with Exonuclease III and Recircularization of the DNA with T4 DNA Ligase (Figure 8.3)

1. Label 15 to 30 microcentrifuge tubes (0.5 ml) depending on the size of insert to be deleted, and add 7.5 to 8.0 µl of nuclease S1 mixture to each tube. Store on ice until use.

2. Dissolve or dilute 5 to 6 µg linearized DNA in a total 60 µl of 1X exonuclease III buffer.

3. Warm the sample at an appropriate temperature, start digestion by adding 250 to 550 units of exonuclease III, and mix as quickly as possible. Transfer 2.5 to 3.0 µl of the reaction mixture at 0.5 min intervals to the labeled tubes containing nuclease S1 (step 1). Quickly mix by pipetting up and down several times and place on ice until use.

*Notes:* *(1) The amount of DNA, the volume of reaction, and the amount of exonuclease III should be optimized experimentally depending on the size of insert to be deleted. (2) The digestion rate depends on the reaction mixture temperature as shown in Table 8-13. (3) One does not need to change buffer between exonuclease III and nuclease S1 since the S1 buffer contains zinc cations and has a low pH, which can inactivate the exonuclease III. But some exonuclease III buffer will not inhibit the activity of nuclease S1.*

4. After all of the samples have been taken, transfer all of the tubes to room temperature and carry out nuclease S1 digestion of single strands by incubating at room temperature for 30 to 35 min.

5. Terminate the reaction by adding 1 µl of S1 stopping buffer and heat at 70°C for 10 min.

6. Check the efficiency of digestion by preparing a 1% agarose gel using 2 to 3 µl of sample from each time point.

7. Carry out blunting of termini by adding 1 µl of Klenow mixture to each tube and incubating at 37°C for 4 min. Add 1 µl of dNTP mixture to each tube and incubate at 37°C for 5 to 6 min.

8. Recircularize the DNA by adding 40 µl of ligase mixture to each tube, mixing and incubating at room temperature for 60 to 70 min.

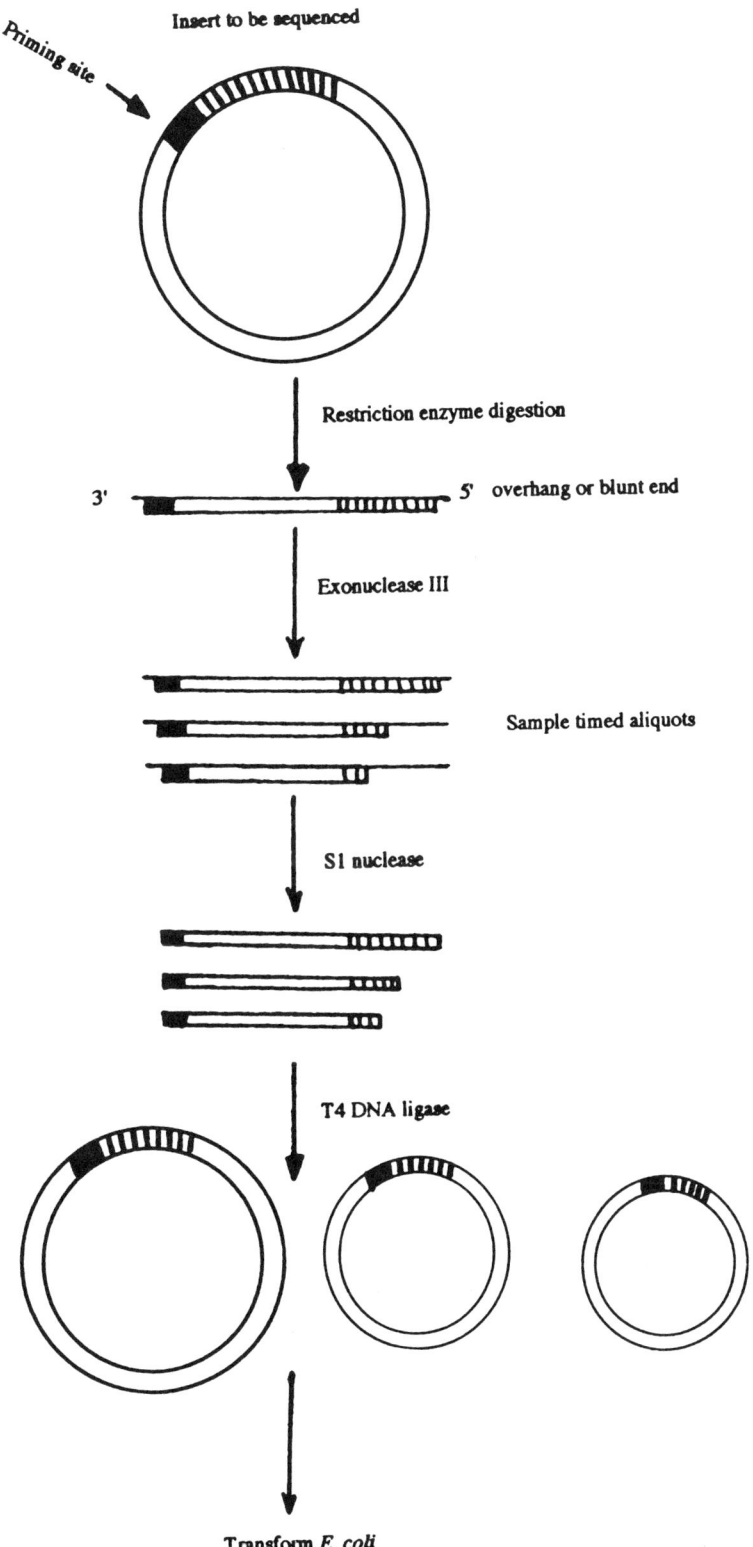

**FIGURE 8.3**

Diagram showing progressive deletions of a DNA fragment to be sequenced.

**Table 8–13 Dependence of Digestion
Rate on Temperature**

| Temperature (°C) | Digestion rate (bp/min) |
|---|---|
| 4 | 25–30 |
| 22 | 80–85 |
| 25 | 90–100 |
| 30 | 200–220 |
| 37 | 455–465 |
| 45 | 600–630 |

## Reagents Needed

### 10X Exonuclease III Buffer

0.66 $M$ Tris-HCl, pH 8.0
6.6 m$M$ MgCl$_2$

### 1X Exonuclease III Buffer

Dilute 10X exonuclease III buffer in dd.H$_2$O.

### S1 Buffer

2.5 NaCl
300 m$M$ Potassium acetate, pH 4.6
10 m$M$ ZnSO$_4$
50% Glycerol

### Nuclease S1 Mixture (Fresh)

54 µl of S1 buffer
0.344 ml dd.H$_2$O
120 units Nuclease S1

### S1 Stopping Buffer

300 m$M$ Tris base
50 m$M$ EDTA, pH 8.0

### Klenow Buffer

20 m$M$ Tris-HCl, pH 8.0
100 m$M$ MgCl$_2$

### Klenow Mixture

6 to 12 units Klenow DNA polymerase
60 µl Klenow buffer

## dNTP Mixture

> 0.13 m$M$ dATP
> 0.13 m$M$ dTTP
> 0.13 m$M$ dGTP
> 0.13 m$M$ dCTP

## 10X T4 DNA Ligase Buffer

> 0.5 $M$ Tris-HCl, pH 7.6
> 0.1 $M$ MgCl$_2$
> 10 m$M$ ATP

## T4 DNA Ligase Mixture

> 0.1 ml 10X Ligase buffer
> 0.79 ml dd.H$_2$O
> 0.1 ml 50% PEG
> 10 µl of 0.1 $M$ DTT
> 5 units T4 DNA ligase

## Protocol 5: Performing Transformation of Recircularized Plasmids for each Time Point to Appropriate Bacterial Hosts

The general procedures of transformation and selection of transformants are described in detail under the section on subcloning in Chapter 7.

## Protocol 6: Estimation of the Sizes of Deletion Subclones by a Fast Screening Method

1.  Mark 5 to 10 microcentrifuge tubes (0.5 ml) for each time point of deletions and the transformant colonies to be analyzed in the same order.

2.  Individually pick a portion (approximately two thirds) of each of 5 to 10 transformant colonies for each time point of the deletions using sterile toothpicks or pipette tips, and transfer each to the bottom of appropriate tubes that were labeled in step 1. Wrap the plates containing partial colonies of transformants and store at 4°C until use.

3.  Add 50 µl of 10 m$M$ EDTA (pH 8.0) to each tube and suspend the cells by vortexing for 1 min.

4.  Add 50 µl of 2X lysis buffer [8% (v/v) 2.5 $M$ NaOH, 5% (v/v) of 10% SDS, 20% (w/v) sucrose] to each tube, vortex for 1 min, and incubate at 70°C for 5 min. Cool to room temperature.

5.  Add 3 µl of 2 $M$ KCl and 0.5 µl of 0.2% bromophenol blue and vortex. Incubate on ice for 5 min.

6.  Centrifuge at 9000 × $g$ for 4 min at 4°C. Transfer the supernatant to fresh tubes.

7.  Load appropriate supercoiled DNA markers and 50 to 60 µl of each supernatant onto a 0.7% agarose gel containing ethidium bromide (10 µl of 10 mg/ml ethidium bromide per 100 ml of gel mixture). Allow electrophoresis to occur until the dye reaches three quarters the length of the gel.

8.  Photograph the gel and estimate the size of the plasmids from each time point of deletions.

9.  Culture appropriate colonies from the plates stored at 4°C, which contain the expected sizes after deletions, in an appropriate volume of LB medium with appropriate antibiotics, and isolate plasmids for nucleotide sequencing.

*Protocol 7: Identification of Overlapping Sequences of Different Deletions and Performing a Full Nucleotide Sequence Analysis*

1.  For each subclone, set up a T-track analysis using only the ddTTP reaction. The band patterns will reveal the overlapping sequences of multiple deletions.

2.  Carry out a full nucleotide sequence analysis for each deleted insert. The entire sequence of the insert will be obtained. The detailed procedures for sequencing are described in Section I.

# References

1.  **Cullmann, G., Hubscher, U., and Berchtold, M. W.,** A reliable protocol for dsDNA and PCR product sequencing, *BioTechniques,* 15, 578, 1993.

2.  **Sanger, F., Nicken, S., and Coulson, A. R.,** DNA sequencing with chain termination inhibitors, *Proc. Natl. Acad. Sci. U.S.A.,* 74, 5463, 1977.

3.  **Church, G. M. and Gilbert, W.,** Genomic sequencing, *Proc. Natl. Acad. Sci. U.S.A.,* 81, 1991, 1984.

4.  **Bishop, M. J. and Rawlings, C. J.,** *Nucleic Acid and Protein Sequence Analysis: A Practical Approach,* IRL Press, Oxford, 1987.

5.  **Wiemann, H. V., Grothues, D., Sensen, C., Zimmermann, C. S., Stegemann, H. E., Rupp, T., and Ansorge, W.,** Automated low-redundancy large-scale DNA sequencing by primer walking, *BioTechniques,* 15, 714, 1993.

6.  **Reynolds, T. R., Uliana, S. R. B., Floeter-Winter, L. M., and Buck, G. A.,** Optimization of coupled PCR amplification and cycle sequencing of cloned and genomic DNA, *BioTechniques,* 15, 462, 1993.

Chapter

# PCR Techniques and Applications

## Contents

The polymerase chain reaction (PCR) is powerful technique used for *in vitro* amplification of specific DNA sequences using appropriate primers.[1,2] PCR is a rapid, sensitive, and inexpensive procedure to amplify specific DNA of interest, and is a major breakthrough technology in the analysis of DNA and RNA, DNA cloning, genetic diagnosis, detection of mutations, and genetic engineering.[3-9] In the process, *Taq* DNA polymerase purified from *Thermus aquaticus* is a heat-stable enzyme and carries out the synthesis of a complementary strand of DNA from 5′ to 3′ direction by the primer extension reaction. The primers are designed so that primer A directs the synthesis of DNA towards the other, primer B, and vice versa, resulting in the synthesis of the region of DNA flanked by the two primers. Because the *Taq* DNA polymerase is high-temperature (94°C) stable, it allows the target sequences of interest to be amplified many cycles using excess primers and commercial thermocycler apparatus. The present chapter describes the procedures of PCR-directed dsDNA amplification, semiquantitative PCR for analysis of gene expression, RT-PCR cDNA cloning, detection of genetic diseases and mutations by PCR, and PCR sequencing.

# I.    General Amplification of Double-Strand DNA by PCR

*Protocol*

1.   Carry out PCR amplification.

   a.  Design oligonucleic primers that are complementary to two different regions of the DNA molecule or DNA fragment of interest. For example, two oligonucleotide primers can be designed as follows for the amplification of actin cDNA.

   | | |
   |---|---|
   | Primer 1 | 5′ ATGGATGACGATATCGCTG 3′ |
   | Primer 2 | 5′ ATGAGGTAGTCTGTCAGGT 3′ |

*Note:*   *These primers can generate a product of 568 base pairs.*

   b.  In a 0.5-ml microcentrifuge tube on ice, set up a standard reaction by adding the following in the order listed:
   * 10X Amplification buffer (5 μl)
   * dd.H$_2$O (10 μl)
   * Mixture of four dNTPs (1.25 m$M$ each) (8.5 μl)
   * Primer 1 (50 to 100 ng) in dd.H$_2$O (5 μl)
   * Primer 2 (50 to 100 ng) in dd.H$_2$O (5 μl)
   * dsDNA in TE buffer (100 to 250 ng) (2 μl)
   * *Taq* DNA polymerase (5 units/μl) (0.5 μl)
   * Add dd.H$_2$O to a final volume of 50 μl.

   c.  Overlay the mixture with 50 μl of light mineral oil (Sigma or equivalent) to prevent evaporation of the sample.

   d.  Carry out 25 to 35 cycles of PCR amplification in a PCR machine, which is programed as

   | Cycle | Denaturation | Annealing | Polymerization |
   |---|---|---|---|
   | First | 4 min at 94°C | 2 min at 50°C | 3 min at 72°C |
   | Subsequent | 1 min at 94°C | 2 min at 50°C | 3.5 min at 72°C |
   | Last | 1 min at 94°C | 2 min at 50°C | 10 min at 72°C |

   Finally, hold at 4°C until the sample is removed.

   e.  Carefully remove the reaction mixture from the mineral oil using a pipette with a relatively long tip attached. Slowly insert the tip into the bottom of the tube and then carefully take the sample until the oil phase. Withdraw the tip from the tube, wipe the outside of the tip with a clean paper towel, and transfer the sample into a fresh tube.

2.   Purify the amplified cDNA by agarose gel elution.

   a.  Add DNA loading buffer to the amplified cDNA sample and DNA standard markers, and load the sample into 1% agarose gel (low-melting point) in 1X TBE buffer, which contains EtBr for staining. Carry out electrophoresis at 7.5 V/cm for 4 to 5 h.

   b.  Under UV light, slice the individual sharp band(s) using a clean razor blade and trim away excess agarose gel as much as possible. Place the band slice into a clean Eppendorf tube and add 2 volumes of TE buffer to the tube.

    c.  Completely melt the gel slice at 75°C for 5 min and quickly chill on dry ice or at –80°C for 30 min.

    d.  Remove the frozen tube to room temperature and thaw the gel mixture by vigorously tapping the tube until a liquid suspension appears.

    e.  Centrifuge at $10,000 \times g$ for 6 min and carefully transfer the liquid containing the DNA into a fresh tube. Discard the gel pellet.

    f.  Add 0.15 volume of 3 $M$ sodium acetate buffer (pH 5.2) and 2.5 volumes of chilled 100% ethanol and allow precipitation at –80°C for 30 min.

    g.  Centrifuge at $12,000 \times g$ for 10 min and briefly rinse the pellet with 1 ml of 70% ethanol. Dry the pellet under vacuum for 10 min and dissolve the DNA in 20 µl TE buffer. Measure the concentration of the DNA (see Chapter 1) and store at –20°C until use.

3.    Carry out Southern blot or dot blot hybridization to verify whether the eluted DNA is the amplified DNA of interest using a DNA template as a probe (see Chapter 5 for details).

## Reagents Needed

### 10X Amplification Buffer

      100 m$M$ Tris-HCl, pH 8.3
      500 m$M$ KCl
      15 m$M$ MgCl$_2$
      0.1% BSA

### TE Buffer

      10 m$M$ Tris-HCl, pH 8.0
      1 m$M$ EDTA

### TE-Saturated Phenol/Chloroform

      Thaw crystals of phenol at 65°C and mix with an equal volume of TE buffer. Mix well and allow the phases to separate at room temperature for 30 min. Take 1 part of the lower, phenol phase and mix with 1 part of chloroform:isoamyl alcohol (24:1). Mix well, allow the phases to separate, and store at 4°C until use.

### 7.5 M Ammonium Acetate

### Ethanol (100% and 70%)

## II.    Analysis of Gene Expression by Semiquantitative PCR

Northern blot hybridization and *in situ* hybridization of mRNA are techniques that are commonly used for analysis of gene expression. The disadvantage of these approaches, however, is the difficulty in obtaining the desired result(s) when analyzing nonconstitutively expressed, cell- or tissue-specific genes having low-abundance mRNAs or when limiting amounts of mRNA are available due to degradation. This drawback is overcome by quantitative PCR. The principle of this technology is that total RNAs or mRNAs are isolated from different samples of cell types, or tissue types, or developmental stages including chemically treated (+) and untreated (−), and time-course treatments to study the kinetic expression of specific genes. The same amount of total RNAs or mRNAs from each sample is reverse transcribed into cDNAs under the same conditions. The particular cDNA of interest in each of the samples is then amplified by PCR using specific primers. The same amount of the amplified cDNA from each sample is then analyzed by dot blotting or Southern blotting using the target sequence between primers as a probe. In this case, different amounts of mRNA expressed by specific genes under different conditions will generate different amounts of cDNA. By comparing the measurements of the hybridized signal intensities of different samples, the different levels of gene expression can be detected with ease. At the same time, a control mRNA (usually an appropriate housekeeping gene transcript) such as actin mRNA is reverse transcribed and amplified by PCR, using specific primers under exactly the same conditions, in order to monitor equivalent reverse transcription to cDNA and equivalent amplification in PCR.

## *Protocol*

1. Isolate total RNAs or purify mRNAs from different cell or tissue types of interest, or the same cell or tissue type at different development stages or under different treatments for the analysis of gene expression. The detailed procedures are described in Chapter 2.

2. Synthesize the first-strand cDNAs from the isolated total RNAs or mRNAs using reverse transcriptase and oligo(dT) as a primer.

   a. Anneal 1 μg total RNA or 50 ng mRNA template with 2 μg oligo(dT) primer in a sterile RNase-free microcentrifuge tube. Add nuclease-free dd.H$_2$O to a total volume of 15 μl. Heat the tube at 70°C for 5 min and allow it to slowly cool to room temperature to finish annealing. Briefly spin down the mixture to the bottom.

   b. To the annealed primer/template, add the following in the order shown. To prevent precipitation of sodium pyrophosphate when it is added to the buffer components, the buffer should be preheated at 40 to 42°C for 4 min before the addition of sodium pyrophosphate and AMV reverse transcriptase. Gently mix well after each addition.

      • First-strand 5X buffer (5 μl)

      • rRNasin ribonuclease inhibitor (25 units/μg RNA) (50 units)

      • 40 m*M* Sodium pyrophosphate (2.5 μl)

      • AMV reverse transcriptase (15 units/μg RNA) (30 units)

      • Add nuclease-free dd.H$_2$O to total volume of 25 μl.

      **Functions:** The 5X reaction buffer contains components required for cDNA synthesis. Ribonuclease inhibitor inhibits RNase activity and protects mRNA template. AMV reverse transcriptase catalyzes the synthesis of the first-strand cDNA from the mRNA template based on the rule of complementary bases.

   c. Set up a tracer reaction by removing 5 μl of the mixture to a fresh tube containing 1 μl of 4 μCi [α-$^{32}$P]dCTP (>400 Ci/mmol, less than one week old). The first-strand synthesis will be measured by trichloroacetic acid (TCA) precipitation and alkaline agarose gel electrophoresis using the tracer reaction.

*Caution:* *[α-$^{32}$P]dCTP is a dangerous isotope. Gloves should be worn when working on the tracer reaction, TCA assay, and gel electrophoresis (see Chapter 6, cDNA libraries). Waste material such as contaminated gloves, tips, solutions, and filter papers should be put in a special container.*

d. Incubate both reactions at 42°C for 1.5 h and place on ice. At this stage, the synthesis of the first-strand cDNA is completed. To the tracer reaction, add 50 m$M$ EDTA up to a total volume of 100 μl and store on ice for TCA incorporation assays or for alkaline agarose gel analysis after extraction.

e. Inactivate the reactions by heating at 70°C for 10 min, add 20 μl dd.H$_2$O to the unlabeled reaction, and store at –20°C until amplification by PCR.

*Notes:*   *(1) Small amounts of total RNA (1 to 1.5 μg) or mRNA (20 to 50 ng) should be used for semiquantitative analysis of specific mRNA. (2) The conditions for reverse transcription should be the same for each of the comparison samples and actin control including all the components, volume, temperature, and reaction time.*

3.   Amplify the specific cDNA of interest by PCR using specific primers.

a. Design oligonucleic primers based on conserved amino acid sequences in two different regions of the specific gene product (protein) of interest. For example, two oligonucleotide primers can be designed as follows for the two amino acid-sequence regions, NDPNG and DPCEW, of invertase.

   Primer 1      5′ AAC(T)GAT(C)CCIAA(C)TGGI 3′ for NDPNG

   Primer 2      3′ GGTGAGCGTCCCTAG_5′        for DPCEW

*Notes:*   *I stands for the third position of the codon, which can be any of T, C, A, G. The primers can be synthesized by a DNA synthesizer. The primers can generate a product of 554 base pairs.*

For amplification of actin cDNA, the primers can be designed as follows:

   Primer 1      5′ ATGGATGACGATATCGCTG 3′

   Primer 2      5′ ATGAGGTAGTCTGTCAGGT 3′

*Note:*   *These primers can generate a product of 568 base pairs.*

b. Carry out PCR amplification.

   i.   In a 0.5-ml microcentrifuge tube on ice, add the following in the listed order for amplification of one sample:

   • 10X Amplification buffer (10 μl)

   • dd.H$_2$O (20 μl)

   • Mixture of four dNTPs (1.25 m$M$ each) (17 μl)

   • Primer 1 (100 to 110 pmol) in dd.H$_2$O (4 μl)

   • Primer 2 (100 to 110 pmol) in dd.H$_2$O (4 μl)

   • cDNA synthesized previously (0.1 to 1 μg) (6 μl)

   • Add dd.H$_2$O to a final volume of 100 μl.

   **For no cDNA control:**

   • 10X Amplification buffer (10 μl)

   • dd.H$_2$O (20 μl)

   • Mixture of four dNTPs (1.25 mM each) (17 μl)

   • Primer 1 (100 to 110 pmol) in dd.H$_2$O (4 μl)

   • Primer 2 (100 to 110 pmol) in dd.H$_2$O (4 μl)

   • Add dd.H$_2$O to a final volume of 100 μl.

ii.   Repeat until all samples, including the control actin cDNA, have been set up. Add 2.5 units of *Taq* DNA polymerase (5 units/μl, Perkin Elmer Cetus) to each of the tubes. Gently mix well.

iii.  Overlay the mixture with 50 μl of light mineral oil (Sigma or equivalent) to prevent evaporation of the sample.

iv.   Carry out PCR amplification for 35 to 40 cycles in a PCR cycler, which is programmed as follows:

| Cycle | Denaturation | Annealing | Polymerization |
|---|---|---|---|
| First | 4 min at 94°C | 2 min at 50°C | 3 min at 72°C |
| Subsequent | 1 min at 94°C | 2 min at 50°C | 4 min at 72°C |
| Last | 1 min at 94°C | 2 min at 50°C | 10 min at 72°C |

*Note:*  *The conditions for amplification should be exactly the same for every sample, whether with actin cDNA or no cDNA.*

v.    Starting after 15 cycles, carefully push a pipette tip through the oil and remove 15 μl of the reaction from each of the samples and from the controls every 5 cycles at the end of appropriate cycle (at 72°C extension phase). Place tubes containing the 15 μl reactions at 4°C until use.

*Note:*  *Sampling should be done in 2 min for all samples. There should be six samplings for each sample in 40 cycles. Oil should be avoided as much as possible.*

vi.   Use the 15 μl of amplified cDNA from each of the samples and controls to carry out dot blot hybridization using the appropriate [32]P-labeled internal oligonucleotides or the target sequence (e.g., cDNA fragment) between the two primers as a probe. The detailed procedures for dot blot hybridization are described in Chapter 5.

vii.  Measure the signal intensities of the dot blots with a densitometer and analyze the data. For cDNA controls, no signal should be visible. For actin cDNA, the signal intensities should be increased as amplified cycles, but no differences should be seen between any cell or tissue types or any developmental stages, or treated and untreated tissues. In contrast, if the expression of a specific gene is indeed nonconstitutive, cell- or tissue-specific, or chemical-treatment induced, clear signal patterns of amplified cDNA dot blot hybridization can be seen. The signal for each sample should be stronger as amplified cycles (Figure 9.1). For example, if the signal of amplified invertase cDNA in gibberellin-treated tissue is much stronger than that of untreated tissue, and if the signal is increased from cycle 15, to 20, 25, 30, 35, and to 40, the expression of invertase mRNA, which was used for the synthesis of the cDNA blotted, is potentially induced by gibberellin treatment.

viii. To determine the size of PCR products, repeat amplification of 6 μl cDNA synthesized previously, and carry out 1.0 to 1.4% agarose gel electrophoresis or Southern blot hybridization using the target sequence as a probe (see Southern blotting in Chapter 5).

## Reagents Needed

### First-Strand 5X Buffer

250 m*M* Tris-HCl, pH 8.3 (42°C)
50 m*M* MgCl$_2$
250 m*M* KCl
2.5 m*M* Spermidine
50 m*M* DTT
5 m*M* Each of dATP, dCTP, dGTP, and dTTP

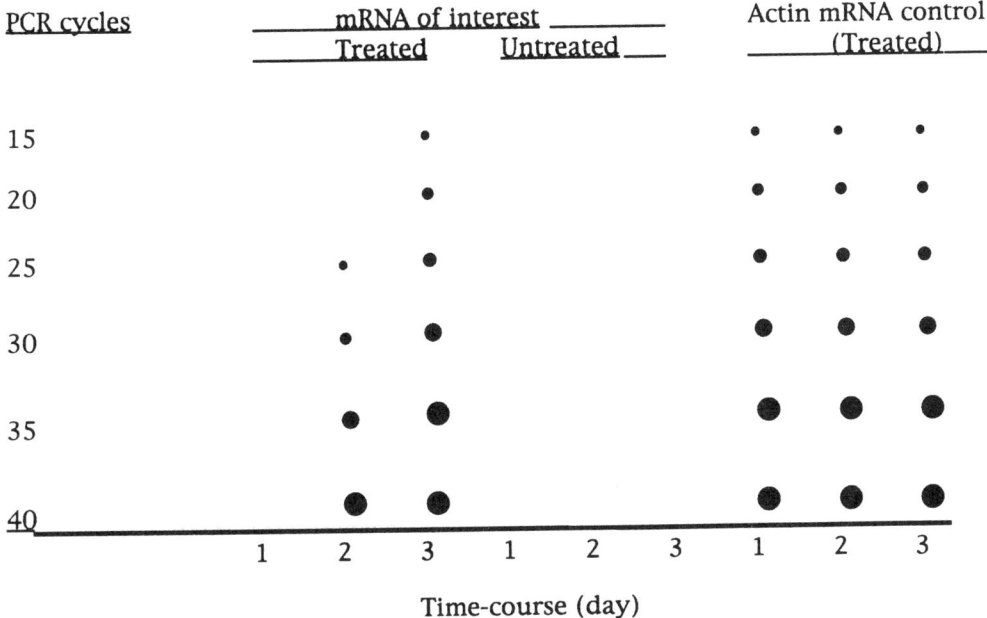

**FIGURE 9.1**
Diagram showing an analysis of a given mRNA of interest and actin mRNA used as a control obtained by the use of a quantitative PCR.

*10X Amplification Buffer*
> 100 m$M$ Tris-HCl, pH 8.3
> 500 m$M$ KCl
> 15 m$M$ MgCl$_2$
> 0.1% BSA

*Mixture of Four dNTPs (1.25 mM each)*

*rRNasin Ribonuclease Inhibitor*

*200 mM Sodium Pyrophosphate*

*AMV Reverse Transcriptase*

*Nuclease-Free dd.H$_2$O*

*[$\alpha$-$^{32}$P]dCTP (>400 Ci/mmol)*

*50 mM EDTA*

*Ethanol (100% and 70%)*

*Trichloroacetic Acid (5% and 7%)*

# III.    cDNA Cloning by RT-PCR

RT-PCR is a powerful technique used for cDNA cloning as long as specific primers are available. It is a relatively fast, simple, and inexpensive procedure as compared with other strategies described above. Total RNA or mRNA is first reverse transcribed into cDNA that is then amplified by PCR — the RT-PCR technique. Briefly, specific primers are designed according to conserved amino acid motifs in two different regions of known proteins or enzymes. The primers will be annealed to the first-strand cDNA synthesized from mRNA templates of particular organisms of interest by reverse transcription. Double-strand cDNAs can be generated and amplified using *Taq* DNA polymerase. There are, however, two major disadvantages of applying the PCR cloning strategy: (1) the cDNA obtained is usually of partial length ranging from 200 to 1000 base pairs, depending on the distance between the two primers designed; and (2) the annealing of the primer to template, sometimes, is not specific and amplify nonspecific sequences called "artifacts". Therefore, we strongly recommend that the PCR products be verified by dot blot or Southern blot hybridization using the targeting sequence as probe, and that positive PCR products be sequenced and compared with other known sequences.

## *Protocol*

1.   Carry out isolation of total RNA and purification of poly(A)+RNA from tissue or cell lines of interest (see Chapter 2).

2.   Design oligonucleic primers based on conserved amino acid sequences in two different regions of known proteins or enzymes. Each primer is designed with one specific restriction enzyme site for subcloning of the forthcoming double-strand cDNAs.

   For example, two amino acid sequence regions, NDPNG and DPCEW, of invertase are conserved from prokaryotes to eukaryotes, which have been characterized and published in professional journals. The first one is close to the N'-terminal, the other is toward the C-terminal. Two oligonucleotides can be designed according to the two conserved amino acid regions. In order to subclone the forthcoming ds cDNAs into a specific vector for sequencing, a *BamH* I restriction site is designed at the N-terminal of the first primer, and a *Hind* III site follows the C-terminal of the second primer. The design is as follows:

   Primer 1     5′ <u>GGATCC</u>AAC(T)GAT(C)CCIAA(C)TGGI3′ for NDPNG

   Primer 2     3′GGTGAGCGTCCCTAGTTCGAA5′          for DPCEW

   The sequences underlined are the *BamH* I site at the 5′ end and the *Hind* III site at the 3′ end of the forthcoming ds cDNAs. I stands for the third position of the codon, which can be any of TCAG. The synthesis of the primers can be done with a DNA synthesizer.

   Alternatively, primer 1, 5′<u>GGATCC</u>AAC(T)GAT(C)CCIAA(C)TGGI3′, can be designed for NDPNG, which is close to the N'-terminal of the protein, and primer 2 can be oligo(dA)<u>TTCGGA</u>. The length of the forthcoming ds cDNA will be longer than that made by the first choice of the two primers.

3.   Synthesize the first-strand cDNA followed by the second-strand cDNA as described previously (see Chapter 6). However, if two primers are available, PCR amplification can be immediately carried out directly after the first-strand cDNA.

4.  Carry out PCR amplification of ss cDNA or ds cDNA.

    a.  In a microcentrifuge tube on ice, add the following in the order listed:

        • 10X Amplification buffer (10 µl)

        • dd.H$_2$O (20 µl)

        • Mixture of four dNTPs (1.25 m*M* each) (17 µl)

        • Primer 1 (100 to 110 pmol) in dd.H$_2$O (5 µl)

        • Primer 2 (100 to 110 pmol) in dd.H$_2$O (5 µl)

        • ss cDNA or ds cDNA in TE buffer (1 to 2 µg)

        • *Taq* DNA polymerase (5 units/µl) (1 µl)

        • Add dd.H$_2$O to a final volume of 100 µl.

    b.  Overlay the mixture with 100 µl of light mineral oil (Sigma or equivalent) to prevent evaporation of the sample.

    c.  Carry out PCR amplification in a PCR machine, which is programed as follows:

| Cycle | Denaturation | Annealing | Polymerization |
|---|---|---|---|
| First | 4 min at 94°C | 2 min at 52°C | 3 min at 70°C |
| Subsequent | 1 min at 94°C | 2 min at 52°C | 3.5 min at 70°C |
| Last | 1 min at 94°C | 2 min at 52°C | 10 min at 70°C |

    Finally, hold at 4°C until the sample is removed.

    d.  Carefully remove the reaction mixture from the mineral oil using a pipette with a relatively long tip attached. Slowly insert the tip into the bottom of the tube and then carefully take the sample until the oil phase. Withdraw the tip from the tube, wipe the outside of the tip with a clean paper towel, and transfer the sample into a fresh tube.

5.  Purify the amplified cDNA by agarose gel elution.

    a.  Add DNA loading buffer to the amplified cDNA sample and DNA standard markers, and load the sample into 1% agarose gel (low-melting point) in 1X TBE buffer, which contains EtBr for staining. Carry out electrophoresis at 7.5 V/cm for 4 to 5 h (Figure 9.2).

    b.  Under UV light, slice the individual sharp band(s) using a clean razor blade and trim away excess agarose gel as much as possible. Place the band slice into a clean Eppendorf tube and add 2 volumes of TE buffer to the tube.

    c.  Completely melt the gel slice at 75° for 5 min and quickly chill on dry ice or at −80°C for 30 min.

    d.  Remove the frozen tube to room temperature and thaw the gel mixture by vigorously tapping the tube until a liquid suspension appears.

    e.  Centrifuge at 10,000 × *g* for 6 min and carefully transfer the liquid containing the DNA into a fresh tube. Discard the gel pellet.

    f.  Add 0.15 volume of 3 *M* sodium acetate buffer (pH 5.2) and 2.5 volumes of chilled 100% ethanol and allow precipitation at −80°C for 30 min.

    g.  Centrifuge at 12,000 × *g* for 10 min and briefly rinse the pellet with 1 ml of 70% ethanol. Dry the pellet under vacuum for 10 min and dissolve the DNA in 20 µl TE buffer. Measure the concentration of the DNA (see Chapter 1) and store at −20°C until use.

6.  Ligate the cDNA to an appropriate vector.

    a.  Digest the cDNA and vector such as pGEM-11Zf(+) (Promega Corporation) or equivalent with *BamH* I and *Hind* III to generate sticky ends for subcloning.

        • cDNA or vector (2 to 5 µg)

        • *Hamd* III 10X buffer (5 µl)

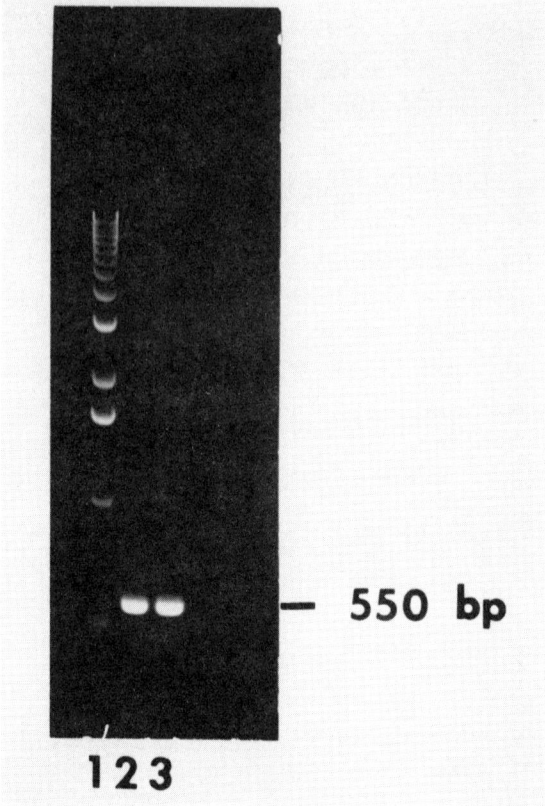

**FIGURE 9.2**

Amplification of a partial-length cDNA (550 bp) after 30 cycles of the PCR. The products were subjected to 1% agarose gel electrophoresis, stained with ethidium bromide and photographed.

- BSA (optional) (2.5 μl)
- *Hind* III (20 units)
- *BamH* I (20 units)
- Add dd.H$_2$O to a final volume of 50 μl.

b. Incubate at 37°C for 2.5 h and extract once by adding 1 volume of TE-saturated phenol/chloroform. Vortex for 1 min.

c. Centrifuge at 11,000 × g for 5 min and carefully remove the top, aqueous phase to a fresh tube.

d. Extract the supernatant once with 1 volume of chloroform:isoamyl alcohol (24:1) and centrifuge as in step c.

e. Add 0.5 volume of 7.5 M ammonium acetate to the supernatant, mix, and add 2 volumes of chilled 100% ethanol. Allow precipitation at –20°C for 2 h.

f. Centrifuge at 12,000 × g for 10 min and briefly rinse the pellet with 1 ml of cold 70% ethanol. Dry the cDNA-linker under vacuum for 10 min and resuspend the pellet in 20 μl of TE buffer. Store the sample at –20°C until use.

g. Set up the ligation reaction as follows:
- Ligase 10X buffer (2 μl)
- Vectors (2 μg)
- cDNA inserts (0.5 μg)
- T4 DNA ligase (10 Weiss units)
- Add dd.H$_2$O to a final volume of 20 μl.

h. Incubate the ligation reactions at room temperature (22 to 24°C) for 4 h and proceed to subcloning as described in Chapter 7. At this stage, the cDNA inserts are inserted in the same direction in the vector.

i. Carry out sequencing of the amplified cDNA as described in Chapter 8.

j. Compare the nucleotide sequence and the deduced amino acid sequence with known protein/enzyme from which the primers were designed. The homology should be 40 to 99% if the right cDNA was cloned.

k. Label the putative cDNA as a probe which can be used for Southern blot, Northern blot and screen cDNA library to obtain full-length cDNA.

## Reagents Needed

### 10X Amplification Buffer

100 m$M$ Tris-HCl, pH 8.3
500 m$M$ KCl
15 m$M$ MgCl$_2$
0.1% BSA

### Ligase 10X Buffer

300 m$M$ Tris-HCl, pH 7.8
100 m$M$ MgCl$_2$
100 m$M$ DTT
10 m$M$ ATP

### Hind III 10X Buffer

300 m$M$ Tris-HCl, pH 7.8
100 m$M$ MgCl$_2$
500 m$M$ NaCl
100 m$M$ DTT

### TE Buffer

10 m$M$ Tris-HCl, pH 8.0
1 m$M$ EDTA

### TE-Saturated Phenol/Chloroform

Thaw crystals of phenol at 65°C and mix with an equal part of TE buffer. Mix well and allow the phases to separate at room temperature for 30 min. Take 1 part of the lower, phenol phase and mix with 1 part of chloroform:isoamyl alcohol (24:1). Mix well, allow the phases to separate, and store at 4°C until use.

### 7.5 M Ammonium Acetate

### Ethanol (100% and 70%)

# IV.    Detection of Genetic Disease and Mutations by PCR, and Direct Sequencing of PCR Products

It has been demonstrated that many human and plant genetic diseases such as sickle-cell anemia and defective X-linked gene are caused by mutation in the coding regions of a single gene. Several approaches such as cDNA and genomic DNA cloning, sequencing, and Southern and Northern blot analyses have been developed to attempt to detect all mutations. But the techniques are quite complicated, time-consuming, expensive, and cannot identify all mutations. These disadvantages have been overcome by recent PCR technology. The basic principles of the PCR approach is that primers are designed for individual exons of the gene of interest based on the known sequence of the gene, such as the ornithine transcarbamylase (OTC) gene whose deficiency/mutations result in the most common urea cycle disorder in man. Multiple exons of the gene can be simultaneously amplified *in vitro* using a multiplex reaction of PCR. The PCR products are analyzed by agarose gel electrophoresis, and deletions affecting part or all of the gene can be detected as missing band(s) within the multiplex PCR. Furthermore, the PCR products from the mutant exon(s) can be directly sequenced, using appropriate primer(s), to identify the mutant point(s) or small rearrangements, including the splice sites flanking the mutant exon(s), as compared with wild-type gene sequences.

## Protocol A: Detection of Mutation at the Exon Level in the Coding Region of the Gene of Interest

1.  Isolate and purify genomic DNA for amplification and screening from blood cells, cell culture/ suspension, or tissue from the individual or species of interest. The general procedures are described in Chapter 1.

2.  Design multiplex PCR primers according to the known sequence of the gene of interest, and synthesize and purify the primers with an automatic synthesizer and purification column(s).

*Notes:*    *(1) Two oligonucleotide primers should be complementary to the target sequence of each exon. One primer (5′PCR primer) corresponds to the sense strand of the exon DNA, the other (3′PCR primer) to the opposite (antisense) strand of the DNA. The 3′ ends of the two primers should face each other so that the target sequence between the two primers can be amplified by PCR, generating a major product. (2) If two exons are very close to each other, a pair of primers flanking these two exons are sufficient for amplifying the two exons in a single fragment of PCR product. (3) The primers should be 24 to 25 nucleotides long with 50% G/C-rich, or a long A/T-rich, or shorter G/C-rich segment, depending on the individual gene of interest. (4) In order to obtain the entire sequence of the individual exon including the splicing site, the oligonucleotide primers should be designed based on the near intron sequences that flank the exon. (5) If sequencing is desired, a third oligonucleotide primer (sequencing primer) can be designed to correspond to the antisense strand near to the exon of interest in order to sequence the sense strand of the exon.*

3.  Carry out multiplex amplification of multiple exons in a single PCR reaction.

    a.  In a 0.5-ml microcentrifuge tube on ice, set up a standard reaction by adding the following in the order listed:

    •  10X Amplification buffer (5 µl)

    •  dd.H$_2$O (10 µl)

    •  Mixture of four dNTPs (10 m$M$ each) (8.5 µl)

- 15 to 25 pmol Each of primers designed (16 µl) (total)
- dsDNA template in dd.H$_2$O (200 to 250 ng) (2 µl)
- *Taq* DNA polymerase (5 units/µl) (1 µl)
- Add dd.H$_2$O to a final volume of 50 µl.

*Note:*    *The amplification of exons of normal (wild-type) and mutant genes should be carried out under the same conditions.*

c.  Overlay the mixture with 50 µl of light mineral oil (Sigma or equivalent) to prevent evaporation of the sample.

d.  Carry out 25 to 30 cycles of PCR amplification in an automatic thermocycler, which is programed as follows:

| Cycle | Denaturation | Annealing | Polymerization |
|---|---|---|---|
| First | 4 min at 94°C | 2 min at 50°C | 3 min at 68°C |
| Subsequent | 1 min at 94°C | 1 min at 60°C | 3 min at 68°C |
| Last | 1 min at 94°C | 1 min at 60°C | 5 min at 68°C |

Finally, hold at 4°C until the sample is removed.

e.  Carefully remove 15 µl from the reaction mixture through the paraffin oil. Using a pipette with a relatively long tip attached, slowly insert the tip into the bottom of the tube and then carefully take the sample until the oil phase. Withdraw the tip from the tube, wipe the outside of the tip with a clean paper towel, and transfer the sample into a fresh tube. Carry out electrophoresis on 1.4% agarose gel in 1X TBE buffer and run in 0.5X TBE buffer (see section on electrophoresis in Southern blotting, Chapter 5). Control PCR products amplified from exons of normal genes (wild type) should be loaded next to the potentially mutant PCR products for comparison.

f.  Analyze the PCR products and identify the deleted/mutant exon(s) in the gene of interest. If the exon is partially or completely deleted, PCR amplification cannot be performed, resulting in the missing band(s) as compared with the PCR product bands of wild-type genes.

4.  Prepare the single-strand PCR products for sequencing.

a.  For each exon to be sequenced, chose only one primer corresponding to the sense strand of the DNA (5′PCR primer) in order to synthesize a complementary antisense strand (PCR product). Thus, when the antisense strand of DNA is used as a template for sequencing, the sequence obtained by synthesizing a new strand DNA is the sequence of the original, sense strand DNA.

b.  In a 0.5-ml microcentrifuge tube on ice, set up a reaction by adding the following in the order listed:

- 5X Single-strand amplification buffer (20 µl)
- dd.H$_2$O (20 µl)
- Mixture of four dNTPs (10 m*M* each) (8 µl)
- 25 pmol of Appropriate 5′PCR primer designed (3 µl)
- Multiplex PCR product or dsDNA template in dd.H$_2$O (200 ng, in 2 µl) (0.5 µl)
- *Taq* DNA polymerase (5 units/µl) (1 µl)
- Add dd.H$_2$O to a final volume of 100 µl.

*Note:*    *The amplification of exons of normal (wild type) and mutant genes should be carried out under the same conditions.*

c.  Overlay the mixture with 100 µl of light mineral oil (Sigma or equivalent) to prevent evaporation of the sample.

d. Carry out 30 cycles of PCR amplification in an automatic thermocycler, which is programed as follows:

| Cycle | Denaturation | Annealing | Polymerization |
|---|---|---|---|
| First | 4 min at 94°C | 2 min at 50°C | 3 min at 68°C |
| Subsequent | 1 min at 94°C | 1 min at 60°C | 3 min at 68°C |
| Last | 1 min at 94°C | 1 min at 60°C | 5 min at 68°C |

Finally, hold at 4°C until the sample is removed.

e. Dilute the PCR products with 1 volume of dd.H$_2$O and precipitate the DNA by adding 0.5 volume of 7.5 $M$ ammonium acetate, mix, and add 2.5 volumes of chilled 100% ethanol. Allow to precipitate at –70°C for 30 min.

f. Centrifuge at $12,000 \times g$ for 15 min at room temperature. Carefully discard the supernatant, resuspend the pellet in 0.2 ml dd.H$_2$O, and repeat ammonium acetate precipitation as in step e.

g. Centrifuge as in step f. Briefly rinse the pellet with 0.4 ml of 70% ethanol and dry the pellet under vacuum for 30 min. Dissolve in 10 μl dd.H$_2$O immediately before the sequencing reaction.

## Buffer Needed

### 5X Single-Strand Amplification Buffer

0.25 $M$ KCl

50 m$M$ Tris-HCl, pH 8.8

7.5 m$M$ MgCl$_2$

5. Perform sequencing of single-strand PCR product.

a. Design a sequencing primer (17 to 18 nucleotides long) that hybridizes to the antisense strand DNA at 20 to 70 bp downstream of each exon to be sequenced. This allows sequencing of the splice sites flanking the exon.

b. Set up an end-labeled primer as follows:

- 5X Kinase buffer (10 μl)
- Sequencing primer (30–50 pmol) (<2 μl)
- [γ-$^{32}$P]dATP (6000 Ci/mmol) (65 μCi)
- T4 Polynucleotide kinase (30 units)
- Add dd.H$_2$O to a final volume of 50 μl.

c. Incubate at 37°C for 50 min.

d. Purify the labeled primer by passing through a NENsorb column (NEN/DuPont). Precipitate the primer with ammonium acetate and wash the pellet with 70% ethanol as described in step 4. After completely drying under vacuum, dissolve the primer in 10 μl dd.H$_2$O prior to use.

e. Assemble the following in microcentrifuge tube:

- Single-strand template (PCR product prepared in step 4) (10 μl)
- End-labeled sequencing primer (3 μl)
- 5X POL buffer (2 μl)

f. Heat at 95°C for 10 min, cool to room temperature, and briefly spin down.

g. Transfer 2.5 μl of the reaction mixture to each of four microcentrifuge tubes labeled A, T, G, and C.

h. Add 2 μl of the appropriate deoxy:dideoxy-terminator/Sequenase mixture to the appropriate tube in step g immediately before use. The mixture is composed of 80:8 μ$M$ deoxynucleotide (dNTP):dideoxynucleotide (ddNTP) that is mixed with 0.5 unit/μl Sequenase (United States Biochemical Corporation).

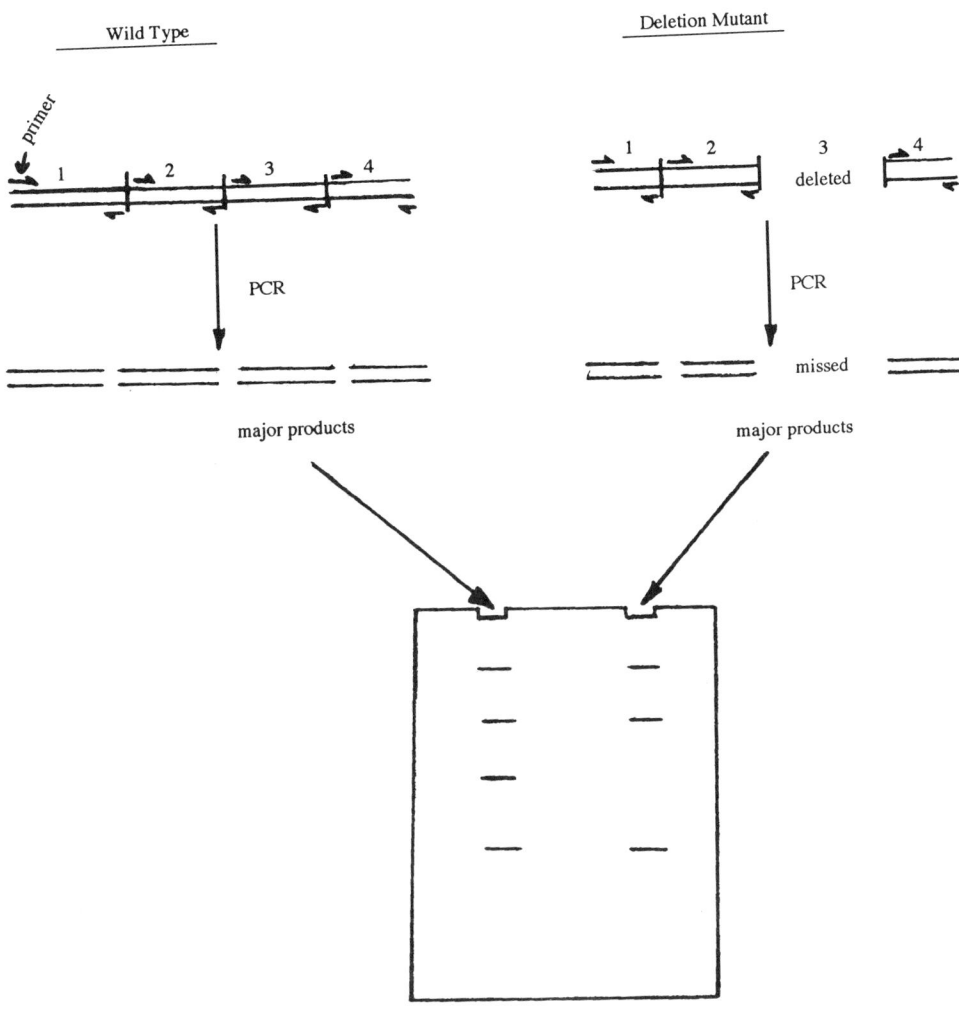

**FIGURE 9.3**
Diagram illustrating the multiplex PCR strategy. It shows a comparison of PCR products from a wild-type gene and a deletion mutant.

i.  Incubate all tubes at 50°C for 10 min, briefly spin down, and add 4μl of stopping solution. The mixture can be heated to 75°C for 2 to 3 min prior to being loaded into a standard sequencing gel or stored at –20°C until use. The procedures of electrophoresis and autoradiography are described in Chapter 8.

j.  Compare the sequences of individual exons of the mutant gene with those of the normal (wild-type) gene and identify the mutant sequence(s) (Figure 9.3).

## Buffers Needed

### 5X Kinase Buffer

  50 m$M$ Tris-HCl, pH 7.6
  50 m$M$ MgCl$_2$
  5 m$M$ Dithiothreitol (DTT)

### 5X POL Buffer

> 0.25 $M$ Tris-HCl, pH 8.3 (37°C)
> 40 m$M$ MgCl$_2$
> 0.15 $M$ KCl
> 50 m$M$ DTT

### Stop Solution

> 95% (v/v) Formamide
> 20 m$M$ Na$_2$EDTA
> 0.1% (w/v) Bromophenol blue
> 0.1% (w/v) Xylene cyanol

### Protocol B: Detection of Mutant Point at cDNA Level of a Gene of Interest

The principle of this technique is that wild-type and mutant cDNAs are respectively synthesized from wild-type or mutant mRNA by reverse transcription. One of the strands of wild-type cDNA is radioactively end-labeled by incorporation of a radiolabeled oligonucleotide primer during the PCR reaction. Wild-type and mutant cDNAs are then mixed and denatured. When the denatured single-stranded cDNAs are rehybridized, heteroduplexes may be formed with some mismatch point(s). On exposure to hydroxylamine or osmium tetroxide, mismatched C or T nucleotides, respectively, are modified. If treated with piperidine, the mismatched sites in the heteroduplexes are cleaved. After electrophoresis and autoradiography, the cleaved site can be identified by comparing the denatured cDNA with wild-type cDNA.

1.  Isolate total RNA or purify mRNA from wild-type and mutant individuals of interest. The procedures are described in Chapter 2.

2.  Synthesize the first-strand cDNAs from wild-type and mutant mRNA or total RNA, respectively. Precipitate and dissolve the cDNA in dd.H$_2$O. The procedures are given in Section III, cDNA cloning by RT-PCR.

3.  Design primers for amplification. The primers should lie outside the coding region of the cDNA. 5′PCR primer should correspond to the sense strand in order to anneal with the cDNA; 3′PCR primer corresponds to the strand opposite to the 5′PCR primer.

4.  Prepare the primers.

    a. End-label one of the primers as follows:
      • 5X Kinase buffer (10 μl)
      • Sequencing primer (30 to 50 pmol) (<2 μl)
      • [γ-$^{32}$P]dATP (6000 Ci/mmol) (65 μCi)
      • T4 Polynucleotide kinase (30 units)
      • Add dd.H$_2$O to a final volume of 50 μl.

    b. Incubate at 37°C for 50 min.

    c. Purify the labeled primer by passing through a NENsorb column (NEN/DuPont). Precipitate the primer with ammonium acetate and wash the pellet with 70% ethanol as described in Protocol A, step 4. After completely drying under vacuum, dissolve the primer in 10 μl dd.H$_2$O prior to use.

5.  Carry out PCR amplification of cDNA.

    a. In a 0.5-ml microcentrifuge tube on ice, set up a reaction by adding the following in the order listed:

- 10X Amplification buffer (10 μl)
- dd.H$_2$O (20 μl)
- Mixture of four dNTPs (10 m$M$ each) (10 μl)
- 50 pmol of each of two primers designed (4 μl) (total)
- ss cDNA template in dd.H$_2$O (0.3 to 1 μg) (2 μl)
- *Taq* DNA polymerase (5 units/μl) (1 μl)
- Add dd.H$_2$O to a final volume of 100 μl.

*Note:*  *The amplification of wild-type cDNA and mutant cDNA should be carried out under the same conditions.*

b. Overlay the mixture with 100 μl of light mineral oil (Sigma or equivalent) to prevent evaporation of the sample.

c. Carry out 30 cycles of PCR amplification in automatic thermocycler, which is programed as follows:

| Cycle | Denaturation | Annealing | Polymerization |
|---|---|---|---|
| First | 4 min at 94°C | 2 min at 45°C | 3.5 min at 68°C |
| Subsequent | 1 min at 94°C | 1 min at 45°C | 3.5 min at 68°C |
| Last | 1 min at 94°C | 1 min at 45°C | 10 min at 68°C |

Finally, hold at 4°C until the sample is removed.

d. Dilute the PCR products with 1 volume of dd.H$_2$O and precipitate the DNA by adding 0.5 volume of 7.5 $M$ ammonium acetate, mix, and add 2.5 volumes of chilled 100% ethanol. Allow to precipitate at –70°C for 30 min.

e. Centrifuge at 12,000 × $g$ for 15 min at room temperature. Carefully discard the supernatant, resuspend the pellet in 0.2 ml dd.H$_2$O, and repeat ammonium acetate precipitation as in step d.

f. Centrifuge as step e. Briefly rinse the pellet with 0.4 ml of 70% ethanol and dry the pellet under vacuum for 30 min. Dilute the PCR products to 5 ng/μl (wild-type cDNA) or 50 ng/μl (mutant cDNA) in dd.H$_2$O. At this stage, the primers and unincorporated nucleotides are removed by repeated (once) precipitation.

g. Measure the activity of wild-type cDNA (cpm/μg) and concentration of unlabeled mutant cDNA. The procedures are described in Chapters 1 and 4. Store at –20°C until use.

6. Prepare the wild-type and mutant cDNA heteroduplexes.

a. Mix 20 ng of labeled wild-type cDNA as a probe (100,000 cpm) with 300 ng unlabeled mutant cDNA in 30 μl of 5X hybridization buffer.

b. Boil the mixture for 10 min and quickly chill on ice for 5 min in order to denature ds cDNAs.

c. Allow rehybridization for 2 to 5 h at 42°C.

d. Precipitate the DNA as described above, and dissolve the DNA in 30 μl dd.H$_2$O.

## *Buffer Needed:*

### *5X Hybridization Buffer*

   3 m$M$ Tris-HCl, pH 7.7
   300 m$M$ NaCl
   3.5 m$M$ MgCl$_2$

7.  Carry out chemical cleavage.

    a.  Label four tubes for each PCR product: one hydroxylamine and one osmium tetroxide treatment for each of two strands. Add 6 μl of the appropriate heteroduplex to each tube.

*Note:*    *The osmium tetroxide tubes should be placed on ice.*

    b.  **For osmium tetroxide treatment** — On ice, add 2.5 μl of 10X osmium tetroxide buffer and 15 μl of 2% osmium tetroxide solution to the tubes. Mix well (yellow) and incubate at 37°C for 5 min with frequent agitation. Place on ice and add 0.2 ml of stopping buffer to it. Precipitate by adding 0.75 ml chilled 100% ethanol and chilling at –70°C for 30 min. Centrifuge at 12,000 × *g* for 15 min, rinse with 0.5 ml of 70% ethanol, and dry under vacuum for 30 min.

       **For hydroxylamine treatment** — Add 20 μl of hydroxylamine solution to the tubes, mix well, and incubate at 37°C for 20 min with frequent agitation. Place on ice and add 0.2 ml of stopping buffer to it. Precipitate by adding 0.75 ml chilled 100% ethanol and chilling at –70°C for 30 min. Centrifuge at 12,000 × *g* for 15 min, rinse with 0.5 ml of 70% ethanol, and dry under vacuum for 30 min.

    c.  To cleave the piperidine, add 50 μl of 1 *M* piperidine to each tube and mix well. Incubate at 90°C for 30 min. Place on ice, and add 50 μl of 0.6 *M* sodium acetate buffer (pH 5.2) and 0.25 ml of chilled ethanol. Place at –70°C for 30 min. Centrifuge at 12,000 × *g* for 15 min, rinse with 0.5 ml of 70% ethanol, and dry under vacuum for 30 min.

    d.  Completely redissolve the pellet in 15 μl of 100% formamide; check the counts with a Geiger counter.

    e.  Heat at 90°C for 4 min and chill on ice.

    f.  Carry out electrophoresis on a 4%, denaturing polyacrylamide gel, and perform autoradiography.

    g.  Analyze the results by comparing the band pattern of mutant cDNA and that of wild-type cDNA (Figure 9.4). Additional small bands in the mutant pattern should result from point mutation(s).

## *Buffers Needed*

### *Osmium Tetroxide Solution*

Make 2% stock solution by dilution of a commercial 4% stock. Store at 4°C for up to 3 months.

### *Hydroxylamine Solution*

Dissolve 4.17 g hydroxylamine chloride in 4.8 ml $H_2O$ under warm water, add 3 ml diethylamine, and adjust the pH to 6.0 with diethylamine. Store at 4°C for up to 9 days.

### *Stopping Solution*

300 m*M* Sodium acetate, pH 5.2
0.5 m*M* EDTA
Autoclave and add 25 μg/ml tRNA.
Clean the tRNA before using in the phenol/chloroform extraction and the sodium acetate/ethanol precipitation. Dissolve in dd.$H_2O$ and heat at 95°C for 10 min. Store at –20°C until use.

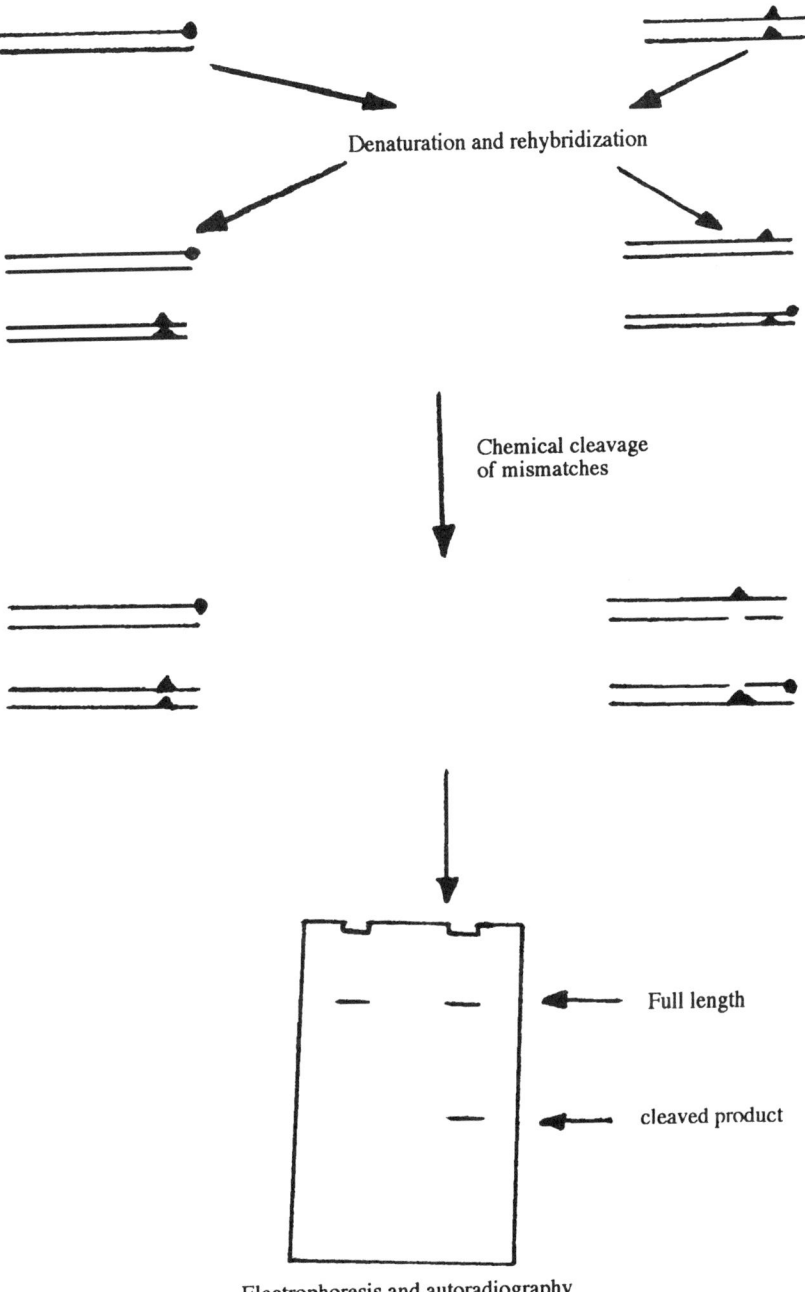

Denaturation and rehybridization

Chemical cleavage
of mismatches

Full length

cleaved product

Electrophoresis and autoradiography

**FIGURE 9.4**

Diagram of the mismatch-chemical cleavage strategy. It shows the difference between wild-type cDNA and
mutant cDNA after PCR amplification, denaturation, rehybridization, and chemical cleavage of the mismatches.

### *10X Osmium Tetroxide Buffer*

0.1 $M$ Tris-HCl, pH 7.7
10 m$M$ EDTA
15% (v/v) Pyridine

# References

1. **Erlich, H. A.,** *PCR Technology: Principles and Applications for DNA Amplification,* Stockton Press, New York, 1989.

2. **Innis, M. A., Gelfand, D. H., Sninsky, J. J., and White, T. J.,** *PCR Protocols: A Guide to Methods and Applications,* Academic Press, New York, 1989.

3. **Wang, A. M., Doyle, M. V., and Mark, D. F.,** Quantitation of mRNA by the polymerase chain reaction, *Proc. Natl. Acad. Sci. U.S.A.,* 86, 9717, 1989.

4. **Wu, L.-L., Song, I., Karuppiah, N., and Kaufman, P. B.,** Kinetic induction of oat shoot pulvinus invertase mRNA by gravistimulation and partial cDNA cloning by the polymerase chain reaction, *Plant Mol. Biol.,* 21, 1175, 1993.

5. **Arnold, C. and Hodgson, I. J.,** Vectorette PCR: a novel approach to genomic walking, *PCR Methods Appl.,* 1, 39, 1991.

6. **Barany, F.,** Genetic disease detection and DNA amplification using cloned thermostable ligase, *Proc. Natl. Acad. Sci. U.S.A.,* 88, 189, 1991.

7. **Lu, W., Han, D.-S., Yuan, J., and Andrieu, J.-M.,** Multitarget PCR analysis by capillary electrophoresis and laser-induced fluorescence, *Nature,* 368, 269, 1994.

8. **Bloomquist, B. T., Johnson, R. C., and Mains, R. E.,** Rapid isolation of flanking genomic DNA using biotin-RAGE: a variation of single-sided polymerase chain reaction, *DNA Cell Biol.,* 10, 791, 1992.

9. **Edwards, J. B. D. M., Delort, J., and Mallet, J.,** Oligodeoxyribonucleotide ligation to single-stranded cDNAs: a new tool for cloning 5′ ends of mRNAs and for constructing cDNA libraries by *in vitro* amplification, *Nucleic Acids Res.,* 19, 5227, 1991.

# Chapter 10

# DNA Fingerprinting

## Contents

DNA fingerprinting, a powerful technique, is commonly used in genetic studies such as restriction fragment length polymorphic genetic disease diagnosis, mutation analysis, as well as genomic mapping. It is a simple, fast, and relatively accurate method that is widely applied. The basic principle of DNA fingerprinting is that genomic DNA, which is isolated from individuals, or from different cell or tissue types such as normal and tumor cells, is subjected to restriction enzyme digestion and then to Southern blot analysis using a labeled, known gene fragment of interest as a probe. DNA from each cell or tissue type or individual may show a unique band pattern (fingerprinting) when it hybridizes with the probe.[1,2] Any slight differences can be determined.

## I. Protocols

1.  Isolate high-molecular weight genomic DNA from a given cell or tissue type, or from different individuals (e.g., normal person vs. a patient with a genetic disease, or wild-type vs. mutant plants). The general procedures are described in Chapter 1.

2.  Carry out restriction enzyme digestion of the isolated genomic DNA as follows:

    •  DNA in TE buffer or dd.$H_2O$ (20 to 30 μg genomic or 2 to 10 μg plasmid or λDNA) (2 to 8 μl)

    •  Appropriate restriction 10X buffer (4 μl)

    •  Appropriate restriction enzyme (3.4 units/μg DNA)

    •  Add dd.$H_2O$ to a final volume of 40 μl.

    Incubate at optimal temperature for the enzyme (usually 37°C) for 2 to 3 h and store at 4°C until electrophoresis is performed.

*Notes:*  *If the amount of DNA in each well is not enough, the band(s) to be detected will be quite weak. However, if too much DNA is loaded in one well, the enzyme digestion may not be complete and electrophoresis may be not good, thus affecting hybridization and detection. After digestion, the sample without extraction and precipitation can be directly loaded into the gel and electrophoresis performed with no side effects.*

3.   Prepare an agarose mixture in a clean bottle or a beaker as follows:

| Components | Mini gel | Medi gel | Big gel |
|---|---|---|---|
| 1X TBE or TAE buffer | 30 ml | 75 ml | 120 ml |
| Ultrapure agarose | 0.3 g | 0.75 g | 1.2 g |
| (1% w/v) | | | |

*Notes:   1X TBE or TAE buffer may be diluted from 5X TBE or 10X TAE stock solution with dd.$H_2O$. Since the agarose is only 1%, its volume can be ignored when calculating the final volume.*

4.   Slightly mix and melt the agarose by gently boiling in a microwave oven for 1 to 3 min depending the gel volume. Alternatively, put a magnetic stir bar in the bottle or beaker and heat on a stirring hotplate gently boiling until agarose dissolves. Gently mix and place at room temperature to cool to 50 to 60°C.

5.   While the gel mixture is being cooled, seal the air-dried gel tray at the two open ends with a tape or gasket and insert the comb. Add 10 μl of 10 mg/ml ethidium bromide (EtBr) to 100 ml of agarose gel solution (50 to 60°C), gently mix, and slowly pour into the assembled gel tray. Allow the gel to harden for 20 to 30 min at room temperature.

*Caution:   EtBr is a mutagen and a potential carcinogen. Gloves should be worn when working with this material. The gel running buffer containing EtBr should be collected in a special container.*

*Notes:   EtBr is used to stain DNA or RNA molecules, which interlaces between the complementary strands of a double-strand DNA or in the regions of secondary structure in the case of ssDNA or RNA and it fluoresces orange when illuminated with UV light. The merit of adding EtBr in gel is that DNA bands can be stained and monitored with a UV lamp during electrophoresis. The drawback, however, is that running buffer and gel apparatus are contaminated with EtBr. An alternative way is to carry out electrophoresis without EtBr. The gel is then stained with EtBr for 10 to 30 min following electrophoresis.*

6.   **Carefully** remove the comb and sealing tape or gasket from the gel tray. Place the gel tray in the electrophoresis tank and add enough 0.5X TBE or TAE buffer to the tank until the gel is covered to a depth of 1.5 to 2 mm above the gel.

*Notes:   The comb should be slowly and vertically removed from the gel, since any cracks inside the wells of the gel will cause leaking when the sample is loaded. The well-side of the gel must be placed at the negative pole end since the negatively charged DNA will migrate toward the positive pole. When the gel is covered with running buffer, each well should be flushed with the buffer using a small pipette tip to flush the buffer up and down inside the well several times. The purpose of doing this is to remove any potential bubbles that will adversely influence the loading of the samples and the electrophoresis.*

7.  Prerun the gel at a constant voltage (5 to 10 V/cm) for 10 min, but this is optional.

8.  Add 5X or 10X loading buffer to the DNA sample and DNA standard marker to a final concentration of 1X. Mix well. Carefully load the commercial DNA standard markers (0.2 to 2 µg per well) in the very left or the right well, or in both left and right wells of the gel. Leave one well blank from the DNA marker well and carefully load the samples, one by one, into the empty wells in the submerged gel.

*Notes:* *(1) Do not insert the pipette tip all the way to the bottom of the well, otherwise, it may break the well and cause sample leaking. (2) For genomic DNA, the sample may be heated to 60°C for 5 min prior to loading. (3) The loading buffer must be at least 1X to final concentration, or the sample may float out of the well. (4) For nonisotopic detection, we recommend that the DNA markers be prelabeled, precipitated, and directly loaded into the gel without denaturation. The labeled markers are still double-strand DNA fragments. The labeling methods are described in Chapter 4. The prelabeled DNA standard markers allow easy and accurate estimation of the sizes of detected bands in the DNA samples after being exposed onto X-ray film or after being color developed. For isotopic detection, however, we do not recommend prelabeled DNA markers due to potentially massive contamination during various steps. If necessary, the hybridized filter can be stripped off the hybridized probe for the DNA samples and reprobed with labeled DNA standard markers, and the marker positions can be compared with those of the detected bands of interest. For estimation of the size of the bands, a photograph may be taken, with ruler markers under UV light, of EtBr stained DNA standard markers and DNA samples and used to compare the sizes of detected bands of interest.*

9.  Measure the length of the gel between the two electrodes and apply power to 5 to 10 V/cm. Allow the electrophoresis to run for 5 to 6 h or until the first blue dye reaches to 2 cm from the end of the gel. If overnight running of the gel is desired, the total voltage should be 20 to 25 V.

10. Stop the electrophoretic running of the gel and remove the gel to observe under UV light. Photograph the gel with a Polaroid™ camera and directly proceed to the blotting procedure.

*Caution:* *Remember to wear safety glasses and gloves for protection from UV light.*

*Notes:* *Pictures are recommended to be taken with a ruler at short exposure to obtain sharp band(s) and clear background and at longer exposure to visualize some very weak band(s). For genomic DNA, the picture should be smeared in each lane with some weak visible bands. For plasmid and lambda DNA containing inserts of interest, the picture should display some sharp bands without a smeared background.*

## Materials Required

*Ultrapure Agarose*

*Gel Casting Tray*

*Gel Combs*

*Electrophoresis Apparatus*

## DC Power Supply

## 5X TBE Buffer

>   600 ml dd.$H_2O$
>   0.45 *M* Tris base (54 g)
>   0.45 *M* Boric acid (27.5 g)
>   0.01 *M* EDTA (20 ml 0.5 *M* EDTA, pH 8.0)
>   Dissolve well after each addition. Add dd.$H_2O$ to 1 liter.
>   Autoclave.

## 10X TAE Buffer

>   600 ml dd.$H_2O$
>   0.4 *M* Tris-acetate (48.4 g tris base, 11.42 ml glacial acetic acid)
>   10 m*M* EDTA (20 ml 0.5 *M* EDTA, pH 8.0)
>   Dissolve or mix well after each addition. Add dd.$H_2O$ to 1 liter.
>   Sterile-filter.

## Ethidium Bromide (EtBr)

>   10 mg/ml in dd.$H_2O$
>   Dissolve well and keep in a dark or brown bottle at 4°C.

## 5X Loading Buffer

>   50% Glycerol
>   1 m*M* EDTA
>   0.25% Bromophenol blue
>   0.25% Xylene cyanol
>   Dissolve well and store at 4°C.

11.   Carry out the procedures (Capillary or Vacublot Method) given below for blotting DNA onto nylon or nitrocellulose membranes.

12.   Carry out the hybridization procedure given below.

## Capillary Method

1.   Right after photographing the electrophoresed gel, soak the gel for 5 to 8 min in 0.25 *N* HCl to depurinate DNA. This treatment can increase the efficiency of transferring of DNA fragments that are more than 8 Kb in length.

2.   Place the gel in 500 to 1000 ml of denaturing solution to denature the dsDNA molecules for later hybridization to the probe. Allow the denaturation to occur for 40 to 45 min at room temperature with slow shaking at 60 rpm.

3.   While denaturation of the gel takes place, cut a piece of nylon or nitrocellulose membrane the same size as the gel, a piece of 3MM Whatman filter paper the same size as the membrane and a relatively large piece of 3MM Whatman filter paper. The membrane filter should be marked at the upper left or the right corner with a pencil.

*Notes:*   *(1) Gloves should be worn when cutting the membrane filter. (2) Nylon membrane is stronger and better than nitrocellulose membrane, which is easily broken. Positively charged nylon membrane is better than neutral nylon membrane to interact with negatively charged DNA. However, neutral nylon membrane has a lower background as compared with positively charged nylon filter following detection. (3) For reprobing the hybridized membrane filter, a nylon membrane is strongly recommended. It can be reprobed several times without any significant decrease in the signals. Nitrocellulose membrane, however, cannot endure repeated probing due to its physical condition, and the signal usually decreases significantly during the procedure of stripping. (4) For subsequent reprobing of the hybridized membrane filter, the filter must never become dry (even partially dry) during and after hybridization, washing, and exposure.*

4. Quickly rinse the denatured gel with 500 to 1000 ml distilled water and place in 500 to 1000 ml of neutralization buffer to partially neutralize the gel but not renature the DNA molecules. Allow the neutralization to occur for 40 to 45 min at room temperature with slow shaking at 60 rpm.

5. While neutralization of the gel takes place, soak the membrane filters in 10X or 20X SSC or SSPE solution for at least 20 min.

6. Set up a clean blotting tray, put four test tube caps in the tray to serve as columns, and place an appropriate size of plate on those columns, or equivalent Plexiglas blotting plate with standing legs, or use an upside-down gel-forming tray as the plateform. Fill the tray with 10X SSC or 10X SSPE up to the edges of the flat plate. **Do not allow fluid to come up over the plate.**

7. Assemble the blotting apparatus in the order listed below (see Figure 5.2 in Chapter 5):

   a. Place the soaked large 3MM Whatman filter on the flat plate prepared in step 6. The two ends of the filter should hang into the 10X SSC solution to serve as wicks. Smooth out any bubbles by lifting one end of the filter and slowly laying it down on the plate.

   b. Carefully place the gel, starting from one side, on the filter with the well-side facing down. Gently press the gel to remove any bubbles underneath the gel.

   c. Carefully overlay the gel with the membrane filter starting from one side of the gel, then proceeding to the other end. The marked side and left or right orientation should be facing the gel. Wet the filter with some 10X SSC buffer and remove any bubbles between the membrane and the gel by lifting and laying down the membrane. **Do not press the membrane;** otherwise, bubbles that are not visible may be produced underneath the gel.

   d. Gently overlay the membrane with a same-sized piece of 3MM Whatman filter paper with no bubbles.

   e. Carefully place a strip of SaranWrap™ of appropriate size against each side of the gel to prevent any potential flow of the 10X SSC buffer to the paper towels.

   f. Gently place a precut stack of paper towels (5 to 10 cm thick) of relatively the same size as the membrane filter on top of the 3MM Whatman paper. **Do not distribute the filters lying underneath;** this will produce bubbles.

   g. Place a glass plate, or equivalent, on top of the paper towel stack. Wrap the whole apparatus with SaranWrap™ (optional). Put a bottle or beaker containing 500 ml water or equivalent on the glass plate to serve as a weight.

8. Allow the DNA to transfer to the membrane filter by capillary action overnight or for 16 h.

9. Remove the paper towel stack, the Whatman filter, and the membrane filter. Mark the wells on the membrane filter. The filter is then subjected to UV-induced cross-linking for 30 to 60 s at the optimal setting. Check the blotted gel under UV light. An efficient transfer should have no visible DNA staining left in the gel.

10. Air dry the membrane filter for about 20 min. If the membrane is not subjected to UV cross-linking, one should dry the membrane in a oven under vacuum at 80°C for 2 h to immobilize the DNA on the membrane filter.

11. Wrap the membrane filter with aluminum foil and store at 4°C until hybridization is performed.

*Notes:*     *Gloves should be worn and changed during the above procedures. Blotting should be handled properly without any bubbles that block local DNA transference onto the membrane. If this happens, it should be visible from the blotted membrane filter and may affect the detection of bands of interest.*

## Vacublot Method

This is a fast but a relatively expensive method of blotting DNA to membrane. The gel to be blotted does not need to be predenatured and neutralized.

1. Cut a nylon membrane to the size of the gel and a plastic mask smaller than the gel.
2. Place a piece of 3MM Whatman™ filter on the metal grid of a vacublot apparatus and wet it with distilled water.
3. Place the nylon membrane on the 3MM Whatman™ filter, then the mask and then the gel (well-side facing up) to form a seal between the gel and nylon membrane.
4. Apply a vacuum (2.5 in. Hg).

*Note:*     *You should be able to feel with your finger tips that the vacuum causes the gel to form a seal along the edges of the mask.*

5. Pour 5 to 10 ml of depurination solution to cover the surface of the gel and let it permeate the gel for 5 to 7 min.
6. Remove the depurination solution with a pipette and replace it with 20 ml of denaturing solution. Allow the solution to permeate the gel for 8 to 15 min.
7. Remove the denaturing solution and replace it with 20 ml of neutralizing solution. Allow the solution to permeate the gel for 8 to 15 min.
8. Remove the neutralizing solution and replace it with 20 ml of 20X SSC or 20X SSPE solution. Allow the solution to permeate the gel for 40 min.
9. After the transfer has finished, mark the wells and remove the membrane. Place the membrane on SaranWrap™ or filter paper and subject it to UV cross-linking at the optimal setting for 20 s.
10. Air dry the filter and place it into a heat-sealable bag and seal the bag about 1 to 2 cm from the edge of the membrane. Store the membrane at this stage at –20°C until hybridization is performed.

## Materials Needed

### Nylon Membranes or Nitrocellulose Membranes

### Heat-Sealable Bags

### 3MM Whatman Paper

### Baking Oven

### Vacublot Apparatus

### Depurination Solution
         0.25 *M* HCl

### Denaturing Solution

0.5 $M$ NaOH
1.5 $M$ NaCl

### Neutralization Solution

1.5 $M$ NaCl
1 $M$ Tris-HCl, pH 7.5

### 20X SSC Solution

175.3 g NaCl
88.4 Sodium citrate
Adjust the pH to 7.5 with HCl.

### 20X SSPE Solution

3 $M$ NaCl
0.2 $M$ NaH$_2$PO$_4$
20 m$M$ EDTA, pH 7.4

## Hybridization to $^{32}$P-Labeled Probe

1.  Immerse the filters in 5X SSC for 5 min at room temperature to equilibrate the filters.

*Note:* *Do not let the filters dry out during subsequent steps. Otherwise, a high background and/or anomalous results will show up.*

2.  Place the filters in the prehybridization solution and carry out prehybridization for 2 to 4 h with slow shaking at 60 rpm.

*Notes:* *(1) We strongly recommend not using plastic hybridization bags because, in using them, it is usually not easy to get rid of air bubbles nor can they be well sealed. This causes leaking and contamination. An appropriate size of plastic beaker or tray is the best type of hybridization container to use for this purpose. (2) The prehybridization temperature depends on the prehybridization buffer. The temperature should be set at 42°C if the buffer contains 30 to 40% (for low-stringency conditions) or 50% formamide (for high-stringency conditions). If the buffer, on the other hand, does not contain formamide, the temperature is set at 65°C. Low-stringency conditions will help identify cDNAs of a potential multigene family. High-stringency conditions help prevent nonspecific cross-linking hybridization. (3) We strongly recommend the use of a regular culture shaker with a cover and temperature control as the hybridization chamber. Such a chamber is easily handled by placing the hybridization beaker or tray containing the filters and buffer on the shaker in the chamber. A commercial hybridization oven may be difficult to operate because the filters must be covered with a matrix and put inside the hybridization bottle, during which time air bubbles are easily generated. (4) If many filters are to be used for hybridization, one beaker should not contain more than three filter disks. Too many filters in one beaker may cause weak hybridization to occur. (5) The volume of prehybridization solution should be 15 ml per 100-cm² filter disk.*

3.   Denature the labeled double-stranded DNA probe contained in a microcentrifuge tube in boiling
     water for 10 min and immediately chill on ice for 5 min to denature the probe for hybridization.
     Briefly spin down prior to use with a microcentrifuge.

*Notes:*   *(1) This is a critical step. If the probes are not completely denatured, a weak or no
           hybridization signal will occur. Single-strand oligonucleotide probes, however, usu-
           ally do not require denaturation. (2) The DNA used for labeling can be a specific
           gene (usually a conserved partial-length fragment), an oligonucleotide (where syn-
           thesis is based on the conserved regions of known DNA), or a specific cDNA (partial
           or full length) from other organisms. The DNA is labeled with [α-$^{32}$P]dCTP and is
           ready for hybridization (see DNA-labeling protocols in Chapter 4). (3) The labeled
           probe should be separated from the unincorporated nucleotides by use of a G-50
           column (see Chapter 4 for details). Otherwise, nonspecific black spots will appear
           on the filter. (4) It is recommended, but not required, to calculate the cpm number
           of the labeled probe prior to hybridization.*

*Caution:*   *[α-$^{32}$P]dCTP is a dangerous isotope. A lab coat and gloves should be worn when
             working with this isotope. Gloves should be changed often and put in a special
             container. Waste liquid, pipette tips, and papers contaminated with the isotope
             should be collected in labeled containers. After finishing, a radioactive contami-
             nation survey should be performed and recorded.*

4.   Dilute the purified probe with 1 ml of hybridization solution and add the probe at 2 to $10 \times 10^6$
     cpm/ml to the hybridization buffer. Mix well and carefully transfer the prehybridized filters to the
     hybridization solution. Allow hybridization to proceed overnight or up to 19 h.

*Note:*   *For prehybridization, notes are the same as for hybridization.*

5.   Wash the hybridized filters according to the following conditions:
     a.   High-stringency conditions
          i.    Wash the filters in a solution (50 ml per filter) containing 2X SSC and 0.1% SDS (w/v)
                for 15 min at room temperature with slow shaking. Repeat once.
          ii.   Wash the filters in a fresh solution (50 ml per filter) containing 2X SSC and 0.1% SDS
                (w/v) for 20 min at 65°C with slow shaking. Repeat two to four times.
          iii.  Air dry the filters at room temperature for about 40 min and proceed with autoradiography.
     b.   Low-stringency conditions
          i.    Wash the filters in a solution (50 ml per filter) containing 2X SSC and 0.1% SDS (w/v)
                for 10 min at room temperature with slow shaking. Repeat once.
          ii.   Wash the filters in a fresh solution (50 ml per filter) containing 2X SSC and 0.1% SDS
                (w/v) for 15 min at 50 to 55°C with slow shaking. Repeat once.
          iii.  Air dry the filters at room temperature for 40 min and proceed with autoradiography.

6.   Wrap the filters, one by one, with SaranWrap™ and place in an exposure cassette. In a dark room
     with the safe light on, cover the filters with a piece of X-ray film and place the cassette with an
     intensifying screen at –80°C for 2 to 24 h prior to their being developed.

## Reagents Needed ——————————————————————————————

### 20X SSC Solution (1 l)

          3 *M* NaCl
          0.3 *M* Na$_3$ citrate (trisodium citric acid)
          Autoclave. Adjust the pH to 7.0 with HCl.

*5X SSC Solution*

        Dilute 20X SSC solution four times with sterile water.

*50X Denhardt's Solution*

        1% (w/v) BSA (bovine serum albumin)
        1% (w/v) Ficoll (Type 400, Pharmacia)
        1% (w/v) PVP (polyvinylpyrrolidone)
        Dissolve well after each addition, adjust the final volume into 500-ml aliquots with distilled water and sterile-filter. Divide the solution into 50-ml portions and store at –20°C. Dilute tenfold into prehybridization and hybridization buffers.

*Prehybridization Buffer*

        5X SSC
        0.5% SDS
        5X Denhardt's reagent
        0.2% Denatured and sheared salmon sperm DNA

*Hybridization Buffer*

        5X SSC
        0.5% SDS
        5X Denhardt's reagent
        0.2% Denatured salmon sperm DNA
        [$\alpha$-$^{32}$P]-Labeled DNA probe

# II.  Applications of DNA Fingerprinting

1.  **Detection of alterations of bands by comparing different tissues from different individuals** — For example, in order to identify DNA fingerprinting changes between normal tissue and tumor tissue of the same patient or different patients, DNA can be isolated individually from normal and tumor tissues and restriction enzyme digestion performed. Southern blot analysis is then performed as described previously using the specific gene of interest as a probe. As shown in Figure 10.1, the patient has a unique DNA fingerprinting pattern with addition or deletion of bands occurring for the tumor tissue as compared with the normal tissue. This is a very useful technique in cancer diagnosis because relevant losses of alleles are characteristic of tumor cells.

2.  **Comparison and verification of pedigree DNA fingerprinting in genetic studies** — DNA can be isolated from the blood of father, mother, and offspring. As described previously, the DNA is subjected to restriction enzyme digestion and Southern blot analysis using a specific probe such as minisatellite DNA. After autoradiography, the true biological paternity of the offspring can be determined with ease by comparing the three DNA fingerprinting patterns. According to Mendelian inheritance, all of the bands in the offspring's DNA fingerprinting pattern should be identified in the fingerprinting of either the father or the mother, and approximately 50% of the bands in the offspring's fingerprinting should come from the father. If the fingerprinting pattern of the offspring is totally different, or more than 70% of the bands are different from that of the father, a different biological paternity may be indicated.

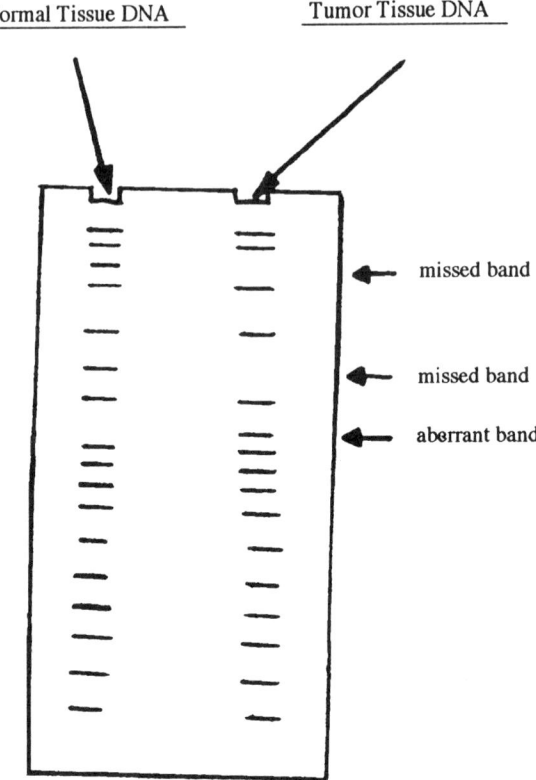

**FIGURE 10.1**
Diagram of DNA fingerprinting in order to compare the differences between normal and mutant tissues.

# References

1. **Sambrook, J., Fritsch, E. F., and Maniatis, T.,** *Molecular Cloning: A Laboratory Manual,* 2nd ed., Cold Spring Harbor Press, Cold Spring Harbor, NY, 1989.

2. **DeMot, R. and Vanderleyden, J.,** Application of two-dimensional protein analysis for strain fingerprinting and mutant analysis of *Azospirillum* species, *Can. J. Microbiol.,* 35, 960, 1989.

Chapter

# DNA *In Vitro* Mutagenesis

## Contents

DNA *in vitro* mutagenesis is a very useful technology that plays an important role in the expression of gene regulation and in the identification of transcriptional regulatory sequences such as promoters, enhancer elements, and silencer elements.[1-6] There are several mutagenesis techniques that have been developed. The present chapter describes two well-established methods: DNA detetions and oligonucleotide site-directed mutagenesis. The general principles and procedures are

1.  The DNA of interest is first mutated *in vitro* by an appropriate method followed by being ligated to an appropriate vector. If blunt ends or compatible cohesive termini are available after mutagenesis or restriction enzyme digestion, the mutated DNA and vector can be directly ligated to each other by DNA ligase. However, in most cases, if the DNA and vector ends are incompatible, the Klenow fragment of *E. coli* DNA polymerase I, or bacteriophage T4 DNA polymerase, mung-bean nuclease, or nuclease S1 must be used to generate blunt ends of the molecules so that they can then be ligated or be recircularized with DNA ligase.

2.  The recombinant DNA constructs are transferred into a particular cell line or whole organism.

3.  The effects or functions of the mutated gene or DNA fragment can then be analyzed from the transfected cells or from transgenic animals or plants.[3–10]

# I.   Mutagenesis by Deletion

Deletion is the simplest and most widely used method of mutagenesis. Nested series of deletions can be produced, using exonuclease *Bal* 31 or exonuclease III, by progressively degrading sequences from one end of the target DNA of interest, generating different lengths of mutants (Figure 11.1). If the recombinant DNA is a plasmid, the plasmid should first be linearized with an appropriate restriction enzyme that cleaves at one end of the target sequence based on a known sequence of the DNA of interest. The linearized DNA is then subjected to time-course digestions by *Bal* 31 or exonuclease III. The overhanging ends are usually generated as blunt termini for efficient ligation. The mutant DNA can be ligated to an appropriate vector with T4 DNA ligase and transformed to *E. coli*. The following procedures are for *Bal* 31 deletion of recombinant plasmids. *Bal* 31 has exonuclease activity that can progressively remove double-stranded DNA from the 3′ end. It also has weaker singly stranded endonucleolytic activity.

## *Protocol A: Linearizing Recombinant Plasmids to Completion Using Appropriate Endonuclease(s)*

1.  Set up, on ice, a standard restriction enzyme digestion as follows:

    **Single-restriction enzyme digestion**
    *   Plasmid DNA (20 μg)
    *   10X Appropriate restriction enzyme buffer (10 μl)
    *   1 mg/ml Acetylated BSA (optional) (10 μl)
    *   Appropriate restriction enzyme (3.4 units/μg DNA)
    *   Add dd.H$_2$O to a final volume of 100 μl.

    **Double-restriction enzyme digestion**
    *   Plasmid DNA or DNA to be inserted (20 μg)
    *   10X Appropriate restriction enzyme buffer (10 μl)
    *   1 mg/ml Acetylated BSA (optional) (10 μl)
    *   Appropriate restriction enzyme A (3.4 units/μg DNA)
    *   Appropriate restriction enzyme B (3.4 units/μg DNA)
    *   Add dd.H$_2$O to a final volume of 100 μl.

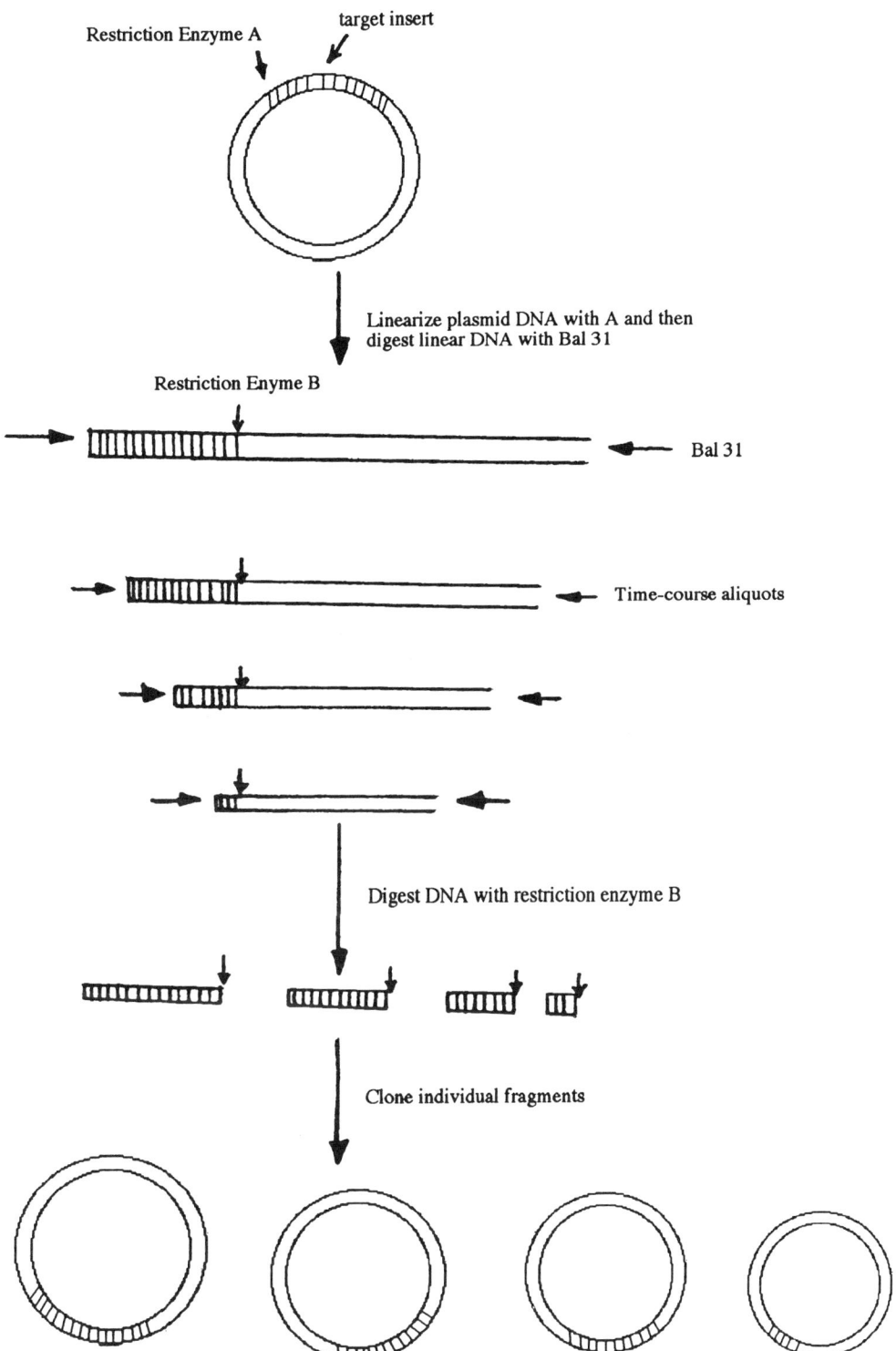

**FIGURE 11.1**
Mutagenesis strategy by nested deletions with *Bal* 31.

*Note:*     *For double-restriction enzyme digestions, the appropriate 10X buffer containing a higher NaCl concentration than the other buffer may be chosen for the double-enzyme digestion buffer.*

2.    Incubate at the appropriate temperature depending on the appropriate restriction enzyme (e.g., 37°C) for 2 to 3 h. For single-enzyme-digested DNA, proceed to step 3. For double-enzyme-digested DNA, proceed to step 5.

*Notes:*     *(1) The digestion efficiency can be checked by loading 1 µg of the digested DNA (10 µl) with loading buffer onto a 1% agarose mini-gel. In the meantime, undigested vectors and DNA (1 µg) and standard DNA markers should be loaded in the adjacent wells. After electrophoresis, the undigested plasmid DNA may reveal multiple bands because of different amounts of supercoiled DNA. However, one band will be visible for a complete single-enzyme digestion; one major band and one small band will be visible after digestion by two different restriction enzymes. The digested DNA to be transferred should be distinguished from undigested DNA. (2) After completion of restriction enzyme digestion, calf intestinal alkaline phosphatase (CIAP) treatment should be carried out for the above single-restriction enzyme digestion of the plasmid vector. This treatment removes 5′-phosphate groups, thus preventing recircularization of the vector during ligation. Otherwise, the efficiency of ligation between vector and insert DNA will be very low. For double-restriction enzyme-digested vectors, the CIAP treatment is not necessary.*

3.    Carry out CIAP treatment by adding the following directly to the single-enzyme-digested plasmid vector DNA sample (90 µl).
   - 10X CIAP buffer (15 µl)
   - CIAP diluted in 10X CIAP buffer (0.01 unit/pmol ends)
   - Add dd.H$_2$O to a final volume of 150 µl.

*Notes:*     *(1) CIAP and 10X CIAP should be kept at 4°C. CIAP treatment should be set up at 0°C. (2) Calculation of the amount of ends is as follows: There is a 9 µg digested DNA left after taking 1 µg of 10 µg digested DNA for checking on an agarose gel. If the vector is 3.2 Kb, the amount of ends can be calculated by the formula below:*

$$\text{pmol of ends} = \frac{\text{amount of DNA}}{\left(\text{base pairs} \times 660/\text{base pair}\right)} \times 2$$

$$= \frac{9}{3.2 \times 1000 \times 660} \times 2$$

$$= 4.2 \times 10^{-6} \times 2$$

$$= 8.4 \times 10^{-6} \ \mu M$$

$$8.4 \times 10^{-6} \times 10^{-6} = 8.4 \text{ pmol of ends}$$

4.    Incubate at 37°C for 1 h and add 2 µl of 0.5 *M* EDTA buffer (pH 8.0) to stop the reaction.

5. Extract with 1 volume of TE-saturated phenol/chloroform. Mix well by vortexing for 1 min and centrifuge at $11,000 \times g$ for 5 min at room temperature.

6. Carefully transfer the top, aqueous phase to a fresh tube and add 1 volume of chloroform:isoamyl alcohol (24:1) to the supernatant. Mix well and centrifuge as in step 5.

7. Carefully transfer the upper, aqueous phase to a fresh tube and add 0.1 volume of 3 $M$ sodium acetate buffer (pH 5.2) or 0.5 volume of 7.5 $M$ ammonium acetate to the supernatant. Briefly mix and add 2 to 2.5 volumes of chilled 100% ethanol to the supernatant. Allow to precipitate at –70°C for 1 h or at –20°C for 2 h.

8. Centrifuge at $12,000 \times g$ for 10 min and carefully decant the supernatant. Briefly rinse the DNA pellet with 1 ml of 70% ethanol and dry the pellet under vacuum for 20 min. Dissolve the DNA pellet in 100 µl dd.H$_2$O. Take 4 µl of the sample to measure the concentration of the DNA at 260 nm. Store the sample at –20°C until use.

*Note:* *Adding 0.5 volume of 7.5* M *ammonium acetate to the supernatant at step 7 yields a higher amount of DNA precipitation than by adding 0.1 volume of 3* M *sodium acetate buffer (pH 5.2).*

## Reagents Needed

*Appropriate Enzymes*

*10X Appropriate Restriction Enzyme Buffer*

*1% Agarose Mini-Gel*

*TE-Saturated Phenol/Chloroform*

*Chloroform:Isoamyl Alcohol (24:1)*

*3 M Sodium Acetate Buffer, pH 5.2*

*7.5 M Ammonium Acetate*

*Ethanol (100%, 70%)*

*0.5 M EDTA, pH 8.0*

*Calf Intestine Alkaline Phosphatase (CIAP)*

*TE Buffer*

*10X CIAP Buffer*
> 0.5 $M$ Tris-HCl, pH 9.0
> 10 m$M$ MgCl$_2$

1 m$M$ ZnCl$_2$

10 m$M$ Spermidine

## Protocol B: Performing Bal 31 Deletions

1.  Add 90 µl of the linearized recombinant plasmid to a microcentrifuge tube containing 90 µl of 2X *Bal* 31 buffer, mix well, and incubate at 30 or 37°C.

2.  Label six microcentrifuge tubes (numbered 1 to 6) at room temperature and add 70 µl of TE-saturated phenol:chloroform:isoamyl alcohol (25:24:1) to each tube.

3.  Transfer 30 µl of the DNA mixture from step 1 into tube numbers 1 and mix by vortexing. This represents zero time point for the enzyme degradation. Meanwhile, add 3 µl of diluted *Bal* 31 (0.2 unit/µl to the remaining DNA mixture in step 1.

4.  Transfer 30 µl of the *Bal* 31/DNA mixture from step 1 into tube numbers 2 to 6 at intervals of 2 to 5 min, respectively. Immediately vortex the tubes to stop the *Bal* 31 activity.

5.  Centrifuge all tubes at 11,000 × *g* for 5 min at room temperature.

6.  Carefully transfer the top, aqueous phase to fresh tubes and add 1 volume of chloroform:isoamyl alcohol (24:1) to the supernatant. Mix well and centrifuge as in step 5.

7.  Carefully transfer the upper, aqueous phase to fresh tubes and add 0.1 volume of 3 $M$ sodium acetate buffer (pH 5.2) or 0.5 volume of 7.5 $M$ ammonium acetate to each supernatant. Briefly mix and add 2 to 2.5 volumes of chilled 100% ethanol to the supernatant. Allow precipitation to occur at –70°C for 1 h or at –20°C for 2 h.

8.  Centrifuge at 12,000 × *g* for 10 min and carefully decant the supernatant. Briefly rinse the DNA pellet with 1 ml of 70% ethanol and dry the pellet under vacuum for 20 min. Dissolve the DNA pellet in 20 µl dd.H$_2$O. Store the samples at –20°C until use.

9.  Take 5 µl from each sample corresponding to each of the time points and analyze the extent of deletion by agarose gel electrophoresis as described previously.

## Reagents Needed

### 2X Bal 31 Buffer

> 40 m$M$ Tris-HCl, pH 8.0
> 24 m$M$ CaCl$_2$
> 24 m$M$ MgCl$_2$
> 400 m$M$ NaCl
> 2 m$M$ EDTA, pH 8.0

### Bal 31 Dilution Buffer

> 20 m$M$ Tris-HCl, pH 6.8
> 0.1 $M$ NaCl
> 5 m$M$ CaCl$_2$
> 5 m$M$ MgCl$_2$

### Diluted Bal 31 Solution

> 0.2 unit/µl in *Bal* 31 dilution buffer
> *Note:* The diluted enzyme should be stored at 4°C instead of –20°C.

## Protocol C: Blunt-Ending of the Deleted Plasmids by Filling in the 5′ Overhangs

1. Set up the 5′ overhangs reaction for each deletion in a microcentrifuge tube on ice as follows:
   - Linearized and deleted plasmid (10 μl)
   - 10X 5′ Overhang buffer (10 μl)
   - 2 m$M$ dNTP mixture (10 μl)
   - DNA polymerase I (Klenow fragment) (10 units)
   - Add dd.H$_2$O to a final volume of 100 μl.
2. Incubate at room temperature for 20 min and heat at 70°C for 5 min to stop the reaction.
3. Add 1 volume of TE buffer (pH 7.5) and carry out phenol:chloroform extraction and precipitation as described previously.

## Reagents Needed

### 10X 5′ Overhang Buffer

> 500 m$M$ Tris-HCl, pH 7.2
> 100 m$M$ MgSO$_4$
> 1 m$M$ DTT
> 500 μg/ml BSA

### 2 mM dNTP Mixture

> 2 m$M$ Each of dATP, dCTP, dGTP, and dTTP in 20 m$M$ Tris-HCl, pH 7.5

### DNA Polymerase I (Klenow Fragment)

> 1 unit/μl

## Protocol D: Synthesis of Phosphorylated, Double-Stranded Oligonucleotide Linkers Containing Appropriate Restriction Enzyme Sites

This is an optional step but highly recommended. The purpose of using linkers in the following recircularization of recombinant plasmid constructs is to digest the recircularized plasmids with the appropriate restriction enzyme for further analysis.

1. Design two complementary oligonucleotides containing 1 to 3 restriction enzyme sites. The ends of the oligonucleotides may be designed as *Sma* I termini for later blunt-ends ligation of plasmids. The oligonucleotides can be readily synthesized by a DNA synthesizer. The oligonucleotides should be purified afterwards.
2. Add 100 pmol of each of the two complementary oligonucleotides in a total volume of 20 μl in a microcentrifuge tube.
3. Add 1 μl of 20X annealing buffer to the tube, incubate at 90°C for 6 min followed by slow cooling to room temperature, which can be done by placing the heated tube in a rack in a beaker filled with an appropriate volume of 90°C water. The cooling procedure should take 20 to 30 min for complete annealing.
4. Add phosphate groups to the 5′ ends of the oligonucleotides by the kinase reaction.
   a. Set up a reaction mixture as follows in a microcentrifuge tube on ice:
      - The annealed oligonucleotides (15 μl)
      - 10X Polynucleotides kinase buffer (3 μl)

- 10 m$M$ ATP (1.5 μl)
- 100 m$M$ DTT (1.5 μl)
- T4 Polynucleotide kinase (15 units)
- Add dd.H$_2$O to a final volume of 30 μl.

b. Incubate at 37°C for 40 min and heat at 70°C for 10 min to inactivate the enzyme.

c. Add 1 volume of TE buffer (pH 7.5) and carry out phenol/chloroform extraction and ethanol precipitation as described previously.

## Reagents Needed

### 20X Annealing Buffer

200 m$M$ Tris-HCl, pH 7.9
40 m$M$ MgCl$_2$
1 $M$ NaCl
20 m$M$ EDTA, pH 8.0

### T4 Polynucleotide Kinase

1 unit/μl

### 10X Polynucleotide Kinase Buffer

500 m$M$ Tris-HCl, pH 7.5
100 m$M$ MgCl$_2$
50 m$M$ DTT
1 m$M$ Spermidine
1 m$M$ EDTA

### 10 mM ATP

## Protocol E: Recircularization of the Deleted Plasmid Constructs in the Presence of the Phosphorylated Oligonucleotide Linker

1. On ice, set up the following ligation:
   - Linearized, deleted plasmid DN (4 μg)
   - Oligonucleotide linker (0.3 μg)
   - 10X Ligase buffer (3 μl)
   - T4 DNA ligase (10 Weiss units)
   - Add dd.H$_2$O to a final volume of 30 μl.

2. Incubate the reactions at 4°C for 12 to 24 h, or 16°C for 4 to 6 h, or at room temperature (22 to 25°C) for 1 to 2 h.

*Note:*     *After ligation is finished at the above temperatures, the mixture can be stored at 4°C until later use.*

3.   The efficiency of ligation/recircularization can be checked by 1% (w/v) agarose gel electrophoresis. The ligated plasmids can be readily distinguished from unligated/uncircularized linkers and linear plasmids.

## Reagents Needed

### Dephosphorylated, Deleted Linear Plasmid Constructs

### Phosphorylated Oligonucleotide Linkers

### 5X Ligase Buffer
> 0.25 $M$ Tris-HCl, pH 7.5
> 50 m$M$ MgCl$_2$
> 5 m$M$ ATP
> 5 m$M$ DTT
> 25% (v/v) Polyethylene glycol (PEG) 8000

### T4 DNA Ligase
> 2 Weiss units/µl

## Protocol F: Transformation of Appropriate Strains of E. coli *with the* Ligated, Recombinant Plasmids and Isolation of the Plasmids

1.   Prepare the LB medium, LB plates, and competent cells as described previously. These should be done before ligation.
2.   Carry out transformation of individual mutant plasmids into bacteria in order to amplify the recombinant plasmids. The detailed procedures for the preparation of competent cells, transformation, selection of transformants, and verification of transformants are described under the section on subcloning in Chapter 7.
3.   Isolate recombinant plasmid DNA for sequencing of individual deletions by equilibrium centrifugation in CsCl-ethidium bromide gradients or by the alkaline mini-prep method. The detailed procedures are described under the section on isolation and purification of plasmid DNA procedures in Chapter 1.

## Protocol G: End-Labeling of the Plasmid DNA for Sequencing

1.   Linearize mutant plasmids isolated from recombinant colonies of bacteria to generate 5′ overhangs using appropriate restriction enzymes.
2.   Set up the 5′ overhangs reaction in a microcentrifuge tube on ice as follows:
     • Linearized plasmids (2 µg)
     • 10X 5′ Overhang buffer (2 µl)
     • 2 m$M$ dNTP mixture (2 µl)
     • [α-$^{32}$P]dNTP (4 µl)
     • DNA polymerase I (Klenow fragment) (3 to 4 units)
     • Add dd.H$_2$O to a final volume of 20 µl.

3.    Incubate at room temperature for 20 min and heat at 70°C for 5 min to stop the reaction.

4.    Add 1 volume of TE buffer (pH 7.5) and carry out phenol:chloroform extraction and precipitation as described previously.

## Reagents Needed

### 10X 5' Overhang Buffer

> 500 m$M$ Tris-HCl, pH 7.2
> 100 m$M$ MgSO$_4$
> 1 m$M$ DTT
> 500 µg/ml BSA

### 2 mM dNTP Mixture

> 2 m$M$ Each of dATP, dCTP, dGTP, and dTTP in 20 m$M$ Tris-HCl, pH 7.5

### DNA Polymerase I (Klenow Fragment)

> 1 unit/µl

### [α-³²P]dNTP

> 3000 Ci/mmol, 10 µCi/µl

### Protocol H: DNA Sequencing of the End-Labeled Plasmids to Map the Precise End Point of Each Deletion Using the Maxam and Gilbert Chemical Sequencing or Dideoxynucleotide Sequencing Method

The detailed procedures are described elsewhere in this book.

# II.    Oligonucleotide Site-Directed Mutagenesis

Oligonucleotide site-directed mutagenesis is a powerful technique that has been widely used to generate site-specific mutations in any cloned gene or cDNA of interest. The basic principle is that a mismatched oligonucleotide is synthesized, annealed to a specific region of single-stranded target DNA, and elongated by DNA polymerase, generating a mutant second strand of DNA containing the modified sequence in the region of interest. The target, single-stranded DNA can be prepared by subcloning the target DNA fragment into a single-stranded phage vector such as M13. An alternative way is to use a commercial Bluescript-based series of pCAT vectors containing the origin of replication of phage, called phagemids. The plasmid can generate single-stranded F1-packaged phage with the aid of helper phage following transformation into *E. coli* strains harboring an F1 episome. (Figure 11.2)

### Protocol A: Generation of Single-Stranded Phagemid DNA

1.    Plate out the transformants of *E. coli* that have the recombinant phagemids and the specific region of DNA of interest on an LB plate. Invert the plate and incubate at 37°C to obtain single colonies.

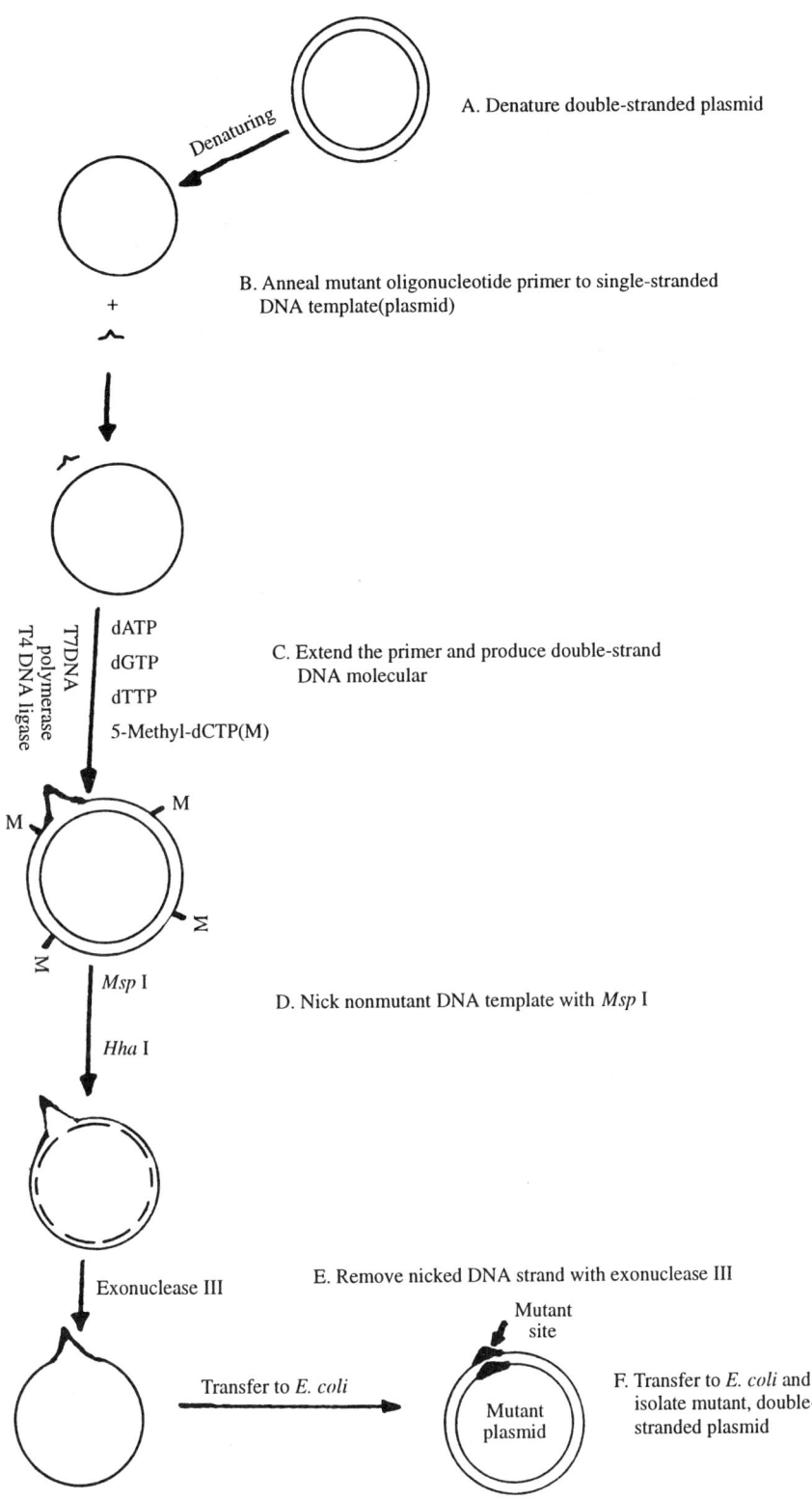

A. Denature double-stranded plasmid

B. Anneal mutant oligonucleotide primer to single-stranded
    DNA template(plasmid)

C. Extend the primer and produce double-strand
    DNA molecular

D. Nick nonmutant DNA template with *Msp* I

E. Remove nicked DNA strand with exonuclease III

F. Transfer to *E. coli* and
    isolate mutant, double-
    stranded plasmid

**FIGURE 11.2**
Strategy for generating site-specific mutations by oligonucleotide-directed *in vitro* mutagenesis.

2.  Inoculate a single colony in 5 ml of LB medium containing 50 µg/ml ampicillin and 12 µg/ml tetracycline. Incubate at 37°C overnight with shaking at 150 rpm.

3.  Add 0.3 ml of the overnight culture to 3 ml of superbroth in a 50-ml conical tube. Incubate at 37°C with shaking at 150 rpm for 2 to 3 h.

4.  Add 8 µl of helper phage R408 (pfu. = $1 \times 10^{11}$, available from Stratagene) to the culture at step 3 and continue to incubate for 8 to 10 h.

5.  Transfer 1.5 ml of the culture to each of two microcentrifuge tubes and centrifuge at $11,000 \times g$ for 2 min.

6.  Transfer 1.2 ml of the supernatant from each tube into a fresh tube and add 0.3 ml of PEG precipitation buffer containing 3.5 $M$ ammonium acetate buffer (pH 7.5) and 20% (w/v) PEG to the supernatant. Vortex for 1 min and leave at room temperature for 20 min.

7.  Centrifuge at $12,000 \times g$ for 15 min and decant the supernatant completely.

8.  Resuspend the PEG pellet in 0.3 ml of TE buffer (pH 8.0) and extract it with 1 volume of TE-saturated phenol:chloroform:isoamyl alcohol (25:24:1). Mix by vortexing and centrifuge at $11,000 \times g$ for 5 min.

9.  Transfer the top, aqueous phase to a fresh tube and repeat extraction as in step 8 twice.

10. Precipitate the single-stranded DNA by adding 0.5 volume of 7.5 $M$ ammonium acetate (pH 7.5) and 2.5 volumes of chilled 100% ethanol to the supernatant. Place at –70°C for 30 min.

11. Centrifuge at $12,000 \times g$ for 20 min at 4°C, decant the supernatant, and briefly rinse the DNA pellet with 1 ml of 70% ethanol. Dry the pellet under vacuum for 30 min and dissolve the DNA in 10 µl of TE buffer (pH 7.6). Combine the two DNA samples into one tube and take 2 to 4 µl to measure the concentration of the single-stranded DNA at 260 and 280 nm. Store the sample at –20°C until use.

## Reagents Needed

### Superbroth Medium

> Bacto-tryptone (12 g)
> Bacto-yeast extract (24 g)
> 0.4% (v/v) Glycerol
> Dissolve in a total volume of 900 ml of dd.$H_2O$ and autoclave. Cool to about 50°C and add 100 ml of phosphate buffer containing 170 m$M$ $KH_2PO_4$ and 720 m$M$ $K_2HPO_4$. Autoclave again.

### PEG Precipitation Buffer

> 3.5 $M$ Ammonium acetate, pH 7.5
> 30% (w/v) PEG, 8000

### Protocol B: Performing Site-Directed Mutagenesis

After the single-stranded DNA is obtained, the DNA is phosphorylated by T4 polynucleotide kinase. The synthetic primer containing the mutant, mismatched sequence is annealed to the single-stranded DNA as a template. The second-strand DNA can be synthesized using T7 DNA polymerase and T4 ligase in the presence of 5-methyl-deoxycytosine triphosphate, generating a methylated new strand DNA containing methylcytosine residues. However, the DNA template is not methylated and can be nicked at multiple points with the methylation-sensitive restriction endonuclease *Msp* I. Incubation with exonuclease III results in the degradation of the nicked strand. The mutant, unnicked strand DNA can then be used to transform an appropriate nonrestrictive *E. coli* host (e.g., SDM cells). For efficient site-directed mutagenesis, the USB T7 *in vitro* mutagenesis kit is strongly recommended.

1.  Set up the phosphorylation reaction of the mutant, mismatched oligonucleotides as follows in a microcentrifuge tube on ice:
    - Mutant oligonucleotides (200 pmol)
    - 10X Polynucleotides kinase buffer (2 μl)
    - 10 m$M$ ATP (4 μl)
    - T4 Polynucleotide kinase (1 unit)
    - Add dd.H$_2$O to a final volume of 20 μl.

2.  Incubate at 37°C for 40 min and heat at 70°C for 10 min to inactivate the enzyme.

3.  Set up the annealing reaction in a microcentrifuge tube as follows:
    - Single-stranded DNA template (6 μg)
    - Phosphorylated mutant oligonucleotide (15 pmol)
    - 5X Annealing buffer (1.5 μl)
    - Add dd.H$_2$O to a final volume of 20 μl.

4.  Incubate at 65°C for 5 min and allow to cool slowly to room temperature. This can be done by placing the rack containing the tube in a beaker filled with 65°C water. The cooling procedure takes about 30 min.

5.  Synthesize the mutant, methylated new strand by setting up the following reaction:
    - Annealed template and oligonucleotide primer (10 μl)
    - 10X Synthesis mixture (5 μl)
    - T7 DNA polymerase (12.5 units)
    - T4 DNA ligase (25 units)
    - Add dd.H$_2$O to a final volume of 50 μl.

6.  Incubate at 3°C for 60 to 80 min and heat at 70°C for 10 min to terminate the reaction.

7.  Carry out the nicking reaction of the DNA template (unmethylated strand) by adding 20 units of *Msp* I and 20 units of *Hha* I to the mixture and incubate at 37°C for 50 min.

8.  Remove the nicked strand by adding 150 units of exonuclease III and incubate at 37°C for 50 min.

9.  Stop the reaction by heating at 70°C for 10 min.

10. Take 1 μl of the reaction mixture to transform competent SDM *E. coli* cells using the electroporation method (see the section on subcloning in Chapter 7). It is recommended that one carry out at least five samples of transformation.

11. Pick up at least 15 colonies for preparation of single-stranded phagemid DNA and carry out sequencing by the dideoxy method (see Chapter 8).


## *Reagents Needed*

### *5X Annealing Buffer*

> 200 m$M$ Tris-HCl, pH 7.5
> 100 m$M$ MgCl$_2$
> 250 m$M$ NaCl


### *T4 Polynucleotide Kinase*

1 unit/µl

### 10X Polynucleotide Kinase Buffer

500 m*M* Tris-HCl, pH 7.5
100 m*M* MgCl$_2$
50 m*M* DTT
1 m*M* Spermidine
1 m*M* EDTA

### 10 mM ATP

### 10X Synthesis Mixture

0.1 *M* Tris-HCl, pH 7.5
5 m*M* dATP
5 m*M* dGTP
5 m*M* dTTP
5 m*M* 5-Methyl-dCTP
20 m*M* DTT
10 m*M* ATP

### T7 DNA Polymerase (USB)

### T4 DNA Ligase (USB)

### Msp I, Hha I, and Exonuclease III (USB)

# References

1. **Sambrook, J., Fritsch, E. F., and Maniatis, T.,** *Molecular Cloning: A Laboratory Manual,* 2nd ed., Cold Spring Harbor Press, Cold Spring Harbor, NY, 1989.

2. **Dag, A. G., Bejarano, E. R., Buck, K. W., Burrell, M., and Lichtenstein, C. P.,** Expression of an antisense viral gene in transgenic tobacco confers resistance to the DNA vu-us tomato golden mosaic virus, *Proc. Natl. Acad. Sci. U.S.A.,* 88, 6721, 1991.

3. **Della-Cioppa, G., Bauer, S. C., Taylor, M. L., Rochester, D. E., Klein, B. K., Haughn, G. W., Smith, J., Mazur, B. J., and Somerville, C.,** Transformation with a mutant *Arabidopsis* acetolactate synthase gene renders tobacco resistant to sulfonylurea herbicides, *Mol. Gen. Genet.,* 211, 266, 1988.

4. **Przibilla, E., Heiss, S., Johanningmeicr, U., and Trebst, A.,** Site-specific mutagenesis of the D1 subunit of photosystem 11 in wild type *Chlamydomonas, Plant Cell,* 3, 169, 1991.

5. **Fobert, P. R., Miki, B. L., and Iyer, V. N.,** Detection of gene regulatory signals in plants revealed by T-DNA-mediated fusions, *Plant Mol. Biol.,* 17, 837, 1991.

6. **Morrison, H. G. and Desrosiers, R. C.,** A PCR-based strategy for extensive mutagenesis of a target DNA sequencing, *BioTechniques,* 15, 454, 1993.

7. **Coldschmidt-Clermont, M.,** Transgenic expression of aminoglycoside adenine transferase in the chloroplast: a selectable marker for site-directed transformation of *Chlamydomonas, Nucleic Acids Res.,* 15, 4083, 1991.

8. **Bloch, C. A. and Ausubel, F. M.,** Paraquat-mediated selection for mutations in the manganese-superoxide dismutase gene *sodA, J. Bacteriol.,* 168, 795, 1986.

9. **Watkins, B. A., Davis, A. E., Cocchi, F., and Reitz, M. S.,** A rapid method for site-specific mutagenesis using larger plasmids as templates, *BioTechniques,* 15, 700, 1993.

10. **Klein, R., Silos-Santiago, I., Smeyne, R. J., Lira, S. A., Brambilla, R., Bryant, S., Zhang, L., Snider, W. D., and Barbacid, M.,** Disruption of the neurotrophin-3 receptor gene *trk*C eliminates Ia muscle afferents and results in abnormal movements, *Nature,* 368, 249, 1994.

# Inhibition of Gene Expression by Antisense DNA and RNA

## Contents

It has been demonstrated that gene expression can be down-regulated or completely knocked out by either synthetic antisense oligonucleotides or antisense orientation of the gene of interest under an appropriate promoter.[1-4] Antisense DNA and RNA approaches have been widely applied in eukaryotes in which the antisense oligonucleotides or antisense RNA transcripts expressed successfully inhibit the expression of specifically targeted sense RNA.

Although the detail mechanisms of antisense DNA/RNA inhibition are not very well known, the antisense oligonucleotides or antisense RNA can interact with the complementary sequence or sense RNA transcripts by hydrogen bonding, thus, blocking the processing or translation of the sense RNA. The duplex of antisense RNA and the sense RNA may be rapidly degraded. There is no evidence, however, the antisense RNA directly affects the target gene or DNA sequence. Therefore, the basic principle of using antisense DNA and antisense RNA is to partially or completely knock out the expression of the target gene or some specific DNA sequence.[2-7] The antisense approach has been used to address fundamental questions in molecular and cellular biology in animals, human diseases, and plants. The present chapter focuses on techniques of using exogenous antisense oligonucleotides and antisense gene to control the expression of specific genes.

# I.    Antisense Oligonucleotides

Antisense oligomers can be chemically synthesized based on any target sequence of DNA or RNA of interest. Consequently, gene expression might be regulated at different steps such as DNA replication and transcription; retroviral RNA reverse transcription; pre-mRNA splicing and polyadenylation; tRNA translation; and mRNA transport, degradation, and translation. This section, however, primarily focuses on the blocking or inhibition of translation of specific mRNA.

## A.    Synthesis of Antisense Oligonucleotides

### Protocol

1.   Design appropriate antisense oligomers based on the target sequence of mRNA (sense RNA) or its first-strand cDNA. For example, if the target sequence is 5′-CAUGCCCCUCAACGUUAGC-3′, the antisense oligonucleotides should be designed as 3′-GTACGGGGAGTTGCAATCG-5′. They are complementary to each other and can form a hybrid as follows:

     5′-CAUGCCCCUCAACGUUAGC-3′    mRNA

     3′-GTACGGGGAGTTGCAATCG-5′    antisense oligomer

   a.  **Target region(s)**

       i.    The upstream region of the initiation codon AUG including the first codon AUG should be chosen as the most effective target site when designing an antisense oligomer. The hybridization of the antisense oligomers may prevent the ribosome from interacting with the upstream region of the initiating codon AUG, blocking the initiation of translation.

       ii.   Antisense oligonucleotides can also be designed to be complementary to the coding regions such as conserved motif(s) to block the chain elongation of the polypeptide of interest. However, it is less effective as compared with (i).

       iii.  The terminal region of translation can be chosen as a target to block the termination of translation. The partial-length polypeptide may not be folded properly and be easily degraded.

       iv.   The effective way is to design antisense oligomers that are complementary to the upstream region including the first codon AUG and to the terminal site of the coding region. The efficiency of inhibition of specific gene expression can be 95 to 100%.

    b. **Oligomer modification** — Conventional or unmodified oligonucleotides are usually substrates for nucleases (DNases) such as RNase H and cannot effectively block the expression of the gene of interest. To prevent from being degraded by nucleases, the antisense oligomers should be chemically modified prior to use. Active compounds such as cross-linking and cleaving reagents can be coupled to the oligomers that will then be much more stable, cause irreversible damage to the target sequence, and subsequently result in a very effective blocking of translation. For example, oligomers can be conjugated to an alkylating group or to photosensitizers. An antisense oligomer linked to either nuclease or chemical reagents can generate sequence-dependent cleavage of the target RNA following the binding of the oligomer to its complementary sequence. The cleavage site in the target RNA is usually in the area surrounding the modified antisense oligomer, preventing the translation of the target RNA. Chemical groups linked to the end of the oligomers can introduce strand breaks either directly (e.g., metal chelates and ellipticine) or indirectly (e.g., photosensitizers) after alkaline treatment. Metal complexes such as $Fe^{2+}$-EDTA, $Cu^+$-phenanthroline and $Fe^{2+}$-porphyrin can produce hydroxyl radicals (OH) that can oxidize the sugar and result in phosphodiester bond cleavage. Phosphodiester internucleoside linkages can be substituted with phosphotriester, methylphosphonate, phosphorothioate, or phosphoroselenoate. Oligomers that are modified to phosphorothioates and phosphoroselenoates are resistant to nuclease (RNase H) and can induce cleavage of the target RNA. For this reason, phosphorothioate analogs have been widely applied in intact cells.

    c. **Oligomer size** — 15 to 26-mer depending on particular sequences.

    d. **Oligomer amount** — 10 to 100 μg in total depending on particular experiments.

2.    Synthesize and purify appropriate antisense oligonucleotides, based on their designs, by DNA synthesizer according to the instructions.

## B.    Treatment of Cultured Cells Using the Antisense Oligomers and Determination of the Optimium Dose of the Oligomers

### *Protocol*

1.    Under sterile conditions, culture cells of interest in prewarmed, appropriate medium such as modified DMEM containing 10 to 15% (v/v) heat-inactivated fetal bovine serum (FBS) or 5 to 10% (v/v) heat-inactivated newborn calf serum with appropriate antibiotics such as $10^5$ units/l penicillin and 0.01 and 0.05% (w/v) streptomycin. Incubate the cells at 37°C in 5% $CO_2$ and 95% air and saturated with water vapor. Maintain the cells in logarithmic phase.

2.    Spin down the cells at 100 to 1000 × *g* for 5 min under sterile conditions. In a sterile cabinet, carefully remove the supernatant and resuspend the cells into minimum volume of fresh medium. Determine the viability and density of the cells by counting Trypan blue-excluding cells in a hemocytometer.

3.    Carry out dose-response of synthetic antisense oligomers as follows in order to optimize the amount of the oligomers to be used to block the expression of gene of interest:

    a. Resuspend each sample of $2 \times 10^6$ cells in 2 to 5 ml fresh medium contained in 24-well culture plates or T-flasks.

    b. Add the oligomers to the medium for the following samples:

       • Sample no. 1: no antisense oligomer as a negative control

       • Sample no. 2: 6 μ*M* antisense oligomers

       • Sample no. 3: 4 μ*M* antisense oligomers

       • Sample no. 4: 2 μ*M* antisense oligomers

- Sample no. 5: 1 $\mu M$ antisense oligomers
- Sample no. 6: 0.5 $\mu M$ antisense oligomers
- Sample no. 7: 0.25 $\mu M$ antisense oligomers

*Note:*  *The initiation concentration of appropriate oligomers varies with the particular experiment.*

c. Incubate the cells for 2 to 6 days, add the same amount of the oligomers to appropriate samples every day, and gently mix in the medium.

d. Centrifuge at $1000 \times g$ for 5 to 10 min. If the protein of interest is a secretory protein, the supernatant should be used for extraction of proteins. If not, the pellet cells of each sample should be washed with appropriate buffer, lysed, and extracted for proteins analysis. The general procedures of protein extraction from cultured cells are described in Chapter 3. The protein of interest may be purified from the total proteins with a specific antibody against the protein by the immunoprecipitation method. We recommend using total proteins for inhibition analysis, using the antibody against appropriate housekeeping gene products as a control.

e. Analyze the inhibition of the protein of interest by Western blotting, whose detailed procedures are given in Chapter 5. Antibody against actin or tubulin should be used as the control. The example data are as follows with the optimium dose of oligomers at 2 $\mu M$.

## C. Analysis of Inhibition of Gene Expression Using the Optimium Dose of the Oligomers

### *Protocol*

1. Resuspend each sample of $2 \times 10^6$ healthy cells in 2 to 5 ml fresh medium contained in 24-well culture plates or T-flasks.

2. Add the optimium amount of antisense oligomers (e.g., 2 $\mu M$) to the medium of each sample and culture for appropriate period of time as follows:
   - Samples nos. 1 to 8: no oligomer as controls (0, 0.5, 1, 2, 3, 4, 5, and 6 day)
   - Sample no. 9: 2 $\mu M$ antisense oligomers for 0.5 day
   - Sample no. 10: 2 $\mu M$ antisense oligomers for 1 day
   - Sample no. 11: 2 $\mu M$ antisense oligomers for 2 days
   - Sample no. 12: 2 $\mu M$ antisense oligomers for 3 days
   - Sample no. 13: 2 $\mu M$ antisense oligomers for 4 days
   - Sample no. 14: 2 $\mu M$ antisense oligomers for 5 days
   - Sample no. 15: 2 $\mu M$ antisense oligomers for 5 days
   - Sample no. 16: 2 $\mu M$ antisense oligomers for 5 days

3. Harvest the cells at the appropriate time point as indicated above by centrifuging at $1000 \times g$ for 10 min. If the protein of interest is a secretory protein, the supernatant should be used for extraction of proteins. If not, the pellet cells of each sample should be washed with appropriate buffer, lysed, and extracted for proteins analysis. The general procedures of protein extraction from cultured cells are described in Chapter 3. The protein of interest may be purified from the total proteins with a specific antibody against the protein by the immunoprecipitation method. We recommend using total proteins for inhibition analysis, using the antibody against appropriate housekeeping gene products as a control.

4.  Analyze the inhibition of the protein expressed by the target mRNA using Western blotting. Antibody against actin or tubulin should be used as the control.

# II. Antisense Orientation of Gene of Interest

Synthetic antisense oligonucleotides can effectively inhibit the expression of a specific gene. However, the effect of inhibition is only temporary and is usually applied to cultured cells. To continuously down-regulate the gene expression, it is necessary to keep adding synthetic antisense oligonucleotides, which is time-consuming and expensive. In order to produce stable transformants that can stably express antisense RNA, antisense orientation of the specific gene should be generated. Generally, the target DNA or gene is first mutated *in vitro* either by complete inversion of the target sequence (antisense orientation) or by transferring an appropriate promoter from upstream to downstream of the coding region (behind the promoter). The mutated DNA is ligated to the appropriate vector (usually a plasmid-based vector) by ligase and propagated in an appropriate *E. coli* strain for amplification. Recombinant plasmids will then be isolated from transformants and the DNA fragments containing the target sequence can be removed from the plasmids (Figure 12.1). The sense and antisense orientations of plasmids can be selected on the basis of appropriate restriction enzyme(s) digestion patterns (Figure 12.2). The mutated constructs will then be transferred into appropriate cell lines or totipotency cells that can generate transgenic animals or plants in which expression of the target gene should be significantly inhibited by the overexpressed antisense RNA. The transfer and analysis of the expression of the transferred gene are described in Chapters 15 and 16.

## A. Preparation of Target DNA, Strong Promoters, Enhancers, Poly(A) Signals, and Vectors

### 1. Preparation of Target DNA

The target DNA should already be characterized and subcloned. The DNA sequence and restriction enzyme sites are well known. For eukaryotic genes, the target DNA selected is usually the coding region of the cDNA of interest in order to prevent potential interference of intron sequences. The target DNA should first be removed from the subcloned plasmids using appropriate restriction enzyme(s) that cannot cut within the target sequence.

*Protocol*

1.  Set up, on ice, a standard restriction enzyme digestion reaction as follows:
    **Single-restriction enzyme digestion**
    - Plasmids (20 µg)
    - 10X Appropriate restriction enzyme buffer (10 µl)
    - 1 mg/ml Acetylated BSA (optional) (10 µl)
    - Appropriate restriction enzyme (3.4 units/µg DNA)
    - Add dd.$H_2O$ to a final volume of 100 µl

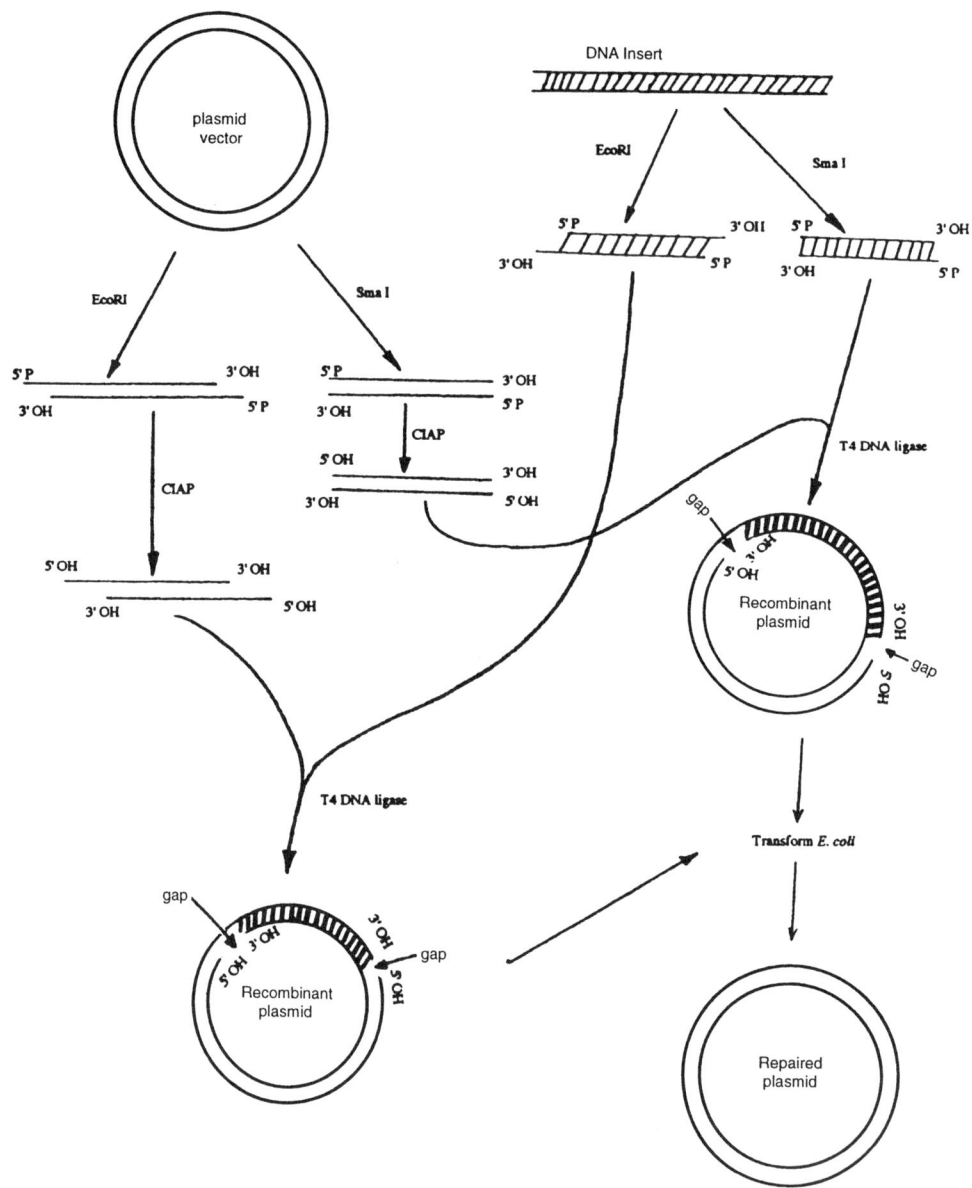

**FIGURE 12.1**

Schematic representation of strategies for DNA recombination and transformation of *E. coli* with recombinant constructs.

**Double-restriction enzyme digestion**

- Plasmid DNA or DNA to be inserted (20 μg)
- 10X Appropriate restriction enzyme buffer (10 μl)
- 1 mg/ml Acetylated BSA (optional) (10 μl)
- Appropriate restriction enzyme A (3.4 units/μg DNA)
- Appropriate restriction enzyme B (3.4 units/μg DNA)
- Add dd.H$_2$O to a final volume of 100 μl.

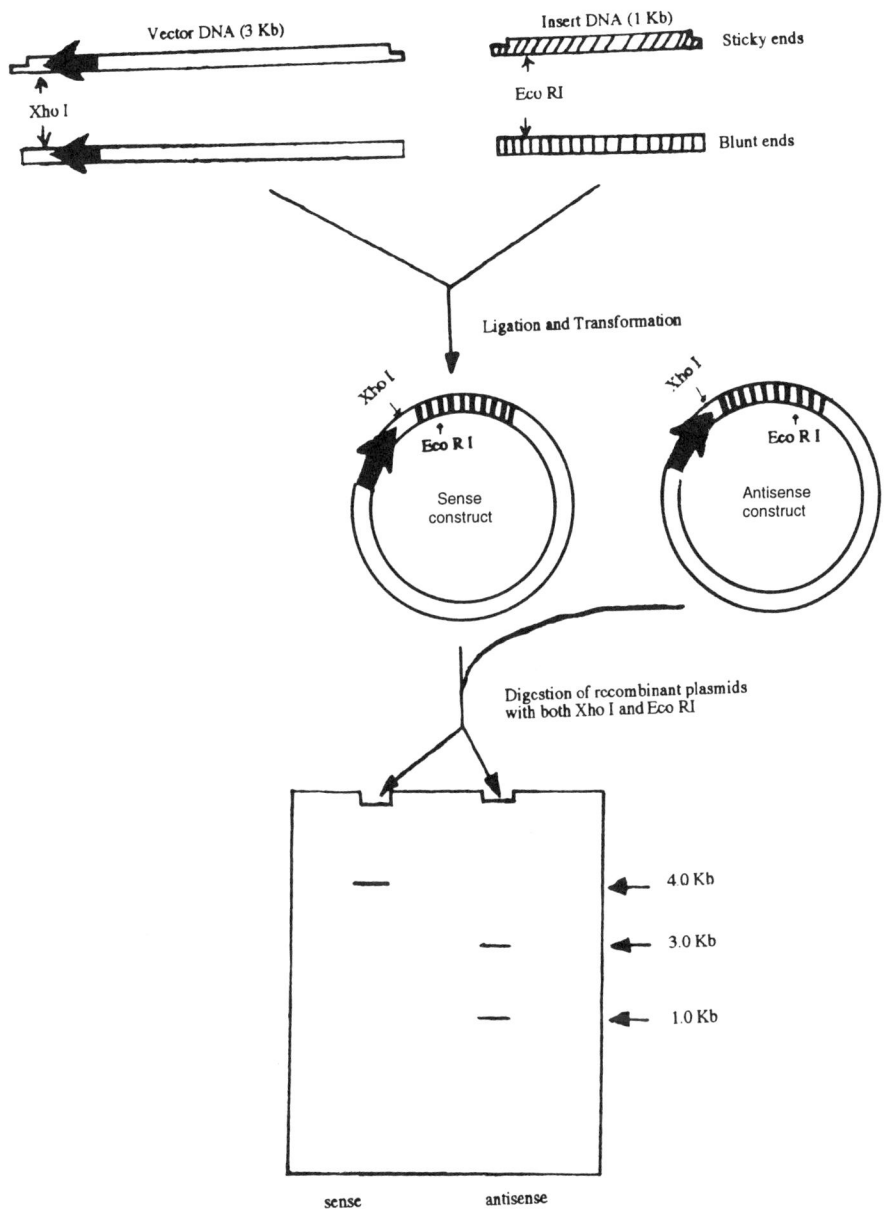

**FIGURE 12.2**

Strategies for making both sense and antisense DNA constructs and for identifying both constructs by patterns of appropriate restriction enzyme digestions.

*Note:* *For double-restriction enzyme digestions, the appropriate 10X buffer containing higher NaCl concentration than the other buffer may be chosen for the double-enzyme digestion buffer.*

2. Incubate at an appropriate temperature depending on the appropriate restriction enzyme (e.g., 37°C) for 2 to 3 h.

3. Purify the target DNA from the digested vector by agarose gel electroelution.

   a. Carry out electrophoresis as described in Southern blotting except use 1% low-melting point
      agarose instead of normal agarose. Monitor the separation of the target fragment and vector
      as compared with known DNA sizes.

   b. Transfer the gel onto a long-wavelength UV transilluminator (305 to 327 nm) to visualize the
      DNA bands. Excise the band(s) of interest with a clean razor blade and place the slices in a
      microcentrifuge tube.

*Notes:*  *(1) The gel should not be placed on a short-wavelength (e.g., <270 nm) UV
          transilluminator as the UV light may cause internal breaking of the DNA fragment,
          and thus significantly inhibit the subcloning of the DNA insert of interest. (2) UV light
          is harmful to the human body. Protective glasses, gloves, and a lab coat should be
          worn when working with UV light. (3) The gel slices containing the DNA bands of
          interest should be trimmed of unstained gel regions as much as possible.*

   c. Add 2 volumes of TE buffer to the gel slices and completely melt the gel in a 60 to 70°C water
      bath.

*Notes:*  *The gel slices can be directly melted without adding any TE buffer. The DNA
          concentration is usually high, but the total yield of the eluted DNA fragment is much
          lower than when TE buffer is added.*

   d. Immediately chill the melted gel solution on dry ice and place the tube at –70°C for at least
      20 min.

   e. Thaw the gel mixture by tapping the tube vigorously. It takes about 5 to 10 min to thaw the
      gel to obtain a resuspension.

   f. Centrifuge at $11,000 \times g$ for 5 min at room temperature.

   g. Carefully transfer the liquid phase containing the elution DNA fragment to a fresh tube. The
      DNA solution can be directly used for labeling except at low concentration.

   h. Extract EtBr three to four times with 3 volumes of water-saturated *n*-butanol.

   i. Precipitate the DNA by adding 0.1 volume of 3 *M* sodium acetate buffer (pH 5.2) and 2.5
      volumes of chilled 100% ethanol to the DNA solution. Allow to precipitate at –70°C for 1 h
      and centrifuge at $12,000 \times g$ for 5 min at room temperature.

   j. Discard the supernatant and briefly rinse the pellet with 1 ml of 70% ethanol. Dry the DNA
      under vacuum for 15 min and dissolve the DNA in an appropriate volume of dd.$H_2O$ or TE
      buffer. Store at –20°C until use.

4. Purify the target DNA using NA45 DEAE membranes.

   a. Carry out electrophoresis as described for Southern blotting except use 1% low-melting point
      agarose instead of normal agarose. Use a long-wavelength of UV lamp to monitor the
      separation of the target fragment and the vector as compared with known sizes of DNA.

   b. Soak a membrane (Scheicher and Schuell) in 10 m*M* EDTA buffer (pH 7.6) for 15 min at room
      temperature and store at 4°C until use.

   c. During the electrophoresis, monitor the migration of the DNA bands stained by EtBr in the gel
      using a long-wavelength UV lamp. Make an incision below the DNA band of interest and
      insert a piece of the prepared membrane into the incision in the gel.

   d. Continue electrophoresis, while monitoring the migration of the stained band, until the DNA
      fragment migrates onto the membrane.

   e. Remove the membrane strip from the gel and place it in a solution containing 20 m*M* Tris-HCl
      (pH 8.0), 0.15 *M* NaCl, and 0.1 m*M* EDTA. Shake it for 1 min to remove any agarose.

   f. Transfer the strip to 0.2 to 0.3 ml of elution buffer containing 20 m*M* Tris-HCl (pH 8.0), 1*M*
      NaCl, and 0.1 m*M* EDTA.

g. Incubate the strip at 55 to 68°C for 30 to 60 min with shaking at 60 rpm.

h. Rinse the strip with 0.1 ml of elution buffer and pool the eluate together.

i. Extract EtBr four times with 3 volumes of water-saturated *n*-butanol.

j. Precipitate and dissolve the eluted DNA as previously described.

## 2. Selections of Promoter, Enhancer, Poly(A) Signal, and Appropriate Vectors

In order to obtain highly efficient inhibition of the target RNA transcripts, the antisense RNA transcripts should be overexpressed using an appropriate strong promoter. An enhancer may be used by inserting it upstream of the promoter. Poly(A) is used to provide 3′ polyadenylation of the transcript. For animal systems, an SV40 promoter, an SV40 enhancer, an SV40 poly(A) signal, and retrovirus long-terminal repeats (LTRs) containing both a promoter and an enhancer are recommended. For plant systems, the cauliflower mosaic virus (CaMV) 35S promoter and a poly(A) signal for the nopaline synthase gene of *agrobacterium tumefaciens* Ti plasmid are recommended. These regulatory fragments are well characterized, subcloned in plasmids, and available from Promega Corporation, CLONTECH Laboratories, Inc., or other companies and laboratories. The DNA fragment of interest can be purified from the recombinant plasmid by using appropriate restriction enzyme(s), but this is not necessary since the appropriate promoter, enhancer, or even a reporter gene are already built into the commercial plasmid vectors such as pBI121 and pBI221 for plant systems (CLONTECH Laboratories, Inc.) and retrovirus-based plasmid vectors for animal systems. Generally, there is a unique restriction enzyme site such as *Sma* I (pBI121 and pBI221) or *Hind* III (pGL2-promoter vector) downstream of the promoter, which allows the insertion of antisense- or sense-oriented DNA with ease. In that case, the plasmids can be linearized with *Sma* I or *Hind* III for the ligation with mutated DNA. In addition, some commercial vectors contain the gene *neo*r, or equivalent, as a selectable marker of stable transformants by using appropriate antibiotics (e.g., G418) in culture medium.

*Protocol*

1. Set up, on ice, a standard enzyme digestion as follows:

   *Sma* **I**
   - Plasmid vectors or DNA to be inserted (10 μg)
   - 10X *Sma* I buffer (10 μl)
   - 1 mg/ml Acetylated BSA (optional) (10 μl)
   - *Sma* I (30 units)
   - Add dd.H$_2$O to a final volume of 100 μl.

   *Hind* **III**
   - Plasmid vectors or DNA to be inserted (10 μg)
   - 10X *Hind* III buffer (10 μl)
   - 1 mg/ml Acetylated BSA (optional) (10 μl)
   - *Hind* III (30 units)
   - Add dd.H$_2$O to a final volume of 100 μl.

2. Incubate at 37°C for 2 to 3 h.

*Notes:*   *The digestion efficiency can be checked by loading 1 μg of the digested DNA (10 μl)*
*and undigested plasmids with loading buffer to a 1% agarose mini-gel, respectively.*
*After electrophoresis, the undigested plasmid DNA may reveal multiple bands be-*
*cause of different supercoiled DNA. However, one band will be visible after* Sma I
*digestion.*

3.   Carry out CIAP treatment, as shown below, to remove the 5′ phosphate group in order to prevent
the recircularization of the linearized plasmid.

•  Linearized plasmids (90 μl)

•  10X CIAP buffer (15 μl)

•  CIAP diluted in 10X CIAP buffer (0.01 unit/pmol ends)

•  Add dd.H$_2$O to a final volume of 150 μl.

*Notes:*   *(1) CIAP and 10X CIAP should be kept at 4°C. CIAP treatment should be set up at*
*0°C. (2) The calculation of the amount of ends is as follows: There is 9 μg digested*
*DNA left after taking 1 μg of 10 μg digested DNA for checking on an agarose gel.*
*If the vector is 3.2 Kb, the amount of ends can be calculated by use of the formula*
*below:*

$$\text{pmol ends} = \frac{\text{amount of DNA}}{\left(\text{base pairs} \times 660/\text{base pair}\right)} \times 2$$

$$= \frac{9}{3.2 \times 1000 \times 660} \times 2$$

$$= 4.2 \times 10^{-6} \times 2$$

$$= 8.4 \times 10^{-6} \ \mu M$$

$$8.4 \times 10^{-6} \times 10^{-6} = 8.4 \text{ pmol ends}$$

4.   Incubate at 37°C for 1 h and add 2 μl of 0.5 $M$ EDTA buffer (pH 8.0) to stop the reaction.

5.   Extract with 1 volume of TE-saturated phenol/chloroform. Mix well by vortexing for 1 min and
centrifuge at $11,000 \times g$ for 5 min at room temperature.

6.   Carefully transfer the top, aqueous phase to a fresh tube and add 1 volume of chloroform:isoamyl
alcohol (24:1) to the supernatant. Mix well and centrifuge as in step 5.

7.   Carefully transfer the upper, aqueous phase to a fresh tube and add 0.1 volume of 3 $M$ sodium
acetate buffer (pH 5.2) or 0.5 volume of 7.5 $M$ ammonium acetate to the supernatant. Briefly mix
and add 2 to 2.5 volumes of chilled 100% ethanol to the supernatant. Allow precipitation to occur
at −70°C for 1 h or at −20°C for 2 h.

8.   Centrifuge at $12,000 \times g$ for 10 min and carefully decant the supernatant. Briefly rinse the DNA
pellet with 1 ml of 70% ethanol and dry the pellet under vacuum for 20 min. Dissolve the DNA
pellet in 20 to 40 μl dd.H$_2$O. Take 4 μl of the sample to measure the concentration of the DNA
at 260 nm. Store the sample at −20°C until use.

*Note:*   *Addition of 0.5 volume of 7.5 M ammonium acetate to the supernatant in step 7 yields*
*a greater amount of DNA precipitation than by adding 0.1 volume of 3 M sodium*
*acetate buffer (pH 5.2).*

*Reagents Needed:*—————————————————————————

*Appropriate Enzymes*

*10X Appropriate Restriction Enzyme Buffer*

*1% Agarose Mini-Gel*

*TE-Saturated Phenol/Chloroform*

*Chloroform:Isoamyl Alcohol (24:1)*

*3 M Sodium Acetate Buffer, pH 5.2*

*7.5 M Ammonium Acetate*

*Ethanol (100%, 70%)*

*0.5 M EDTA, pH 8.0*

*Calf Intestinal Alkaline Phosphatase (CIAP)*

*TE Buffer*

*10X Sma I Buffer*
        0.1 $M$ Tris-HCl, pH 7.8
        0.5 $M$ KCl
        70 m$M$ MgCl$_2$
        10 m$M$ DTT

*10X Hind III Buffer*
        60 m$M$ Tris-HCl, pH 7.5
        0.5 $M$ NaCl
        100 m$M$ MgCl$_2$
        10 m$M$ DTT

*10X CIAP Buffer*
        0.5 $M$ Tris-HCl, pH 9.0
        10 m$M$ MgCl$_2$
        1 m$M$ ZnCl$_2$
        10 m$M$ Spermidine

## 3.     Preparation of Blunt-End DNA

If both vectors and mutated, digested DNA fragments have sticky or overhanging ends at the right sites, they can be readily ligated to each other. If not, it is usually necessary to generate blunt ends on the digested DNA fragments prior to their being ligated to the blunt vectors. If the vectors have sticky ends generated by *Hind* III, the vectors may be also changed to blunt ends before CIAP treatment. Blunt ends can be made by filling in 5′ overhangs with DNA polymerase I (Klenow fragment).

### *Protocol*

1.  Set up the 5′ overhang reaction in a microcentrifuge tube on ice as follows:
    - DNA fragment or vector DNA (1 to 4 μg in 4 μl)
    - 10X 5′ Overhang buffer (2 μl)
    - 2 m$M$ dNTP mixture (0.5 to 2 mM each) (2 μl)
    - DNA polymerase I (Klenow fragment) (2 to 3 units)
    - Add dd.H$_2$O to a final volume of 20 μl.
2.  Incubate at room temperature for 20 min and heat at 70°C for 5 min to stop the reaction.
3.  Add 1 volume of TE buffer (pH 7.5) and carry out phenol:chloroform extraction and precipitation as previously described.

### *Reagents Needed*

#### *10X 5′ Overhang Buffer*

> 500 m$M$ Tris-HCl, pH 7.2
> 100 mM MgSO$_4$
> 1 m$M$ DTT
> 500 μg/ml BSA

#### *2 mM dNTP Mixture*

> 2 m$M$ Each of dATP, dCTP, dGTP, and dTTP in 20 m$M$ Tris-HCl, pH 7.5

#### *DNA Polymerase I (Klenow Fragment)*

> 1 unit/μl

## 4.     Preparation of Insert DNA Lacking the 5′ Phosphate Group

If the insert DNA lacks the 5′ phosphate group, the following reaction should be carried out prior to carrying out ligation to blunt-ended and dephosphorylated vectors.

### *Protocol*

1.  Add the phosphate group to the 5′ end of the insert DNA by the following kinase reaction:
    - Blunt-end, insert DNA (15 μl)
    - 10X Polynucleotides kinase buffer (3 μl)

- 10 m$M$ ATP (1.5 µl)
- 100 m$M$ DTT (1.5 µl)
- T4 Polynucleotide kinase (15 units)
- Add dd.H$_2$O to a final volume of 30 µl.

2. Incubate at 37°C for 40 min and heat at 70°C for 10 min to inactivate the enzyme.

3. Add 1 volume of TE buffer (pH 7.5) and carry out extraction and precipitation as previously described.

## Reagents Needed

### T4 Polynucleotide Kinase

1 unit/µl

### 10X Polynucleotide Kinase Buffer

500 m$M$ Tris-HCl, pH 7.5
100 m$M$ MgCl$_2$
50 m$M$ DTT
1 m$M$ Spermidine
1 m$M$ EDTA

### 10 mM ATP

## B.    Blunt-End Ligation of Plasmid Vectors and Inserts of Antisense and Sense Orientation

To achieve optimal ligation, the ratio of vector to insert DNA (1:1, 1:2, 1:3, and 3:1 molar ratios) is strongly recommended to be optimized by using a small-scale reaction. The following reaction is standard for the ligation of 3.2 Kb plasmid vector and 3.0 Kb insert DNA.

### Protocol

1. Calculate molar weight of vector and insert DNA:

$$1 \ M \text{ plasmid vector} = 3.2 \times 1000 \times 660 = 2.112 \times 10^6$$

$$1 \ M \text{ insert DNA} = 3 \times 1000 \times 660 = 1.98 \times 10^6$$

2. Calculate the molar ratio of vector to insert DNA as follows:

| Ratio of vector DNA:insert DNA | Amount of DNA (µg) | |
|---|---|---|
| | Vector | Insert |
| 1:1 | 1 | 0.792 |
| 1:2 | 1 | 1.584 |
| 1:3 | 1 | 2.376 |
| 3:1 | 1 | 0.264 |

3.  Set up the ligations on ice:

| Components | Ligation reactions | | | |
|---|---|---|---|---|
| | 1 (1:1) | 2 (1:2) | 3 (1:3) | 4 (3:1) |
| Plasmid DNA as vector (µg) | 1 | 1 | 1 | 1 |
| Insert DNA (µg) | 0.792 | 1.584 | 2.376 | 0.244 |
| 10X Ligase buffer (µl) | 1 | 1 | 1 | 1 |
| T4 DNA ligase (Weiss units) | 4 | 4 | 4 | 4 |
| Add dd.H$_2$O to (µl) | 10 | 10 | 10 | 10 |

*Notes:* *Restriction enzyme-digested plasmid (vector) and insert DNA should be dissolved in dd.H$_2$O (nuclease-free) to 0.5 to 1.0 µg/µl. If the DNA is less than 0.4 µg/µl, it is recommended that the DNA be precipitated in order to dissolve in about 1 µg/µl.*

4.  Incubate the reactions at 4°C for 12 to 24 h, or 16°C for 4 to 6 h, or at room temperature (22 to 25°C) for 1 to 2 h.

*Note:* *After ligations are completed at one of the above temperatures, the mixture can be stored at 4°C until later use.*

5.  Check the efficiency of the ligations by performing 1% agarose gel electrophoresis. When the electrophoresis is complete, photograph the EtBr-stained gel under UV light. As compared with unligated vector or insert DNA wells, a highly efficient ligation should show less than approximately 10% unligated vector and insert DNA by estimation of the intensity of fluorescence. Approximately 90% of the vector and insert DNA are ligated to each other and show strong band(s) with molecular weight shifts compared to the vector and insert DNA sizes. By comparing the efficiency of ligations using different molar ratios, the optimal conditions for carrying out the ligation can be determined with ease. These can then be used as a guide for large-scale ligation.

*Note:* *The above small-scale ligation is optional, but is strongly recommended.*

6.  Large-scale ligation of vector and insert DNA is based on determinations of the optimal conditions by small-scale ligations. For example, if one uses a 1:2 molar ratio of plasmid DNA:insert DNA as the optimal condition for ligation, a large-scale ligation can be carried out as follows:
    - Plasmid DNA as the vector (3 µg)
    - Insert DNA (4.75 µg)
    - 10X Ligase Buffer (3 µl)
    - T4 DNA ligase (15 Weiss units)
    - Add dd.H$_2$O to a final volume of 30 µl.
    - Incubate the ligation mixture at 4°C for 12 to 24 h, or 16°C for 4 to 6 h, or at room temperature (22 to 25°C) for 1 to 2 h. Store at 4°C until use. Proceed now to transformation.

## Reagents Needed

### Dephosphorylated Linearized Plasmid Vectors

### Phosphorylated DNA Fragments

### 5X Ligase Buffer

> 0.25 $M$ Tris-HCl, pH 7.5
> 50 m$M$ MgCl$_2$
> 5 m$M$ ATP
> 5 m$M$ DTT
> 25% (v/v) PEG, 8000

### T4 DNA Ligase

> 2 Weiss units/μl

## C.  Transformation of Appropriate Strain of Bacteria to Amplify the Recombinant Plasmids

The general procedures for transformation of bacteria and selection of transformants are described under the section on subcloning of plasmids in Chapter 7.

## D.  Selection of Plasmids with Antisense and Sense Orientations

During the blunt-end ligations, each of the antisense or sense orientations of DNA has a 50% probability to be ligated to the vectors. However, they can be separated from each other on the basis of appropriate restriction enzyme(s) digestion patterns. For example, if *Xba* I is the closest enzyme to the blunt-end *Xba* I, and if the first restriction enzyme at the 5′ end of the coding region of the sense orientation gene is *EcoR* I, the recombinant plasmids can be digested with both *Xba* I and *EcoR* I. After electrophoresis and comparison with vector and insert DNA as controls, the plasmids containing the sense-oriented DNA will show one large band that is almost the size of the vector and inserted DNA. However, the plasmids containing the antisense-oriented DNA should show two bands. One is almost the same size as the vector DNA, while the other is approximately the same size as the insert DNA.

### Protocol

1. Isolate the recombinant plasmid DNA for transfection by the DNA mini-prep method. The detailed procedures are described under the section on isolation and purification of plasmid DNA in Chapter 1.

2. Set up a double-restriction enzyme digestion as follows:
   - Plasmid DNA or DNA to be inserted (20 μg)
   - 10X Appropriate restriction enzyme buffer (10 μl)
   - 1 mg/ml Acetylated BSA (optional) (10 μl)
   - Appropriate restriction enzyme A (3.4 units/μg DNA)
   - Appropriate restriction enzyme B (3.4 units/μg DNA)
   - Add dd.H$_2$O to a final volume of 100 μl.

*Note:* *For double-restriction enzyme digestions, the appropriate 10X buffer containing a higher NaCl concentration than the other buffer may be chosen for the double-enzyme digestion buffer.*

3. Incubate at the appropriate temperature depending on the appropriate restriction enzyme (e.g., at 37°C) for 2 to 3 h.

4. Carry out electrophoresis on a 1% agarose gel as described previously and identify antisense-oriented, recombinant plasmids. Culture the bacteria containing the antisense-oriented plasmids under large-scale conditions. Isolate the plasmids and store at –20°C until use for gene transfer.

## E.     Gene Transfer and Expression of Antisense RNA

After the recombinant constructs containing the antisense orientation of the gene of interest and selectable marker gene (e.g., *neo*<sup>r</sup>) have been obtained, the next procedure is to transfer the constructs into specific cell lines, animals, or plants of interest. Stable transformant cell lines, transgenic animals, or transgenic plants can be selected by using an appropriate antibiotics (e.g., *neo*<sup>r</sup>). The antisense RNA transcripts expressed in transformants can hybridize with sense RNA. The translation of mRNA will then be blocked. The detail protocols of gene transfer and expression are described in Chapters 15 and 16. The stable transformants should be analyzed as follows:

1. Evidence of successful integration of antisense gene in the host genome by Southern blot analysis using the transferred gene as a probe.

2. Evidence of the expression of antisense RNA by Northern blot analysis using the transferred gene as a probe.

3. Evidence of the inhibition of the translation of the target sense RNA at the protein level by Western blot analysis using the antibody raised against the gene product. This is the most critical proof of the inhibition of gene expression by antisense RNA.

4. Functional evidence altered by antisense RNA.

# References

1. **Robert, L. S., Donaldson, P. A., Ladaigue, C., Altosaar, L., Arnison, P. G., and Fabijanski, S. F.,** Antisense RNA inhibition of β-glucuronidase gene expression in transgenic tobacco can be transiently overcome using a heat-inducible β-glucuronidase gene construct, *Biotechnology,* 8, 459, 1990.

2. **Rezaian, M. A., Skene, K. G. M., and Ellis, J. G.,** Anti-sense RNAs of cucumber mosaic virus in transgenic plants assessed for control of the virus, *Plant Mol. Biol.,* 11, 463, 1988.

3. **Hemenway, C., Fang, R. -X., Kaniewski, W. K., Chua, N. -H., and Tumer, N. E.,** Analysis of the mechanism of protection in transgenic plants expressing the potato virus X coat protein or its antisense RNA, *EMBO J.,* 7, 1273, 1988.

4. **Dag, A. G., Bejarano, E. R., Buck, K. W., Burrell, M., and Lichtenstein, C. P.,** Expression of an antisense viral gene in transgenic tobacco confers resistance to the DNA vu-us tomato golden mosaic virus, *Proc. Natl. Acad. Sci. U.S.A.,* 88, 6721, 1991.

5. **Zelenin, A. V., Titomirov, A. V., and Kolesnikov, V. A.,** Genetic transformation of mouse cultured cells with the help of high-velocity mechanical DNA injection, *FEBS Lett.,* 244, 65, 1989.

6. **Perlak, F. J., Fuchs, R. L., Dean, D. A., McPherson, S. L., and Fischhoff, D. A.,** Modification of the coding sequence enhances plant expression of insect control protein genes, *Proc. Natl. Acad. Sci. U.S.A.,* 88, 3324, 1991.

7. **Morrison, H. G. and Desrosiers, R. C.,** A PCR-based strategy for extensive mutagenesis of a target DNA sequencing, *BioTechniques,* 15, 454, 1993.

# Chapter 13

# DNA Footprinting and Gel Retardation Assay

## Contents

Footprinting or DNA–protein interaction test is a powerful technique that is used to detect specific DNA fragment(s) or sequences bound by specific protein(s) such as transcriptional factors or equivalent. A particular dsDNA is labeled at the 5′ end with $^{32}$P or its equivalent and allowed to interact with particular proteins of interest.[1,2] The DNA–protein complex is then subject to a DNA endonuclease digestion such as *DNase* I. However, the nicking by *DNase* I is so brief that no DNA molecule gets more that one single-strand cut. The specific DNA fragment bound by protein should be resistant to *DNase* I nicking. If the DNA contains $n$ base pairs in which there are $y$ base pairs bound by the protein, $n$ different sizes of DNA pieces would exist in non-protein-interacted DNA molecules as a control.[2-4] Since the protein binding prevents the specific DNA fragment from endonuclease digestion, only $n-y$ different sizes of DNA fragments are present in protein-binding DNA molecules. These fragments from both control and DNA–protein-interacted samples can be separated by gel electrophoresis and their sizes can be determined by the distance from the labeled 5′ end to the nicks. The protein-binding region of the DNA can then be determined with ease. Furthermore, the actual points of the bound region of the DNA fragment can be identified by the dimethyl sulfate protection technique. Dimethyl sulfate only methylates A and G but not C and T. The methylation should not occur in the actual protein-contact region of the DNA fragment. The actual size of the regions of contact by the protein can then be identified by the positions of nonmethylated A and G in the entire bound region of the DNA fragment. The DNA–protein complex can be isolated and denatured for DNA and protein sequencing in order to identify the specific sequence bound by the protein.

The gel retardation assay or the gel mobility shift assay, on the other hand, are methods used to detect the specific protein or protein fraction that binds to a specific DNA sequence. The DNA molecule is highly negatively charged and migrates very fast toward a positive electrode in an electric field. During a native or nondenaturing polyacrylamide gel electrophoresis, DNA molecules are separated based on their sizes. Smaller molecules move faster than larger ones. When a DNA molecule or DNA fragment is bound by protein(s), it moves more slowly through the gel. The larger the protein that is bound, the more slowly the DNA–protein complex migrates. Based on this principle of the gel retardation assay, DNA molecules or fragments of interest (usually its length and sequence are known) are radioactively or nonradioactively labeled and mixed with a specific protein or protein fraction for a specified period of time. The mixture is loaded onto a native polyacrylamide gel and subject to electrophoresis followed by autoradiography or its equivalent. As compared with control DNA fragments without the protein mixture, the free DNA molecules or fragments will run rapidly close to the bottom of the gel, but those molecules or fragments bound by protein(s) are retarded above the free DNA bands. All these bands are revealed on the developed X-ray film or an equivalent membrane.

# I.     Footprinting Protocols

## A. Preparation of Single-End (3′ or 5′ end) Labeled DNA Probe

DNA purification is described in Chapter 1. Radioactively labeled DNA fragments of interest are prepared by single-end labeling in order to achieve high specific activity. The labeling procedures are described in Chapter 4. The probe DNA is purified by extraction and precipitation with ethanol. The DNA is dissolved in an appropriate volume of TE buffer, and then the cpm of the probe is measured. The sample is stored at –20°C until use.

## B. Preparation of Denaturing Polyacrylamide Gels

1. Thoroughly clean an appropriate vertical gel electrophoresis apparatus including glass plates (15 × 60 cm or equivalent), spacers (0.4 mm), tanks, and comb (0.4 mm). Siliconize one of the glass plates with 1 to 1.5 ml of Sigmacote (Sigma Chemicals), wash it with ethanol, then wipe dry. Assemble the sandwich according to the instructions.

2. In a clean beaker, prepare 100 ml of the gel mixture as follows:

| Stock solution | Polyacrylamide gel concentration (%) | | |
|---|---|---|---|
|  | 5 | 6 | 7 |
| 5X TBE solution (ml) | 20 | 20 | 20 |
| 40% Acrylamide (ml) | 12.5 | 15 | 17.5 |
| Ultrapure urea (g) | 48 | 48 | 48 |
| dd.H$_2$O (ml) | 32.5 | 30 | 27.5 |

3. Gently heat and stir the gel mixture until the urea is dissolved and filter or degas (optional).

4. Add 250 μl of 10% ammonium persulfate and 40 μl of TEMED to the gel mixture, gently mix well, and then immediately pour the gel. Insert the comb and allow the gel to polymerize for 1 to 1.5 h.

*Notes:* *The glass sandwich can be laid on a flat table or bench without sealing the bottom edge. The gel mixture can be poured horizontally by capillary action. If the gel is poured vertically or at an angle, the side and bottom edges must be sealed tightly to prevent leaking. The polymerized gel can be wrapped with wet paper towels at the top and bottom and covered with SaranWrap™ to prevent the gel from drying out. The gel can be left overnight until use.*

## C. DNase I Protection Assay

1. Prepare the DNA mixture as follows:
   - Single-end-labeled DNA fragments (2 to $4 \times 10^5$ cpm) (5 to 10 μl)
   - 20% Polyvinyl alcohol (10 μl)
   - Poly(dI-dC) (1 mg/ml) (1 μl)
   - Add TE buffer to a final volume of 50 μl.
2. Prepare the protein mixture as follows:
   - 2X Binding buffer (25 μl)
   - DNA-binding protein (50 to 1500 ng) (1 to 20 μl)
   - Add dd.H$_2$O to a final volume of 50 μl.

*Notes:* *The extraction and purification of proteins (cytosolic and nucleic proteins) are described in Chapter 3. The protein used for the assay can be a specific protein or protein fraction of a total protein mixture.*

3. Set up a DNA–protein interaction by combining the protein mixture and DNA mixture into 100 μl and incubating the reaction mixture at room temperature for 20 min. Meanwhile, set up a free DNA sample by mixing another 50 μl of the DNA mixture with 25 μl of 2X binding buffer and 25μl dd.H$_2$O.
4. Add 1 volume of salt solution to each tube and incubate for 1 to 2 min.
5. Add 10 μl of the diluted *DNase* I solution to each tube and incubate at room temperature for 1 min. Stop the reaction by adding 200 μl stopping solution followed by incubating at 37°C for 20 to 30 min.

*Notes:* *The DNase I should be prediluted from 1:1000 to 1:10,000 with 2X binding buffer and salt solution. A series of DNase I protection assays should be carried out using the same amount of DNA in order to find the optimal dilution. Use the amount of DNase I that digests about one half of the labeled DNA probe in reaction step 5.*

6. Extract by adding 200 μl or 1 volume of phenol:chloroform:isoamyl alcohol (24:24:1) to each tube. Mix well by vortexing for 15 s.
7. Centrifuge at $11,000 \times g$ for 5 min and carefully transfer the upper, aqueous phase to a fresh tube and add 2.5 volumes of chilled 100% ethanol to the tube. Allow precipitation to occur at –70°C for 30 min or at –20°C for 2 h.
8. Centrifuge at $12,000 \times g$ for 5 to 8 min, decant the supernatant, and briefly rinse the pellet with 1 ml of cold 70% ethanol. Dry the pellet under vacuum for 15 min. Dissolve the pellet in 10 μl of formamide-loading dye and retain until loading of sample into the gel.

*Note:* *The DNA–protein complex sample and control (free DNA sample) should be loaded adjacent to each other in order to easily identify the missed bands that were bound by proteins.*

9. Determine the total radioactivity recovered for each sample by counting for 1 min in a scintillation counter.

## D. Electrophoresis

1. Carefully remove the comb from the gel sandwich and rinse the wells with distilled water. Attach the sandwich to the apparatus, fill the tanks with 1X TBE solution, and flush the wells with buffer using a pipette.

2. Preelectrophorese the gel for 30 to 70 min at 45 to 70 W constant power depending on the size of the gel. Circulate the buffer between tanks during the electrophoresis.

3. Stop the preelectrophoresis and heat the prepared samples (step 8 above) at 95°C for 2 to 4 min and immediately load the same amounts of radioactivity (6000 to 10,000 cpm per lane) onto the preelectrophoresed gel.

4. Connect the electrophoresis apparatus to the power supply and run the electrophoresis in 1X TBE buffer at the same constant power as for preelectrophoresis for 2 to 4 h or until the bromophenol blue is about 2 cm from the bottom of the gel.

5. Stop the electrophoresis, remove the gel plates from the tank, and allow to cool for 10 min at room temperature.

6. Slowly separate the glass plates starting from one corner near the spacer with a spatula or its equivalent, and set the top plate aside.

*Notes:* *(1) As long as the top glass plate starts to separate from the other plate, keep going and never allow the two glass plates to become attached again; otherwise, the gel may be torn apart. (2) The gel usually sticks on the bottom plate. If the gel sticks to the top glass plate, carefully turn the plate over and lay it on the table or bench.*

7. Carry out autoradiography.

   a. Lay one piece of 3MM Whatman filter paper on the gel and gently press it onto the filter using an appropriate amount of presser to make the gel adhere to the filter.

   b. Starting from one corner of the gel, carefully peel off the gel, together with the filter paper, and lay it on the table or bench. Cover the gel with SaranWrap™ and remove any air bubbles between the gel and the SaranWrap™.

   c. Dry the gel at 80°C for 30 to 60 min under vacuum.

   d. Remove the SaranWrap™ from the gel and place it in the exposure cassette. Lay an X-ray film on the gel in the dark room and cover the cassette. Allow exposure to take place at room temperature or at –70°C with an intensifying screen for desired time.

   e. Develop the X-ray film and analyze the data. When comparing with the control lane, the missing bands in DNA–protein-interacted sample are the regions bound by protein (Figure 13.1).

## Reagents Needed

### 5X TBE Buffer

> Tris base (54 g)
> Boric acid (27.5)
> 0.5 $M$ EDTA (pH 8.0) (20 ml)
> Add dd.$H_2O$ to a final volume of 1000 ml.
> Autoclave.

### 1X TBE Buffer

> Dilute 5X TBE buffer with dd.$H_2O$.

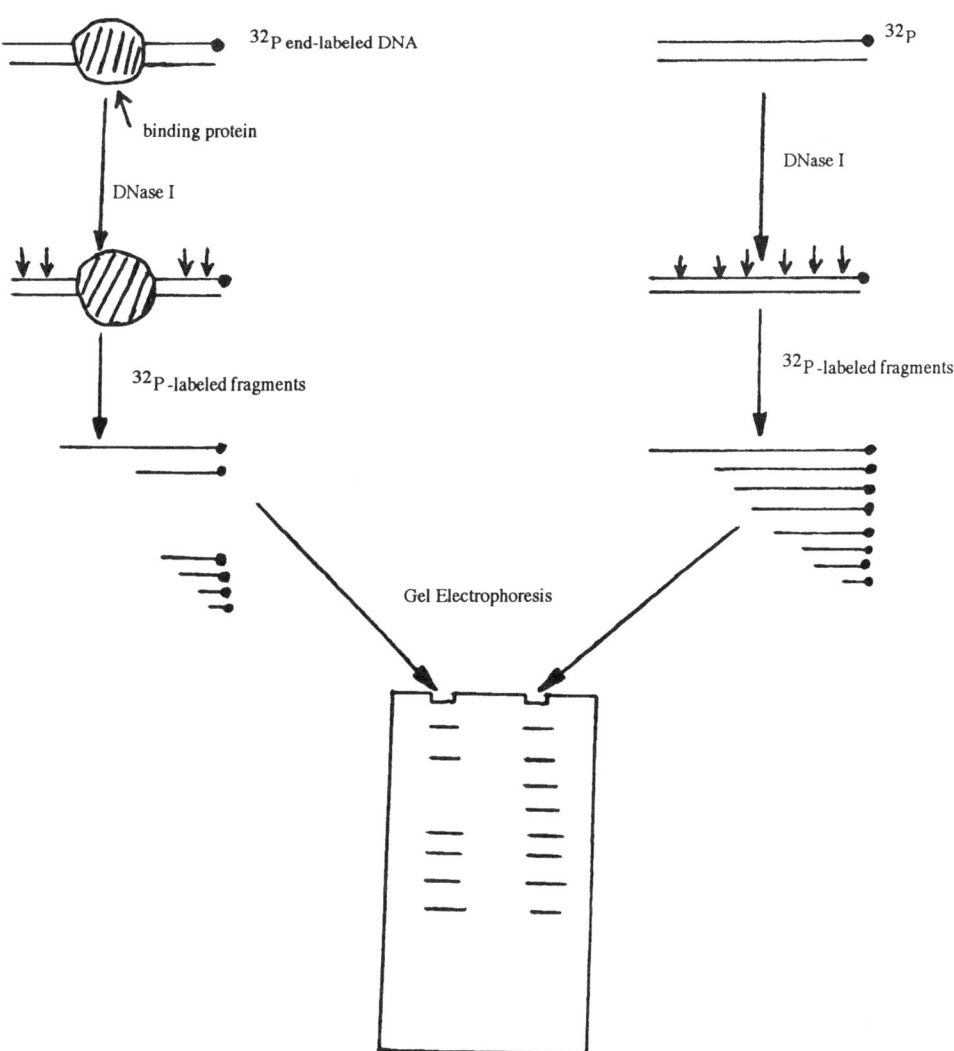

**FIGURE 13.1**
Diagram showing the nuclease protection assay of DNA–protein interactions. Following gel electrophoresis and autoradiography, the missing bands represent the regions that are bound by specific protein(s).

## *40% Acrylamide Stock Solution*

Acrylamide (76 g)

*N,N'*-Methylene-*bis*-acrylamide (2 g)

Add dd.H$_2$O to a final volume of 200 ml.

Heat at 37°C to dissolve each component completely after each addition and sterile-filter through a 2 μm-membrane and store at 4°C in an aluminum foil-wrapped bottle for a period of up to 5 weeks.

## 10% Ammonium Persulfate Solution (Fresh)

Ammonium persulfate (0.1 g)
Add dd.$H_2O$ to a final volume of 1 ml.

## 2X Protein-Binding Buffer

40 m$M$ Tris-HCl, pH 7.9
2 m$M$ EDTA
20% Glycerol
0.2 ml Nonidet P-40
4 m$M$ $MgCl_2$
2 m$M$ DTT

## Salt Solution

10 m$M$ $MgCl_2$
5 m$M$ $CaCl_2$

## 20% Polyvinyl Alcohol

(Sigma Chemicals)

## TE Buffer

10 m$M$ Tris-HCl, pH 8.0
1 m$M$ EDTA

## Phenol:Chloroform:Isoamyl Alcohol

24:24:1 (v/v/v)

## Ultrapure Urea

## DNase I Stopping Solution

0.2 $M$ NaCl
40 m$M$ EDTA
1% SDS
125 µg/ml tRNA
100 µg/ml Proteinase K (add prior to use)

## Formamide-Loading Dye Buffer

95% Formamide
4% EDTA, pH 8.0
0.2% Bromophenol blue
0.2% Xylene cyanol

# II.   Gel Retardation Assay

## *Protocol*

1.  Prepare DNA fragments by digesting the DNA molecules of interest with appropriate restriction enzyme(s) that create a 3′- or 5′-end overhang.

*Notes:*   *DNA purification is described in Chapter 1. DNA containing the binding site of interest can come from cDNA, genomic DNA, plasmid DNA, or a synthetic oligonucleotide. The sizes and/or sequences of the DNA fragments used for gel retardation assay should be known.*

2.  Radioactively label the DNA fragments of interest by end labeling (3′ or 5′ end) to achieve a high specific activity. The labeling procedures are described in Chapter 4. Purify the probe DNA by extraction and precipitation with ethanol. Dissolve the DNA in an appropriate volume of TE buffer and measure the cpm of the probe. Store the sample at –20°C until use.

3.  Prepare the protein sample for the assay. The extraction and purification of proteins (cytosolic and nucleic proteins) are described in Chapter 3. The protein used for the assay can be a specific protein or a protein fraction of a total protein mixture.

*Notes:*   *The protein or protein fraction must be nondenatured. The tertiary structure of the protein(s) is important for binding to the DNA fragment. For optimal binding, the amount of proteins should be in excess of the amount of DNA amount, usually 5 to 15 μg per lane in the gel for protein fraction or 2 to 5 μg per lane for a purified protein.*

4.  Thoroughly clean and assemble a vertical electrophoresis apparatus according to the instructions for the apparatus. The gel size is normally $14 \times 19$, $15 \times 20$, or $20 \times 26$ cm using 1- or 1.5-mm spacers. Insert an appropriate comb into the sandwich. Check if leaking occurs by using distilled water.

*Notes:*   *The comb can be inserted into the sandwich after pouring the gel mixture. However, air bubbles may be trapped around the teeth; this should be avoided.*

5.  Prepare the native gel mixture containing 4 to 8% acrylamide in the order listed below (for 100 ml gel, 4% polyacrylamide):
    *   5X TBE buffer (10 ml)
    *   30% (w/v) Acrylamide stock solution (13.3 ml)
    *   dd.$H_2O$ (75 ml)
    *   Mix well after each addition and degas (optional).
    *   Add 0.9 ml of freshly prepared 10% (w/v) ammonium persulfate (AP) solution and 100 μl TEMED to the gel mixture. Gently mix and immediately pour the solution carefully into the glass sandwich or assembled mold.

6.  Allow the gel to polymerize for about 1 h at room temperature and carefully remove the comb. Rinse the wells ($5 \times 3 \times 1$ or 1.5 mm) thoroughly with dd.$H_2O$ several times.

7.  Attach the gel sandwich properly to the electrophoresis tanks. If a notched glass plate is used, it should face inward toward the buffer reservoir.

8. Fill the the buffer tanks with 0.5X TBE running buffer. Flush out the wells to remove any air bubbles with a Pasteur pipette or its equivalent.

9. Connect to the power supply (positive electrode should be connected to the bottom tank) and run the preelectrophoresis for 1.5 h at 200 V at room temperature or at 350 V for 1.5 h at 4°C.

*Notes:*   *Native or nondenaturing polyacrylamide gels should run at 3 to 8 V/cm. Too high a voltage may cause overheating that may result in "bowing" of the DNA bands. Whenever possible, it is strongly recommended that recirculation of the running buffer between the top and bottom buffer tanks be maintained by using a pump. This will help maintain uniform pH and ionic strength throughout the system during the electrophoresis.*

10. While the gel is prerunning, set up the DNA and protein-binding reactions in microcentrifuge tubes as indicated below:
    - 2X DNA–protein-binding buffer (10 µl)
    - Competitor DNA [1 mg/ml poly(dI-dC) or poly(dA-dT)] (1.5 µl)
    - DNA-binding protein (5.0 µl)
    - Add dd.H$_2$O to a final volume 20 µl.

*Notes:*   *(1) The nonspecific competitor DNA serves to provide a large excess of low affinity binding sites that absorb nonspecific DNA-binding proteins, thus increasing the detection of the specific protein–DNA complexes. (2). The normal DNA competitors are synthetic copolymers such as poly(dI-dC) or poly(dA-dT).*

11. Incubate the reaction at 22 to 25°C or desired temperature for 20 to 30 min.

12. Add 1 to 2 ml of labeled DNA fragments (5 to 10 pmol; $5 \times 10^4$ to $5 \times 10^6$ cpm if $^{32}$P-labeled) to the reaction and continue incubation at the same temperature for 10 to 30 min.

13. Add 2 ml of 10X loading buffer to the binding reaction mixture and store at room temperature until use.

14. Stop the preelectrophoresis and carefully load the reaction mixture onto the gel (10 µl per well).

15. Start the electrophoresis at the same voltage as for pre-electrophoresis for 1 to 2 h or until the bromophenol blue reaches a distance of 1 cm from the bottom of the gel. Maintain the running buffer circulation between tanks during electrophoresis.

16. Stop the electrophoresis, remove the gel plates from the tank, and allow to cool for 10 min at room temperature.

17. Slowly separate the glass plates starting from one corner near the spacer with a spatula or its equivalent. Remove the top plate and put it aside.

*Notes:*   *(1) As long as the top glass plate starts to separate from the other plate, keep going and never allow the two glass plates to become attached again; otherwise, the gel may be torn apart. (2) The gel usually sticks to the bottom plate. If the gel sticks to the top glass plate, carefully turn the plate over and lay it on the table or bench.*

18. Carry out autoradiography.
    a. If the bands of the DNA–protein complex are to be recovered from the gel, the gel should not be fixed or dyed.

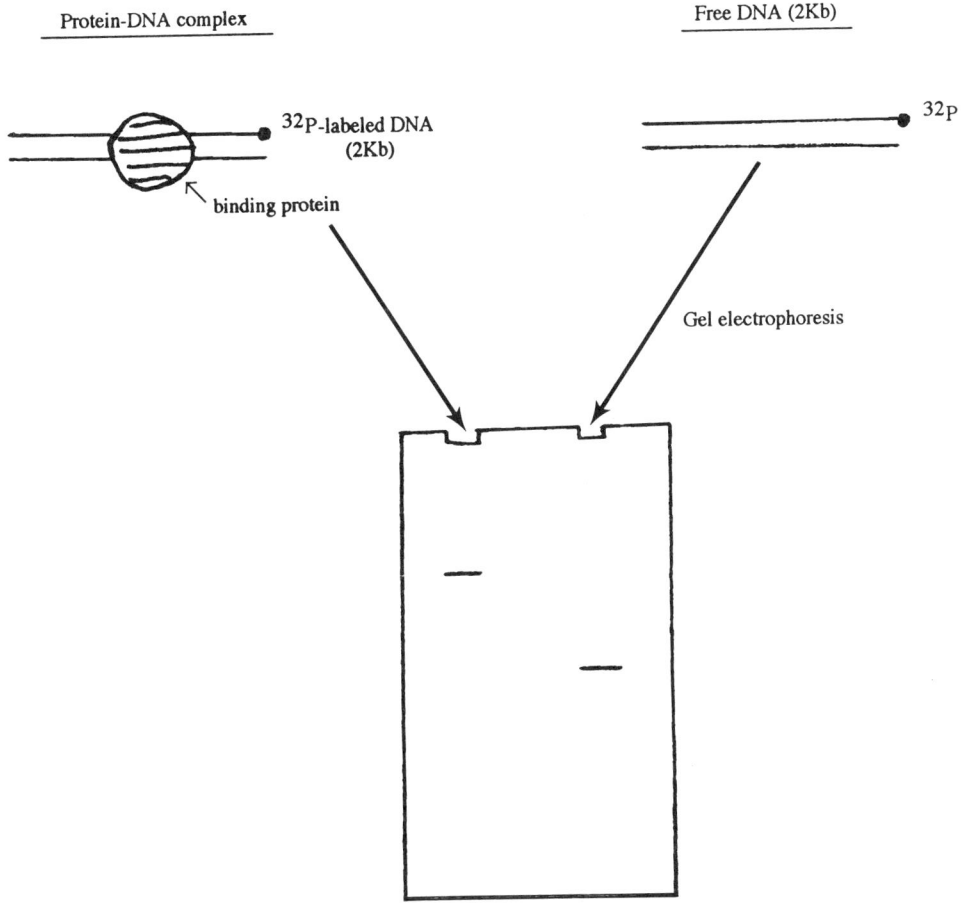

**FIGURE 13.2**
Schematic representation of gel retardation. DNA bound by specific protein(s) will be retarded during gel electrophoresis as compared with the free DNA control.

    i.    Cover the gel properly without air bubbles present and wrap it together with the supporting glass plate using SaranWrap™.

    ii.    In a dark room with a safety light, place an X-ray film on the gel and stick the edges to the plate with tape. Wrap the whole apparatus with light-tight aluminum foil and invert the gel so that the glass plate on its top serves as a weight. Allow exposure to occur at room temperature or at −70°C with an intensifying screen for several hours or until desired bands are visible.

*Note:*    *Never use a metal film cassette because it may break the glass plate and crush the gel.*

    iii.    Develop the film and analyze the data. Free DNA fragments should be visulized as black bands down near the bottom of the gel, but DNA–protein-complex bands are retarded in their movement and thus occur at the upper part of the gel or above the free DNA bands. The more proteins that are bound to the DNA, the stronger is the signal, and the smaller the amount of the free DNA that is present.

    b. Dry the gel and carry out autoradiography.

        i. Lay one piece of 3MM Whatman™ filter paper on the gel and gently press it onto the filter using an appropriate amount of pressure to make the gel to stick to the filter.

        ii. Starting from one corner of the gel, carefully peel off the gel, together with the filter paper, and lay it on the table or bench. Cover the gel with SaranWrap™ and remove any air bubbles that occur between the gel and the SaranWrap™.

        iii. Dry the gel at 80°C for 30 to 60 min under vacuum.

        iv. Remove the SaranWrap™ from the gel and place it in an exposure cassette. Lay an X-ray film on the gel in the dark room and cover the cassette. Allow exposure to take place at room temperature or at −70°C with an intensifying screen for the desired time.

        v. Develop the X-ray film and analyze the data (Figure 13.2).

## Reagents Needed

### 5X TBE Buffer

    Tris base (54 g)
    Boric acid (27.5)
    0.5 $M$ EDTA (pH 8.0) (20 ml)
    Add dd.$H_2O$ to a final volume of 1000 ml.
    Autoclave.

### 1X TBE or 0.5X TBE Buffer

    Dilute 5X TBE buffer with dd.$H_2O$.

### 30% Acrylamide Stock Solution

    Acrylamide (58 g)
    $N,N'$-Methylene-*bis*-acrylamide (1 g)
    Add dd.$H_2O$ to a final volume of 200 ml.
    Heat at 37°C to dissolve each component completely after each addition and sterile-filter through 2-μm membrane. Store at 4°C in an aluminum foil-wrapped bottle for up to 5 weeks.

### 10% Ammonium Persulfate Solution (Fresh)

    Ammonium persulfate (0.1 g)
    Add dd.$H_2O$ to a final volume of 1 ml.

### 0.5X Running Buffer

    Dilute 5X TBE buffer with dd.$H_2O$.

### 2X Protein-Binding Buffer

    40 m$M$ Tris-HCl, pH 7.9
    100 m$M$ NaCl
    20% Glycerol
    0.2 m$M$ DTT

### 10X Loading Buffer

    50% Glycerol
    0.25% Bromophenol blue
    0.25% Xylene cyanol

## III.  Footprinting Following the Gel Retardation Assay Using the Copper–Phenanthroline Method

This method combines and simplifies the above two techniques and allows one to obtain detailed nucleotide sequence information. The 1,10-phenanthroline–copper ion is able to cleave DNA when the DNA–protein complex is embedded within the nondenaturing gel matrix following the gel retardation assay. The cleaved DNA is eluted from the gel and resolved by denaturing gel electrophoresis.

### Protocol

1. Carry out the gel retardation experiment as described previously except that the standard reaction should be scaled up 6- to 12-fold and loaded in several wells in the gel.

2. After electrophoresis, carefully remove the top glass plate and place the gel with its supporting glass plate in a tray or its equivalent containing 200 ml of 10 m$M$ Tris-HCl (pH 8.0) to equilibrate the gel for 10 to 14 min at room temperature.

3. While the gel is equilibrating, prepare the following mixture:

   - 9 m$M$ $CuSO_4$ (1ml)

   - 40 m$M$ 1,10-phenanthroline (1 ml)

   - Let the mixture stand for 1 to 1.5 min at room temperature to turn blue, and then add 18 ml of dd.$H_2O$.

4. Add the 20 ml of the $CuSO_4$–phenanthroline solution to the equilibrated gel and initiate the reaction by adding 0.1 ml of 3-mercaptopropionic acid diluted in 20 ml dd.$H_2O$. The solution turns brown.

5. Allow the cleavage reaction to proceed at room temperature for 1 min without any shaking.

*Note:*   *The optimal time of incubation of the cleavage reaction should be predetermined.*

6. Add 10 ml 1.2% 2,9-dimethyl-1,10-phenanthroline in 100% ethanol to stop the reaction for 2 min. The solution should change to a yellow color.

7. Decant the solution and rinse the gel once with distilled water. Partially drain excess water and wrap the gel together with its supporting glass plate with SaranWrap™. Carry out autoradiography with the wet gel as described for the gel retardation protocol.

8. Slice out the region of the gel containing the DNA–protein complex and the same area of the gel of the free DNA probe, respectively. Place the gel slices into a microcentrifuge tube containing 0.5 ml of gel elution buffer containing 0.2 $M$ NaCl, 20 m$M$ EDTA (pH 8.0), 1% SDS, and 1 mg/ml tRNA.

9. Crush the gel slices with a clean spatula and allow the elution to proceed at 37°C for at least 6 h or overnight.

10. Briefly spin down to remove the gel pellet. Add 1 volume of phenol:chloroform:isoamyl alcohol (24:24:1) to the supernatant.

11. Centrifuge at 11,000 × g for 5 min and carefully transfer the supernatant to a fresh tube. Add 2.5 volumes of chilled 100% ethanol to the supernatant. Allow to precipitate at –70°C for 30 min.

12. Centrifuge at 12,000 × g for 5 min and rinse the pellet with 1 ml of 70% ethanol. Dry the pellet for 15 min under vacuum and dissolve it in 10 to 20 µl of formamide-loading dye.

13. Denature the sample at 90 to 95°C for 4 min and load the same amount of radioactivity (cpm) onto a preelectrophoresed urea denaturing gel or an appropriate sequencing gel (see preelectrophoresis and electrophoresis in footprinting protocol described previously).

*Notes:* *The free and bound DNA samples should be loaded in the adjacent lanes. An A + G ladder or DNA probe cleaved with guanosine-specific or cytidine- and thymidine-specific reagents, respectively, for chemical sequencing should be loaded in lanes to the left or right of the samples for identification of the sequence in the protein-bound region of the DNA.*

14. Dry the gel and carry out autoradiography to visualize the DNA band profile as described in the footprinting section.

# References

1. **Cartwright, I. L. and Kelly, S. E.,** Probing the nature of chromosomal DNA-protein contacts by *in vivo* footprinting, *BioTechniques,* 11, 188, 1991.

2. **Saluz, H. P. and Jost, J. P.,** A simple, high-resolution procedure to study DNA methylation and *in vivo* DNA–protein interactions on a single-copy gene level in higher eukaryotes, *Proc. Natl. Acad. Sci. U.S.A.,* 86, 2602, 1989.

3. **Mueller, P. R. and Wold, B.,** Ligation-mediated PCR: applications to genomic footprinting, *Methods,* 2, 20, 1991.

4. **Mueller, P. R. and Wold, B.,** *In vivo* footprinting of a muscle specific enhancer by ligation mediated PCR, *Science,* 246, 780, 1989.

# *In Vitro* Translation of mRNA(s) and Analysis of Proteins by Gel Electrophoresis

## Contents

Purified mRNAs from total RNAs or mRNA transcribed *in vitro* from cloned cDNA inserts in expressional vectors can be translated in cell-free extracts to produce protein(s). The synthesized protein(s) may be then analyzed by SDS-PAGE, 2-D gel electrophoresis, immunoprecipitation, or biological activity assay depending on the particular research interest.[1–5] The present chapter describes, step by step, the protocols for *in vitro* translation of mRNA(s) using commercial lysates of rabbit reticulocytes and wheat germ extracts, and protocols for analysis of the products by SDS-PAGE. The basic principle is that commercial translational kits contain translational machinery including all the components such as amino acids, tRNA, rRNA, and factors (initiation, elongation, and termination) for the *in vitro* synthesis of either radioactively labeled or unlabeled polypeptide(s).

# I.    *In Vitro* Translation of mRNA(s)

## Protocol A: Synthesis of Proteins from Purified mRNAs Using Rabbit Reticulocyte Lysates

1.    Isolate the total RNAs from cell cultures or tissues of interest (see Chapter 2).
2.    Purify poly(A)+RNA from the total RNA (see Chapter 2).
3.    Thaw the reagents, on ice, from the commercial translation kit stored at –20°C or –70°C.

*Note:*    *We recommend that the translation kit components not be thawed at room temperature or at 37°C. This will decrease the efficiency of* in vitro *translation.*

4.    Add purified mRNAs (0.1 to 0.2 µg in 2 to 4 µl) in a sterile microcentrifuge tube. Denature the secondary structures of the mRNA by heating at 65 to 67°C for 10 min and quickly chilling the tube on ice for 5 min. Spin down briefly.
5.    Set up a standard reaction on ice as follows:

   **Sample**
   - Denatured mRNAs (4 µl)
   - dd.$H_2O$ (14 µl)
   - RNasin ribonuclease inhibitor (40 units/µl) (2 µl)
   - Micrococcal nuclease-treated rabbit reticulocyte lysate (70 µl)
   - 1 m$M$ Amino acid mixture (minus methionine) (2 µl)
   - [$^{35}$S]Methionine (1200 Ci/mmol) at 10 mCi/ml (8 µl)
   - Total volume of 100 µl.

*Caution:*    *[$^{35}$S]Methionine is radioactive. Care should be taken.*

   **Background control (no mRNA)**
   - RNasin ribonuclease inhibitor (40 units/µl) (1 µl)
   - Micrococcal nuclease treated rabbit reticulocyte lysate (35 µl)
   - 1 m$M$ Amino acid mixture (minus methionine) (1 µl)
   - dd.$H_2O$ (9 µl)
   - [$^{35}$S]Methionine (1200 Ci/mmol) at 10 mCi/ml (4 µl)
   - Total volume of 50 µl.

*Note:*    *After use, the components should be quickly stored at –70 or –20°C.*

6.    Incubate the reactions at 30°C for 1 h and proceed to analysis of the protein products (see Protocol E).

## Protocol B: Synthesis of Proteins from Purified mRNAs Using Wheat Germ Extract

1.    Preparation of Wheat Germ Extract

All steps should be carried out on ice or in a cold room.

1. Grind wheat germ (3 to 4 g) in liquid $N_2$ in a chilled mortar or blender. Add more liquid $N_2$ to keep the powder from thawing and repeat grinding until a fine powder is obtained.

*Note:*    *Gloves should be worn because liquid $N_2$ can burn the skin.*

2. Briefly warm the powder at room temperature for about 2 min and add 10 ml of cold homogenization buffer to the mortar. Repeat grinding for 2 min or until a fine homogenate is visible.

3. Transfer the homogenate, using a clean spatula, into a sterile 30-ml Corex centrifuge tube on ice and centrifuge at 20,000 × *g* for 10 min at 4°C.

4. Carefully transfer the supernatant to a fresh tube.

*Note:*    *Be careful to prevent any pellet material from entering the supernatant.*

5. Prepare a Sephadex G-25 column (2 × 30 cm) in a cold room (see Chapter 3 for general column preparation).

6. Pass the supernatant through the column. Check for the occurence of precipitation of a few drops of effluent in a test tube containing 10% TCA. As long as precipitation occurs, collect the effluent in a sterile tube until approximately an equal volume of the supernatant loaded is collected.

7. Centrifuge the supernatant in a Corex tube at 20,000 × *g* for 10 min at 4°C.

8. Divide the extract into 0.4-ml portions in microcentrifuge tubes. Immediately freeze them in liquid $N_2$ for 4 min and then store at –70°C until use.

*Notes:*    *The cell-free extract prepared from the wheat embryos contains tRNA, rRNA, soluble factors for initiation, elongation, and termination needed for* in vitro *translation of mRNA. The endogenous mRNA level is relatively low.*

## 2.    *In Vitro* Translation of mRNAs Using Wheat Germ Extract

1. Thaw commercial reagents on ice.

*Notes:*    *We recommend that the components not be thawed at room temperature or at 37°C. This will decrease the efficiency of* in vitro *translation.*

2. Add purified mRNAs (0.1 to 0.2 μg in 2 to 4 μl) in a sterile microcentrifuge tube. Denature the secondary structures of the mRNA by heating at 65 to 67°C for 10 min and quickly chilling the tube on ice for 5 min. Briefly spin down.

3. Set up a standard reaction on ice as follows:

**Sample**

- Denatured mRNAs in dd.$H_2O$ (4 μl)
- dd.$H_2O$ (24 μl)
- RNasin ribonuclease inhibitor (40 units/μl) (2 μl)
- Wheat germ extract (60 μl)
- 1 m*M* Amino acid mixture (minus methionine) (2 μl)
- [$^{35}$S]Methionine (1200 Ci/mmol) at 10 mCi/ml (8 μl)
- Total volume of 100 μl.

**Background control (no mRNA)**

- RNasin ribonuclease inhibitor (40 units/µl) (1 µl)
- Wheat germ extract (30 µl)
- 1 m$M$ Amino acid mixture (minus methionine) (1 µl)
- dd.H$_2$O (14 µl)
- [$^{35}$S]Methionine (1200 Ci/mmol) at 10 mCi/ml (4 µl)
- Total volume of 50 µl.

4.   Incubate the reactions at 25°C for 1 h and proceed to products analysis (Protocol E).

## Reagents Needed

### Homogenization Buffer

> 40 m$M$ HEPES/KOH, pH 7.6
> 1 m$M$ Magnesium acetate
> 0.1 $M$ Potassium acetate
> 2 m$M$ CaCl$_2$
> 4 m$M$ DTT

### Column Buffer

> 40 m$M$ HEPES/KOH, pH 7.6
> 0.1 $M$ Potassium acetate
> 5 m$M$ Magnesium acetate
> 4 m$M$ DTT

### 10% (w/v) TCA Solution

> 20 g trichloroacetic acid in 200 ml dd.H$_2$O

## Protocol C: Synthesis of Proteins from mRNA Transcribed In Vitro

### 1.    In Vitro Trascription of mRNA from a Subloned cDNA Insert

A number of plasmid vectors contain polycloning sites downstream from the powerful bacteriophage promoters SP6, T7, or T3. The cDNA or genomic DNA insert of interest can be cloned at the polycloning site between promoters SP6 and T7 or T3, forming a recombinant plasmid. The cDNA or genomic DNA inserted can be transcribed *in vitro* into single-strand sense RNA from a linear plasmid DNA under the control of promoter SP6, T7, or T3.

1.   In a microcentrifuge tube on ice, add the following in the order listed below:
- 5X Transcription buffer (20 µl)
- 0.1 $M$ DTT (8 µl)
- rRNasin ribonuclease inhibitor (100 units)
- Mixture of ATP, GTP, CTP and UTP (2.5 m$M$ each) (20 µl)
- Linearized DNA template (1 to 2.5 µg/µl) (2 µl)
- SP6, T7, or T3 RNA polymerase (15 to 20 units/µl) depending on the specific promoter (5 µl)
- Add dd.H$_2$O to a final volume of 100 µl.

2.   Incubate the reaction mixture at 37 to 40°C for 1 to 2 h.

3.  Add RNase-free *DNase* I to a concentration of 1 unit/μg DNA template.

4.  Incubate for 15 min at 37°C.

5.  Extract the enzyme by adding 1 volume of TE-saturated phenol/chloroform. Mix well by vortexing for 1 min and centrifuging at 11,000 × g for 5 min at room temperature.

6.  Transfer the top, aqueous phase into a fresh tube and add 1 volume of chloroform:isoamyl alcohol (24:1). Mix well by vortexing and centrifuging at 11,000 × g for 5 min.

7.  Carefully transfer the upper, aqueous phase to a fresh tube. Add 0.5 volume of 7.5 *M* ammonium acetate solution and 2.5 volumes of chilled 100% ethanol. Allow to precipitate at –70°C for 30 min or at –20°C for 2 h

8.  Centrifuge at 12,000 × g for 5 min. Carefully discard the supernatant and briefly rinse the pellet with 1 ml of 70% ethanol and dry the pellet under vacuum for 15 min.

9.  Dissolve the RNA probe in 20 to 50 μl of dd.H$_2$O and store at –20°C until use.

*Note:*    *The quantity and quality of the labeled RNA can be checked by denaturing agarose gel electrophoresis using 4 to 5 μl of the sample (see Northern blotting).*

## 2.    *In Vitro* Translation of the Purified mRNA
The procedures are described in Protocols A and B.

### Protocol D: TCA Assay for Amino Acid Incorporation

1.  Remove 5 μl of the translation reaction mixture from each of the samples and directly spot the aliquots onto individual glass-fiber filters. Air dry the filters and transfer the filter to a vial containing an appropriate volume of liquid scintillation "cocktail" to cover the filter. Use a liquid scintillation counter to record the total counts present in the translation reaction mixture.

2.  Transfer 2 μl of the translation reaction mixture from the background control (no *in vitro* mRNA translation), 2 μl from the *in vitro* translation using rabbit reticulocyte lysate, and 5 μl from the sample translated *in vitro* using wheat germ extract into microcentrifuge tubes. Dilute the aliquot to a total volume of 0.25 ml with 1 *N* NaOH/H$_2$O$_2$ solution for the reticulocyte lysate sample or 1 *N* NaOH for the wheat germ sample.

3.  Incubate at 37°C for 10 to 15 min.

*Notes:*    *NaOH hydrolyzes aminoacyl tRNAs, thus preventing their inclusion in the labeled tRNA in the isotope incorporation calculation. H$_2$O$_2$ removes the red color from the reticulocyte lysate reaction prior to scintillation counting.*

4.  Add 1 ml of precipitation solution containing 25% (w/v) TCA and 2% (w/v) casamino acids as carriers to each tube and incubate for 30 min on ice.

5.  Prewet Whatman GF/A glass-fiber filters with 3 ml of cold 5% TCA and filter the sample(s) under vacuum onto the filter. Wash the filter three times with 3 ml of cold TCA solution and rinse the filter once with 3 ml of acetone.

6.  Completely dry the filter(s) at room temperature and transfer each filter into a vial.

7.  Add enough scintillation "cocktail" mixture to cover the filter (5 to 10 ml per vial) and carry out cpm counting in a liquid scintillation counter.

8.  Calculate the efficiency of *in vitro* translation as follows:

$$\text{Total radioactivity per microliter} = \frac{\text{amount of radioactivity determined}}{\text{microliters of reaction filtered}}$$

For example:

$$\frac{5 \times 10^6}{2\mu l} = 2.5 \times 10^6 \, \text{cpm}/\mu l$$

Total radioactivity in the translation reaction $= \dfrac{\text{amount of radioactivity}}{\text{microliters of reaction filtered}}$

$$= 2.5 \times 10^6 \, \text{cpm}/\mu l \times 100 \, \mu l$$

$$= 2 \times 10^8 \, \text{cmp}$$

Radioactivity incorporated into protein $= \dfrac{\text{total radioactivity in translation reaction}}{\text{microliters of reaction precipitated by TCA}}$

$$= 2 \times 10^8 \, \text{cpm}/2\mu l$$

$$= 1 \times 10^8 \, \text{cpm}/\mu l$$

Total radioactivity incorporated into protein in a standard reaction (volume $= 100 \, \mu l$)

$$= \text{radioactivity incorporated into protein} \times \text{total volume}$$

$$= 1 \times 10^8 \, \text{cpm}/\mu l \times 100 \, \mu l$$

$$= 5 \times 10^7 \, \text{cpm}$$

Percent incorporation $= \dfrac{\text{total radioactivity incorporated into protein}}{\text{total radioactivity in the reaction}} \times 100$

$$= \frac{5 \times 10^7}{2 \times 10^8} \times 100$$

$$= 25\% \text{ incorporation}$$

## Protocol E: Analysis of Labeled Proteins by SDS-PAGE

SDS is an anionic detergent that denatures proteins and makes them negatively charged by wrapping around the polypeptides, thus giving equal charge densities per unit length. SDS-PAGE can separate and determine the molecular weights (MW) of proteins using standard protein size markers. There is a linear relationship between the log of the MW of a polypeptide and its $R_f$, which is the ratio of the distance from the top of the gel to the polypeptide divided by the distance from the top of the gel to the dye front. A standard curve can then be generated by plotting the $R_f$ of each standard polypeptide marker as the abscissa and the $\log_{10}$ of its MW as the ordinate. The MW of an unknown protein can be determined with ease by finding its $R_f$ that vertically crosses on the standard curve and reading the $\log_{10}$ MW that horizontally crosses to the ordinate. The antilog of the $\log_{10}$ MW is the actual MW of the protein.

1.     Preparation of the Separating Gel

    1.    Thoroughly clean glass plates and spacers with detergent, wash with tap water, rinse with distilled water several times, and air dry.

2.  Wear gloves and wipe dry the glass plates and spacers with 100% ethanol. Assemble the vertical slab gel unit, such as the SE 600 Vertical Slab Gel Unit (Hoeffer), in the casting mode according to the instructions. Place one spacer (1.5 mm thick) at each of the two sides between the two glass plates and fix in place with clamps, forming a sandwich. Repeat for the second sandwich. The size of the separating gel varies with individuals. A standard size is $120 \times 140 \times 1.5$ mm.

*Note:*    *Spray a little grease oil on the spacer areas at both the top and bottom ends of the sandwiches to prevent potential leaking.*

3.  Check potential leaking by pipetting some distilled water into the sandwich and then drain away the water by inverting the unit. Set the unit at an even level using a water balancer or its equivalent.

4.  Prepare the separating gel solution in a clean 100-ml beaker or a 125-ml flask in the order shown below (for two pieces of standard-sized gels):
    *   20 ml Monomer solution
    *   15 ml Running gel buffer
    *   0.6 ml 10% SDS
    *   24.1 ml dd.$H_2O$
    *   Mix well after each addition.

5.  Degas the solution to remove any bubbles by use of vacuum for 5 min (optional). Add 0.25 ml of freshly prepared 10% ammonium persulfate (AP) and 20 µl TEMED and mix by gently swirling. **Do not generate air bubbles.**

6.  Immediately pipette the mixture with a 50-ml syringe into the assembled sandwich up to a level that is 4 cm from the top.

7.  Take up 0.8 ml of dd.$H_2O$ in a 1-ml syringe equipped with a 2-in. 22-gauge needle or its equivalent and carefully load 0.3 ml of the water, starting from one top corner next to the spacer, onto the surface of the acrylamide gel solution. Repeat on the other side of the slab next to the other spacer. The water layer will evenly flow across the surface of the gel mixture. The purpose of applying water is to make the surface of the gel very even. A very sharp gel–water interface can be visible after the gel has polymerized.

8.  Drain away the water layer by gently tilting the casting unit and rinse once with 2 ml of overlay solution. Add 1 ml of overlay solution on top of the gel and allow the gel to sit for 1 h or overnight by covering the top of the gel with a piece of Parafilm to prevent evaporation.

## 2.     Preparation of the Stacking Gel

1.  Prepare the stacking gel solution in a clean 50-ml beaker or a 50-ml flask in the order shown below (for two pieces of standard-sized gels):
    *   2.66 ml Monomer solution
    *   5 ml Stacking gel buffer
    *   0.2 ml 10% SDS
    *   12.2 ml dd.$H_2O$
    *   Mix well after each addition.

2.  Drain the overlay solution on the separating gel by tilting the casted unit and rinse with 2 ml of the stacking gel mixture. Drain away the stacking gel mixture and insert the comb (1.5 mm thick) into the glass sandwich according to the instructions for the apparatus.

*Notes:*   *The stacking gel solution may be filled prior to inserting the comb into the sandwich. However, air bubbles are easily generated and trapped around the teeth of the comb. We recommend that the comb be inserted into the sandwich prior to filling with the stacking gel mixture.*

3.   Add 100 µl of freshly prepared 10% AP and 10 µl TEMED to the stacking gel solution, and mix by gently swirling. **Do not generate air bubbles.**

4.   Immediately and slowly pipette the solution, from both sides next to the spacers, into the sandwich up to the top. Allow the gel to polymerize and sit for 30 min.

## Reagents Needed

### Monomer Solution

> Acrylamide (116.8 g)
> N,N-Methylene-bis-acrylamide (3.2 g)
> Dissolve well after each addition in 300 ml dd.$H_2O$.
> Add dd.$H_2O$ to a final volume of 400 ml.
> Wrap the bottle with aluminum foil and store at 4°C in the dark.

*Caution:*   *Acrylamide is neurotoxic. Gloves should be worn when handling this chemical.*

### Running Gel Buffer

> 1.5 *M* Tris (72.6 g)
> Dissolve well in 200 ml dd.$H_2O$.
> Adjust pH to 8.8 with 2 *N* HCl.
> Add dd.$H_2O$ to a final volume of 400 ml.
> Store at 4°C.

### Stacking Gel Buffer

> 0.5 *M* Tris (6 g)
> Dissolve well in 50 ml dd.$H_2O$.
> Adjust pH to 6.8 with 2 *N* HCl.
> Add dd.$H_2O$ to a final volume of 100 ml.
> Store at 4°C.

### 10% SDS

> SDS (10 g)
> Dissolve well in 100 ml dd.$H_2O$.
> Store at room temperature.

### 10% Ammonium Persulfate (AP)

> AP (0.2 g)
> Dissolve well in 2.0 ml dd.$H_2O$.
> Store at 4°C for up to 10 days.

### Overlay Buffer

Running gel buffer (25 ml)
10% SDS solution (1 ml)
Add dd.H$_2$O to 100 ml.
Store at 4°C.

## 3.  Loading the Samples and Protein Standard Markers onto the Gel

1.  While the stacking gel is polymerizing, prepare samples and protein standard markers. Add 1 volume of 2X denaturing buffer to each of the samples and to the protein standard markers. Cap the tubes and place them in boiling water for 2 to 4 min. Immediately chill the tubes on ice for 4 min. Briefly spin down, add 10% (v/v) of bromophenol blue solution to the sample, and place on ice until use.

*Notes:*  *At this stage, the proteins and markers are denatured, which is important for efficient electrophoresis. The isolation and purification of proteins are described in Chapter 3. The amount of proteins loaded into one well should be 10 to 35 µg in a total of 5 to 15 µl for Coomassie Blue staining and Western blotting, and 5 to 15 µg in 5 to 10 µl for highly sensitive silver staining. The amount of protein standard markers for one well should be 2 to 10 µg.*

2.  Slowly and vertically pull the comb straight up from the gel. Rinse each well with running buffer using a syringe or pipette to add the running buffer. Carefully invert the casting stand to drain the wells and repeat twice.

3.  Position the casting stand upright and fill each well with 10 µl of running buffer. Take up sample or markers into a syringe equipped with a 2-in. 22-gauge needle or a pipette with an equivalent tip attached. Carefully insert the needle into the running buffer in each well and **underlay** the sample or standard marker in the well. Repeat until all of the samples are loaded.

*Notes:*  *The volume loaded into each well should be less than 40 µl for a standard-sized gel or less than 15 µl for a mini-gel. Overloading may cause samples to float out and cause contamination among wells. The markers are recommended to be loaded into the front left or right well, or both wells, leaving a blank well between markers and samples. The volume and order of each sample loaded in the two sheets of gels should be identical. One gel will be used for staining in the determination of the MW of the proteins. The other gel will be used in Western blotting.*

4.  Carefully overlay the wells with running buffer, using a syringe or pipette, until the top of the gel is covered.

*Notes:*  *The samples and markers containing blue dye should be visible at the bottom of each well. Mark the orientation of samples and record in lab notebook.*

## Reagents Needed

### 2X Denaturing Buffer

Stacking gel buffer (5 ml)
10% SDS solution (8 ml)
Glycerol (4 ml)

2-Mercaptoethanol (2 ml)
Add dd.H$_2$O to a final volume of 20 ml.
Divide in aliquots and store at –20°C.

### Bromophenol Blue Solution

Bromophenol blue (0.05 g)
Sucrose (20 g)
Dissolve well after each addition in 15 ml dd.H$_2$O.
Add dd.H$_2$O to a final volume of 50 ml, aliquot, and store at –20°C.

### Running Buffer

0.25 *M* Tris (12 g)
Glycine (57.6 g)
10% SDS solution (40 ml)
Add dd.H$_2$O to a final volume of 4 l.

## 4.      Electrophoresis

1.  Carefully put the upper buffer chamber in place according to the instructions. Remove lower clamps and clamp the sandwiches to the bottom of the upper buffer chamber. **Do not disturb the wells.** Check for any potential leaking by adding 10 to 20 ml of running buffer into the chamber.

*Note:*   *Apply a little grease oil to the spacer areas at the top corners of the sandwiches to prevent potential leaking.*

2.  Carefully transfer the assembled unit into the lower buffer chamber according to the instructions. Place the entire tank on a magnetic stirrer. Slowly fill the lower chamber with 3 l of running buffer and add 1 liter to the upper chamber, or use an appropriate volume depending on the sizes of the chambers.

*Notes:*   *The volume of bottom-chamber buffer should be sufficient to cover two thirds of the slabs; otherwise, the heat generated during electrophoresis will not be distributed evenly, thus causing distortion of the band patterns on the gel. For the upper chamber, gently fill with the running buffer, adding it from one corner into the chamber. Do not pour the buffer into the well areas to avoid washing the samples out.*

3.  Put a stir bar in the lower chamber and stir the buffer for 1 min to remove any air bubbles trapped under the ends of the sandwiches.

4.  Add a few drops of dye solution such as phenol red or equivalent into the upper-chamber buffer if the dye in the samples is not sufficient.

5.  Place the lid, or its equivalent, on the unit and connect the power supply. The cathode (negative pole) should be connected to the upper buffer chamber and the anode (positive pole) should be connected to the bottom buffer chamber. Proteins that become negatively charged by SDS will migrate from the cathode to the anode and will separate according to their MW.

6.  Set the power supply at a constant power or current and turn it on. Adjust the current at 25 to 30 mA/1.5-mm thick standard-sized gel. The voltage will increase during the running process. Allow the gel to run electrophoretically for several hours until the dye reaches the bottom of the gel.

7. Turn off the power supply and disconnect the power cables or their equivalent. Loosen and remove the clamps at both sides of the sandwiches. Place the gel sandwiches on the table or bench. Slowly remove the spacers and use an extra spacer to carefully separate the two glass plates starting from one corner. Remove the top plate; the gel should be on the bottom glass plate. Make a small cut at the upper left or right corner of the gel to record the orientation by using a razor blade. Use one gel for staining to determine the MW of different proteins. The other gel is used for Western blotting.

8. Dry the gel under vacuum at 60 to 80°C and carry out autoradiography by exposing the dried gel on X-ray film for 6 to 24 h or for appropriate period of time.

9. Analyze the results.

# References

1. **Sambrook, J., Fritsch, E. F., and Maniatis, T.,** *Molecular Cloning: A Laboratory Manual,* 2nd ed., Cold Spring Harbor Press, Cold Spring Harbor, NY, 1989.

2. **Smith, B. J.,** SDS polyacrylamide gel electrophoresis of proteins, in *Methods in Molecular Biology, Proteins,* Vol. 1, Walker, J. M, Ed., Humana Press, Clifton, NJ, 1984, chap. 6.

3. **Merril, C. R.,** Gel-staining techniques, in *Guide to Protein Purification. Methods in Enzymology,* Vol. 182, Deutscher, M. P., Ed., Academic Press, San Diego, 1990, chap. 36.

4. **Knudsen, K. A.,** Proteins transferred to nitrocellulose for use as immunogens, *Anal. Biochem.,* 147, 285, 1985.

5. **Kyhse-Anderson, J.,** Electroblotting of multiple gels: a simple apparatus without buffer tank for rapid transfer of proteins from polyacrylamide to nitrocellulose, *J. Biochem. Biophys. Methods,* 10, 203, 1984.

# Chapter 15

## Gene Transfer and Expression in Animals

## Contents

Gene engineering technology has been well established in mammalian cells and animals (e.g., mice) as a major revolutionary approach to address a vast spectrum of fundamental biological questions. These include activity assays of promoters, enhancers, and silencers of interest; analysis of expression of mutated genes in their coding regions; analysis of the expression of reporter genes; and applications of gene therapy, using, for example, an antisense approach for knocking out the expression of cancer genes or introducing normal genes into patients containing specific mutant genes. Therefore, gene transfer[1–7] has a broad range of applications in animal systems. It consists of highly involved techniques that include specific gene isolation and characterization, gene recombination, gene transfer, and analysis of the gene expression in transformed cells or transgenic animals.[2,4,8–12] This technology constitutes one of the major advances in medicine and molecular genetics in recent years.[1,7,13–15] It has allowed the possibility of treating some human diseases resulting from defects in single genes by transferring normal genes into some of the cells of the patients, generating gene therapy. Some genes that are involved in tumor or cancer formation have been isolated and characterized. The present chapter describes detailed procedures for gene transfer and analysis of gene expression in mammalian cells and mice. The techniques can also be adapted for other animal systems.

# I.     Gene Transfer and Analysis of Expression in Mammalian Cells

Two types of mammalian cell gene transfer systems have been developed. One is the **transient expression system** in which exogenous genes are transferred into cells, and the introduced genes are allowed to be expressed for 1 to 3 days. The transfected cells are lysed and the gene products are analyzed. The object of transient transfection is usually to obtain a burst in the expression of the transferred gene(s), but such transfection is not suitable for the selection of transformed, stable cell lines. The other is the **stable transformation system** in which foreign genes are introduced into cells and stably integrated into the chromosomes or genomes of the host cells. The integrated DNA replicates efficiently and is maintained during cell division. The gene expression can be analyzed from successive generations of divided cells, establishing genetically stable transformed cell lines. Both gene transfer systems involve the construction of chimeric genes (e.g., promoter, enhancer, vector, interesting gene to be transferred, and/or reporter gene), selection of cell lines, gene transfer into cells, cell culture, and analysis of the expression of the introduced genes in the transformed cells.

## A.     Selection of Vectors, Promoters, and Enhancers for Gene Transfer and Expression

Selection of an appropriate vector, including promoters, enhancers, selectable markers, and poly(A) signals, is very important for the success of gene transfer and analysis of gene expression. A number of vectors have been developed and each vector has its own strengths and weaknesses. We recommend and describe the following vectors that have been well established and widely used in mammalian systems.

### 1.     SV40-Based Vectors

The simian virus 40 (SV40) is a small double-stranded circular DNA tumor virus with a 5.2 Kb genome, which has been well studied. This genome contains an early region encoding the tumor (T) antigen, a late region encoding the viral coat proteins, the origin of replication (ori), and enhancer elements near the ori. The ori and early region play essential roles in the expression of genes. Two divergent transcription units are produced from a single complex promoter/replication region. These viral transcripts are referred to as *early* and *late* due to the time of maximal expression during infection. Both transcripts contain introns and are polyadenylated.

DNA manipulation technology has made it possible to mutate the SV40 genome and fuse it with a plasmid such as pBR322, generating a series of valuable vectors that have been used for gene transfer and expression.

#### a.     Expression Vectors for Testing the Activity of the Promoter of Interest

These plasmid vectors are constructed so as to contain an appropriate reporter gene-coding region, the chloramphenicol acetyl transferase (CAT) gene-coding region including the start and stop sites, the SV40 small T antigen region for intron and polyadenylation signal, multiple cloning sites for the insertion of a foreign promoter of interest, β-lactamase Amp$^r$-coding region, and the origin of replication. In order to test the activity of a promoter of

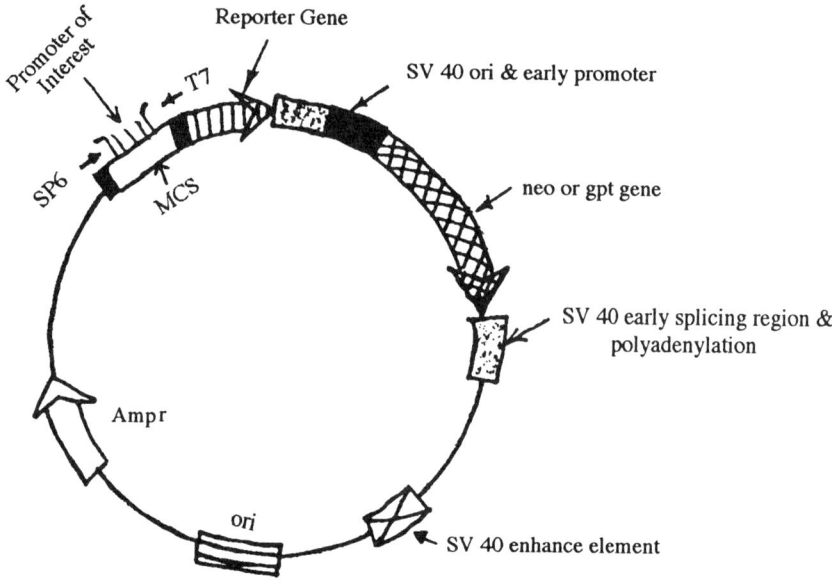

**FIGURE 15.1**

General map of a desirable expression vector used for insertion and analysis of a given promoter of interest.

interest, the promoter can be inserted into the multiple cloning sites (MCS) upstream from the reporter gene. To enhance and test putative promoter activity, the SV40 enhancer elements can be inserted. For the selection of eukaryotic transformants, an appropriate antibiotic-resistant gene (e.g., the *E. coli neo*[r] gene and the *gpt* gene) should be inserted between SV40 ori. and early promoter and SV40 early splicing region and polyadenylation (Figure 15.1).

b.     Vectors for Monitoring the Expression of Introduced Genes

A typical expression vector should contain a strong promoter such as SV40 or LTR and other elements as shown in Figures 15.2 and 15.3.

There are three widely used reporter genes as follows:

1.   **CAT gene:** A bacterial gene encoding for CAT has proven to be useful as a reporter gene for monitoring the expression of transferred genes in transformed cells or transgenic animals, since eukaryotic cells contain no endogenous CAT activity. The CAT gene was isolated from the *E. coli* transposon, Tn9, and its coding region was fused to an appropriate promoter. CAT enzyme activity can be assayed readily by incubating the cell extracts with acetyl CoA and [14]C-chloramphenicol. The enzyme acylates the chloramphenicol, and products can be separated by thin layer chromatography (TLC) on silica gel plates, followed by autoradiography.

2.   **Luciferase gene:** The luciferase gene, which encodes for firefly luciferase, has been isolated and widely used as a highly effective reporter gene. As compared with the CAT assay, the assay of luciferase activity is more than 100-fold more sensitive. It is much simpler, rapid, and relatively inexpensive. Luciferase is a small, single polypeptide with a molecular weight of 62 kDa, and it does not need posttranslational modification for the activity. Other advantages of using the luciferase gene as a reporter gene are that mammalian cells do not have endogenous luciferase activity, and that luciferase can produce chemiluminescent light with very high efficiency, which can be readily detected.

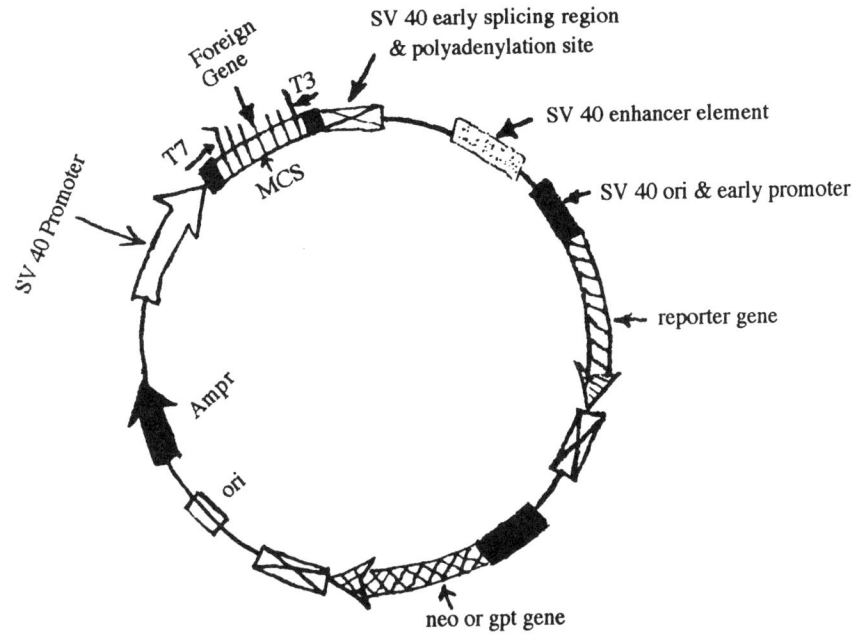

**FIGURE 15.2**
Structural map of an expression vector with an SV40 promoter that can be used for cloning and analysis of a foreign gene of interest.

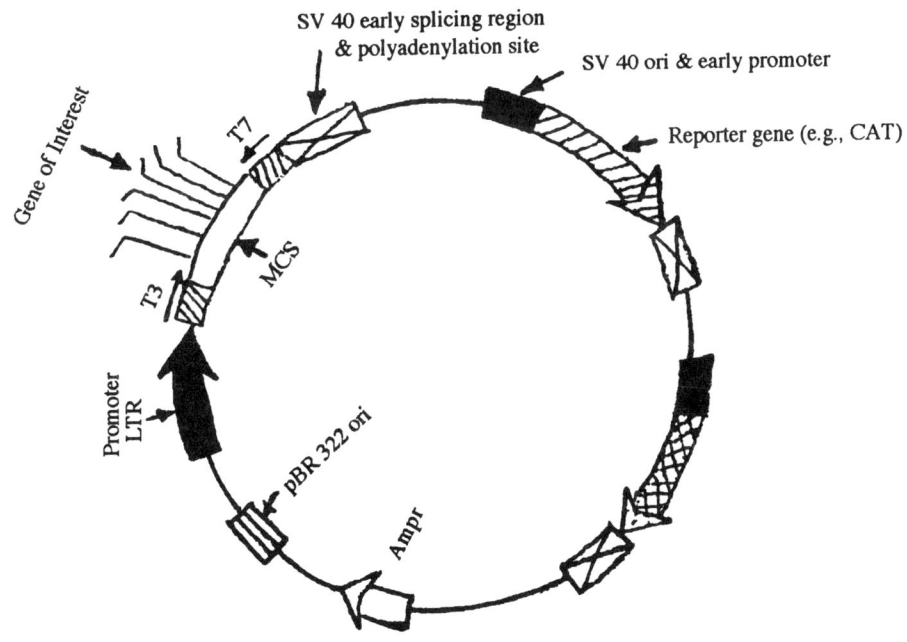

**FIGURE 15.3**
Structural map of an overexpression vector with a retroviral promoter, LTR, that can be used for cloning and analysis of a foreign gene of interest.

Some investigators have used *uidA* or the gusA gene from *E. coli,* which encodes for β-glucuronidase (GUS), as a reporter gene used for gene transfer. However, a disadvantage here is that mammalian tissues contain endogenous GUS activity, thus making the enzyme assay more difficult. The GUS reporter gene, on the other hand, is an excellent reporter gene widely used in higher plant systems because plants do not contain detectable endogenous GUS activity.

3.    **β-Galactosidase gene:** *Lac* Z encoding for β-galactosidase is also widely used as a reporter gene. The cell extracts of transformants can be directly assayed for β-galactosidase activity with a spectrophotometric method.

## 2.    Retrovirus Vectors

It has been demonstrated that retroviruses can be used as effective vectors for gene transfer in mammalian system. There are several advantages over SV40-based plasmid vectors, which include the following: (1) the retroviral genome can stably integrate into the host chromosomes of infected cells and be passed from generation to generation, thus providing an excellent vector for stable transformation; (2) retroviruses have a broad infectivity and expression-host range for any animal cells via viral particles; (3) integration is site-specific with respect to the viral genome at long terminal repeats (LTRs), which can easily preserve the structure of the gene to remain intact after integration; and (4) the viral genomes are very plastic and manifest a high degree of natural size manipulation. Before we describe the construction of retrovirus-based vectors for mediating gene transfer, it is necessary to briefly elaborate the life cycle of retroviruses.

Retroviruses, such as Rous sarcoma virus (RSV) and Moloney murine leukemia virus (MoMLV), are RNA viruses that cause a variety of diseases including tumors in human beings. The viruses contain two tRNA primer molecules, two copies of genomic RNA (38S), reverse transcriptase, RNase H, and integrase, which are packaged with an envelope. The viral envelope contains glycoproteins that serve to determine the host range of infection. When a virus or virion attaches to a host cell, the viral glycoproteins in the envelope bind to specific receptors in the plasma membrane of the host cell. The bound complex then facilitates the internalization of the virus that is now uncoated as it passes through the cytoplasm of the host cell. In the cytoplasm, reverse transcriptase contained in the viral genome catalyzes the formation of a double-stranded DNA molecule from the single-stranded virion RNA. The DNA molecule circularizes, enters the nucleus, and integrates into the chromosome of the host cell, forming a provirus. Subsequently, the integrated provirus serves as a transcriptional template for both mRNAs and virion genomic RNA. Such transcription is catalyzed by the host RNA polymerase II. The mRNAs undergo translation to produce viral proteins and enzymes using the host machinery. All of these components are packaged into viral core particles. The particles move through the cytoplasm, attach to the inner side of the plasma membrane, and bud off. This cycle of infection, reverse transcription, transcription, translation, virion assembly, and budding is repeated over and over again, thus causing infection of new host cells.

The best-understood retrovirus is RSV. The mechanism of synthesis of a double-stranded DNA intermediate (provirus) from viral RNA is unique and complex. The nucleotide base sequence of the DNA molecule is different from that of the viral RNA. The sequence U3-R-U5, which is a combination of the 5' r-u5 segment and the 3' u3-r segment of the RNA, is present at both 5' and 3' ends of the double-stranded DNA molecule. The U3-R-U5 is called the long terminal repeat (LTR). The complete scheme can be divided into eight steps.

Step 1. One of the proline tRNA primer anneals to the pbs region in the 5′ r-u5 of the viral genome RNA. Reverse transcriptase catalyzes the extension of the tRNA primer from its 3′-OH end to the 5′ end, generating a DNA fragment called 3′ R′-(U5)′-tRNA.

Step 2. RNase H removes the cap and poly(A) tail from the viral RNA, and the viral r-u5 segment in the double-stranded region.

Step 3. The 3′ R′-(U5)′-tRNA separates from the pbs region, jumps to the 3′ end of the viral RNA and forms an R′/r duplex.

Step 4. The 3′ R′-(U5)′-tRNA elongates up to the pbs region of the viral RNA by reverse transcriptase, producing the minus (–) strand of DNA.

Step 5. RNase H removes the u3-r from the 3′ end of the viral RNA, followed by synthesis of DNA from the 3′ end of the RNA by reverse transcriptase, forming the first LTR (U3-R-U5) that contains the promoter sequence. This is part of the plus (+) strand DNA.

Step 6. All RNA including the tRNA primer are removed by RNase H.

Step 7. The U3-R-U5 (PBS) is separated from (U3)′R′(U5)′, jumps to the 3′ end of the complementary strand of DNA, and forms a (PBS)/(PBS)′ complex. This is an important process for the virus since the "act" of jumping brings the promoter in the U3 region from the 3′ end to the 5′ end of the plus strand DNA. The promoter is now upstream from the coding region for *gag-pol-env-src* in the 38S RNA genome.

Step 8. Reverse transcriptase catalyzes the extension of DNA from the 3′ termini of both strands, generating a double-stranded DNA (provirus). The LTRs at both ends of the DNA contain promoter and enhanced elements for transcription of the virus genome. The double-stranded DNA molecule integrates into the chromosomes of the infected cell.

Because retroviruses can cause tumors in animals and human beings, we obviously do not want the whole virus genome to be used for gene transfer and as an expression vector. Recent DNA recombination technology makes it possible to modify the retrovirus genome so that it is an efficient tool for gene transfer. The simplest type of gene transfer system is that in which all or most of the *gag, pol,* and *env* genes in the provirus are deleted. However, all the *cis*-active elements such as the 5′ and 3′ LTRs, PBS(+), PBS(–), and psi (ψ) are left intact. A selectable gene such as *neo, gpt, dhfr,* or *hprt* is inserted at the initiating ATG site for *gag*. The expression of the inserted gene is controlled by the 5′ LTR. This manipulated vector is then fused to a plasmid fragment (e.g., pBR322) containing the origin of replication (ori) and an antibiotic-resistant gene. The recombinant plasmids are propagated in *E. coli*. The vectors are then used as a standard DNA-mediated transfection of suitable recipient cells and stable transformants can be selected by using an appropriate antibiotic chemical such as G418. The partial viral RNA transcribed from the transformants can be further packaged into retroviral particles to become an infectious recombinant retrovirus. This can be done by cotrainfecting the host cells with a helper virus that can produce *gag* and *env* proteins, which can recognize the psi site on the recombinant transcript and which become packaged to form particles. The disadvantage of this strategy is that the culture supernatant will contain recombinant and wild-type viruses. Some vectors are constructed by deleting the psi region, replacing the 3′ LTR with a SV40 terminator, the poly(A) signal, and fusing it with the backbone of pBR322. Another recombinant expression vector is constructed by deleting *gag* and *env* genes and by inserting a selectable marker gene (e.g., *neo*) or a reporter gene, which is followed by polylinker sites for the insertion of the foreign gene or cDNA of interest, at the ATG site for *gag* downstream from the 5′ LTR. The 3′ LTR can be replaced with the SV40 poly(A) signal downstream from the introduced gene. The recombinant vector is then fused with the pBR322 backbone containing the origin of replication and the Amp$^r$ gene. (Delete the CAT gene and insert a recombinant viral vector at the polylinker sites where a foreign gene can be inserted downstream from the promoter in the 5' LTR.) The recombinant plasmids are cloned and amplified in *E. coli*. The foreign gene or cDNA of interest can be inserted at the polylinker site downstream from the 5' LTR of recombinant vectors, and gene transfer can be carried out by a standard plasmid DNA-mediated procedure. There are also other recombinant viral vectors used for gene transfer depending on individual research groups. A typical expression vector is shown in Figure 15.3.

## B.     Purification of Plasmid Vectors and DNA Fragments or Genes to be Used for Transfer

The DNA to be studied in transfection experiments may be a potential or putative regulatory sequence, such as a promoter or an enhancer element, a cDNA-coding region, a complete gene, or the truncated/mutated, partial-length segment of a gene of interest. Whatever the source, the DNA fragment or gene to be transferred should be partially characterized. This should include having some known information on DNA size, nucleotide sequence, and restriction enzyme sites. The DNA fragment or gene is then separated and purified from the rest of the DNA in order to be ligated to an appropriate vector for transfection. The general procedures and protocols are described below.

*Protocol 1: Purification of Plasmid Vectors, DNA Fragments, cDNA or Drug-Selectable Marker Gene by Restriction Enzyme Digestion followed by Elution from an Electrophoresed Gel*

### a.     Preparation of Vectors

Select appropriate plasmid vectors such as pCAT and pGL2 series (Promega Corporation) and other pSV-based vectors or retrovirus-based vectors depending on the particular experiments you wish to carry out.

1.     For the transfer and over- and underexpression of the gene or cDNA of interest, a standard vector may have: (1) a strong promoter, such as the SV40 promoter or a retroviral LTR; (2) multiple cloning sites downstream from the promoter, which can be used for the insertion of the exogenous gene; (3) SV40 small T intron (early splice) and poly(A) signal downstream from the foreign gene; (4) an enhancer element from SV40 or a retrovirus can be upstream from the promoter or upstream from the polylinker sites in either orientations; (5) SV40 ori early promoter followed by the coding region (initiation at AUG) of a reporter gene (e.g., CAT gene, or luciferase gene) downstream from the SV40 promoter; (6) SV40 ori early promoter followed by the coding region (initiation at AUG) of selectable marker gene (e.g., *neo*$^r$); (7) the Col E1 ori and F1(+) ori of replication for high copy number; and (8) the Amp$^r$ gene for the selection of antibiotic resistance bacteria.

2.     For testing the activity of the promoter of interest, the basic characteristics of the vector are almost the same as described above except that there is no promoter upstream from the reporter gene, and that the polylinker sites are located upstream from the reporter gene for the insertion of the promoter of interest.

3.     For testing the activity of enhancer or silencer of interest, the basic backbone of the vector may be similar to that described above except that there is no enhancer. The enhancer element of

interest can be cloned upstream from the promoter or downstream from the reporter gene at the polylinker sites in either orientation. The silencer of interest is usually linked upstream from the promoter.

### b.    Restriction Enzyme Digestion of Plasmid Vectors and DNA of Interest for Ligation

The selection of restriction enzymes should be based on what kinds of restriction enzyme sites are present in the polylinker sites of the vector and what restriction enzyme sites are in the DNA to be inserted. Choose a unique restriction enzyme(s) for the vector and for DNA, which can cut the inserted DNA outside of the region of interest (e.g., cDNA-coding region, promoter, enhancer, or mutated and recombinant genes). The general procedures are described as follows:

1.   Set up, on ice, a standard restriction enzyme digestion as follows:

**Single-restriction enzyme digestion**

- Plasmid vectors or DNA to be inserted (10 µg)
- 10X Appropriate restriction enzyme buffer (10 µl)
- 1 mg/ml Acetylated BSA (optional) (10 µl)
- Appropriate restriction enzyme (3.3 units/µg DNA)
- Add dd.H$_2$O to a final volume of 100 µl.

**Double-restriction enzyme digestion**

- Plasmid DNA or DNA to be inserted (10 µg)
- 10X Appropriate restriction enzyme buffer (10 µl)
- 1 mg/ml Acetylated BSA (optional) (10 µl)
- Appropriate restriction enzyme A (3.3 units/µg DNA)
- Appropriate restriction enzyme B (3.3 units/µg DNA)
- Add dd.H$_2$O to a final volume of 100 µl.

*Notes:*   *(1) The restriction enzyme(s) used for the vector and insert DNA digestions should be the same to ensure the optimal conditions for ligation. (2) For directional cloning, the plasmid and insert DNA may be digested using two different restriction enzymes. The double-enzyme digestion of DNA may be set up in a single reaction at the same time or carried out with two single-enzyme digestions performed at different times. (3) For double-restriction enzyme digestions, the appropriate 10X buffer containing a higher NaCl conentration than for the other buffer may be chosen as the double-enzyme digestion buffer.*

2.   Incubate at an appropriate temperature depending on the appropriate restriction enzyme (e.g., 37°C) for 2 to 3 h. For single-enzyme-digested DNA, proceed to step 3. For double-enzyme-digested DNA, proceed to step 5.

*Notes:*   *(1) To ensure an optimal ligation to the insert DNA, both vectors and DNA should be completely digested. The digestion efficiency can be checked by loading 1 µg of the digested DNA (10 µl) with loading buffer to a 1% agarose mini-gel. In the meantime, undigested vectors and DNA (1 µg) and standard DNA markers should be loaded in the adjacent wells. After electrophoresis, the undigested plasmid DNA may reveal multiple bands because of different levels of supercoiling. However, one band should be visible for a complete single-enzyme digestion; one major band and one small band should be visible after digestion with two different restriction enzymes. One should be able to distinguish the digested DNA to be transferred from the undigested DNA. (2) After completion of the restriction enzyme digestion, calf intestinal alkaline phosphatase (CIAP) treatment should be carried out for the above single-restriction enzyme digestion of the plasmid vector. This treatment removes 5′-phosphate groups, thus preventing recircularization of the vector during ligation. Otherwise, the efficiency of ligation between vector and insert DNA will be very low. For double-restriction enzyme-digested vectors, the CIAP treatment is not necessary.*

3. Carry out CIAP treatment by adding the following directly to the single-enzyme-digested plasmid vector DNA sample (90 μl).
   - 10X CIAP buffer (15 μl)
   - CIAP diluted in 10X CIAP buffer (0.01 unit/pmol ends)
   - Add dd.$H_2O$ to a final volume of 150 μl.

*Notes:* *(1) CIAP and 10X CIAP should be kept at 4°C. CIAP treatment should be set up at 0°C. (2) Calculation of the amount of ends is as follows. There is 9 μg digested DNA left after taking 1 μg of 10 μg digested DNA for checking on agarose gel. If the vector is 3.2 Kb, the amount of ends can be calculated by the formula below:*

$$\text{pmol ends} = \frac{\text{amount of DNA}}{\left(\text{base pairs} \times 660/\text{base pair}\right)} \times 2$$

$$= \frac{9}{3.2 \times 1000 \times 660} \times 2$$

$$= 4.2 \times 10^{-6} \times 2$$

$$= 8.4 \times 10^{-6} \; \mu M$$

$$8.4 \times 10^{-6} \times 10^{-6} = 8.4 \text{ pmol ends}$$

4. Incubate at 37°C for 1 h and add 2μl of 0.5 *M* EDTA buffer (pH 8.0) to stop the reaction.

5. Extract with 1 volume of TE-saturated phenol/chloroform. Mix well by vortexing for 1 min and centrifuge at $11,000 \times g$ for 5 min at room temperature.

6. Carefully transfer the top, aqueous phase to a fresh tube and add 1 volume of chloroform:isoamyl alcohol (24:1) to the supernatant. Mix well and centrifuge as in step (5).

7. Carefully transfer the upper, aqueous phase to a fresh tube and add 0.1 volume of 3 *M* sodium acetate buffer (pH 5.2) or 0.5 volume of 7.5 *M* ammonium acetate to the supernatant. Briefly mix and add 2 to 2.5 volumes of chilled 100% ethanol to the supernatant. Allow to precipitate at −70°C for 1 h or at −20°C for 2 h.

8. Centrifuge at $12,000 \times g$ for 10 min and carefully decant the supernatant. Briefly rinse the DNA pellet with 1 ml of 70% ethanol and dry the pellet under vacuum for 20 min. Dissolve the DNA pellet in 20 to 40 μl dd.$H_2O$. Take 4 μl of the sample to measure the concentration of the DNA at 260 nm. Store the sample at −20°C until use.

*Note:* *Adding 0.5 volume of 7.5 M ammonium acetate to the supernatant at step 7 yields a higher amount of DNA precipitated than adding 0.1 volume of 3 M sodium acetate buffer (pH 5.2).*

## *Reagents Needed* ————————————————————————

### *Appropriate Enzymes*

> 10X Appropriate Restriction Enzyme Buffer
> 1% Agarose Mini-Gel
> TE-Saturated Phenol/Chloroform
> Chloroform:Isoamyl Alcohol (24:1)
> 3 *M* Sodium Acetate Buffer, pH 5.2
> 7.5 *M* Ammonium Acetate
> Ethanol (100%, 70%)
> 0.5 *M* EDTA, pH 8.0
> Calf Intestinal Alkaline Phosphatase (CIAP)
> TE Buffer

### *10X CIAP Buffer*

> 0.5 *M* Tris-HCl, pH 9.0
> 10 m*M* $MgCl_2$
> 1 m*M* $ZnCl_2$
> 10 m*M* Spermidine

c.        Electrophoresis and Elution of DNA

### *Method i: Elution of DNA by Agarose Gel Slices* _____

1.   Electrophorese the digested vectors and DNA to be inserted on 0.8 to 2% agarose depending on the DNA sizes as described under Southern blotting procedures in Chapter 5, except that one should use low-melting point agarose instead of normal agarose.

2.   Transfer the gel onto a long-wavelength UV transilluminator (305 to 327 nm) to visualize the DNA bands. Excise the band(s) of interest with a clean razor blade and place the slices into a microcentrifuge tube. The DNA can be eluted from the agrose gel slices by one of two procedures below.

*Notes:*   *(1) The gel should not be placed on a short-wavelength (e.g., <270 nm) UV transilluminator since UV light may cause breakage inside the DNA fragment. This significantly inhibits the subcloning of the insert of interest. (2) UV light is toxic to the human body. Protective glasses, gloves, and a lab coat should be worn when using UV light. (3) The gel slices containing the DNA bands of interest should be trimmed of excess unstained gel area as much as possible.*

3.   Carry out elution of DNA by using one of the following methods:
   - **Freezing and thawing method**
   a.   Add 2 volumes of TE buffer to the gel slices and completely melt the gel in a 60 to 70°C water bath.

*Notes:*   *The gel slices can be directly melted without adding any TE buffer. The DNA concentration is usually high but total yield of elution of the DNA fragment is much lower than when TE buffer is added.*

b. Immediately chill the melted gel solution on dry ice and place the tube at –70°C for at least 20 min.

c. Thaw the gel mixture by tapping the tube vigorously. It takes about 5 to 10 min to thaw the gel into a resuspension status.

d. Centrifuge at $11,000 \times g$ for 5 min at room temperature.

e. Carefully transfer the liquid phase containing the elution DNA fragment to a fresh tube.

f. Extract EtBr with 3 volumes of water-saturated $n$-butanol.

g. Precipitate the DNA by adding 0.1 volume of 3 $M$ sodium acetate buffer (pH 5.2) and 2.5 volumes of chilled 100% ethanol into the DNA solution. Allow to precipitate at –70°C for 1 h and centrifuge at $12,000 \times g$ for 5 min at room temperature.

h. Discard the supernatant and briefly rinse the pellet with 1 ml of 70% ethanol. Dry the DNA under vacuum for 15 min and dissolve the DNA in an appropriate volume of dd.$H_2O$ or TE buffer. Measure the concentration of the eluted DNA and store at –20°C until use.

- **Phenol/chloroform extraction method**

a. Add 1 volume of Tris buffer-equilibrated phenol to the gel slices, vortex the microcentrifuge tube for 2 min, and place the tube in a dry ice–methanol bath for 20 min.

b. Centrifuge at $12,000 \times g$ at room temperature for 15 min and transfer the aqueous phase to a fresh tube.

c. Reextract the phenol phase with 0.3 ml dd.$H_2O$, mix, and repeat step b.

d. Extract the pooled aqueous phases with 1 volume of phenol:chloroform:isoamyl alcohol (25:24:1) and vortex for 1 min.

e. Repeat step b and precipitate the DNA as described in step g above.

## Method ii: Elution of DNA Fragments in Wells of Agarose Gel

1. Carry out DNA digestion and gel electrophoresis as described in Method i, except that one should add running buffer up to the upper edges of the gel instead of covering the gel.

2. During electrophoresis, monitor the separation of the DNA bands stained by EtBr in the gel using a long-wavelength UV lamp. Stop electrophoresis and use a razor blade or a spatula to make a well in front of the band of interest. Add 20 to 60 µl of running buffer to the well.

3. Continue electrophoresis until the band migrates into the well. This is done by monitoring the band which is stained by EtBr.

4. Stop the electrophoresis and transfer the solution containing the DNA of interest from the well to a fresh tube.

5. Extract EtBr with 3 volumes of water-saturated $n$-butanol.

6. Precipitate and dissolve the DNA as described in Method i.

### d.    Preparation of Blunt-End DNA Fragments

If both the vector and the digested DNA fragment have sticky or overhanging ends at the right sites, they can be readily ligated to each other. If not, it is usually necessary to trim the DNA ends prior to their being inserted into a blunt-ended site such as the *Sma* I site of the vector. This can be done either by filling-in the 5′ overhangs with DNA polymerase I (Klenow fragment) (see Method i below) or by digesting the 3′ overhangs with exonuclease T4 DNA polymerase (see Method ii below), or by mung bean nuclease that digests single-stranded DNA at both 5′ and 3′ overhangs (see Method iii below).

## Method i: Filling in 5′ Overhangs

1. Set up the 5′ overhang reaction in a microcentrifuge tube on ice:
   - DNA fragment or vector DNA (2 to 4 µg in 4 µl)
   - 10X 5′ Overhang buffer (2 µl)
   - 2 m$M$ dNTP mixture (2 µl)
   - DNA polymerase I (Klenow fragment) (2 to 3 units)
   - Add dd.H$_2$O to a final volume of 20 µl.
2. Incubate at room temperature for 20 min and heat at 70°C for 5 min to stop the reaction.
3. Add 1 volume of TE buffer (pH 7.5) and carry out phenol:chloroform extraction and precipitation as described previously.

## Reagents Needed

### 10X 5′ Overhang Buffer

500 m$M$ Tris-HCl, pH 7.2
100 m$M$ MgSO$_4$
1 m$M$ DTT
500 µg/ml BSA

### 2 mM dNTP Mixture

2 m$M$ Each of dATP, dCTP, dGTP, and dTTP in 20 m$M$ Tris-HCl, pH 7.5

### DNA Polymerase I (Klenow Fragment)

1 unit/µl

## Method ii: Removal of 3′ Overhangs

1. Set up a standard reaction in a microcentrifuge tube on ice as follows:
   - DNA fragment or vector DNA (2 to 4 µg in 4 µl)
   - 10X 3′ Overhang buffer (2 µl)
   - 2 m$M$ dNTP mixture (2 µl)
   - T4 DNA polymerase (2 to 3 units)
   - Add dd.H$_2$O to a final volume of 20 µl.
2. Incubate at 12°C for 20 min and heat at 70°C for 5 min to stop the reaction.
3. Add 1 volume of TE buffer (pH 7.5) and carry out phenol:chloroform extraction and precipitation as described previously.

## Reagents Needed

### 10X 3′ Overhang Buffer

300 m$M$ Tris-acetate, pH 7.9
660 m$M$ Potassium acetate

100 m$M$ Magnesium acetate
5 m$M$ DTT
1 µg/µl BSA

### 2 mM dNTP Mixture

2 m$M$ Each of dATP, dCTP, dGTP, and dTTP in 20 m$M$ Tris-HCl, pH 7.5

### T4 DNA Polymerase

## Method iii: Using Mung Bean Nuclease

1.  Set up a standard reaction in a microcentrifuge tube on ice as follows:
    *   DNA fragment or vector DNA (2 µg in 2 µl)
    *   10X Mung bean nuclease buffer (2 µl)
    *   Mung bean nuclease (5 units)
    *   Add dd.H$_2$O to a final volume of 20 µl.
2.  Incubate at room temperature for 20 min and heat at 70°C for 10 min to stop the reaction.
3.  Add 1 volume of TE buffer (pH 7.5) and carry out phenol:chloroform extraction and precipitation as described previously.

*Note:*    *Mung bean nuclease is better than S1 due to the lower intrinsic activity on double-stranded DNA.*

## Reagents Needed

### 10X Mung Bean Nuclease Buffer

300 m$M$ Sodium acetate, pH 5.0
500 m$M$ NaCl
10 m$M$ ZnCl$_2$
50% (v/v) Glycerol

### Mung Bean Nuclease

Immediately before use, the enzyme is diluted to 5 units/µl in mung bean nuclease dilution buffer.

### Mung Bean Nuclease Dilution Buffer

10 m$M$ Sodium acetate, pH 5.0
0.1 m$M$ Zinc acetate
1 m$M$ Cysteine
0.1% (v/v) Triton X-100
50% (v/v) Glycerol

## C. Preparation of Double-Stranded Oligonucleotides for Cloning

### Procedure

1. Design two complementary oligonucleotides of interest, which can be synthesized by a DNA synthesizer. The oligonucleotides should be purified afterwards.

2. Add 100 pmol of each of the two complementary oligonucleotides in a total volume of 20 µl in a microcentrifuge tube.

3. Add 1 µl of 20X annealing buffer to the tube, incubate at 90°C for 6 min, followed by slow cooling to room temperature. This can be done by placing the heated tube in a rack in a beaker filled with an appropriate volume of 90°C water. The cooling procedure should take 20 to 30 min for complete annealing.

4. Add phosphate groups to the 5′ ends of the oligonucleotides by the kinase reaction.

   a. Set up a reaction as follows in a microcentrifuge tube on ice:
      - The annealed oligonucleotides (15 µl)
      - 10X Polynucleotides kinase buffer (3 µl)
      - 10 m$M$ ATP (1.5 µl)
      - 100 m$M$ DTT (1.5 µl)
      - T4 Polynucleotide kinase (15 units)
      - Add dd.H$_2$O to a final volume of 30 µl.

   b. Incubate at 37°C for 40 min and heat at 70°C for 10 min to inactivate the enzyme.

   c. Add 1 volume of TE buffer (pH 7.5) and carry out extraction and precipitation.

### Reagents Needed

### 20X Annealing Buffer

> 200 m$M$ Tris-HCl, pH 7.9
> 40 m$M$ MgCl$_2$
> 1 $M$ NaCl
> 20 m$M$ EDTA, pH 8.0

### T4 Polynucleotide Kinase

> 1 unit/µl

### 10X Polynucleotide Kinase Buffer

> 500 m$M$ Tris-HCl, pH 7.5
> 100 m$M$ MgCl$_2$
> 50 m$M$ DTT
> 1 m$M$ Spermidine
> 1 m$M$ EDTA

### 10 mM ATP

## D.    Construction of Chimeric Genes for Transfection

After the necessary components for gene transfer are purified as generally described above, chimeric gene constructs may be constructed, which are either in a circular plasmid form or in a linear form. Generally, standard chimeric gene constructs for gene transfer, gene expression, and selection of transformants should include a variety of components: vectors, promoters, enhancers, SV40 introns, poly(A) signals, reporter gene-coding region, cDNA or genomic DNA of interest, and a drug-selection marker gene such as aminoglycoside 3'-phosphotransferase *(aph)* gene, thymidine kinase *(tk)* gene, hygromycin B phosphotransferase gene, and tryptophan synthetase gene. For each gene or cDNA or drug-selection marker gene to be transferred, its coding region should be downstream from appropriate promoter, followed by the splicing intron (e.g., SV40 intron) and poly(A) signal. An enhancer element may be inserted upstream from the promoter or downstream from the poly(A) signal. Commercial plasmid vectors are available from Promega Corporation or other companies. Plasmids contain most of the necessary components including polylinker sites for insertion of the foreign DNA of interest. However, those plasmid vectors such as the pCAT series and the pGL-2 series do not contain a drug-selection marker gene for the selection of stable transformants. Therefore, an appropriate marker gene should be fused to the recombinant gene constructs for transfection. An alternative way is cotransfect a plasmid bearing the marker gene and a plasmid containing the DNA of interest into the cells. But we recommend that all necessary components be fused together in a single plasmid DNA for efficient transfection, analysis of the expression of the introduced genes, and selection of stable transformants. However, it has been reported that a chimeric gene containing bacterial DNA could affect the expression of the introduced gene. Therefore, the gene of interest can be purified and cotransfect the cells with separate DNA containing a drug-selectable gene depending on the particular gene transfer experiment. A typical construction of chimeric genes involves ligation, one by one, recircularization of the recombinant plasmid, transfer into an appropriate bacterial host for amplification, selection of transformants, and purification of the recombinant plasmids for transfection.

### *1. Ligation of Plasmid Vector and Insert DNA*

To achieve optimal ligation, the ratio of vector to insert DNA (1:1, 1:2, 1:3 and 3:1 molar ratio) is strongly recommended to be optimized by using a small-scale reaction series. The following reaction is standard for the ligation of 3.2 Kb plasmid vector and 3.0 Kb insert DNA.

1.  Calculate the molar weight of the vector and insert DNA as follows:

$$1\ M \text{ plasmid vector} = 3.2 \times 1000 \times 660 = 2.112 \times 10^6$$

$$1\ M \text{ insert DNA} = 3 \times 1000 \times 660 = 1.98 \times 10^6$$

2.  Calculate the molar ratio of vector to insert DNA as follows:

|  | Amount of DNA (µg) | |
| --- | --- | --- |
| Vector DNA:insert DNA | Vector | Insert |
| 1:1 | 1 | 0.792 |
| 1:2 | 1 | 1.584 |
| 1:3 | 1 | 2.376 |
| 3:1 | 1 | 0.264 |

3. Set up the following ligations on ice:

| Components | Ligation reactions | | | |
| --- | --- | --- | --- | --- |
| | 1 (1:1) | 2 (1:2) | 3 (1:3) | 4 (3:1) |
| Plasmid DNA as vector ($\mu$g) | 1 | 1 | 1 | 1 |
| Insert DNA ($\mu$g) | 0.792 | 1.584 | 2.376 | 0.244 |
| 10X Ligase buffer ($\mu$l) | 1 | 1 | 1 | 1 |
| T4 DNA ligase (Weiss units) | 4 | 4 | 4 | 4 |
| Add dd.$H_2O$ to ($\mu$l) | 10 | 10 | 10 | 10 |

*Notes:* *The restriction enzyme-digested plasmid (vector) and insert DNA should be dissolved in dd.$H_2O$ (nuclease-free) to 0.5 to 1.0 $\mu$g/$\mu$l. If the DNA is less than 0.4 $\mu$g/$\mu$l, the DNA should be precipitated so as to dissolve in about 1 $\mu$g/$\mu$l.*

4. Incubate the reactions at 4°C for 12 to 24 h, or 16°C for 4 to 6 h, or at room temperature (22 to 25°C) for 1 to 2 h.

*Note:* *After ligations are finished at the above temperatures, the mixture can be stored at 4°C until use.*

5. Check the efficiency of ligation by 1% agarose electrophoresis. When the electrophoresis is complete, photograph the gel stained with EtBr under UV light. As compared with the unligated vector or insert DNA wells, a highly efficient ligation should allow one to visualize less than approximately 10% unligated vector and insert DNA by estimation of the intensity of fluorescence. Approximately 90% of the vector and insert DNA are ligated to each other and show strong band(s) with molecular weight shift compared to the size of vector and insert DNA. By comparing the efficiency of ligation using different molar ratios, the optimal conditions can be determined with ease. These can be used as a guide for large-scale ligations.

*Note:* *The above small-scale ligation is optional but it is strongly recommended that one carry it out before attempting large-scale ligations.*

6. Carry out large-scale ligation of vector and insert DNA based on the optimal conditions determined by small-scale ligations. For example, if one uses a 1:2 molar ratio of plasmid DNA:insert DNA as being the optimal for the ligation, a large-scale ligation can be carried out as follows:
   - Plasmid DNA as vector (3 $\mu$g)
   - Insert DNA (4.75 $\mu$g)
   - 10X Ligase buffer (3 $\mu$l)
   - T4 DNA ligase (15 Weiss units)
   - Add dd.$H_2O$ to a final volume of 30 $\mu$l.

   Incubate the ligation at 4°C for 12 to 24 h, or 16°C for 4 to 6 h, or at room temperature (22 to 25°C) for 1 to 2 h. Store at 4°C until use. Proceed to transformation.

## Reagents Needed

### Dephosphorylated Linearized Plasmid Vectors Phosporylated DNA Fragments

*5X Ligase Buffer*

> 0.25 $M$ Tris-HCl, pH 7.5
> 50 m$M$ MgCl$_2$
> 5 m$M$ ATP
> 5 m$M$ DTT
> 25% (v/v) PEG, 8000

*T4 DNA Ligase*

> 2 Weiss units/μl

## 2. Transformation of an Appropriate Strain of E. coli with the Ligated, Recombinant Plasmids and Isolation of the Plasmids for Transfection

1.  Prepare the LB medium and LB plates as described previously. This should be done before ligation.

2.  Prepare the competent cells as previously described. This should be completed before ligation.

3.  Prepare competent cells and carry out transformation of bacteria in order to amplify the recombinant plasmids. The competent cells (such as DH5αF′, JM109, and HB 101; for transformation, selection of transformants, and verification of transformants) are commercially available. The detailed procedures for preparation of competent cells are described under the section on plasmid transformation in Chapter 7.

4.  Isolation of recombinant plasmid DNA for transfection is carried out by equilibrium centrifugation in CsCl–ethidium bromide gradients. The detailed procedures are described under the section on isolation and purification of plasmid DNA in Chapter 1.

## E.    Transfection of Reporter DNA Constructs into Cultured Cells

There are a number of methods that have been developed for the transfer of foreign genes into mammalian cells. Each method has advantages and disadvantages. The most widely used methods are compared in Table 15–1.

### Method 1: Calcium Phosphate Transfection

The principle of calcium phosphate-mediated transfection is that the DNA to be transferred is mixed with CaCl$_2$ and phosphate buffer to form a fine calcium phosphate precipitation containing the DNA. The precipitate is then placed on a cell monolayer. It binds/attaches to the plasma membrane and is taken into the cell by endocytosis. This is the most widely used method for both transient and stable transfection. It is carried out as follows:

1.  Carry out trypsinizing to remove adherent cells for subculturing or cell counting as well as for monolayer cells preparation.

    a.  Remove the media from cell-culture or tissue-culture dish and wash the cells twice with 15 ml of PBS buffer or other calcium- and magnesium-free salt solution.

    b.  Remove the washing solution and add 1 ml of trypsin solution per 100-mm dish. Quickly rock the dish to evenly distribute the solution over the cells and incubate at 37°C for 1 to 2 min depending on the cell type (e.g., NIH 3T3, 1 min; CHO, 1 min; HeLa, 1.5 min; COS, 2 min).

## Table 15–1 Advantages and Disadvantages of Various Methods of Gene Transfer

| Method | Advantages | Disadvantages | Efficiency |
|---|---|---|---|
| Calcium phosphate co-precipitation | Simple and widely used | Low efficiency | 10–15% |
| DEAE dextran-mediated transfection | Simple and high efficiency | DNA may be mutated, not for stable transfection | 50–80% |
| Electroporation | Simple, suitable for transient and stable transfection | Up to 50% of cells will die due to high-voltage electric pulse | Up to 50% |
| Protoplast fusion | High efficiency, suitable for table and transient transfection | Complicated procedure and time-consuming | High |
| DNA microinjection | High efficiency | Needs expensive equipment and special techniques and skills | High |
| SV40 vectors-mediated transfer | High efficiency in COS cells | Not for stable transfection and DNA rearrangement occurs | High |
| Retroviral vector-mediated transfer | High efficiency, wide host range, works well for both stable and transient transfection | Low protein level due to splicing of RNA | 100% |
| Adenovirus vector-mediated transfer | High efficiency | | High |

    c. Closely monitor the cells, once they begin to detach, immediately remove the trypsin solution and strike the bottom and sides of the culture to loosen the remaining adherent cells.

    d. Add culture medium containing 10% serum (e.g., PBS) to the cells to inactivate the trypsin. Gently shake the culture dish. The cells are then ready for subculture and counting.

2.    Prepare cell monolayers the day before the transfection experiment. The cell monolayers should be at 50 to 70% confluence. Remove the old culture medium and feed the cells with 5 to 10 ml of fresh culture medium.

*Notes:*   *(1) It is recommended that this procedure be carried out in a sterile laminar flow hood. Gloves and a mask should be worn in order to prevent contamination of the cells. (2) The common cell lines used for transfection are NIH 3T3, CHO, HeLa, and COS. Other mammalian cells of interest can also be transfected in the same way. (3) The density of plating of cells depends on the particular cell line. A general guideline is $7 \times 10^6$ cells per 100-mm culture dish. (4) All of the solutions and buffers should be sterilized by autoclaving or by sterile-filtration. All the components should be warmed to room temperature prior to their being used.*

3.  Prepare a transfection mixture for each 100-mm dish as follows:
    - 2.5 $M$ $CaCl_2$ solution (50 μl)
    - Recombinant reporter DNA construct (10 to 30 μg)
    - Add dd.$H_2O$ to a volume of 0.5 ml.

    Mix well, and then slowly add the DNA mixture dropwise to 0.5 ml of well-suspended 2X HBS buffer with continuous mixing by vortexing. Incubate the mixture at room temperature for 30 to 40 min to allow coprecipitation to occur.

*Notes:* (1) The DNA to be transferred can be in a circular form (e.g., recombinant plasmids) or be directly constructed as linear chimeric genes. (2) After the DNA mixture has been added into the 2X HBS buffer, the mixture should be slightly opaque due to the formation of a fine coprecipitate of DNA with calcium phosphate.

4.  Slowly add the precipitated mixture dropwise to the dish containing the cells while continuously swirling the plate to achieve a good mixing. For each 100-mm dish containing the cells, add 1 ml of the DNA mixture. For each 60-mm dish, add 0.5 ml.

5.  Transfer and incubate the plate containing the cells overnight in a clean 37°C $CO_2$/air (2 to 5% $CO_2$) incubator.

6.  Remove the medium and wash the monolayer once with 10 ml of serum-free DMEM medium and once with 10 ml of PBS. Add fresh culture medium and return the Petri dishes to the incubator. Normally, cells can be harvested 48 to 72 h after transfection for transient transfer analysis.

## Reagents Needed

*Note:* All solutions, buffers, and media must be sterilized.

### 10X HBS Buffer

> 8.18% (w/v) NaCl
> 5.94% (w/v) HEPES, pH 7.1
> 4-(2-Hydroxyethyl)-1-piperazine ethanesulfonic acid
> 0.2% (w/v) $Na_2HPO_4$

### 2X HBS Buffer

> Dilute the 10X HBS in dd.$H_2O$.
> Adjust the pH to 7.1 with 1 $N$ NaOH solution.

### DNA to be Transferred

> Circular or linear recombinant DNA constructs in 50 μl of TE buffer, pH 7.5

### 2.5 M CaCl$_2$

### DMEM Medium

> Available from Gibco/BRL or prepare it as follows:

Dissolve one bottle of Dublecco's Modified Eagle's Medium (DMEM) powder in 800 ml dd.$H_2O$; add 3.7 g $NaHCO_3$; adjust the pH to 7.2 with 1 $N$ HCl; add dd.$H_2O$ to 1 liter. Sterile-filter medium (do not autoclave) in a laminar flow hood. Divide the medium into 200- to 500-ml aliquots in sterile bottles. Store at 4°C until use. Prior to use, add 10 to 20% FBS, depending on the particular cell type, to the medium and warm to room temperature.

## PBS Buffer

0.137 $M$ NaCl
2.7 m$M$ KCl
4.3 m$M$ $Na_2HPO_4$
1.47 m$M$ $KH_2PO_4$
The final pH is 7.2 to 7.4.

## 1X Trypsin–EDTA Solution

0.05% (w/v) Trypsin
0.53 m$M$ EDTA, pH 7.6
Dissolve in PBS buffer, pH 7.4 to 7.6.

## Method 2: Transfection by DEAE-Dextran

This is a highly efficient method used for transient transfection, but it is not suitable for stable transformants. The basic principle underlying this method is that DNA constructs to be transferred bind to DEAE-dextran, which is then applied to cells so as to allow it to enter the cells by endocytosis.

1. Carry out steps 1 to 2 in Method 1.
2. Prepare a transfection mixture for each 100-mm dish as follows:
   - 1X PBS buffer (0.54 ml)
   - DEAE-Dextran (10 mg/ml) (40 µl)
   - Recombinant reporter DNA constructs (10 to 30 µg in 20 µl)
   - **Total volume: 0.6 ml**
3. Wash the cells twice with 10 ml of serum-free medium or PBS buffer.
4. Add the transfection mixture over the monolayer cells. Dispense evenly over the cells.
5. Incubate the cells in the Petri dishes in a clean 37°C $CO_2$ (2 to 5%) incubator for 4 h.
6. Remove the solution from the dishes and add 1 ml per dish of 25% glycerol. Leave for 1 min in order to enhance uptake of DNA by cells.
7. Remove the glycerol, wash the cells twice with serum-free medium (10 ml per dish), and add 10 to 15 ml per dish culture medium (DMEM). Return the dishes to the incubator. Normally, cells can be harvested 48 to 72 h after transfection.

## Reagents Needed

*Note:    All solutions, buffers, and media must be sterilized.*

### DNA to be Transferred

Circular or linear recombinant DNA constructs in 50 μl of TE buffer, pH 7.5

### 2.5 M CaCl₂

### DMEM Medium

Available from Gibco/BRL or prepare it as follows:

Dissolve one bottle of Dublecco's Modified Eagle's Medium (DMEM) powder in 800 ml dd.H₂O; add 3.7 g NaHCO₃; add 10 ml of ABAM as an antibiotic; adjust the pH to 7.2 with 1 $N$ HCl; add dd.H₂O to 1 liter. Sterile-filter the medium (do not autoclave) in a laminar flow hood. Divide the medium into 200- to 500-ml aliquots in sterile bottles. Store at 4°C until use. Prior to use, add 10 to 20% FBS to the medium if it is required, and warm to room temperature.

### PBS Buffer

0.137 $M$ NaCl
2.7 m$M$ KCl
4.3 m$M$ Na₂HPO₄
1.47 m$M$ KH₂PO₄
The final pH is 7.2 to 7.4.

### DEAE-Dextran Solution

10 mg/ml DEAE-dextran ($M_r$ 5 × 10⁵, chloride form, Sigma Chemical Co.) in dd.H₂O
Autoclave. Store at 4°C. Warm up to 37°C prior to use.

## Method 3: Transfection by Electroporation

This is a widely used transformation technique that is simple, fast, and effective. The principle underlying this method is that cells and constructed DNA to be transferred are mixed and subjected to a high-voltage electric pulse that creates pores in the plasma membranes of the cells. This allows the DNA to enter the cells and become integrated into the genome. The pores are closed by incubation at 0°C after electroporation. This method is suitable for both transient and stable transformation experiments. However, the pulse conditions should be adjusted empirically for the particular cell type.

1.   Harvest the cells by gentle trypsinization as described in step 1 in Method 1. Centrifuge at 1000 × $g$ for 1 min and resuspend the cells in PBS buffer at about 10⁶ cells/ml.

2.   Add 0.6 ml of the cell suspension and 10 to 20 μl constructed DNA (15 to 25 μg) to be transferred in a disposable microelectroporation chamber (BRL) on ice. Mix by pipetting and incubate on ice for 15 min (optional).

3.   Carry out electroporation using an apparatus such as that of BioRad's Gene Pulser (with a pulse controller and capacitance extender):

*Note:*   *The pulse is set at the appropriate capacitance and the voltage settings need to be adjusted experimentally.*

    a. Connect the power cable to a BRL-Porator pulse control + power supply apparatus, to the BRL-Porator voltage booster, and between the two units.

    b. Set up the pulse control as follows:
- Power: charge
- Capacitance (μF): 330
- High ohm/low ohm: low ohm
- Charge rate: fast
- Set the voltage booster at 4 kV.

    c. Add ice water into the chamber safe up to four fifths of the volume, and place the chamber rack in the chamber safe.

    d. Transfer the cuvette/chamber to the chamber of the electroporation apparatus. Gently place the cuvette into the cell in the safe-rack, cover the chamber safe and turn the electric-shock pointer toward the cell containing the bacterial cell chamber.

    e. Connect the power from the voltage booster to the chamber safe, and turn on power for both the pulse control and the voltage booster.

    f. Press the charge button on the pulse control up to 365. When the DC voltage goes down to 345 to 350, turn the button from "charge" to "arm". Quickly push the tigger button for 1 s. The voltage goes down to <10 and the voltage booster should read 1.9 to 2.0 kV.

    g. Turn off the power and carefully take the chamber from the cell. The transformed cell suspension should still be positioned between the positive and negative electrode poles.

4. Place the cuvette on ice for 15 min to allow the membrane pores to close.

5. Transfer the cells to 10 ml of culture medium and incubate at 37°C in a $CO_2$ incubator overnight.

6. Remove the old medium containing the dead cells and add fresh culture medium. Place in incubator until cells are harvested.

*Notes:* *For optimal transfection under appropriate pulse conditions, the percentage of dead cells should not be more than 50%. Using the trypan blue staining method, the extent of cell death can be readily monitored. Suspend the cells in PBS buffer at approximately $5 \times 10^5$ cells/ml. Mix 0.5 ml of the cell suspension and 0.5 ml of 0.4% (w/v) trypan blue solution in PBS. Incubate for 10 to 15 min at room temperature. Transfer 1 to 2 drops of the trypan blue cell suspension to a haemocytometer and count the stained (dead cells) and unstained cells under a light microscope. Calculate the percentage of surviving cells in total stained and unstained cells.*

## F.    Analysis of Expression of Reporter Gene or Inserted Gene of Interest in Transfected Cells

### 1. Preparation of Cytoplasmic Extract from Transfected Cells for Enzyme Assay

1. Approximately 48 to 72 h after transfection, remove the culture medium from the dish and wash the cells twice with 10 ml of PBS.

2. Remove the washing PBS buffer and add 1 ml of fresh PBS. Loosen and resuspend the cells to the buffer using a sterile rubber policeman.

3. Transfer the cells into a fresh microcentrifuge tube, centrifuge at 100 to 500 × g for 5 min and decant the supernatant.

*Note:*   *Do not centrifuge at top speed; otherwise, cells may rupture.*

4. Resuspend the cells in 0.25 ml of 250 m*M* Tris-HCl, pH 8.0.

5. Lyse the cells by 3 to 4 cycles of freezing (dry ice or dry ice–methanol bath, 5 min) and thawing (37°C, 5 min).

6. Centrifuge at 10,000 ×*g* for 10 min in a microcentrifuge, carefully transfer the supernatant (extract) to a fresh tube. Store the extract at –70°C until use.

## 2. Chloramphenicol Acetyl Transferase (CAT) Enzyme Assay

The CAT enzyme expressed by the CAT reporter gene, which originally comes from bacteria, catalyzes the conversion of [$^{14}$C]chloramphenicol to its 1-acetyl and 3-acetyl derivatives. The derived products are separated from the unconverted compound by thin-layer chromatography (TLC) on plates coated with silica gel, which in turn are then exposed to X-ray film for autoradiography.

1. Add the following components in the order shown below to a microcentrifuge tube:
   - Prepared cell extract (100 μl)
   - 500 m*M* Tris-HCl, pH 8.0 (60 μl)
   - [$^{14}$C]Chloramphenicol (0.025 mCi/ml) (10 μl)
   - *n*-Butyryl coenzyme A (5 mg/ml) (10 μl)
   - **Total volume: 180 μl**

*Note:*   *A standard curve for CAT activity includes one blank with the CAT enzyme.*

2. Incubate at 37°C for the optimum time period (0.5 to 20 h) as determined from prior experiments.

3. Terminate the reaction by adding 0.5 ml of ethyl acetate for the TLC assay and vortex for 1 min.

4. Centrifuge at 10,000 × *g* for 3 min and transfer the upper, organic phase to a fresh tube and dry down the ethyl acetate in a vacuum evaporator.

5. Resuspend the residue in 20 μl of ethyl acetate. Spot 10 μl of each sample onto a silica gel TLC plate and air dry.

6. Place the TLC plate in a tank containing approximately 100 ml of solvent (chloroform:methanol, 95:5). When the solvents migrate up to about 1 cm from the top of the plate, remove the plate and air dry.

7. Expose the plate to X-ray film overnight at room temperature. In order to carry out a quantitative assay, for each sample, cut a square corresponding to the monoacetylated form from the silica plate and place in vial. Add 5 ml of scintillation fluid to the vial and count in a scintillation spectrometer.

## 3. Luciferase Assay

Luciferase encoded by the reporter luciferase gene is based on an oxidation reaction mediated by luciferyl-CoA. Light is produced as a result of this reaction.

1. For each sample, add 20 to 30 μl of the cell extract as described previously to a vial containing 100 to 150 μl of luciferase assay reagent at room temperature.

2. Quickly place the reaction in a luminometer and measure light produced for a period of 10 s over a 2-min period depending on the sensitivity.

*Note:*    *The luciferase activity decreases very rapidly.*

## Reagent Needed

### Luciferase Assay Reagent

> 20 m$M$ Tricine
> 1.07 m$M$ $(MgCO_3)_4Mg(OH)_2)5H_2O$
> 2.67 m$M$ $MgSO_4$
> 0.1 m$M$ EDTA, pH 7.8
> 33.3 m$M$ DTT
> 0.27 m$M$ Coenzyme A
> 0.47 m$M$ Luciferin
> 0.53 m$M$ ATP
> Adjust the pH to 7.8.

## 4. β-Galactosidase Assay

1.  For each sample, add the following components into a microcentrifuge tube in the order shown below:
    - Magnesium solution (3 μl)
    - 0.1 $M$ Sodium phosphate buffer (200 μl)
    - CPRC (4 mg/ml) (66 μl)
    - Cell extract prepared (31 μl)
    - **Total volume: 300 μl**

    **For blank**
    - Magnesium solution (3 μl)
    - 0.1 $M$ Sodium phosphate buffer (200 μl)
    - CPRC (4 mg/ml) (66 μl)
    - 0.25 $M$ Tris-HCl, pH 8.0 (31 μl)
    - **Total volume: 300 μl**

2.  Incubate at 37°C for 30 to 60 min and add 0.7 ml of 1 $M$ $Na_2CO_3$ to stop the reaction.

3.  Measure the absorbance, against blank reference, at $A_{574}$.

## Reagents Needed

### Magnesium Solution

> 0.1 $M$ $MgCl_2$
> 5 $M$ 2-Mercaptoethanol

### 0.1 M Sodium Phosphate Buffer, pH 7.3

### CPRG Solution

> 4 mg/ml chlorophenol red-β-D-galactopyranoside in 0.1 $M$ sodium phosphate buffer

## 5. Isolation of RNA from Transfected Cells for Transcripts Analysis

In order to obtain molecular proof for transformed cells, the expression of transferred genes should be analyzed at the mRNA level in addition to reporter enzyme assay as described above.

1.  Harvest transfected cells by centrifugation at 200 × g for 4 min. Wash the cells and repeat centrifuging three times with PBS buffer.
2.  Resuspend the cells in 5 ml of 4 M guanidinium isothiocyanate buffer and vigorously vortex for 0.5 to 1 min to completely disrupt the cells.
3.  Add 1 volume of water-saturated phenol:chloroform. Shake vigorously for 20 s.
4.  Centrifuge at 9000 × g for 10 min at 4°C. Carefully transfer the top, aqueous phase to a fresh tube.
5.  Repeat phenol:chloroform extraction once as in steps 3 to 4.
6.  Precipitate RNA by adding 0.1 volume of 3 M sodium acetate buffer and 2.5 volumes of chilled 100% ethanol. Place the tube at –20°C for 2 h.
7.  Centrifuge at 10,000 × g for 10 min at 4°C and discard the supernatant. Briefly rinse the pellet with 1 ml of 70% ethanol and dry the pellet under vacuum for 10 min.
8.  Resuspend the RNA in 200 μl of TE buffer.
9.  Add 200 μl of 8 M LiCl solution to the RNA solution and place at –20°C for 3 to 5 h to precipitate the RNA. Centrifuge at 11,000 × g for 15 min at 4°C and carefully decant the supernatant. Briefly rinse the pellet with 2 ml of 70% ethanol and dry the pellet under vacuum for 15 min. Dissolve the RNA in 100 μl of water, 0.1 volume of 3 M sodium acetate buffer, and 3 volumes of chilled ethanol. Place at –20°C for more than 2 h. Repeat step 7. Dissolve the RNA in 50 μl of TE buffer. Take 5 μl of the sample to measure the concentration and quality (see Chapter 2). Store the sample at –20°C until use. (This is an optional step.)

## Reagents Needed

### 4 M Guanidinium Thiocyanate Buffer

> 4 M Guanidinium thiocyanate
> 25 mM Sodium citrate, pH 7.0
> 0.5% Sarcosyl
> 0.7% (v/v) β-Mercaptoethanol
> Sterile-filter and store in a lighttight bottle at room temperature up to 3 months.

### 8 M LiCl Solution

> 8 M LiCl in DEPC-treated dd.H$_2$O
> Sterile-filter.

## 6. Analysis of RNA by Northern or Dot Blot Hybridization

Total RNA isolated as above can be used to carry out dot blot hybridization or Northern blot hybridization to check the presence and size of mRNA expressed by the introduced gene in transfected cells. The probe should be a gene fragment that was used for transfection. If a hybridization signal shows up, then the introduced gene is indeed expressed in the transformants. This is strong evidence for successful transfection. If not, the transfection may have failed or the introduced gene cannot be expressed due to some adverse influence such as an inhibitor. The detailed procedures for Northern blot and dot blot hybridizations are described in Chapter 5.

## 7. Analysis of the Expression of Transfected Gene and Transcription Initiation Site by Primer Extension

Primer extension is a powerful method used to analyze the expression of gene as well as the initiation site of transcription.

1. Based on the nucleotide sequence information from the introduced gene, synthesize an oligonucleotide primer that will hybridize close to the 5′ end of the transcript.

2. End-labeling the primer:
   a. Set up, on ice, the following reaction in a microcentrifuge tube:
      - Oligonucleotide primer (40 pmol) (1.5 µl)
      - Polynucleotide kinase buffer (7.5 µl)
      - [r-$^{32}$P]ATP (190 µCi) (19 µl)
      - T4 Polynucleotide kinase (30 units) (4 µl)
      - Add dd.H$_2$O to a final volume of 75 µl.
   b. Incubate at 37°C for 40 min and add 75 µl of TE buffer, pH 8.0.
   c. Purify the end-labeled probe by the use of Sephadex G50 spin column subjected to centrifugation. Store the eluate containing the probe at –20°C until use.

3. Anneal the labeled primer with specific mRNA in the total RNA isolated previously.
   a. Set up the annealing reaction as follows:
      - 5X Annealing buffer (4 µl)
      - Total RNA (10 to 15 µg) (6 µl)
      - End-labeled primer (4 µl)
      - Add dd.H$_2$O to a final volume of 20 µl.
   b. Cap the tube and heat to 70°C for 3 min in a heating block. Slowly cool the heating block to 35°C and incubate at 35°C overnight.

4. Carry out primer extension:
   a. Set up the following reaction:
      - Annealed primer/RNA mixture (15 µl)
      - Reverse transcriptase buffer (15 µl)
      - 0.1 *M* DTT (15 µl)
      - 10 m*M* dNTP (8 µl)
      - 1 m*M* Actinomycin D (8 µl)
      - AMV reverse transcriptase (15 units)
      - Add dd.H$_2$O to a final volume of 150 µl.
   b. Incubate at 42°C for 1 to 1.5 h and stop the reaction by adding 8 µl of formamide dye mixture and heating at 95°C for 5 min.

5. Carry out separation of primer and extended products by electrophoresis on a 6% polyacrylamide sequencing gel. The detailed procedure for nucleotide sequencing is described in Chapter 8.

6. Expose the gel to X-ray film at –70°C for an appropriate time period and identify the size of the extended product and the initiation site.

## Reagents Needed

### 10X Polynucleotide Kinase Buffer

500 m*M* Tris-HCl, pH 7.6
100 m*M* MgCl$_2$

50 m$M$ DTT
1 m$M$ Spermidine
1 m$M$ EDTA, pH 7.6

## [r-$^{32}$P]ATP

10 µCi/µl, 3000 Ci/mmol

## TE Buffer

10 m$M$ Tris-HCl, pH 8.0
1 m$M$ EDTA, pH 8.0

## 5X Annealing Buffer

2 $M$ NaCl
0.125 $M$ PIPES
5 m$M$ EDTA, pH 6.8

## Reverse Transcriptase Buffer

1 $M$ Tris-HCl, pH 8.3
500 m$M$ KCl
100 m$M$ MgCl$_2$

## AMV Reverse Transcriptase

50 units/µl

## Formamide Dye Mixture

10 m$M$ NaOH
1 m$M$ EDTA, pH 8.0
0.1% (w/v) Bromophenol blue
0.1% (w/v) Xylene cyanol
80% (v/v) Deionized formamide

# G.  Selection of Stable Transformants

Foreign DNA enter the cells after transfection, but only a small portion of the DNA gets transferred into the nucleus, where some of the DNA might be transiently expressed for a few days. An even smaller portion of the DNA, which is introduced in the nuclei of some transfected cells, can integrate into the chromosomes and be expressed in future generations of the cell population, forming stable transformants. The selection of the stable transformants depends on the drug-selectable marker gene fused in the recombinant plasmid, linear DNA construct, or a separate plasmid that can be cotransfected into the cells with the plasmid containing the foreign gene of interest. Generally, the transfected cells are cultured for 24 to 30 h in the medium lacking specific antibiotics in order to allow these cells to express the selectable marker gene. The cells are then subcultured in a 1:5 dilution in a selective medium containing an appropriate drug. The cells are cultured in the selective medium for 2 to 4 weeks with weekly or frequent changes of medium to remove dead cells and cellular debris. Only those cells that bear and express the drug-selectable marker gene can survive in the selective medium. There are a number of selectable markers that have been used for transfection. The common ones are described below.

## 1.    Resistance to Aminoglycoside Antibiotics

If cells are transfected by recombinant gene constructs that contain a bacterial gene (Tn5 or Tn60) for aminoglycoside 3′-phosphotransferase *(aph)*, the stable, dominant transformants can confer resistance to aminoglycoside antibiotics such as kanamycin *(kan)*, neomycin *(neo)* and geneticin (G418). To select the stable transformants, 0.1 to 0.8 mg/ml of G418 (Sigma Chemicals Co.) can be added to complete media. This is the most widely used selection method.

## 2.    Thymidine Kinase

If cell lines lacking the thymidine kinase gene *(tk⁻)* are transfected by recombinant gene constructs containing the *tk* gene, the stable transformants can be selected by culturing them in a medium containing hypoxanthine, aminopterin, and thymidine (HAT). Untransfected cells cannot metabolize this lethal analog and will die.

The HAT stock solution can be prepared as follows. Dissolve 15 g hypoxanthine and 1 mg aminopterin in 8 ml of 0.1 $N$ NaOH solution. Adjust the pH to 7.0 with 1 $N$ HCl. Add 5 mg thymidine and add dd.$H_2O$ to 10 ml. Sterile-filter it. Dilute the stock HAT to 100-fold in the culture medium.

## 3.    Hygromycin-β-Phosphotransferase

Transformants containing the hygromyxin β-phosphotransferase gene can be selected by culturing the transfected cells in a medium containing the antibiotic hygromycin B, which is a protein-synthesis inhibitor. Hygromycin B stock is available from Sigma Chemicals Co. One uses 10 to 400 µg/ml in the culture medium.

## 4.    Tryptophan Synthetase

If cells are transfected with recombinant gene constructs containing the tryptophan synthetase gene *(trp B)*, the stable, dominant transformants can be selected by culturing the cells in a medium lacking tryptophan, an essential amino acid.

# II.    Gene Transfer and Analysis of Expression in Mice

The mouse is a well-established model organism that has been widely used for gene transfer and analysis of gene expression in mammalian systems. As compared with animal-cell culture systems, transgenic mice allow the gene of interest to be integrated into the germline of the founder mice. Therefore, every cell in future generations of transgenic animals contains the introduced gene(s). Therefore, the expression of the transferred gene(s) can be analyzed during different developmental stages of mice as well as in different generations. In addition, the use of drug-resistant marker gene for selection in cultured cell systems is not necessary in transgenic animals. The drawback of generating transgenic mice, however, is that it requires expensive equipment for oocyte microinjection, is time-consuming, and requires animal-care facilities.

The general procedures include the following: construction of chimeric genes; preparation of oocytes; microinjection; reimplantation of injected eggs; generation of transgenic mice, F1 and future generations; and analysis of the introduced gene in transgenic mice at DNA, mRNA, and protein levels.

## A.   Preparation of Chimeric Genes for Microinjection

The general procedures are similar to those previously described in Section I, except that a drug-selectable marker gene is not necessarily included in the chimeric constructs. The purified gene of interest may be directly transferred into eggs since a chimeric gene containing bacterial DNA might affect the expression of the introduced gene in transgenic mice.

## B.   Preparation of Oocytes

### Procedure

1. Four days (94 to 96 h) before harvesting the eggs, induce superovulation of 10 to 15 female mice (3 to 4 weeks old) by injecting 5 units per female of pregnant mare's serum (PMS) as a stimulating hormone.

2. Inject 5 units per female of human chorionic gonadotrophin (HCG) as a stimulating hormone 75 to 76 h after the injection of PMS. This injection should be made 19 to 20 h before harvesting the eggs.

3. After the injection of HCG, individually cage the females with stud males overnight by maintaining on a 12-h light:12-h dark cycle.

4. Check the females for vaginal plugs. The mated females should have visible, vaginal plugs.

5. Sacrifice the superovulated females that have visible, vaginal plugs by cervical dislocation.

6. Briefly clean the body surface with 70% ethanol, dissect the oviducts, and transfer to disposable Petri dishes containing 30 ml of medium A. Carefully transfer the oocytes (the fertilized eggs) from the oviducts to the medium.

*Notes:  Under a dissecting microscope, the fertilized eggs can be identified as having two nuclei that are called pronuclei. The male pronucleus is bigger than the female pronucleus.*

7. Remove medium A, carefully transfer the oocytes to a fresh Petri dish, and incubate the oocytes with hyaluronidase for an appropriate time period to remove the cumulus cells.

8. Extensively wash the oocytes with four changes of medium A.

9. Carefully transfer the oocytes to a fresh Petri dish containing 10 to 30 ml of medium B. Place the dish in a 37°C incubator with 5% $CO_2$/air.

## *Reagents Needed*

### *Medium A*

4.78 m*M* KCl
94.66 m*M* NaCl
1.71 m*M* CaCl$_2$
1.19 m*M* KH$_2$PO$_4$
1.19 m*M* MgSO$_4$
4.15 m*M* NaHCO$_3$
23.38 m*M* Sodium lactate
0.33 m*M* Sodium pyruvate
5.56 m*M* Glucose
20.85 m*M* HEPES, pH 7.4 (adjust with 0.2 M NaOH)
BSA, 4 g/l
Penicillin G (potassium salt), 100,000 units/l
Streptomycin sulfate, 50 mg/l
Phenol red, 10 mg/l
• Dissolve well after each addition.
• Sterile-filter and store at 4°C.
• Warm up to 37°C prior to use.

### *Medium B*

4.78 m*M* KCl
94.66 m*M* NaCl
1.71 m*M* CaCl$_2$
1.19 m*M* KH$_2$PO$_4$
1.19 m*M* MgSO$_4$
25 m*M* NaHCO$_3$
23.38 m*M* Sodium lactate
0.33 m*M* Sodium pyruvate
5.56 m*M* Glucose
BSA, 4 g/l
Penicillin G (potassium salt), 100,000 units/l
Streptomycin sulfate, 50 mg/l
Phenol red, 10 mg/l
• Dissolve well after each addition.
• Sterile-filter and store at 4°C.
• Warm up to 37°C prior to use.

## C.    Microinjection of DNA Constructs into the Oocytes

The fertilized eggs are incubated in a 37°C incubator. Normally, microinjection of foreign DNA into the oocytes should be carried out during the period of 3 to 8 h after harvesting as most of the pronuclei can be readily seen. Later, the nuclei may start to break down before the first cleavage, which cannot be injected. The equipment for microinjection should include:

1.  An injection microscope on a vibration-free table. For example, Diaphot TMD microscope with Nomarski optics is available from Nikon Ltd.

2.  An inverted microscope for setting up the injecting chamber, holding pipette, and needle. A SMZ-2B binocular microscope is available from Leitz Instruments Co.

3.  Two sets of micromanipulators (Leitz Instruments Ltd.)

4.  Kopf needle puller model 750 (David Kopf Instruments)

5.  A finer needle and pipette holder

6.  A Schott-KL-1500 cold light source (Schott)

## *Procedure*

1.  Thoroughly clean a depression slide with teepol-based detergent and extensively rinse it with running distilled water for 20 min. Rinse it with ethanol and allow to air dry. Add a drop of Medium A on the depression slide followed by a drop of liquid paraffin on the top of a medium drop.

2.  Assemble all parts that are necessary for the injection according to the instructions for the microinjection equipment, and carry out pretesting of the procedure prior to injection.

3.  Carefully transfer the eggs into the medium drop using a handling pipette that can draw the eggs inside by suction under the microscope.

4.  Slowly and carefully fill the injection needle with the DNA sample to be introduced by capillary action.

5.  Under the microscope, hold the egg with the holding pipette and use the micromanipulator to carefully insert the needle into the male pronucleus, which is larger than the female pronucleus. Slowly inject the DNA sample into the pronucleus. The pressure is maintained on the DNA sample in the needle by a syringe connected to the needle holder. As soon as the pronucleus swelling occurs, remove the needle. The volume transferred is approximately 1 to 2 pl.

6.  Transfer the injected egg to one side and repeat injections until all the eggs have been injected. About 40 to 60 eggs can be injected in 1 to 2 h. Some eggs may lyse as a result of the procedure. It is recommended to inject approximately 300 eggs a day, and about two thirds may survive for transfer into the oviduct of the female. Twenty to thirty-five pups are expected and four to six of them are transgenic mice.

7.  Carefully transfer the injected eggs into a fresh dish containing medium B and place it in the incubator.

## D.    Reimplantation of the Injected Eggs into Recipient Female Mice and Generation of Founder Mice

The injected eggs can be reimplanted into recipient female mice right after injection (one-cell stage embryo) or after being incubated in the incubator overnight (two-cell stage embryo). It is recommended that the injected eggs be transferred at the one-cell stage so that the introduced DNA integrates into one of their chromosomes at the same site. Therefore, all the cells in transgenic animals contain the foreign DNA. Usually, the introduced DNA integrates randomly at a single site. In some cases, multiple copies of the DNA are arranged head-to-tail and form in some tandem repeats. The detailed mechanism of integration is not well understood.

## *Procedure*

1. Prepare pseudopregnant females by caging four to eight F1 females (6 to 9 weeks old) with vasectomized males (two females per male) on the day they are needed. Thus, the females will be in the correct hormonal state to allow the introduced embryos to implant, but none of their own eggs can form an embryo. The pseudopregnant females can be distinguished by inspecting the vaginal area for swelling and moistness.

2. Carefully make a small incision in the body wall of a half-day pseudopregnant female (e.g., C57B1/CBA F1), and gently withdraw the ovary and oviduct from the incision. Hold the oviduct in place using a clip placed on the fat-pad attached to the ovary.

3. Under a binocular dissecting microscope, transfer the injected eggs or embryos into the infundibulum using a sterile glass transfer pipette. Generally, 10 to 20 eggs can be injected into each oviduct.

4. Carefully place the oviduct and ovary back in place and seal the incision.

5. Allow the foster mothers to recover by briefly warming them under an infrared lamp (do not overheat), and house two or three to one cage. Live offspring are usually born 18 days after the surgery. These initial transgenic offspring are called the founder animals or the GO in terms of genetics.

## E.    Indentification of Transgenic Mice

In order to identify and prove transgenic mice, at least two experimental analyses of experiments must be carried out. One is genomic DNA analysis by the Southern blot or slot blot hybridization to check whether the transferred gene is integrated into the genome of the mouse, using the introduced DNA as a probe. The other is an analysis of the expression of the introduced gene at the mRNA level by Northern blot hybridization using the DNA as probe. If positive signals are shown by Southern and Northern blot hybridizations, the introduced gene is successfully integrated and expressed in the transgenic mice. For proof of a stably transgenic line of mice, these analyses should be carried out for several generations.

### *1. Isolation of DNA from Mouse Tails*

1. Spray ethyl chloride around the mouse tail to partially freeze the tail.

2. Cut a piece of the tail at approximately 0.5 cm from the end of the tail and place into a microcentrifuge tube containing 700 μl of DNA isolation buffer.

3. Repeat steps 1 to 2 for other mice.

4. Add 25 μl of 10 mg/ml of Proteinase K to each tube. Incubate the tubes at 55°C overnight to digest the tail pieces (homogenizing is an optional procedure).

5. Add 700 μl of phenol:chloroform to each tube and shake the tubes on a shaker at 150 rpm for 15 min.

6. Centrifuge at $11,000 \times g$ for 10 min and carefully transfer the top, aqueous phase to fresh tubes.

7. Repeat steps 5 to 6. (Optional.)

8. Extract with 0.7 ml of chloroform, centrifuge at $11,000 \times g$ for 5 min, and transfer the upper, aqueous phase to fresh tubes.

9. Add 0.8 volume of isopropanol to each tube and allow to precipitate at room temperature or at −20°C for 30 min.

10.  Centrifuge at $11,000 \times g$ for 10 min, decant the supernatant, and rinse the pellet with 1 ml of 70% ethanol. Completely dry the DNA pellet under vacuum for 30 min at room temperature.

11.  Dissolve the DNA in 50 µl of TE buffer. Take 5 µl of the sample to measure the DNA concentration at 260 and 280 nm. Normally, 60 to 100 µg of DNA can be obtained for each sample.

## Reagents Needed

### DNA Isolation Buffer

> 50 m$M$ Tris-HCl, pH 8.0
> 100 m$M$ EDTA, pH 8.0
> 100 m$M$ NaCl
> 1% (w/v) SDS

### Phenol:Chloroform

> One part of TE buffer-saturated (pH 8.0) phenol and one part of chloroform

### TE Buffer

> 10 m$M$ Tris-HCl, pH 8.0
> 1 m$M$ EDTA, pH 8.0

## 2. Isolation of RNA from the Organs (Embryo, Brain, and Liver) of the Transgenic Mice Using Lithium Chloride/Urea

*Note:*   *RNA can also be isolated from lymphocytes of blood taken from the tails of transgenic mice.*

1.  Dissect appropriate tissue from transgenic mice.

2.  Homogenize the tissue in 8 ml of LiCl/urea isolation solution per gram of tissue to a fine homogenate.

3.  Sonicate the sample on ice for 2 min to shear the DNA. Place the sample at 4°C overnight to precipitate RNA.

4.  Centrifuge at $9000 \times g$ for 20 min, decant the supernatant, and resuspend the RNA pellet in 4 ml of LiCl/urea isolation solution.

5.  Repeat step 4.

6.  Dissolve the RNA in 3 ml of TE buffer (pH 7.6) containing 0.4% (w/v) SDS. Add 1 volume of phenol:chloroform to the mixture and mix.

7.  Centrifuge at $11,000 \times g$ for 5 min and carefully transfer the upper, aqueous phase to a fresh tube.

8.  Add 0.1 volume of 3 $M$ sodium acetate buffer (pH 5.2) and two volumes of chilled 100% ethanol the tube and allow precipitation to occur at –20°C for 60 min.

9.  Centrifuge at $11,000 \times g$ for 10 min, decant the supernatant and rinse the pellet with 1 ml of 70% ethanol. Completely dry the DNA pellet under vacuum for 30 min at room temperature.

10.  Dissolve the RNA in 40 µl of dd.$H_2O$. Take 5 µl of the sample to measure the RNA concentration at 260 and 280 nm.

## Reagents Needed

### LiCl/Urea Isolation Solution

> 3 $M$ LiCl
> 6 $M$ Urea

### TE Buffer

> 10 m$M$ Tris-HCl, pH 7.6
> 1 m$M$ EDTA, pH 7.6

### TE-Saturated Phenol:Chloroform (1:1)

## 3. Southern Blot, Slot Blot, and Northern Blot Hybridizations

Follow the detailed procedures as described in Chapter 5, except that the probe should be a partial- or full-length sequence of the gene that has been introduced into the mouse eggs.

## 4. Analysis of the Expression of the Reporter Gene by Staining Embryos from Transgenic Mice

1. Kill females by cervical dislocation and dissect out the uterus that contains embryos.

2. Transfer the uterus to a fresh Petri dish containing 20 ml of PBS. Dissect out embryos using a clean pair of forceps.

3. Wash the embryos with PBS. The washed embryos can be frozen in liquid nitrogen for further RNA isolation or used directly for staining for β-galactosidase expression (as in the following steps).

4. Fix the embryos in cold (4°C) fixative solution for 1 to 1.5 hours.

5. Wash the embryos four times in 40 ml of washing solution at room temperature.

6. Stain the embryos overnight in 20 ml of staining solution at room temperature in the dark. Rinse three times in PBS and store at 4°C in PBS. Stained embryos should be blue in color.

## Reagents Needed

### PBS Solution (1 l)

> NaCl (7.6 g)
> $Na_2HPO_4$ (3.8 g)
> $NaH_2PO_4$ (0.42 g)

### Fixative Solution

> 1% (v/v) Formaldehyde
> 0.2% (v/v) Glutaraldehyde
> 0.02% (v/v) NP-40
> 2 m$M$ $MgCl_2$
> 5 m$M$ EDTA
> Make in PBS

*Washing Solution*

     0.02% (v/v) in PBS

*Staining Solution*

     5 m$M$ $K_3Fe(CN)_6$
     5 m$M$ $K_4Fe(CN)_6 \cdot 3H_2O$
     2 m$M$ $MgCl_2$
     0.01% (v/v) Sodium deoxycholate
     0.02% (v/v) NP-40
     1 mg/ml X-gal (5-Bromo-4-chloro-3-indolyl-β-D-galactopyranoside)   diluted
     from stock solution

## F.   Selection of Transgenic Lines by Breeding the Transgenic Mice

    The sexes of transgenic founders are usually distinguished at approximately 3 weeks after the mice are born. The mice become sexually mature and can breed at about 6 weeks. To establish a transgenic line, allow the founder mice to breed with nontransgenic mice. Males can mate with females within 24 h of the birth of a litter, and the female can continue to produce litters at 3-week intervals. After running Southern blot and Northern blot hybridization analyses as described previously, the mice should be allowed to breed again for up to 4 to 5 generations. If the introduced gene is stably expressed in several generations, a transgenic mouse line has been established.

## References

1. **Yang, N. S., Burkholder, J., Roberts, B., Martinell, B., and McCabe, D.,** *In vitro* and *in vivo* gene transfer to mammalian somatic cells by particle bombardment, *Proc. Natl. Acad. Sci. U.S.A.,* 87, 9568, 1990.

2. **Klein, T. M., Wolf, E. D., Wu, R., and Sanford, J. C.,** High-velocity microprojectiles for delivering nucleic acids into living cells, *Nature,* 327, 70, 1987.

3. **Carmeliet, P., Schoonjans, L., Kiechens, L., Ream, B., Degen, J., Bronson, R., Vos, R. D., van den Oord, J. J. Collen, D., and Mulligan, R. C.,** Physiological consequences of loss of plasminogen activator gene function in mice, *Nature,* 368, 419, 1994.

4. **Koleske, A. J. and Young, R. A.,** An RNA polymerase II holoenzyme responsive to activators, *Nature,* 368, 466, 1994.

5. **Wang, Z., Svejstrup, J. Q., Feaver, W. J., Wu, X., Kornberg, R. D., and Friedberg, E. C.,** Transcription factor b (TFIIH) is required during nucleotide-excision repair in yeast, *Nature,* 368, 74, 1994.

6. **Zelenin, A. V., Titomirov, A. V., and Kolesnikov, V. A.,** Genetic transformation of mouse cultured cells with the help of high-velocity mechanical DNA injection, *FEBS Lett.,* 244, 65, 1989.

7. **Shillito, R. D., Saul, M. W., Paszkowski, J., Muller, M., and Potrykus, L.,** High efficiency direct gene transfer to plants, *BioTechnology,* 3, 1099, 1985.

8. **Johnston, S. A.,** Biolistic transformation: microbes to mice, *Nature,* 346, 776, 1990.

9. **Moore, K. A. and Belmont, J. W.,** Analysis of gene transfer in bone marrow stem cells, in *Gene Targeting: A Practical Approach,* Joyner, A. L., Ed., Oxford University Press, Inc., New York, 1993.

10. **Walmsley, M. E. and Patient, R. K.,** Quantitative and qualitative analysis of exogenous gene expression by the S1 nuclease protection assay, in *Gene Transfer and Expression Protocols,* Murray, E. J., Ed., Humana Press, Clifton, NJ, 1991.

11. **Hasty, P. and Bradley, A.,** Gene targeting vectors for mammalian cells, in *Gene Targeting: A Practical Approach,* Joyner, A. L., Ed., Oxford University Press, Inc., New York, 1993.

12. **Wurst, W. and Joyner, A. L.,** Production of targeted embryonic stem cell clones, in *Gene Targeting: A Practical Approach,* Joyner, A. L., Ed., Oxford University Press, Inc., New York, 1993.

13. **Papaioannou, V. and Johnson, R.,** Production of chimeras and genetically defined offspring from targeted ES cells, in *Gene Targeting: A Practical Approach,* Joyner, A. L., Ed., Oxford University Press, Inc., New York, 1993.

14. **Finer, M. H.,** The RNase protection assay, in *Gene Transfer and Expression Protocols,* Murray, E. J., Ed., Humana Press, Clifton, NJ, 1991.

15. **Gossler, A. and Zachgo, J.,** Gene and enhancer trap screens in ES cell chimeras, in *Gene Targeting: A Practical Approach,* Joyner, A. L., Ed., Oxford University Press, Inc., New York, 1993.

Chapter **16**

# Gene Transfer and Expression in Plants

## Contents

# I.   Introduction

Gene manipulation technology has been widely used in plant systems.[1–10] This has made it possible to select "superior" plants carrying genes for desired traits. In recent years, the application of gene transfer techniques has produced significant achievements in generating transgenic plants with desired traits.[2,4–15] The general reasons for producing transgenic plants can be summarized as follows: (1) to improve the resistance of plants to insect pests, and to fungal and virus diseases; (2) to modify the amino acid composition of storage protein in cereals, potatoes, and legumes to increase their nutritional value; (3) to improve the composition and storage life of fruits and vegetables; (4) to generate transgenic plants that are herbicide-resistant; (5) to increase the capacity of crop plants to fix atmospheric nitrogen; (6) to increase the rate of photosynthesis in green plants; (7) to modify the phenotypic characteristics (color, height, flowering time, and size) of food plants; (8) to reduce photorespiration in $C_3$ plants with resultant increase in C fixation; (9) to introduce new pigment colors in petals of flowers such as blue flower color in roses; and (10) other purposes.

One of the unique characteristics of plant cells, as compared with animal cells, is that plants have cell walls, which act as barriers to the introduction of foreign DNA. As a result, the gene transfer methods used in plants are not identical to those applied to animal systems. The initially used gene transfer technique primarily relies on a natural plant-pathogenic bacterium, *Agrobacterium tumefaciens,* to introduce recombinant plasmid DNA into the plant genome. However, this *Agrobacterium*-mediated gene transfer is largely limited to dicotyledonous flowering plants (dicots). Many of major crops, however, are monocotyledons (monocots), which have been demonstrated not to be readily infected by *Agrobacterium.* For this reason, some other techniques have been developed for gene transfer in monocots. Whichever method is used, however, the general principles are the same. Recombinant DNA carrying the gene(s) of interest is transferred by an appropriate method into topitotent cells or tissues that are capable of regenerating into entire plants with leaves, stems, roots, flowers, and fruits. The foreign DNA construct will integrate into the plant genome, generating transgenic plants in

## Table 16–1 Advantages and Disadvantages of Gene Transfer Techniques

| Method | Advantages | Disadvantages |
|---|---|---|
| *Agrobacterium* (Ti and Ri)-mediated gene transfer<br>  Transfer to leaf discs<br>  Transfer to protoplasts<br>  Transfer to roots | Well established and high efficiency for dicots | Not suitable for monocots |
| Polyethylene (PEG)<br>  Promotes protoplast fusion<br>  and the uptake of DNA | Relatively simple | PEG is toxic to humans |
| Electroporation<br>  The use of short electrical<br>  impulses of high field strength<br>  that increases the permeability<br>  of protoplast membranes | High efficiency and used for both dicots and monocots | Up to 50% cells die due to high-voltage damage to plasma membranes of protoplasts |
| Microinjection<br>  Precisely inject recombinant<br>  DNA into specific compartments<br>  of protoplasts | Specific, highly efficient, and used for all plants | Requires special skills, special equipment, and high cost |
| Particle bombardment<br>  Particle gun shoots coated<br>  recombinant DNA into<br>  protoplasts or cells | High efficiency, and DNA is delivered simultaneously into many cells | Requires special skills, expensive equipment, and high cost |

which the foreign genes are expressed. This chapter describes in detail several well-established gene transfer techniques that are used in transformation of plants with emphasis on dicots. Each method has advantages and disadvantages that are summarized in Table 16–1.

The entire process of genetic engineering in plants can be subdivided into several procedures:

1. Cloning, isolation, characterization, and subcloning of the gene of interest to be transferred
2. Selection and/or purification of promoter, enhancer, poly(A) signal, reporter gene, selectable marker gene, and the gene of interest from recombinant plasmids
3. Construction of chimeric or fusion genes
4. Transformation of plant cells or tissue with the chimeric genes
5. Selection and regeneration of transgenic plants
6. Analysis of the expression of the introduced gene in transgenic plants

# II.    Protocols

## Protocol A: Cloning, Isolation, Characterization, and Subcloning of the Gene of Interest

The general procedures are described in Chapters 6 and 7, and are not included within the scope of the present chapter.

## Protocol B: Purification of the Gene of Interest from the Recombinant Plasmids

The gene to be transferred is usually precharacterized, subcloned in an appropriate plasmid, and propagated in an appropriate bacterial strain. In that case, the gene should be removed from the recombinant plasmid by using an appropriate restriction enzyme(s) depending on what kind of endonuclease(s) are used for subcloning. The general process is as follows:

1. Set up, on ice, a standard restriction enzyme digestion as follows:

   **Single-restriction enzyme digestion**
   - Plasmids (20 μg)
   - 10X Appropriate restriction enzyme buffer (10 μl)
   - 1 mg/ml Acetylated BSA (optional) (10 μl)
   - Appropriate restriction enzyme (3.4 units/μg DNA)
   - Add dd.H$_2$O to a final volume of 100 μl.

   **Double-restriction enzyme digestion**
   - Plasmid DNA or DNA to be inserted (20 μg)
   - 10X Appropriate restriction enzyme buffer (10 μl)
   - 1 mg/ml Acetylated BSA (optional) (10 μl)
   - Appropriate restriction enzyme A (3.4 units/μg DNA)
   - Appropriate restriction enzyme B (3.4 units/μg DNA)
   - Add dd.H$_2$O to a final volume of 100 μl.

*Note:*    *For double-restriction enzyme digestions, the appropriate 10X buffer containing higher NaCl concentration than the other buffer may be chosen for the double-enzyme digestion buffer.*

2. Incubate at appropriate temperature depending on the appropriate restriction enzyme (e.g., 37°C) for 2 to 3 h.
3. Purify the DNA fragment of interest by one of the methods below.

## Method 1: Elution of DNA from Agarose Gel Slices

This is a simple, fast, and effective protocol that is successfully used in our laboratory. The recovery is usually 85 to 95% of the amount of DNA that is loaded in the gel. The DNA eluted by this method is very pure and can be directly used for ligation, cloning and/or labeling.

1. Carry out electrophoresis by using 1% low-melting point agarose instead of normal agarose.
2. Stain the gel with EtBr solution (10 μl of 10 mg/ml of EtBr in 100 ml dd.H$_2$O) for 10 min.
3. Transfer the gel to a long-wavelength UV transilluminator (305 to 327 nm) to visualize DNA bands. Excise the band(s) of interest with a clean razor blade and place the slices in a microcentrifuge tube.

*Notes:*    *(1) The gel should not be placed on a short-wavelength (e.g., <270 nm) UV transilluminator because the UV light may cause breakage within the DNA fragment, and thus significantly inhibit the subcloning of the insert of interest. (2) UV light is toxic to the human body. Protective glasses, gloves, and a lab coat should be worn when using UV light. (3) The gel slices containing the DNA bands of interest should be trimmed of excess unstained gel areas as much as possible.*

3.   Add 2 to 4 volumes of TE buffer to the gel slices and completely melt the gel in a 60 to 65°C water bath.

*Notes:*   *The gel slices can be directly melted without adding any TE buffer. The DNA concentration is usually high but total yield of eluted DNA fragment is much lower than when one adds TE buffer to the gel slice.*

4.   Immediately chill the tube on dry ice or place it at –70°C for at least 20 min. Alternatively, add 1 volume of Tris-buffer-saturated (pH 8.0) phenol (pH 7.0), **not phenol/chloroform,** to the melted gel solution. Vigorously mix by inversion for 2 to 4 min and chill the tube in dry ice or place it at –70°C for at least 20 min.

5.   Centrifuge at $12,000 \times g$ at room temperature for 15 min. An alternative way is to thaw the gel mixture by tapping the tube vigorously. It takes about 5 to 10 min to thaw the gel to a resuspension status followed by centrifuging at $12,000 \times g$ for 5 min at room temperature.

6.   Carefully transfer the liquid phase (supernatant) containing the elution DNA fragment to a fresh tube.

*Notes:*   *(1) Although the DNA solution can be directly used for labeling, except that the concentration is low, we strongly recommend that the DNA be precipitated with ethanol and resuspended in a small volume of TE buffer or dd.$H_2O$. (2) Phenol serves to extract agarose and EtBr from the stained DNA. If one does not add Tris-buffered phenol to the melted gel solution, EtBr should be extracted from the DNA in the supernatant three times with 3 volumes of water-saturated* n-butanol.

7.   Add 0.5 volume of TE buffer or dd.$H_2O$ to the phenol phase, mix well, and centrifuge as in step 5. Transfer the supernatant to the tube in step 6.

8.   Precipitate the DNA by adding 0.05 volume of 3 *M* sodium acetate buffer (pH 5.2) and 2.5 volumes of chilled 100% ethanol to the pooled supernatant. Allow precipitation to occur at –70°C for 1 h and centrifuge at $12,000 \times g$ for 10 min at room temperature.

9.   Decant the supernatant. Briefly wash the DNA pellet with 1 ml of 70% ethanol and centrifuge at $12,000 \times g$ for 5 min. Decant the supernatant and dry the DNA under vacuum for 15 min. Dissolve the DNA in an appropriate volume of dd.$H_2O$ or TE buffer. Store at –20°C until use.

## *Method 2: Purification of the Target DNA Using NA45 DEAE Membranes*

1.   Carry out electrophoresis by using 1% low-melting point agarose containing EtBr (10 μl of 10 mg/ml EtBr in 100 ml of agarose gel solution). Use a long-wavelength UV lamp to monitor the separation of the target fragment and the vector as compared with known sizes of DNA standards.

2.   Soak a membrane (Scheicher and Schuell) in 10 m*M* EDTA buffer (pH 7.6) for 15 min at room temperature and store at 4°C until use.

3.   During the electrophoresis, monitor the migration of DNA bands stained by EtBr in the gel using a long-wavelength UV lamp. Make an incision below the DNA band of interest and insert a piece of the prepared membrane into the incision in the gel.

4.   Continue electrophoresis, while monitoring the migration of the stained band, until the DNA fragment migrates onto the membrane.

5.   Remove the membrane strip from the gel and place it in a solution containing 20 m*M* Tris-HCl (pH 8.0), 0.15 *M* NaCl, and 0.1 m*M* EDTA. Shake it for 1 min to remove any agarose.

6.   Transfer the strip to 0.2 to 0.3 ml of elution buffer containing 20 m*M* Tris-HCl (pH 8.0), 1 *M* NaCl, and 0.1 m*M* EDTA.

7.   Incubate the strip at 55 to 68°C for 30 to 60 min with shaking at 60 rpm.

8. Rinse the strip with 0.1 ml of elution buffer and pool the elution fluids together.

9. Extract EtBr four times with 3 volumes of water-saturated *n*-butanol.

10. Precipitate and dissolve the eluted DNA as described in Method 1.

## Protocol C: Selection of Appropriate Vector, Promoter, Poly(A) Signal, Reporter Gene, and Selectable Marker Gene

Transformation of plants was initiated by *Agrobacterium tumefaciens*-mediated gene transfer methodology. *Agrobacterium tumefaciens* is a gram-negative soil bacterium that can cause crown gall disease in the infected plants. The bacterium contains a unique plasmid called the Ti (tumor-inducing) plasmid that contains oncogenes (tumor-inducing genes). The Ti plasmids are approximately 200 to 250 Kb in size. After infection, a portion of the DNA of the Ti plasmid can transfer into the host cell and integrate into the genome of the host cell. That DNA is called transferred DNA or T-DNA and it contains oncogenes. The T-DNA codes for enzymes that synthesize auxin and cytokinin that cause abnormal plant growth, resulting in the proliferation of undifferentiated wound tissue that forms the crown gall, or tumor. The infected plant cells also produce nitrogenous compounds named opines by opine synthase genes that are present in the T-DNA. There are more than six different classes of opines, but the most well-known Ti plasmids are the nopaline- and octopine-type plasmids. The nopaline-type T-DNA is 23 Kb in length and flanked by 25 bp direct repeats called the left and the right border (LB and RB) sequences. The octopine-type T-DNA is approximately 21 Kb in length and contains 25 bp direct repeat (LB and RB) sequences that are quite similar to the LB and RB sequences of the nopaline-type T-DNA. It has been demonstrated that the border sequences play an essential role in the transfer of T-DNA. In order to make the Ti plasmids nononcogenic and widely useful as vectors for gene transfer into plants, the oncogenes in the T-DNA are deleted except for the LB and RB regions. Such plasmids are said to contain "disarmed" T-DNA. The removed oncogene fragment is replaced with a region of pBR322 or other bacterial plasmid DNA fragment, resulting in modification of the Ti plasmids. Since the Ti plasmids are large plasmids that are usually difficult to transfer, binary vectors have been developed. Later, Ti plasmids are extensively modified and fused as a part of chimeric genes that include a promoter, a selectable marker, a reporter gene, and polylinker sites for the insertion of the gene of interest.

In addition to the modified vectors, other regulatory sequences or genes should be prepared for the gene transfer and analysis of the expression of the introduced gene(s). The cauliflower mosaic virus (CaMV) 35S promoter, the 3′ poly(A) signal from the nopaline synthase gene of the Ti plasmid, the chloramphenical acetyl transferase (CAT) gene or the β-glucuronidase (GUS) gene as a reporter gene, and the neomycin phosphotransferase (NPT II) gene or the hygromycin phosphotransferase (HPT) gene as a selectable marker gene are recommended to be appropriately fused together in the recombinant constructs to be transferred (Figures 16.1 and 16.2). These fragments and/or genes are well characterized, have been subcloned in plasmids, and are commercially available. The fragments can be purified from the recombinant plasmid using appropriate restriction enzyme(s).

1. Set up, on ice, a standard enzyme digestion as follows:
   • Plasmid vectors or DNA to be inserted (10 μg)
   • 10X *Sma* I buffer (10 μl)
   • 1 mg/ml Acetylated BSA (optional) (10 μl)
   • Sma I (30 units)
   • Add dd.H₂O to a final volume of 100 μl.

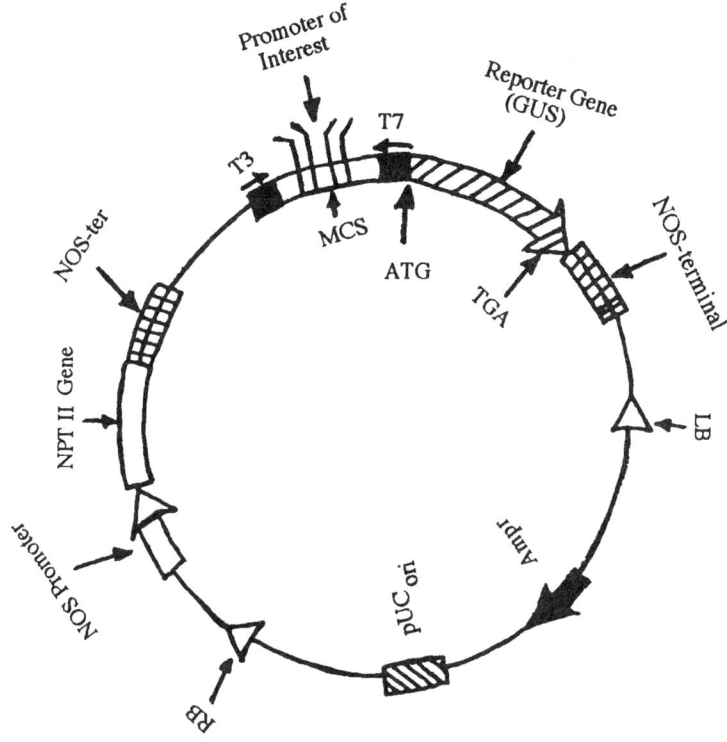

**FIGURE 16.1**
Structural map of an expression vector designed for insertion and analysis of the activity of a specific promoter in plants.

2.   Incubate at 37°C for 2 to 3 h.

*Notes:*   *The digestion efficiency can be checked by loading 1 μg of the digested DNA (10 μl) with loading buffer to a 1% agarose mini-gel. After electrophoresis, the undigested plasmid DNA may reveal multiple bands because of different levels of supercoiled plasmid DNA. However, one band will be visible after* Sma *I digestion.*

3.   Carry out CIAP treatment to remove the 5′ phosphate groups in order to prevent the recircularization of linearized plasmid.
   •   Linearized plasmids (90 μl)
   •   10X CIAP buffer (15 μl)
   •   CIAP diluted in 10X CIAP buffer (0.01 unit/pmol ends)
   •   Add dd.H$_2$O to a final volume of 150 μl.

*Notes:*   *(1) CIAP and 10X CIAP should be kept at 4°C. CIAP treatment should be set up at 0°C. (2) Calculation of the amount of ends is as follows. There is 9 μg digested DNA left after taking 1 μg of 10 μg digested DNA for checking on the agarose gel. If the vector is 3.2 Kb, the amount of ends can be calculated by the use of the formula below:*

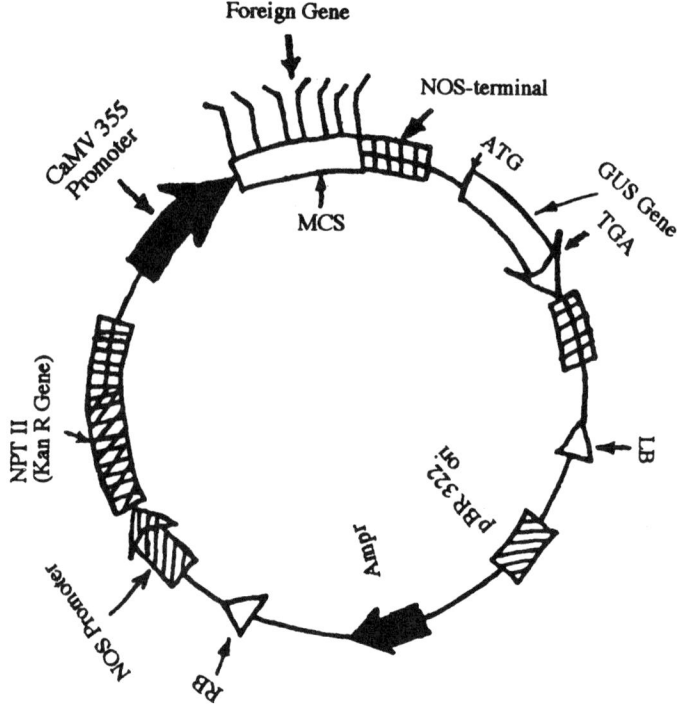

**FIGURE 16.2**
Structural map of an expression vector with a strong promoter, CaMV 35S, which is used for cloning and expression of a foreign gene of interest.

$$\text{pmol ends} = \frac{\text{amount of DNA}}{\left(\text{base pairs} \times 660/\text{base pair}\right)} \times 2$$

$$= \frac{9}{3.2 \times 1000 \times 660} \times 2$$

$$= 4.2 \times 10^{-6} \times 2$$

$$= 8.4 \times 10^{-6} \ \mu M$$

$$8.4 \times 10^{-6} \times 10^{-6} = 8.4 \text{ pmol ends}$$

4.  Incubate at 37°C for 1 h and add 2 μl of 0.5 $M$ EDTA buffer (pH 8.0) to stop the reaction.

5.  Extract with 1 volume of TE-saturated phenol/chloroform. Mix well by vortexing for 1 min and centrifuge at 11,000 × $g$ for 5 min at room temperature.

6.  Carefully transfer the top, aqueous phase to a fresh tube and add 1 volume of chloroform:isoamyl alcohol (24:1) to the supernatant. Mix well and centrifuge as in step 5.

7.  Carefully transfer the upper, aqueous phase to a fresh tube and add 0.1 volume of 3 $M$ sodium acetate buffer (pH 5.2) or 0.5 volume of 7.5 $M$ ammonium acetate to the supernatant. Briefly mix and add 2 to 2.5 volumes of chilled 100% ethanol to the supernatant. Allow to precipitate at –70°C for 1 h or at –20°C for 2 h.

8.  Centrifuge at 12,000 × $g$ for 10 min and carefully decant the supernatant. Briefly rinse the DNA pellet with 1 ml of 70% ethanol and dry the pellet under vacuum for 20 min. Dissolve the DNA pellet in 20 to 40 μl dd.H$_2$O. Take 4 μl of the sample to measure the concentration of the DNA at 260 nm. Store the sample at –20°C until use.

*Note:*    *Adding 0.5 volume of 7.5* M *ammonium acetate to the supernatant at step 7 yields a higher amount of DNA precipitation than adding 0.1 volume of 3* M *sodium acetate buffer (pH 5.2).*

## Reagents Needed

*Appropriate Enzymes*

*10X Appropriate Restriction Enzyme Buffer*

*1% Agarose Mini-Gel*

*TE-Saturated Phenol/Chloroform*

*Chloroform:Isoamyl Alcohol (24:1)*

*3 M Sodium Acetate Buffer, pH 5.2*

*7.5 M Ammonium Acetate*

*Ethanol (100%, 70%)*

*0.5 M EDTA, pH 8.0*

*Calf Intestinal Alkaline Phosphatase (CIAP)*

*TE Buffer*

*10X Sma I Buffer*

        0.1 $M$ Tris-HCl, pH 7.8
        0.5 $M$ KCl
        70 m$M$ $MgCl_2$
        10 m$M$ DTT

*10X CIAP Buffer*

        0.5 $M$ Tris-HCl, pH 9.0
        10 m$M$ $MgCl_2$
        1 m$M$ $ZnCl_2$
        10 m$M$ Spermidine

## Protocol D: Preparation of Blunt-End DNA:

If both the vector and DNA insert have sticky or overhanging ends at the right sites, they can readily ligate to each other. If not, it is usually necessary to generate blunt ends on the DNA fragments prior to their being ligated to the blunt-ended vectors. Blunt ends can be made by filling in 5′ overhangs with DNA polymerase I (Klenow fragment).

1.  Set up the 5′ overhang reaction in a microcentrifuge tube on ice as follows:
    *   DNA fragment or vector DNA (2 to 4 μg in 4 μl)
    *   10X 5′ overhang buffer (2 μl)
    *   2 m$M$ dNTP mixture (2 μl)
    *   DNA polymerase I (Klenow fragment) (2 to 3 units)
    *   Add dd.H$_2$O to a final volume of 20 μl.
2.  Incubate at room temperature for 20 min and heat at 70°C for 5 min to stop the reaction.
3.  Add 1 volume of TE buffer (pH 7.5) and carry out phenol:chloroform extraction and precipitation as described in Protocol C.

## Reagents Needed

### 10X 5′ Overhang Buffer

> 500 m$M$ Tris-HCl, pH 7.2
> 100 m$M$ MgSO$_4$
> 1 m$M$ DTT
> 500 μg/ml BSA

### 2 mM dNTP Mixture

> 2 m$M$ Each of dATP, dCTP, dGTP, and dTTP in 20 m$M$ Tris-HCl, pH 7.5

### DNA Polymerase I (Klenow Fragment)

> 1 unit/μl

## Protocol E: Preparation of Insert DNA Lacking the 5′ Phosphate Groups

If the insert DNA lacks the 5′ phosphate groups, the following reaction should be carried out prior to its being ligated to blunt, dephosphorylated vectors.

1.  Add phosphate groups to the 5′ end of the oligonucleotides by kinase reaction.
    *   Blunt-end, insert DNA (15 μl)
    *   10X Polynucleotides kinase buffer (3 μl)
    *   10 m$M$ ATP (1.5 μl)
    *   100 m$M$ DTT (1.5 μl)
    *   T4 Polynucleotide kinase (15 units)
    *   Add dd.H$_2$O to a final volume of 30 μl.
2.  Incubate at 37°C for 40 min and heat at 70°C for 10 min to inactivate the enzyme.
3.  Add 1 volume of TE buffer (pH 7.5) and carry out extraction and precipitation as described in Protocol C.

## Reagents Needed

### T4 Polynucleotide Kinase

　　　1 unit/μl

### 10X Polynucleotide Kinase Buffer

　　　500 m$M$ Tris-HCl, pH 7.5
　　　100 m$M$ MgCl$_2$
　　　50 m$M$ DTT
　　　1 m$M$ Spermidine
　　　1 m$M$ EDTA

### 10 mM ATP

## Protocol F: Construction of Chimeric Genes by Ligating Plasmid Vector and Insert DNA

In order to achieve optimal ligation, the ratio of vector to insert DNA (1:1, 1:2, 1:3, and 3:1 molar ratio) is strongly recommended to be optimized using a small-scale reaction. The following reaction is standard for the ligation of a 3.2 Kb plasmid vector and a 3.0 Kb insert DNA.

1.　Calculate the molar weight of vector and insert DNA:

$$1\ M \text{ plasmid vector} = 3.2 \times 1000 \times 660 = 2.112 \times 10^6$$

$$1\ M \text{ insert DNA} = 3 \times 1000 \times 660 = 1.98 \times 10^6$$

2.　Calculate the molar ratio of vector to insert DNA:

| Vector DNA:insert DNA | Amount of DNA (μg) | |
|:---:|:---:|:---:|
| | Vector | Insert |
| 1:1 | 1 | 0.792 |
| 1:2 | 1 | 1.584 |
| 1:3 | 1 | 2.376 |
| 3:1 | 1 | 0.264 |

3.　Set up the ligation reactions on ice:

| Components | Ligation reactions | | | |
|---|:---:|:---:|:---:|:---:|
| | 1 (1:1) | 2 (1:2) | 3 (1:3) | 4 (3:1) |
| Plasmid DNA as vector (μg) | 1 | 1 | 1 | 1 |
| Insert DNA (μg) | 0.792 | 1.584 | 2.376 | 0.244 |
| 10X Ligase buffer (μl) | 1 | 1 | 1 | 1 |
| T4 DNA ligase (Weiss units) | 4 | 4 | 4 | 4 |
| Add dd.H$_2$O to (μl) | 10 | 10 | 10 | 10 |

*Notes:*    *Restriction enzyme-digested plasmid (vector) and insert DNA should be dissolved in*
        *dd.H₂O (nuclease-free) to 0.5 to 1.0 µg/µl. If the DNA is less than 0.4 µg/µl, the DNA*
        *should be precipitated to dissolve in about 1 µg/µl.*

4.  Incubate the reactions at 4°C for 12 to 24 h, or 16°C for 4 to 6 h, or at room temperature (22 to 25°C) for 1 to 2 h.

*Note:*    *After the ligations are finished at the above temperatures, the mixture can be stored*
        *at 4°C until use.*

5.  Check the efficiency of ligation by 1% agarose electrophoresis. When the electrophoresis is complete, photograph the gel stained with EtBr under UV light. As compared with unligated vector or insert DNA wells, a highly efficient ligation should allow one to visualize less than approximately 10% unligated vector and insert DNA by estimation of the intensity of fluorescence. Approximately 90% of the vector and insert DNA are ligated to each other and show strong band(s) with molecular weight shifts as compared to vector and insert DNA sizes. By comparing the efficiency of ligation using different molar ratios, the optimal conditions can be determined with ease. This can then be used as a guide for large-scale ligation.

*Note:*    *The above small-scale ligation is optional, but it is strongly recommended.*

6.  Large-scale ligation of vector and insert DNA is based on the optimal conditions determined by the small-scale ligation. For example, if one uses a 1:2 molar ratio of plasmid DNA:insert DNA as the optimal ligation, a large-scale ligation reaction is carried out as follows:
    *   Plasmid DNA as vector (3 µg)
    *   Insert DNA (4.75 µg)
    *   10X Ligase buffer (3 µl)
    *   T4 DNA ligase (15 Weiss units)
    *   Add dd.H₂O to a final volume of 30 µl.

Incubate the ligation reaction at 4°C for 12 to 24 h, or 16°C for 4 to 6 h, or at room temperature (22 to 25°C) for 1 to 2 h. Store at 4°C until use.

## Reagents Needed

### Dephosphorylated Linearized Plasmid Vectors

### Phosphorylated DNA Fragments

### 5X Ligase Buffer

> 0.25 $M$ Tris-HCl, pH 7.5
> 50 m$M$ MgCl$_2$
> 5 m$M$ ATP
> 5 m$M$ DTT
> 25% (v/v) PEG, 8000

### T4 DNA Ligase

> 2 Weiss units/µl

## Protocol G: Transformation of an Appropriate Strain of Bacteria in order to Amplify the Recombinant Plasmids

The general procedures for transformation of bacteria and selection of transformants are as described under the section on subcloning of plasmid DNA in Chapter 7, except for the transformation of *Agrobacteria*.

## Method 1: Introduction of Recombinant Plasmids into Agrobacterium tumefaciens (Biotype 1: Ach5, A6, B6, C58, T37, Bo542, and 15955) by Transformation

1. Inoculate bacteria on a fresh LB plate and incubate at 29°C overnight.

2. Inoculate a single colony of the bacteria in 50 ml of LB medium and culture at 29°C with shaking at 150 rpm.

3. Dilute the culture to an $OD_{660}$ of 0.02 by adding fresh medium and incubating at 29°C for 4 h with shaking at 150 rpm.

4. Centrifuge at $5000 \times g$ for 5 min and wash the cell pellet twice with 5 volumes of 10 m$M$ Tris-HCl buffer (pH 7.5). Repeat centrifugation.

5. Decant the supernatant and resuspend the cells in 0.1 volume of rich medium.

6. Mix 10 µl (3 to 5 µg) of recombinant plasmids with 20 µl of cell suspension and incubate the mixture at –70°C or in a dry ice–ethanol bath for 5 to 8 min.

7. Transfer the mixture to 37°C for 2 min and incubate at room temperature for 30 min.

8. Dilute the suspension with 70 µl of fresh medium and incubate at 29°C for 1 h followed by plating onto a selection medium containing appropriate antibiotics.

9. Select transformants and culture them for isolation of plasmids.

## Method 2: Introduction of Plasmids into Agrobacteria by Conjugation

1. Inoculate the bacteria onto a fresh LB plate and incubate overnight at 29°C.

2. Individually inoculate donor and recipient cells into 10 ml of rich liquid medium in a 125-ml sterile flask and culture overnight at 29°C with shaking at 150 rpm.

3. Dilute the culture to an $OD_{660}$ of 0.05 to 0.1 by adding fresh medium and incubating at 29°C for 4 h with shaking at 150 rpm.

4. Mix equal parts of the donor and recipient cultures.

5. Spread 50 to 100 µl of the mixture onto a sterile membrane filter on top of an agar layer with rich growth medium in a Petri dish.

6. Incubate for 10 to 12 h at 29°C and remove the filter with sterile forceps. Suspend the bacteria in a 0.9% NaCl solution by vigorous shaking.

7. Plate the dilutions of the suspension onto selective medium plate containing appropriate antibiotics.

8. Purify the transconjugants by single-colony isolation and characterize the transconjugants by the plasmid content.

## Reagent Needed

### LB Medium (Rich Medium)

    1% (w/v) Tryptone
    0.5% (w/v) Yeast extract

0.8% (w/v) NaCl
Adjust the pH to 7.0. Autoclave.

## Protocol H: Transformation of Plants with Recombinant Gene Constructs

### Method 1: Gene Transfer by Agrobacterium tumefaciens

Leaf disc transformation, mediated by A. *tumefaciens* containing recombinant Ti plasmids is a classic, well-established method used for gene transfer in plants.

1.  Remove six to eight of the youngest, fully expanded leaves from plants grown in a greenhouse or under sterile conditions. Peel off the lower epidermis using forceps and place the leaf tissue in a Petri dish.

2.  Sterilize the leaves from greenhouse-grown plants by immersing the leaves in 5 to 10% Clorox solution for 5 to 10 min followed by thoroughly rinsing with 40 ml sterile distilled water four to five times to remove the Clorox.

3.  Punch leaf discs with a sterile paper punch or sterile cork borer.

4.  Immerse the leaf discs in overnight-cultured A. *tumefaciens* (1:20 dilution in medium A, $5 \times 10^8$ cells/ml) bearing recombinant Ti plasmids. Allow to inoculate overnight at 28°C followed by gentle blotting to dryness.

5.  Prepare nurse culture plates by adding 1.5 ml of a tobacco cell suspension culture as a "feeder layer" to 25 ml of medium B in a Petri plate. Swirl the plate so that the medium is evenly covered with the cell suspension and cover with an appropriate size of sterile Whatman No. 1 filter paper.

6.  Transfer the inoculated leaf discs upside-down on the nurse culture plates and allow to incubate for 2 to 4 days at 28°C.

7.  Transfer the explants to medium C and incubate at 28°C for several weeks depending on the plants species.

8.  Transfer the entire explants to medium D when shoots appear and incubate at 28°C until young shoots are visible.

9.  Individually cut shoots from the callus and place upright in medium D in order to induce root formation at 28°C.

10. Remove the plantlets from the plates, wash the agar, and plant in sterile soil in 2.5-in. pots. Place the pots in GA7 boxes and close tightly to keep humidity for 8 to 10 days.

11. Transfer the boxes to a greenhouse, slowly open the lid to gradually reduce the humidity, and allow the plants to acclimatize to the ambient humidity.

12. Fertilize plants and allow to grow in a standard greenhouse.

## Reagents Needed

### MS Vitamins (for 100 ml)

Thiamine-HCl (2 mg)
Pyridoxine-HCl (10 mg)
Nicotinic acid (10 mg)
Glycine (40 mg)
Aliquot into appropriate containers and store at −20°C until use.

### B-5 Vitamins (for 100 ml)

Inositol (2000 mg)
Thiamine-HCl (200 mg)

Pyridoxine-HCl (20 mg)
Nicotinic acid (20 mg)

## Preparation of Media (for 1 l)

| Components | Medium | | | |
|---|---|---|---|---|
| | A | B | C | D |
| Inorganic salts (Gibco BRL) | 4.3 g | 4.3 g | 4.3 g | 4.3 g |
| MS vitamins | — | — | 2.5 ml | 2.5 ml |
| B-5 vitamins | 5 ml | 5 ml | — | — |
| myo-Inositol | — | — | 100 mg | 100 mg |
| Sucrose | 30 g | 30 g | 40 g | 40 g |
| Gel-rite | — | — | 1.5 g | 2 g |
| Phytagar | 8 g | 8 g | — | — |
| Hormones | | | | |
| 1-Naphthyleneacetic acid (NAA) (1 mg/ml) | — | 2 ml | — | — |
| 6-Benzylaminopurine (BAP) (1 mg/ml) | — | 0.5 ml | — | — |

*Note:* Dissolve well in 700 ml distilled water, adjust the pH to 5.6 with 1 N KOH for each medium and add distilled water to 1 liter. Autoclave. When the medium has cooled to approximately 55°C, add kanamycin sulfate to each specific medium: 100 mg to media A, B, C, and D.

## Method 2: Direct Gene Transfer to Protoplasts by Electroporation

The plant cell wall is a barrier to the introduction of foreign DNA into plant cells. To increase the efficiency of transformation, electroporation is widely employed to transfer genes of interest into plant protoplasts, which are without cell walls.

### Procedure a: Isolation of Protoplasts

1. Remove six to eight of the youngest, fully expanded leaves from plants grown in a greenhouse or under sterile conditions. Peel off the lower epidermis using a pair of forceps and place the leaf tissue in a Petri dish.

2. Sterilize leaves from greenhouse-grown plants by immersing them in 5 to 10% Clorox solution for 5 to 10 min followed by thoroughly rinsing with 40 ml sterile distilled water four to five times to remove the Clorox.

3. Add 5 to 10 ml of sterile enzyme medium, mix, and incubate in the dark at room temperature (24 to 25°C) for 18 to 20 h without shaking.

*Notes:* At this stage, the cell walls are hydrolyzed by hydrolytic enzymes such as cellulase, hemicellulase, and pectinase. This can be monitored by looking at them under a microscope. Hydrolyzed cells have only protoplasts.

4.  Add 15 ml of washing medium and gently shake to loosen the protoplasts from undigested leaf material.

5.  Filter through a nylon mesh (50 μm pore diameter) to remove undigested materials. The protoplasts are in the filtrate solution.

6.  Centrifuge the protoplasts at $100 \times g$ for 5 min at room temperature and carefully decant the supernatant.

7.  Resuspend the protoplasts in 4 ml of washing medium and centrifuge as in step 6.

8.  Resuspend the protoplasts in 1 ml washing medium, add 1 ml of 18% sucrose as an underlayer in the protoplast suspension, and centrifuge at $120 \times g$ for 5 min at room temperature.

9.  Carefully transfer the protoplasts from the interface, using a wide-bore Pasteur pipette, to a fresh centrifuge tube and add 1 ml of cell suspension buffer.

10. Count the protoplasts under a microscope using a haemacytometer.

11. Centrifuge at $100 \times g$ for 5 min at room temperature and resuspend the protoplasts at approximately $5 \times 10^7$ protoplasts/ml in the cell suspension buffer. Proceed to Procedure b.

## Reagents Needed

### Phosphate-Buffered Saline (PBS)

NaCl (8 g)
KCl (0.2 g)
$Na_2HPO_4$ (1.44 g)
$KH_2PO_4$ (0.24 g)
Dissolve well after each addition in 800 ml dd.$H_2O$.
Adjust pH to 7.4 with 2 N HCl and add dd.$H_2O$ to 1 liter.
Autoclave and store at room temperature.

### Clorox Solution

5 to 10% (v/v) Clorox in dd.$H_2O$

### CPW-Salts Solution (1 l)

$KH_2PO_4$ (27.2 g)
KI (0.16 mg)
$CuSO_4 \cdot 5H_2O$ (0.025 mg)
$KNO_3$ (0.101 g)
$MgSO_4 \cdot 7H_2O$ (0.246 g)

### Enzyme Medium

9% (w/v) Mannitol
3 m$M$ 2-(N-Morpholino)-ethane-sulfonic acid (MES)-KOH, pH 5.8
1% (w/v) Cellulase
0.2% (w/v) Macerozyme
Make up in CPW-salts solution.

### Washing Medium

3 m$M$ MES-KOH, pH 5.8
2% (w/v) KCl
Make up in CPW-salts solution. Autoclave.

### Sucrose Solution

> 18% (w/v) Sucrose
> 3 m$M$ MES-KOH, pH 5.8
> Make up in CPW-salts solution. Autoclave.

### Cell Suspension Buffer

> 10 m$M$ Tris-HCl, pH 7.6
> 100 m$M$ EDTA, pH 8.0
> 20 m$M$ NaCl

## Procedure b: Transfer of Gene Constructs into Protoplasts by Electroporation

1. Measure the resistance of the protoplast suspension by adding 0.35 ml of the suspension to the chamber of the electroporator. Add an appropriate amount of $MgCl_2$ solution to adjust the resistance from an initial 1–4 to 1–1.1 kV.

2. Heat shock the protoplasts at 45°C for 5 min and cool to room temperature, after which they should be placed on ice.

3. Aliquot 250 µl of the protoplast suspension into three to five sterile or disposable tubes. Add 20 µl of DNA solution to be transferred and 125 µl of PEG solution. Mix well and let stand for 10 to 15 min.

4. Transfer the samples to the chamber of the electroporator and pulse three times at 10-s intervals with an initial pulse field strength of 1.4 kV/cm.

5. Transfer each sample to a 6-cm diameter Petri dish in a laminar flow hood and let stand for 10 min.

6. Add 3 ml of a 1:1 mixture of K3 and H media containing 0.6% (w/v) SeaPlaque agarose to the dish. Gently mix the protoplasts in the agarose medium and allow to harden without any disturbance. Proceed to Procedure c.

## Procedure c: Selection and Regeneration of Transgenic Plants

1. Seal the dishes containing the electroporated protoplasts with Parafilm to prevent potential contamination. Incubate at 24°C for 24 h in the dark followed by 4 to 6 h of continuous dim light (500 lux), depending on the plant species.

2. Cut the agarose-containing protoplasts into small quadrants using a clean razor blade and place the agarose blocks from one dish into a culture vessel (10 cm in diameter and 5 cm in depth) containing 50 ml of medium A with 50 µg/ml of kanamycin sulfate for selection of transformants. Culture at 24 to 28°C in continuous dim light (500 lux) with shaking at 80 rpm. Resistant clones will be visible 3 to 4 weeks later, depending on the plant species.

3. When the clones are 2 to 3 mm in diameter (5 to 6 weeks after culture), transfer the clones to medium A containing 0.8% (w/v) agar with 30 g/ml mannitol and 50 µg/ml of kanamycin sulfate. Allow the colonies to grow for 2 to 5 weeks, depending on the plant species.

4. Transfer the colonies to medium A without mannitol and allow to grow for 2 to 3 weeks.

5. Induce the root formation by culturing the colonies in medium A without mannitol but with 20 µg/ml sucrose and 0.25 µg/ml BAP (6-benzylaminopurine) hormone present. Incubate the dishes in the dark for 1 week followed by illumination at 3000 to 5000 lux until shoots are generated from the callus.

6. Cut off the shoots (1 to 2 cm long) from the callus and place on medium B lacking hormones in order to produce roots. When the shoots are 3 to 5 cm long, proceed to step 7.

7. Gently wash away the agar once the root system is established and transfer the plantlets to pots of soil. They can then be gradually grown in a regular greenhouse.

## Reagents Needed

### PEG Solution

24% (w/v) Polyethylene glycol (6000) in 0.4 $M$ mannitol buffer (pH 5.6) containing 20 m$M$ $MgCl_2$

### Prepare media in the order shown below:

| Components | Medium A | B | K3 | H |
|---|---|---|---|---|
| 1. Macroelements (mg/ml) | | | | |
| $KNO_3$ | 1010 | 900 | 2500 | 1900 |
| $KH_2NO_3$ | 136 | 170 | — | 170 |
| $NH_4NO_3$ | 800 | 1650 | 250 | 600 |
| $NaH_2PO_4 \cdot H_2O$ | — | — | 150 | — |
| $CaCl_2 \cdot 2H_2O$ | 440 | 440 | 900 | 600 |
| $MgSO_4 \cdot 7H_2O$ | 740 | 370 | 250 | 300 |
| $(NH_4)_2 \cdot SO_4$ | — | — | 134 | — |
| $NH_4 \cdot$ Succinate | 50 | — | — | — |
| 2. Microelements (µg/ml) | | | | |
| $Na_2EDTA$ | 74.6 | 74.6 | 74.6 | 74.6 |
| $FeCL_3 \cdot 6H_2O$ | 27 | 27 | 27 | 27 |
| $H_3BO_3$ | 3 | 6.2 | 3 | 3 |
| KI | 0.75 | 0.83 | 0.75 | 0.75 |
| $MnSO_4 \cdot H_2O$ | 10 | 16.9 | 10 | 10 |
| $MnSO_4 \cdot 7H_2O$ | 2 | 8.6 | 2 | 2 |
| $CuSO_4 \cdot 5H_2O$ | 0.025 | 0.025 | 0.025 | 0.025 |
| $Na_2MoO_4 \cdot 2H_2O$ | 0.25 | 0.25 | 0.25 | 0.25 |
| $CoCl_2 \cdot 6H_2O$ | 0.025 | — | 0.025 | 0.025 |
| $CoSO_4 \cdot 7H_2O$ | — | 0.03 | — | — |
| 3. Vitamins and organics (µg/ml) | | | | |
| *myo*-Inositol | 100 | 100 | 100 | 100 |
| Vitamins | | | | |
| Biotin | — | — | — | 0.01 |
| Pyridoxine HCl | 1 | | 1 | 1 |
| Thiamine HCl | 10 | 0.04 | 10 | 10 |
| Nicotinamide | — | — | 1 | — |
| Nicotinic acid | 1 | — | 1 | — |
| Folic acid | — | — | 0.4 | — |

| Components | Medium | | | |
|---|---|---|---|---|
| | A | B | K3 | H |
| D-Ca-Pantothenate | — | — | 1 | — |
| p-Aminobenzoic acid | — | — | 0.02 | — |
| Choline chloride | — | — | 1 | — |
| Riboflavin | — | — | 0.2 | — |
| Ascorbic acid | — | — | 2 | — |
| Vitamin A | — | — | 0.01 | — |
| Vitamin D3 | — | — | 0.01 | — |
| Vitamin B12 | — | — | 0.02 | — |
| Glycine | — | — | 0.1 | — |
| 4. Sugars (mg/ml) | | | | |
| Sucrose | 30 | 30 | 103 | 0.25 |
| Glucose | — | — | 68.4 | — |
| Mannitol | 50 | — | 0.25 | — |
| Sorbitol | — | — | 0.25 | — |
| Cellobiose | — | — | 0.25 | — |
| Fructose | — | — | 0.25 | — |
| Mannose | — | — | 0.25 | — |
| Rhamnose | — | — | 0.25 | — |
| Ribose | — | — | 0.25 | — |
| Xylose | — | — | 0.25 | — |
| 5. Hormones (µg/ml) | | | | |
| 2,4-Dichlorophenoxy-acetic acid (2,4-D) | — | — | 0.1 | 0.1 |
| 1-Naphthyleneacetic acid (NAA) | 0.1 | — | 1 | 1 |
| 6-Benzylaminopurine (BAP) | 1 | — | 0.2 | 0.2 |
| 6. Adjust pH | | | | |
| Final pH | 5.7 | 5.8 | 5.8 | 5.8 |

## Method 3: Direct Gene Transfer by Microprojectile Bombardment

Microparticle bombardment, using a biolistic or particle gun, is the most powerful technique used today in direct gene transfer into plant cells. This method involves the acceleration of heavy microparticles coated with recombinant genes. This approach may potentially solve all gene transfer problems in plant systems. It has several major advantages over traditional methods which are as follows: (1) it is relatively simple and easy to handle; (2) one shot can simultaneously transfer genes into many cells; (3) target cells can be cultured cells, pollen, and those in differentiated tissues or meristems of different plant species; and (4) random hits of microparticles can reach competent cells and increase the frequency of stable transformation. The disadvantage, however, is that this method requires expensive bombardment equipment.

## Procedure a: Transfer of Genes into Leaf Tissue

1. Remove two to six fully expanded, young leaves from plants of interest, which have grown under sterile conditions or in a greenhouse. In the latter case, the leaves must be surface-sterilized in 5 volumes of 10% Clorox solution for 10 to 15 min, in a laminar flow hood, followed by being thoroughly rinsed four to five times in 5 volumes of sterile distilled water. Peel off the lower epidermis of the leaf using a pair of jeweler's forceps (optional step).

2. Slice the leaves into approximately $1 \times 0.5$-cm pieces using a clean razor blade. Transfer the excised leaf pieces onto Grade 617 Whatman filter paper in disposable Petri dishes ($60 \times 20$ mm) containing 15 ml callus medium with 100 μg/ml kanacycin (medium A). Orient the leaf pieces in the center of each dish for maximal exposure to bombardment.

3. Carry out macroprojectile bombardment as follows: [steps (a)-(j) should be completed prior to step 3]

   a. Sterilize 100 mg tungsten (1.2-μm microprojectiles) in 1.5 ml of 95% ethanol in a sterile 15-ml centrifuge tube for 5 min.

   b. Sonicate on ice for 10 min with a continuous pulse using a 20% duty cycle at level 2 output.

   c. Transfer the sonicated microprojectiles into a microcentrifuge tube and centrifuge at $12,000 \times g$ for 2 min. Decant the ethanol supernatant and gently resuspend the pelleted microprojectiles in 1.5 ml dd.$H_2O$.

   d. Centrifuge as in step c. Decant the supernatant, add 1.5 ml dd.$H_2O$, and recentrifuge.

   e. Remove the supernatant, resuspend in 1.5 ml dd.$H_2O$, and sonicate the vial. Aliquot 25 μl of the samples into microcentrifuge tubes. Resonicate after every two aliquots to maintain uniform bead concentration for each aliquot.

   f. Add 10 μl of gene constructs (1 μg/μl) to each aliquot of microprojectiles and mix well.

   g. Add 25 μl of 2.5 $M$ $CaCl_2$ to each DNA/microprojectile mixture and mix well.

   h. Add 10 μl of 100 m$M$ spermine to the mixture, mix by vortexing, and let it set for 20 to 30 min.

   i. Centrifuge at $12,000 \times g$ for 2 min and carefully remove the supernatant to a final volume of 30 μl.

   j. Sonicate the DNA/microprojectile mixture and pipette 1.5 to 2.0 μl onto a sterile macroprojectile. Resonicate after every two aliquots.

   k. Place the macroprojectiles in the gun barrel and the power level 1 blank in the chamber. After inserting the stopping plate and tissue sample in place, attach the detonator and draw vacuum. Fire the gun when the vacuum reaches 68 to 71 cm Hg. (See the manufacturer's instructions for details; DuPont Company or GTE Products Corporation).

4. After bombardment, transfer the dishes to a growth room at 28°C with a 12 to 16-h day length at 100 μE m$^{-2}$s$^{-1}$, depending on the plant species, and maintain the bombarded leaf strips on the callus medium (medium A) for 2 to 3 weeks. Callus colonies should be visible.

5. Transfer the leaf strips to regeneration medium (medium B) and allow to grow for 2 to 4 weeks with transfer to fresh medium every 2 weeks. Plantlets will develop from the callus.

6. In a laminar flow hood, separate the plantlets from the callus by cutting their bases with a sterile razor blade. Transfer the plantlets in sterile Flow boxes containing 50 ml of root-inducing medium with 50 μg/ml kanamycin (medium C). Wrap the boxes with parafilm in order to prevent contamination. Allow the plantlets to grow in the boxes until root formation occurs and shoot growth is sufficient for the explants to be transferred to a greenhouse, depending on the particular plant species. Changing to fresh medium is not necessary.

## Procedure b: Selection of Stably Transgenic Plants

1.  Transfer the generated plants (Ro) from the tissue culture environment (Procedure a: step 6) under water to 10-cm pots containing a porous soil mix of Terra-lite/Redi-Earth/Peat-lite mix:Perlite (1:2). Allow the plants to grow and gradually acclimatize by placing them in a mist chamber set at 25 to 28°C day/20°C night with a 12 to 16-h day length at 200 μE $m^{-2}s^{-1}$ light intensity, depending on the particular plant species.

2.  Transfer the established plants to 3.8-l pots and grow in a greenhouse to maturity. Fertilize the plants two to three times a week with fertilizer (e.g., Peters' 20–20–20 N/P/K fertilizer) and feed blood meal once a week.

3.  Allow flower buds from each plant to self-pollinate and others to cross as males or females to control plants. Initiate test cross-pollinations by removing the anthers from the designated female parent and pollen from the designated male parent. Transfer the dehisced anther with forceps and touch it to the receptive stigma.

4.  Harvest seeds from self- and cross-pollinated plants and carry out surface sterilization as previously described.

5.  Allow the seeds (16) to germinate in $100 \times 25$-mm Petri dishes containing 15 to 25 ml of medium E with 200 μg/ml kanamycin. Plant seeds on medium D lacking kanamycin to check germination frequency.

6.  Transfer the dishes to a growth room and allow to germinate over a period of 2 to 3 weeks at 28°C under approximately 200 μE $m^{-2}s^{-1}$ of light and a 12 to 16-h day length.

7.  Score the phenotypes of the germinated seedlings as kanamycin-resistant (e.g., green) and control without kanamycin (e.g., white). Calculate the genetic ratios and analyze the data using $X^2$ (Chi square statistical method). For example, the phenotypic ratios for differing numbers of independently segregating genes can be expected as follows:

    - For self-pollinated plants: one gene, 3 green:1 white; two genes, 15 green:1 white; three genes, 63 green:1 white.

    - For cross-pollinated plants: one gene, 1 green:1 white; two genes, 3 green:1 white; three genes, 15 green:1 white.

## Reagents Needed

### MS Vitamins (for 100 ml)

> Thiamine-HCl (2 mg)
> Pyridoxine-HCl (10 mg)
> Nicotinic acid (10 mg)
> Glycine (40 mg)
> Aliquot and store at −20°C until use.

### B5 Vitamins (for 100 ml)

> Inositol (2000 mg)
> Thiamine-HCl (200 mg)
> Pyridoxine-HCl (20 mg)
> Nicotinic acid (20 mg)

### Preparation of Media (for 1 l)

| | Medium | | | | |
|---|---|---|---|---|---|
| Components | A | B | C | D | E |
| Inorganic salts (Gibco BRL) | 4.3 g | 4.3 g | 4.3 g | 4.3 g | 4.3g |
| MS vitamins | — | — | 2.5 ml | 2.5 ml | 2.5 ml |
| B-5 vitamins | 5 ml | 5 ml | — | — | — |
| *myo*-Inositol | — | — | 100 mg | 100 mg | 100 mg |
| Sucrose | 30 g | 30 g | 40 g | 40 g | 40 g |
| Gel-rite | — | — | 1.5 g | 2 g | 2 g |
| Phytagar | 8 g | 8 g | — | — | — |
| Hormones | | | | | |
|   1-Naphthyleneacetic acid (NAA) (1 mg/ml) | — | 2 ml | — | — | — |
|   6-Benzylaminopurine (BAP) (1 mg/ml) | — | 0.5 ml | — | — | — |

*Note:* Dissolve well in 700 ml distilled water, adjust the pH to 5.6 with 1 $N$ KOH for each medium and add distilled water to 1 liter. Autoclave. When the medium has cooled to approximately 55°C, add kanamycin sulfate to each specific medium: 100 mg to media A and B, 50 mg to medium C, and 200 mg to medium E.

## Protocol I: Proof of Stable Transformation

In order to avoid possible artifacts of transformation, stable transgenic plants should be established for several generations. Some genes transferred into the nucleus of a plant cell can be expressed to produce mRNA and proteins in the first generation of transgenic plants. However, the introduced genes may not be integrated into the genome of plant cell, resulting in the absence of their expression in second or further generations. Therefore, it is necessary to establish stable transformation using appropriate antibiotics in the medium and then find strong evidence to verify the existence of transgenic plants. Generally, the following analyses should be carried out for clear proof of transformation.

## Evidence 1: Phenotypic and/or Functional Examinations

This is a straightforward observation of transgenic and nontransgenic plants. For example, if an insect pest-resistant gene was transferred into crop plants, stable transgenic plants (usually two to four generations) should reveal resistance to attack by the insect as compared with control plants. If a red pigment leaf color gene is introduced into a plant, the stable transgenic plants should have red leaves. In another words, if the phenotype and functions of stable transgenic plants are expected, the gene of interest has been successfully transferred into and expressed in the transgenic plants. However, this is not the only proof because the introduced gene may somehow not be expressed in transgenic plants. In that case, further evidence should be provided.

## Evidence 2: Analysis of the Expression of the Reporter Gene

A gene that encodes for β-glucuronidase (GUS) is usually used as a reporter gene in the chimeric gene constructs for transformation. The enzyme hydrolyzes 4-methyl-umbelliferyl-β-D-glucuronide (MUG) and produces 4-methyl-umbelliferone (4 MeU), generating a blue fluorescence. The enzyme can be assayed by quantitative measurement or by tissue staining.

### a. Quantitative Assay of GUS Activity

1. Homogenize or grind 5 to 10 mg of tissue from transgenic and nontransgenic plants in 0.1 to 0.2 ml of lysis buffer until a fine homogenate is obtained. The lysis buffer is made up of 50 m$M$ sodium phosphate (pH 7.0), 10 m$M$ EDTA, 0.1% (v/v) Triton X-100, 10 m$M$ 2-mercaptoethanol, and 0.1% (w/v) sarcosyl.

2. Transfer the homogenate to a microcentrifuge tube and centrifuge at $12,000 \times g$ for 15 min at 4°C. Transfer the supernatant to a fresh tube and store at 4°C until use.

3. Carry out protein measurement (see Chapter 3).

4. Determine the background fluorescence from nontransgenic tissue, which may reveal nonspecific hydrolysis of the substrate. Determine the maximum amount of protein from the protein extracts, which can be used for the GUS assay using pure GUS as a positive control.

5. Carry out the fluorescence assay for the sample as follows:

   a. Add protein extract to two microcentrifuge wells containing 2 and 4 μg total proteins, respectively.

   b. Add lysis buffer to final volume of 45 μl in each well.

   c. Add 5 μl of MUG to start the reaction.

   d. Cover the plate and incubate at 37°C for 30 min.

   e. Add 150 μl of 0.2 $M$ $Na_2CO_3$ to terminate the reaction.

   f. Read the fluorescence and substract the blank value.

   g. Calculate the specific activity (units/ng protein/min) of GUS in the tissue based on the total proteins and incubation time.

### b: Staining Assay Followed by Light Microscopy

1. Fix whole plant seedlings or slice tissues in 25 m$M$ sodium phosphate buffer (pH 7.0) containing 0.1 to 1% (v/v) glutaraldehyde for 30 min at 4°C or at room temperature.

2. Wash the tissue for 3 min five times in 25 m$M$ sodium phosphate buffer (pH 7.0).

3. Quickly and completely cover the tissue with the substrate mixture and vacuum filter. Incubate in the dark at 37°C overnight or for 1 day. The substrate mixture can be made from stock solution to a working concentration of 10 m$M$ sodium phosphate buffer (pH 7.0), 0.5 m$M$ potassium ferricyanide, 0.5 m$M$ potassium ferrocyanide, and 1 m$M$ X-glucuronide (5-bromo-4-chloro-3-indolyl glucuronide).

4. Rinse the tissue for 5 min twice in sodium phosphate buffer until the tissue shows an intense blue color.

5. Directly observe and photograph the stained tissue under a microscope without further processing. To improve the image, the pigments (chlorophyll) can be removed by passing the tissue through 25, 50, 70, 95, 100, 100, and 100% ethanol with 15 min per step. The tissue can be rehydrated through ethanol series and progressively infiltrated with glycerol. Remove any air bubbles by final vacuum filtration.

## Evidence 3: Analyses of Southern Blotting and Sequencing, and Determination of the Copy Number of the Gene

Genomic DNA should be isolated from transgenic plants and digested with appropriate restriction enzymes. The fragments are separated by agarose gel electrophoresis followed by transfer to a nitrocellulose or a nylon membrane. The membrane is then hybridized with a labeled DNA probe that is the introduced gene or gene fragment. Stable transgenic plants should show a positive signal as compared with nontransgenic plants as control. Further, the positive fragment(s) of DNA may be sequenced. If the sequence shows both host and foreign DNA sequences, the introduced gene has been integrated into the chromosomes of the host cell. This is strong evidence of stable transformation. The detailed procedures for Southern blot hybridization and DNA sequencing are described in Chapters 5 and 8.

## Method a: Determination of the Copy Number of the Gene Integrated in the Genome of Transgenic Plants

1.  Isolate genomic DNA from transgenic plants and measure the concentration of DNA in the sample. The general procedures are described in Chapter 1.

2.  Calculate the quantity of DNA that corresponds to a specific copy number. One copy of the gene per haploid genome is equivalent to the size (bp) of the transferred gene (g) divided by the size (bp) of the genome (G), or g/G. Therefore, if 1 µg of the genomic DNA is used for the experiment, the quantity of the transferred gene will be equal to g/G × 1 µg.

3.  Set up copy number standards of known gene equivalents of the query sequence. If the query sequence (bp), the gene, or cDNA insert (I) is purified from a recombinant plasmid, the amount (N, µg) that is equivalent to one copy per genome of the target sequence can be calculated as: I × N = g/G × 1 µg. If the I (bp) is in the cloning vector such as plasmid (V, bp), the N (µg) can be calculated as: I/V × N = g/G × 1 µg. Prepare a set of copy number standards that is equivalent to 1, 2, 3, 4, 5, 10, and 15 copies of the target or transferred gene.

4.  Carry out restriction enzyme digestion of 1 µg of genomic DNA using an enzyme that cannot cut within the target sequence.

5.  Load the digested genomic DNA and a set of copy number standards into separate wells of an agarose gel. Carry out electrophoresis and Southern blot hybridization using purified, labeled cDNA, the monomeric sequence of a repetitive DNA family, or a linearized plasmid DNA containing the insert as a probe.

6.  Analyze the detected band(s) by scanning with an integrating densitometer. The copy number of the transferred gene in the genome can be estimated by measuring the extent of hybridization signals of genomic bands and comparing the intensity of the signal with those of known gene standards. Thus, the total copy number can be determined with ease. With a family of tandem repeats, the sum of the extent of all hybridized bands is the total number of gene copies, as determined from the copy number standards.

## Method b: Determination of the Copy Number of the Dispersed, Repetitive DNA Sequence

1.  Carry out dot blot or slot blot hybridization as described in Chapter 5.

2.  Scan the autoradiograph with a densitometer.

3.  Calculate the percent (T%) of DNA that is complementary to the probe or gene transferred as follows:

$$T\% = \frac{Sg}{Svi} \times \frac{Avi}{Ag} \times \frac{Lv}{Li} \times 100$$

where Sg represents the signal intensity of genomic DNA; Svi refers to the signal intensity of vector including the DNA insert; Avi is the amount ($\mu$g) of vector plus DNA insert used for the dot blotting; Ag is the amount ($\mu$g) of genomic DNA used for the dot blotting; Lv represents the size (bp) of vector DNA minus the insert; Li is the size (bp) of insert DNA used as the probe in the dot blot hybridization.

4.  Determine the copy number of dispersed, repetitive DNA sequences as follows:

$$\text{Copy number} = \frac{T\% \times \text{genome size (bp)}}{\text{monomer size (bp)}}$$

## Method c: Determination of Multigene Families

1.  Digest the genomic DNA with an appropriate restriction enzyme that does not cut within the target gene or DNA insert used as a probe.

2.  Carry out electrophoresis, Southern blot hybridization, and autoradiography as described in Chapter 5.

3.  Analyze the data as follows:

    a.  If one band is detected, the target gene or DNA exists once in the gemone, or it may be a highly homogeneous multigene family.

    b.  If multiple, discrete bands appear, and if the distance between the bands is larger than the size of the known gene, a multiple gene family may exist. But the multiple bands may not derive from a multiple gene family if the sizes of the bands are very close to each other, or if the restriction enzyme cuts within introns.

    c.  If there are no discrete bands, but rather, a smear appears, the target gene or DNA is a highly repetitive, dispersed DNA family.

    d.  If the sizes of a ladder of bands are equal to multiples of the smallest fragment, the target gene or DNA is composed of a tandemly organized set of repeated units, in which the monomeric length is probably equal to the smallest DNA fragment size.

    e.  If a multiple set of bands is superimposed on a smear, the probe may detect a tandemly repeated gene family that has dispersed units.

## Evidence 4: Northern Blot Analysis

The purpose of gene transfer is to obtain expected gene expression in transgenic plants. This can be carried out by Northern blot hybridization. RNA can be isolated from transgenic plants and separated by formaldehyde-denaturing agarose gel electrophoresis followed by transfer to nitrocellulose or nylon membrane. The membrane is then hybridized with a labeled DNA probe that is the introduced gene or gene fragment. Stable transgenic plants should show a positive signal as compared with nontransgenic control plants. The detailed procedures for Northern blot hybridization analysis are described in Chapter 5.

## Evidence 5: Western Blot Hybridization or Immunocytochemical Localization

This method provides direct evidence for the expression of a transferred gene in transgenic plants. Protein extracts can be prepared from transgenic and nontransgenic plants and separated by SDS-PAGE. The separated proteins are then transferred onto a nitrocellulose membrane followed by being probed with an antibody against the protein encoded by the introduced gene. A positive signal should be evident in stable transgenic plants but not in nontransgenic plants. In addition, the expression of the introduced gene can be checked by immunocytochemical localization *in situ*. That will give detailed information as to where the protein is localized in the cell or the tissue. The detailed procedures for Western blot and immunolocalization analyses are described in Chapters 5 and 19.

# References

1. **Gruber, M. Y and Crosby, W. L.,** Vectors for plant transformation, in *Methods in Plant Molecular Biology and Biotechnology,* Glick, B. R. and Thompson, J. E., Eds., CRC Press, Inc., Boca Raton, Florida, 1993.

2. **Miki, B. L., Fobert, P. F., Charest, P. J., and Iyer, V. N.,** Procedures for introducing foreign DNA into plants, in *Methods in Plant Molecular Biology and Biotechnology,* Glick, B. R. and Thompson, J. E., Eds., CRC Press, Inc., Boca Raton, Florida, 1993.

3. **Greenberg, B. M. and Glick, B. R.,** The use of recombinant DNA technology to produce genetically modified plants, in *Methods in Plant Molecular Biology and Biotechnology,* Glick, B. R. and Thompson, J. E., Eds., CRC Press, Inc., Boca Raton, Florida, 1993.

4. **Potrykus, L.,** Gene transfer to plants: assessment of published approaches and results, *Annu. Rev. Plant Physiol. Plant Mol. Biol.,* 42, 205, 1991.

5. **Gasser, C. S. and Fraley, R. T.,** Genetically engineering plants for crop improvement, *Science,* 244, 1293, 1989.

6. **Heikkila, J. J.,** Use of *Xenopus* oocytes to monitor plant gene expression, in *Methods in Plant Molecular Biology and Biotechnology,* Glick, B. R. and Thompson, J. E., Eds., CRC Press, Inc., Boca Raton, Florida, 1993.

7. **Brunke, K. J. and Meeusen, R. L.,** Insect control with genetically engineered crops, *Tre. Biotechnol.,* 9, 197, 1991.

8. **Saul, M. W., Shillito, R. D., and Negrutiu, L.,** Direct DNA transfer to protoplasts with and without electroporation, in *Plant Molecular Biology Manual,* Vol. A1, Kluwer Academic Publishers, Dordrecht, 1988.

9. **Christou, P., Ford, T. L., and Kofron, M.,** Production of transgenic rice (*Oryza sahva* L.) plants from agronomically important Indica and Japonica varieties via electric discharge particle acceleration of exogenous DNA into immature zygotic embryos, *BioTechnology,* 9, 957, 1991.

10. **Vaeck, M., Regnaerts, A., Hofte, H., Jansens, S., De Beuckeleer, M., Datn, C., Zabeau, M., Van Montagu, M., and Leemans, J.,** Transgenic plants protected from insect attack, *Nature,* 328, 33, 1987.

11. **Cullis, C. A., Rivin, C. J., and Walbot, V.,** A rapid procedure for the determination of the copy number of repetitive sequences in eukaryotic genomes, *Plant Mol. Biol. Rep.,* 2, 24, 1984.

12. **Koncz, C., Martini, N., Mayerhofer, R., Koncz-Kalman, Z., Korber, H., Redei, G. P., and Schell, J.,** High-frequency T-DNA-mediated gene tagging in plants, *Proc. Natl. Acad. Sci. U.S.A.,* 86, 8467, 1989.

13. **Jefferson, R. A.,** Assaying chimeric genes in plants: the GUS gene fusion system, *Plant Mol. Biol. Rep.,* 5, 387, 1987.

14. **Fobert, P. R., Miki, B. L., and Iyer, V. N.,** Detection of gene regulatory signals in plants revealed by T-DNA-mediated fusions, *Plant Mol. Biol.,* 17, 837, 1991.

15. **McCabe, D. E., Swain, W. F., Martinell, B. J., and Christou, P.,** Stable transformation of soybean *(Glycine max)* by particle acceleration, *BioTechnology,* 6, 923, 1988.

# Chapter 17

# Plant Cell Culture

## Contents

Plant cells or tissues can be cultured *in vitro* on agar in the form of dedifferentiated cell masses called callus tissue (see Chapter 18) or in a liquid medium as single cells or small cell clumps (cell suspension cultures). Suspension-cultured plant cells are especially attractive materials for bioreactors to produce valuable medicines and other useful metabolites. The advantages of growing the suspension-cultured cells in a bioreactor over culturing callus are (1) that suspension-cultured cells usually grow more rapidly than callus tissues; (2) that suspension-cultured cells are more homogeneous than cells in a callus; and (3) that every cell in a liquid medium can come in contact with elicitors or precursors for the metabolite production in the medium directly, while only a small portion of cells in calli can be in contact with the agar medium, and thus, major populations of cells in callus tissue can absorb elicitors or precursors only indirectly.

In this chapter, we will describe several basic techniques used for the induction and maintenance of plant suspension cultures. These techniques include: (1) the initiation of suspension cultures; (2) the determination of cell growth; (3) differentiation of suspension-cultured cells; and (4) production of secondary metabolites. Cell suspension cultures are usually initiated from callus tissue or differentiated explants such as hypocotyls. Liquid medium (MS, B5, and LS media) containing 2,4-D (1 to 2 mg · l⁻¹) is worth testing first as an initiation medium if no culture medium is established for the suspension culture of a particular plant species. When the cell suspension culture is to be initiated from callus, one should consider how to break the callus into single cells or small cell clumps. One way to achieve this is to use friable portions of callus tissue. In this case, cells are easily dispersed in the liquid media by one or two subcultures with continuous shaking at about 100 to 125 rpm. A second way to obtain cell suspensions is to culture hard callus in continuously shaken liquid media and keep subculturing it by transfer of the callus tissue to fresh media until it becomes friable. In some cases, the friability of callus tissue increases suddenly after many passages of subculturing. Sometimes, mechanical disruption of callus tissue by the use of a spatula when the callus tissue is subcultured is helpful to speed up the breakage of callus tissue into small cell clumps. Continuous shaking of liquid cultures at about 100 to 125 rpm itself is one way to disrupt callus mechanically due to the shearing forces of swirled liquid. However, if the shaking speed is too high, cells may be damaged due to too much shearing force. Another important attribute of continuous shaking of cell suspension cultures is aeration. Since water contains only limited amounts of oxygen, and its diffusion from the air into the liquid phase is a slow process, stationary liquid-cultured cells will develop under the stress of anoxia (absence or reduced levels of oxygen) that leads to cell death eventually. The initial density of explants or callus is important for successful induction of cell suspension cultures. When a cell suspension is initiated from differentiated tissue, a few grams of explants per 100 ml of liquid media should be inoculated. If cultures are overdiluted (cell density is too low), cell growth becomes very slow.

Once a cell suspension culture is established, it is maintained by regular subculturing once in every one to two weeks. During this maintenance phase, one should keep monitoring cell shape and growth. Cell growth is usually measured by three methods that include counting cell number, measuring dry weight, and measuring packed cell volume.

Cell suspension cultures are used for the regeneration of plantlets and for other purposes, such as the production of valuable secondary metabolites. To achieve these purposes, cultured cells need to be differentiated at the organ level or cellular level. At the organ level, cells undergo a process called somatic embryogenesis, which is one of the two pathways to produce plant regenerants. Organogenesis is briefly discussed in Chapter 18. At the cellular level, cytodifferentiation, such as chloroplast development, is worthwhile to discuss. Since many secondary metabolites are synthesized in plastids, establishing cell cultures with differentiated plastids is necessary for the production of such secondary metabolites. On the other hand some secondary metabolites can be produced by the addition of precursors to cell cultures. In this chapter, several examples of the establishment of photosynthetic cell cultures and the production of secondary metabolites from cultured plant cells are provided.

# I. Initiation of Plant Cell Suspension Cultures

1.  Cell suspension cultures of carrot
    a.  Induce carrot callus to form from sterilized roots, hypocotyls, and cotyledons on agar containing Gamborg's B5 medium supplemented with 0.1 to 1 mg · l-1 2,4-D (Chapter 18).
    b.  After 3 weeks, transfer 5 to 10 calli (about 1 g fresh weight) to 20 ml liquid media of same composition in a 125-ml flask.
    c.  Incubate under continuous light at 26°C with shaking at 100 rpm.
    d.  Subculture regularly (at 2-week intervals) by decanting half of the old medium and replacing it with fresh medium by the use of a sterile pipette. Watch for the formation of a cell suspension.
    e.  After the carrot cell suspension culture is established, continue to subculture by taking 5 ml of cell suspension and inoculating it in 45 ml of fresh medium once every week.
2.  Cell suspension cultures of rice
    a.  At 10 to 20 days after anthesis or shedding of pollen (milk stage, embryos are 0.5 to 1 mm long), remove panicles and husks (palea and lemma) followed by sterilization as described in Chapter 18.

b. Separate embryos from endosperms carefully. Use of a stereo microscope is recommended.

c. Place 10 embryos on agar containing MS media supplemented with 2.5 mg · l⁻¹ 2,4-D (per 9-cm Petri dish) in which the scutellum (cotyledon) surfaces face upwards. Orientation of embryos is important for successful callus induction.

d. Select visually friable embryogenic callus (small clumps whose size is less than 5 mm in diameter). Friable rice callus tissue is yellowish-white with "loose" morphology.

e. Place calli in 10 ml of liquid medium containing N6 media plus 2 mg · l⁻¹ 2,4-D.

f. Incubate callus tissues in the dark at 26°C on the shaker (100 rpm).

g. For the first 2 weeks, subculture once every 2 to 3 days by removing all of the medium and replacing it with fresh media by the use of a sterile pipette.

h. From 2 weeks to about 2 months, subculture twice a week. During this period, selection of fine cell clumps should be achieved by the use of a wide-bore pipette. Care must be taken not to include large clumps and necrotic (brown or black dead) cells.

# II. How to Determine Cell Growth and Viability

1. Counting cell numbers

   a. Take 1 ml of cell suspension culture and mix with 4 ml of 12% (w/v) chromium trioxide ($CrO_3$).

   b. Heat at 70°C for about 20 min until the cells are stained; then disrupt cell clumps by vigorous pumping of the culture with a syringe.

   c. Spin down the cells by the centrifugation at about 3000 rpm for 5 min.

   d. Decant supernatant carefully by the use of a Pasteur pipette and resuspend the pellet of cells in 1 ml of water.

   e. Count the cell number by the use of a hemocytometer and a light microscope.

2. Determining cell dry weight

   a. Take 5 ml from each cell suspension culture and pass through a preweighed filter paper or glass filter.

   b. Wash the filter residue twice with distilled water and dry the residue and the filter overnight at 80°C.

   c. After cooling down to room temperature in a desiccator, measure the weight of the dried cells and filter paper and calculate the dried cell weight by subtracting the weight of the filter paper from the weight of the filter paper plus dried cells.

3. Determining packed-cell volume

   a. Sample 5 ml of cells from a cell suspension culture and put this aliquot of cells in a 15-ml graduated conical centrifuge tube.

   b. Centrifuge the cells down at $2000 \times g$ for 5 min and measure the volume of the packed cells.

*Notes:* *During culture, cell growth should be monitored regularly every 3 to 4 days. If the volume of the cell suspension is too small to harvest cells, one way to measure cell growth is to use a flask with a side arm on the side wall of the flask. When the packed-cell volume is to be measured, cells settle down on the bottom of the side arm for a fixed period of time, and one can thus measure the height of the cell mass by the use of a ruler.*

# III.  Differentiation of Suspension-Cultured Plant Cells

1.  Embryogenesis in carrot cell suspension cultures
    a.  Prepare embryogenic cell suspensions as described above in the section on carrot cell suspension cultures.
    b.  Pass carrot cells through a 125-μm sterile nylon mesh.

*Notes:*   *Sieving cells through a nylon mesh increases the degree of uniformity by excluding larger cell clumps. To further increase this uniformity, it may be helpful to centrifuge the sieved cells in a 16% Ficoll solution containing 2% sucrose.*

    c.  Collect cells by centrifugation at about $100 \times g$ for 10 min. Then wash the cells twice with liquid MS media lacking auxin (MSO media).
    d.  Resuspend the cells in liquid MS media lacking auxin. Adjust the cell density to $1 \times 10^5$ to $1 \times 10^6$ cells per ml. If the cell density is lower than this range, the addition of 0.1 mg · l$^{-1}$ zeatin and 0.1 mg · l$^{-1}$ ABA (abscisic acid) is helpful for facilitating embryogenesis.
    e.  Incubate cells under ambient light conditions with continuous shaking at about 100 rpm. Embryos usually appear in about one week, and 2 to 3 weeks are sufficient for the maturation of the embryos.
    f.  Transfer the mature embryos on the agar containing MSO media and grow under continuous light. Small plantlets that contain several leaves and roots can be transplanted to sterile vermiculite. Alternatively, the embryo can be "packaged" in alginate beads for enhanced survival when shifting them from suspension culture to soil or vermiculite.

*Notes:*   *Before these plantlets are moved to a greenhouse, a hardening process is necessary since they have been grown in flasks with high relative humidity. One way to achieve this is to grow the plantlets under a vinyl or polyethylene film cover with high humidity and then to successively reduce the humidity.*

2.  Photosynthetic and photoautotropic (green) cell cultures
    a.  Rice[2]
        i.   Sterilize rice seeds as described in Chapter 18 and imbibe seeds in sterile water for 2 h.
        ii.  Carefully excise embryos from endosperms and place them on the agar in Petri dishes containing MS medium supplemented with 3 mg · l$^{-1}$ NAA, 6 to 8 mg · l$^{-1}$ kinetin and 0.8% (w/v) phytoagar (about 10 to 16 embryos per Petri dish).
        iii. Culture embryos under continuous light at 28°C. After 2 to 3 weeks, calli are formed from the scutellum (cotyledon) region of embryos.
        iv.  Visually select the green portions of the callus and initiate cell suspension by culturing about 100 green callus tissues in 20 ml of the liquid MS medium supplemented with 10 g · l$^{-1}$ sucrose, 2 mg · l$^{-1}$ NAA, and 4 mg · l$^{-1}$ kinetin under continuous light with shaking (100 rpm).
        v.   Subculture regularly (twice a week) by decanting the old medium using a sterile pipette and replacing it with fresh medium of the same composition until fine and green cell suspension cell lines are established.
    b.  Soybean
        i.   Prepare Petri dishes containing KT medium for callus induction. KT medium contains MS basal salts, 1% (w/v) sucrose, 1 mg · l$^{-1}$ NAA, 0.2 mg · l$^{-1}$ kinetin, vitamins, and 0.8% Phytoagar. Vitamins are supplied from the stock solution (concentrated 1000 times). 100 ml of the stock solution contains 20 mg nicotinamide, 20 mg pyridoxine·HCl, 10 mg

biotin, 10 mg choline chloride, 10 mg calcium panthothenate, 10 mg thiamine·HCl, 5 mg folic acid, 5 mg p-aminobenzoic acid, 5 mg riboflavin, and 0.015 mg cyanocobalamin.

ii. Prepare sterile soybean seedlings by growing chemically sterilized seeds (Chapter 18) on the agar containing MS basal medium in wide-mouth bottles or Magenta™ vessels under continuous light at 25 to 28°C.

iii. Excise cotyledons from 5- to 6-day-old seedlings and place them on the KT media (6 to 10 cotyledons per Petri dish). Seal the Petri dishes with a strip of Parafilm™ and incubate them under continuous light at 26 to 28°C.

iv. After calli emerge, collect greener parts of the callus tissues by visual selection and initiate cell suspension cultures in liquid KT media. For the initiation of cell cultures, put 5 to 6 pieces of callus tissue into 10 ml of KT media in 125-ml flasks. After closing the mouth of each flask with a double layer of sterile aluminum foil, incubate the flasks on a gyratory shaker (100 to 125 rpm) under continuous light. Cell lines established in this way are called SB-M photomixotrophic soybean cell lines.

v. After the SB-M cell line is established, transfer about 400 mg of cells to 80 ml of $KT^0$ medium (KT without sucrose) supplemented with 5 m$M$ HEPES buffer (pH 7.0) in 250-ml flasks and plug flasks with rubber stoppers. In order to supply $CO_2$, two 16- or 17-gauge needles are placed in rubber stoppers and the ends of both needles are connected to small membrane filter units (membrane pore size = 0.45 μm). This rubber stopper-needles-filter unit should be autoclavable. Connect one of the two filter units to a gas humidifier and further to a gas tank containing 5% $CO_2$ by means of polypropylene tube.

vi. Adjust the flow of 5% $CO_2$ at 10 to 15 ml/min per flask while culturing the cells on the gyratory shaker (100 rpm) under continuous light (200 to 300 mE m$^{-2}$s$^{-1}$).

*Note:* *This protocol is based on experiments performed by Widholm's group.*[1]

# IV. Production of Valuable Secondary Metabolites from *Catharanthus roseus* (Periwinkle) Cell Cultures

1. Callus induction

   a. Sterilize *C. roseus* seeds by rinsing them with 95% ethanol for 3 min followed by swirling them in 20% commercial bleach (sodium hypochlorite solution) for 10 min. Seeds should then be extensively washed by sterile water at least three times.

   b. Germinate the seeds on MS basal media containing 0.8% phytoagar in a wide-mouth bottle under sterile conditions with continuous illumination at about 26°C.

   c. Excise cotyledons and chop hypocotyls (0.5 cm in length) from 5- to 6-day-old seedlings and induce callus to develop from these explants on Gamborg's B5 basal media supplemented with 0.8% phytoagar and 3 mg · l$^{-1}$ 2,4-D.

2. Cell suspension cultures and secondary metabolite production

   a. Initiate suspension cell cultures by culturing about 500 mg callus in 20 ml liquid B5 media supplemented with 2 mg · l$^{-1}$ 2,4-D at 26°C in the dark with continuous shaking at about 100 to 125 rpm.

   b. Subculture by decanting half of the old media and replacing it with fresh media using a sterile pipette once a week until a fine cell suspension culture is established.

   c. In order to produce secondary metabolites, add tryptamine (100 mg · l$^{-1}$) and secologanin (100 mg · l$^{-1}$) to 3-week-old suspension cultures and incubate for 1 week.

*Notes:* *Secondary metabolites produced can be extracted with chloroform. After evaporating chloroform under $N_2$, residues are reconstituted in a one-to-one mixture of benzene and methanol. Metabolites are analyzed either by TLC or HPLC (see Chapter 20).*

# References

1. **Horn, M. E., Sherrard, J. H., and Widholm, J. M.** Photoautotropic growth of soybean cells in suspension culture. 1. Establishment of photoautotropic cultures, *Plant Physiol.,* 72, 426, 1983.

2. **Kim, D., Brock, T. G., Karuppiah, N., and Kaufman, P. B.** Green and non-green callus induction from excised rice *(Oryza sativa)* embryos: effects of exogenous plant growth regulators. *Plant Growth Reg. Soc. Am. Q.,* 20(4), 189, 1992.

# Chapter 18

# Plant Tissue Culture

## Contents

Obtaining transgenic plants that contain foreign genes is very important in the field of plant molecular biology and biotechnology. In order to achieve this, suitable plant tissue culture systems should be developed. Plant tissue excised from plants can be cultured *in vitro* and regenerated to whole plants if the culture media contain suitable nutrients and plant growth hormones. This is due to plants having a unique property called *totipotency*. From different parts of plants, dedifferentiated cell masses called callus are formed usually in the presence of 2,4-dichlorophenoxyacetic acid (2,4-D) at the level of few mg · l⁻¹. Differentiation of these calli to whole plants is achieved by changing culture conditions such as levels and compositions of plant growth hormones. Regeneration of plants from callus tissue is usually achieved either by organogenesis or by somatic embryogenesis. In the case of organogenesis, plant organs and tissues, such as shoots, roots, and vascular tissue connecting shoot and root, are formed independently of each other. On the other hand, plant organs regenerated via somatic embryogenesis are thought to originate from a single cell in a callus or from suspension-cultured cells.

Since large numbers of plant species have been cultured and various culture methods have been developed, it is most practical to follow previously reported protocols. However, if such protocols are not available for a particular plant species, one should try to use several well-known culture media with different auxin and cytokinin combinations. Auxin and cytokinin are two key components used in plant tissue culture. The ratio of these plant growth regulators in a culture medium greatly affects the course of development of a plant tissue. For example, high auxin/low cytokinin is usually a favorable condition for root development, while low auxin/high cytokinin promotes shoot development. Therefore, an optimal ratio of plant growth regulators should be determined by experiments with various auxin/cytokinin ratios. The quantity of these plant growth regulators is also important for successful plant tissue culture. For example, 1 mg · l⁻¹ 2,4-D is better for the callus induction from carrot roots than 5 mg · l⁻¹ 2,4-D. On the other hand, callus induction from rice embryos is very poor in the presence of

1 mg · l⁻¹ 2,4-D but excellent with 2.5 mg · l⁻¹ 2,4-D. Other components in tissue culture media that one should consider are quantities and types of carbon and nitrogen sources. Commonly used nitrogen sources are ammonium and nitrate salts, amino acids, casein hydrolysate, and urea. As a carbon source, sucrose is most widely used. However, other carbon compounds such as glucose or even carbon dioxide ($CO_2$) can replace sucrose in certain types of plant tissue culture.

In this chapter, we describe some basic techniques used in plant tissue culture.[1-6] Even though various plant tissue culture methods have been developed, all of these methods are based on general tissue culture techniques. These techniques include the maintenance of sterile conditions, the preparation of culture media, the initiation of plant tissue cultures, and the control of cultured tissue growth and development.

# I.     Basic Tissue Culture Techniques

## A. Sterile Techniques

1.  **Autoclave** — For most tissue culture media and equipment that are heat stable. Items are heated by hot steam (121°C) under pressure for 20 to 30 min.

2.  **Dry oven** — Especially for flasks and glass pipettes. Items are heated by dry air at 200°C for 2 to 3 h.

3.  **Flame sterilization** — Metal equipment such as scalpels and forceps. Items are dipped in ethanol and the ethanol on the items is burned with a flame or infrared heater.

4.  **Filter sterilization** — For liquid media and stock solutions. Solutions are filtered through sterile membrane filters (pore size: usually 0.2 µm). Pressure or vacuum is applied to force solutions to pass through membrane filters.

5.  **Laminar flow cabinet** — This provides a sterile work-bench area for handling plant tissue to be cultured. Once sterilized, plant tissue must be handled with sterile equipment.

6.  **Chemical sterilization** — For plant materials to be cultured.

    a.  Put plant materials in a sterile flask (100 or 250 ml size).

    b.  Add about 50 ml of 70% (v/v) ethanol and shake for 30 s. Then discard ethanol using a pipette.

    c.  Add 20 to 50 % (v/v) commercial bleach (sodium hypochlorite, 3 to 5%) containing a drop of liquid detergent soap sufficient to submerge plant materials completely. Shake for 10 to 20 min on a rotary shaker at about 100 rpm.

*Notes:* *The optimum period of time for the bleach treatment is determined by trial and error. However, general guidelines are as follows: (1) seeds and other hard plant tissue can be treated longer than leaves and other soft tissues; (2) the treatment should be stopped when small bleached spots appear on leaves; and (3) the minimum period of chemical treatment to obtain sterility is the optimum period.*

    d.  Decant the bleach solution using a sterile pipette.

    e.  Wash plant tissues with sterile water extensively (more than three times).

*Notes:* *(1) Every part of the plant body can be sterilized chemically. However, the best way to obtain sterile plant tissue is to grow plant seedlings aseptically from chemically sterilized seeds. (2) Maintaining aseptic culture conditions is critical for successful tissue culture. In order to achieve this, one should be extremely careful not to allow any possibility of bacterial or fungal contamination. Plant tissues to be cultured should be handled with sterile equipment in a sterile laminar flow cabinet and must be placed on a sterile surface. Using antibiotics is often recommended.*

## B. Preparation of Stock Solutions

1. Plant growth regulators
   a. 2,4-D
      - Dissolve 50 mg 2,4-D in a small volume of ethanol.
      - Dilute to 100 ml with distilled water
   b. NAA ($\alpha$-naphthaleneacetic acid), IAA (indole-3-acetic acid)
      - Dissolve 50 mg NAA or IAA in small volume of 1 $N$ NaOH.
      - Dilute to 80 ml with distilled water.
      - Adjust pH to 7.
      - Adjust volume to 100 ml.
   c. Kinetin, BAP (benzyl aminopurine)
      - Dissolve 50 mg kinetin or BAP in a small volume of 1 $N$ NaOH.
      - Dilute to 100 ml with distilled water.
2. Vitamin stock solution [1000 times more concentrated than that used in Murashige–Skoog (MS) media].
   a. Weigh out 50 mg nicotinic acid, 10 mg thiamine·HCl, 50 mg pyridoxine·HCl, and 200 mg glycine.
   b. Dissolve them in distilled water to make 100 ml.
   c. For B5 media, the vitamin stock solution (1000×) contains 10 mg nicotinic acid, 100 mg thiamine·HCl, and 10 mg pyridoxine·HCl in 100 ml.

*Notes:* *(1)For other types of plant tissue culture media, one should consult commercial or published media formulations and prepare vitamin stock solutions according to those formulations. The concentration of the vitamin stock solution is usually 1000 times higher than the working solution. (2) Stock solutions are usually kept in a refrigerator at 4°C.*

## C. Preparation of Culture Media

1. Dissolve MS basal medium mixture for 1 liter in approximately 800 ml of distilled water (B5 basal medium mixture for Gamborg's B5 medium)

*Notes:* *The MS salt mixture supplied by Sigma Chemical Company (St. Louis, MO), sometimes leaves undissolved salt particles. On the other hand, other companies' MS salt mixtures can be dissolved completely. Even though it has not been known whether such undissolved particles severely affect the plant tissue culture, one can purchase a salt mixture suitable for culture purposes.*

2. Add the following chemicals:
   - 30 g Sucrose
   - 1 ml Vitamin stock solution for MS media or B5 media
   - 100 mg *myo*-Inositol
   - Suitable amounts of plant growth hormones

3.   Adjust the pH of the solution to 5.8.

4.   Adjust the volume to 1 liter.

5.   Pour the solution into a 2-l flask.

6.   Add 8 g phytoagar.

7.   Seal the mouth of the flask with two layers of aluminum foil.

8.   Autoclave at 121°C for about 20 min.

*Notes:*   *Sometimes a given medium contains heat-labile compounds. In this case, media without these compounds should be autoclaved first. After step 9, the medium is supplemented with the heat-labile compounds from filter-sterilized stock solutions.*

9.   Cool the solution to about 50°C.

10.   Dispense media into sterile Petri dishes (about 20 to 30 ml per 100-mm dish).

11.   Wrap dishes with SaranWrap™ and store at room temperature until use.

*Note:*   *If the culture medium contains heat-labile compounds, it is better to use freshly prepared media or to keep the medium in a refrigerator.*

# II.     Callus Induction and Maintenance

## A. Callus Induction from Rice Embryos and Rice Callus Subculture

1.   Prepare dehusked rice seeds by removing husks.

2.   Sterilize seeds chemically as described above.

3.   Place sterile seeds on MS media plus 2.5 mg · l⁻¹ 2,4-D in Petri dishes.

*Notes:*   *Rice seeds are submerged halfway into the medium so that the embryo faces upward and is exposed to the air while the endosperm faces downward. Usually 10 to 16 seeds are placed in each Petri dish.*

4.   Close the Petri dish with its cover and seal the unit with a strip of Parafilm™.

5.   Incubate in the dark at 28°C for 2 to 3 weeks and keep watching for the appearance of callus.

*Notes:*   *Callus is formed from the scutellum (cotyledon or "seed-leaf") tissue of the rice embryo. After the appearance of callus, you may find endosperm, coleoptile, and callus tissue with or without roots.*

6.   Excise calli from other tissues carefully by the use of sterile forceps and transfer them to fresh media that contains the same composition of nutrients and plant growth hormones.

7.   Subculture rice calli regularly every 2 to 3 weeks. When transferred, the callus should be subdivided into smaller pieces (about 0.5 cm³).

## B. Callus Induction from Carrot Roots and Carrot Callus Subculture

1.   Remove roots from carrots grown in the greenhouse or in the field, or obtain roots of carrots from a local food market. Collect transverse segments from the middle part of a root by cutting into 5-cm lengths.

2. Wash in a mild soap solution and rinse with distilled water.

3. Sterilize root pieces chemically as described above.

4. Trim away the end 3 to 5 mm of each explant and slice the remaining tissue into 3- to 5-mm lengths.

5. Remove the epidermis with the use of a scalpel.

6. Cut each slice into four to six pieces and place on agar containing B5 medium plus 1 mg · l⁻¹ 2,4-D (six to nine pieces per Petri dish) and seal the Petri dish with Parafilm™.

7. Incubate the explants in the dark at 24 to 26°C and keep watching for the formation of callus tissue.

8. After callus formation has occurred, it should be removed from the explant and transferred to fresh medium.

9. Continue subculturing regularly by transferring callus pieces to fresh medium of the same composition (every 2 to 3 weeks).

### C. Callus Induction from Tobacco Leaves and Callus Subculture

1. Put tobacco seeds in a bag of nylon mesh and sterilize the seeds chemically by submerging the bag in a 20% commercial bleach for 10 min followed by extensive washes with sterile water.

2. Place the seeds on agar containing MS media without any plant growth regulator (MSO media) and let the seeds germinate under continuous white light.

3. Grow tobacco seedlings until young leaves appear and expand.

4. Remove leaves from sterile tobacco plants and cut into pieces of about 0.5 × 0.5 cm.

5. Place the leaf pieces on the agar containing MS medium supplemented with 1 mg · l⁻¹ NAA and 5 mg · l⁻¹ BAP (4 pieces per Petri dish).

6. Seal the Petri dish with Parafilm™ and incubate in the dark at 27°C and watch for the formation of callus from the cut surfaces of the leaf pieces.

7. Collect callus tissue and subculture on fresh medium as described above.

## III.   Plant Regeneration — Organogenesis

### A. Plant Regeneration from Rice Callus

1. Prepare MS16 medium for rice shoot regeneration from callus tissue.

*Notes:*   *MS 16 medium is basically the same as MS medium, as described above, except that the medium contains 2 mg · l⁻¹ kinetin and 1 mg · l⁻¹ NAA as plant growth hormones, and 1.6% (w/v) phytoagar as solid support. The concentration of phytoagar is higher for plant regeneration than for callus induction since it results in a greater amount of plant regeneration and lower vitrification of regenerated plants.*

2. Place 5 callus clumps (about 1 cm³) on the agar per Petri dish.

3. Incubate calli under continuous white light at about 28°C and watch for the appearance of green spots on the surface of the callus and for green shoots.

*Note:*   *Regular 40-W fluorescent lamps give sufficient light energy for shoot induction and greening if they are placed about 50 cm above the cultures.*

4.  Transfer calli containing shoots to MSO media in a bottle or a culture tube.

*Notes:*   *MSO medium is MS medium without plant growth hormones. One can prepare this medium by following the steps described in the section on media preparation, except that no plant growth regulators are added.*

5.  Continue to culture calli under continuous white light and watch for the appearance of roots. One may also find that roots appear in step 4.

6.  When the size of the regenerated plantlets is about 10 cm, plant them in pots containing sterile vermiculite.

7.  While maintaining high humidity by covering the whole unit with a transparent plastic bag, grow plantlets in the culture room. This step is designated as hardening.

8.  After hardening for 1 to 2 weeks, transplant regenerants to pots and transfer to the greenhouse.

## B. Induction of Organogenesis from Cotyledonary-Petioles of Chinese Cabbage

1.  Germinate sterile Chinese cabbage on the agar containing MSO media under white light (continuous or 16:8 h of light:dark cycles).

2.  While waiting for germination and seedling growth, prepare MS media supplemented with 3 mg · l$^{-1}$ IBA (indole-3-butyric acid) and 10 mg · l$^{-1}$ BAP.

3.  Collect cotyledons with petioles from 4- to 5-day-old Chinese cabbage seedlings and place them on the medium (10 to 16 explants per Petri dish).

4.  Culture them under continuous white light while watching for the appearance of shoots and roots.

5.  Transferring plant regenerants, hardening, and transferring plantlets to the greenhouse are essentially the same as described in the section on rice organogenesis.

*Notes:*   *(1) Shoot regeneration directly from explants is especially important in plant transformation since prolonged tissue culture may result in unwanted mutation of plant regenerants. Furthermore, direct shoot-induction is a faster way to obtain plant regenerants. Usually the entire procedure for the plant tissue culture including the callus induction, subculture, and the regeneration of plants from callus takes several months at best. Therefore, many researchers try to induce plant organogenesis directly from explants after transformation by Agrobacterium infection or particle bombardment. (2) Another important consideration in obtaining transgenic plants is the presence of selection pressure. Usually antibiotics such as kanamycin and hygromycin have been used to elicit selection pressure. However, the presence of such antibiotics reduces the culture efficiency.*

# References

1. **Bhojwani, S. S. and Razdan, M. K.,** *Plant Tissue Culture: Theory and Practice,* Elsevier Science, New York, 1983.

2. **Debergh, P. C. and Zimmerman, R. H.,** *Micropropagation. Technology and Application,* Kluwer Academic Publishers, Dordrecht, The Netherlands, 1991.

3. **Evans, D. A., Sharp, W. R., and Ammirato, P. V., Eds.,** *Handbook of Plant Cell Culture, Volume 4, Techniques and Applications,* Macmillan, New York, 1986.

4. **Lindsey, K., Ed.,** *Plant Tissue Culture Manual,* Kluwer Academic Publishers, Dordrecht, The Netherlands, 1991.

5. **Pollard, J. W. and Walker, J. M., Eds.,** *Plant Cell and Tissue Culture, Methods in Molecular Biology,* Vol. 6, Humana Press, Clifton, NJ, 1990.

6. **Vasil, I. K., Ed.,** *Cell Culture and Somatic Cell Genetics of Plants,* Volumes 1–5, Academic Press, 1984–1988.

Chapter

# Microscopy: Light, Scanning Electron, Transmission Electron, and Confocal

## Contents

## I.  Introduction

Major conceptual breakthroughs have occurred in cellular and molecular biology with the development of light microscopy (LM), scanning electron microscopy (SEM), transmission electron microscopy (SEM), and confocal microscopy (CM). For investigators doing studies on *in situ* localization of specific mRNAs or reporter gene expression in animal or plant tissues and organs, light microscopy protocols are required. Immunolocalization of specific gene products using polyclonal antibodies requires a knowledge of transmission electron microscopy and specialized techniques of tissue fixation, embedding, and ultrathin microtome sectioning. Confocal microscopy involves 3-D optical sectioning made possible with powerful computer programs for data and section analyses, dyes, and filter systems that allow investigators to observe fluorescence images of specific proteins, nucleic acids (DNA and RNA), and ions within whole tissues or organs. Scanning electron microscopy allows one to visualize 3-D images of the surfaces of whole organisms, organs, tissues, and cells or parts of cells, made possible by the large depth-of-field capabilities of the SEM. X-ray analysis of the distribution of particular elements present in the cells or tissues can be performed with the SEM as well.

In this chapter, we explore the principles underlying how the LM, SEM, TEM, and CM instruments work, how to operate them, and a number of applications in biology and medicine where these techniques have been used successfully.[1-10]

## II.     Light or Optical Microscopy

Optical microscopy, involving the use of visible or ultraviolet light sources, has undergone a remarkable resurgence of interest in its applications in biology and medicine. Witness the advent of Nomarski, phase contrast and bright-field/dark-field optics to get better contrast images by the use of the polarizing optics with a red plate to determine orientation of cellulose microfibrils in cell walls, digital image processing for contrast enhancement (video-enhanced contrast microscopy or VECM), optical sectioning that has led to development of confocal microscopy (see later in this chapter), and UV fluorescence microscopy. In this section, we cover the basic structural features of light or optical microscopy, illustrate several research light microscopes in use today, then show several applications in the use of these instruments with biological systems.

Figure 19.1 illustrates the basic external and internal components of a commercial research light microscope. Figure 19.2 is a photo of an Olympus Vanox light research microscope in use in the authors' laboratory. It is fitted with photomicrography capabilities with electronic controls for taking photomicrographs; Nomarski, dark-field, and bright-field optics; polarizing filters and red-plate filter; and a UV power supply to do UV fluorescence microscopy. We have successfully used polarized light and the red plate to analyze the orientation of cellulose microfibrils in the guard cells of stomata of plants. We also have used UV fluorescence to study callose (β-1,3-linked glucan, polysaccharide) in pollen tubes, using aniline blue stain and cellulose deposition in developing cell walls that form in isolated plant protoplasts, using Calcofluor, a specific stain for cellulose (β-1,4-linked glucan).

Figures 19.3 to 19.5 illustrate one application of optical light microscopy, namely, analysis of starch-containing chloroplasts in graviresponsive organs (pulvini) in oat shoots. The stain used is iodine-potassium-iodide ($I_2KI$), which is specific for staining starch. From this study, it was discovered by the authors that these starch-containing chloroplasts are the gravireceptors that initiate an upward-bending response in oat shoots as well as shoots of all members of the grass family (Poaceae).

## III.     Scanning Electron Microscope

### A.     Principles of Operation

Figure 19.6 illustrates one of the currently used scanning electron microscopes (SEM). This instrument differs fundamentally from the conventional transmission electron microscope and all other common optical instruments in that no lens acts on the image-forming radiation after it strikes the specimen. Instead, the electron beam is focused to a very small diameter by two or three lenses, which are functionally analogous to the condenser lenses of

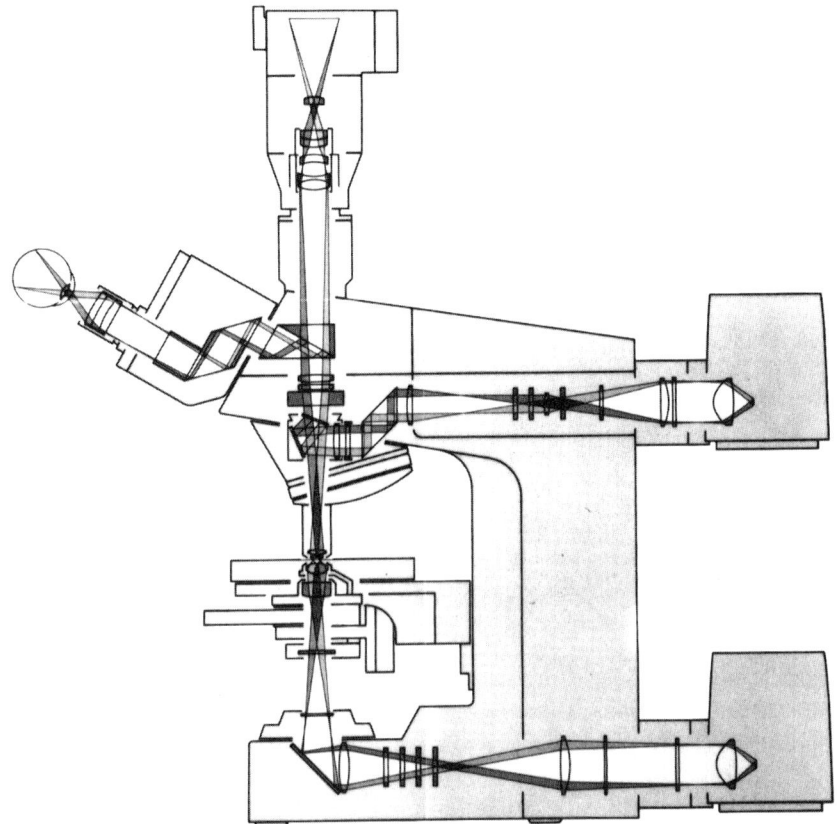

**FIGURE 19.1**
Diagram of an Axioplan Universial Microscope (courtesy of Carl Zeiss Inc., Thornwood, NY) illustrating the light sources and internal optics of their research light microscope. (Poster diagram contributed by Shelly Almburg, Instrument Analyst II, Department of Biology, University of Michigan.)

a TEM, and then caused to scan the specimen in a raster* pattern by a system of beam deflecting coils. The image is then formed in a point-by-point manner using various types of signals generated by interaction of the electron beam with the specimen. As shown in Figure 19.6 B, the image is displayed on a cathode ray tube (CRT) whose electron beam is driven by the same scanning drive generator that scans the beam over the specimen. There is thus a one-to-one correspondence between points on the viewing screen of the cathode ray tube and points on the specimen. Magnification is changed simply by an amplifier, which changes the size of the area scanned on the specimen relative to that of the CRT, since

$$\text{magnification} = (\text{area on CRT})/(\text{area on specimen})$$

Images are produced on the CRT by feeding the signals from the specimen to a video amplifier that modulates the brightness of the beam of the CRT in a manner related to the strength of the signal from the specimen. Several different signals are available for use in image formation; however, the one most frequently used is secondary electrons. These are valence

---

* Raster refers to a sequential line-by-line coverage of an area, usually with quick linear horizontal scans on adjacent top-to-bottom lines. (Used on TV sets.)

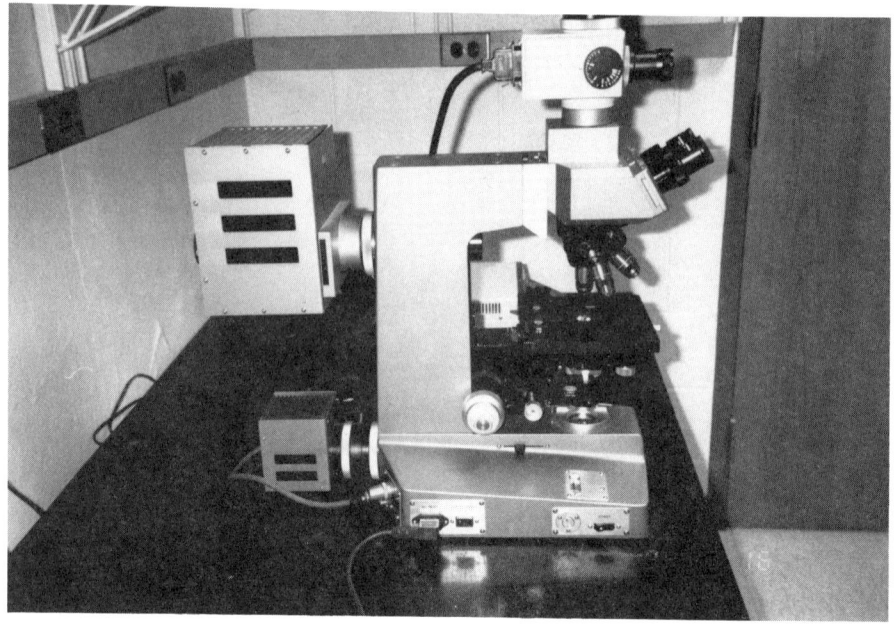

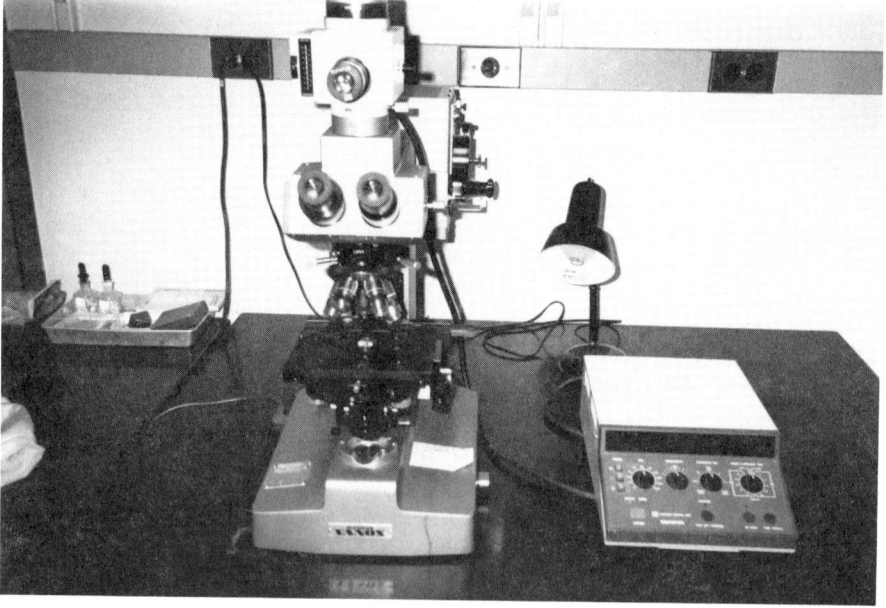

**FIGURE 19.2**

Side view (top) and front view (bottom) of an Olympus Optical Company, Ltd. Vanox research light microscope with ancillary electronic controls. (Photos taken by Peter Kaufman.)

electrons that are dislodged from atoms in the outer 5 to 10 nm of the specimen's surface by the incident electron beam. The secondary electrons emerge from the surface with very low velocities and are generally collected by applying a potential of about +10 kV to the surface of a scintillator. This potential accelerates the secondary electrons enough so that when they

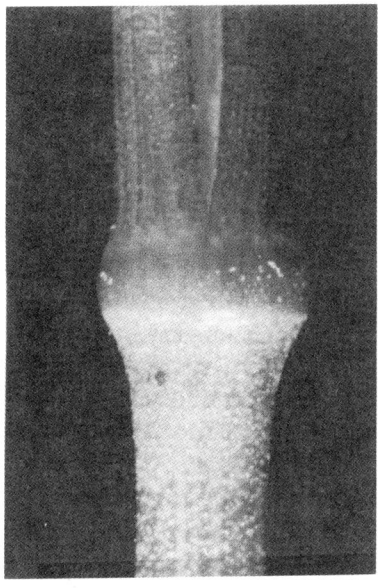

**FIGURE 19.3**

The leaf-sheath pulvinus of oat *(Avena sativa)* from an upright plant. The papery margin of the leaf sheath can be seen above the pulvinus. (From Brock, T. G. and Kaufman, P. B., *The Pulvinus: Motor Organ for Leaf Movement,* 1990, © American Soc. of Plant Physiologists. With permission.)

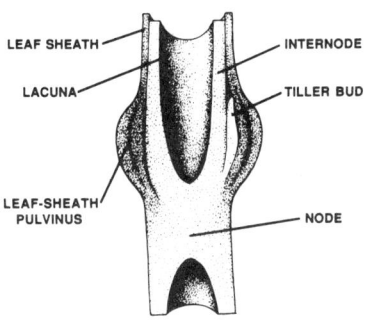

LEAF SHEATH

LACUNA

INTERNODE

TILLER BUD

LEAF-SHEATH
PULVINUS

NODE

**FIGURE 19.4**

Diagram illustrating a medium longitudinal section through the region around the leaf-sheath pulvinus of an oat shoot. (From Brock, T. G. and Kaufman, P. B., *The Pulvinus: Motor Organ for Leaf Movement,* 1990, © American Soc. of Plant Physiologists. With permission.)

strike the scintillator they produce minute flashes of light, which are detected by a photomultiplier tube. The output of the photomultiplier is then used by the video amplifier for image formation. The number of secondary electrons generated at a given point depends basically on the angle at which the beam strikes the specimen surface at that point. More electrons are emitted when the surface is inclined sharply to the beam than when the surface is perpendicular to it. Consequently, image contrast changes as the beam moves over the surface in a manner related to the topography of the specimen surface.

Some of the electrons of the incident beam interact strongly with the atoms in the specimen and are scattered back out of its surface with essentially the same velocity they had upon striking it. These backscattered electrons can also be collected by various types of detectors and the resulting signal amplified for image formation. Likewise, the incident electrons that are not backscattered produce a minute current ($10^{-6}$ to $10^{-12}$ A) of electricity through the sample, which can be amplified for image formation. The fraction of electrons backscattered depends basically on the average atomic number of the atoms in the specimen surface, and ranges from about 10% for an average atomic number of about 10 to nearly 50 for an average atomic number of about 75. Consequently, contrast in backscattered electron and sample current images depends on local variations in composition, although it is very difficult to determine, from image characteristics alone, the elements involved in causing these variations. If the specimen is thin enough, electrons transmitted through it can also be used for

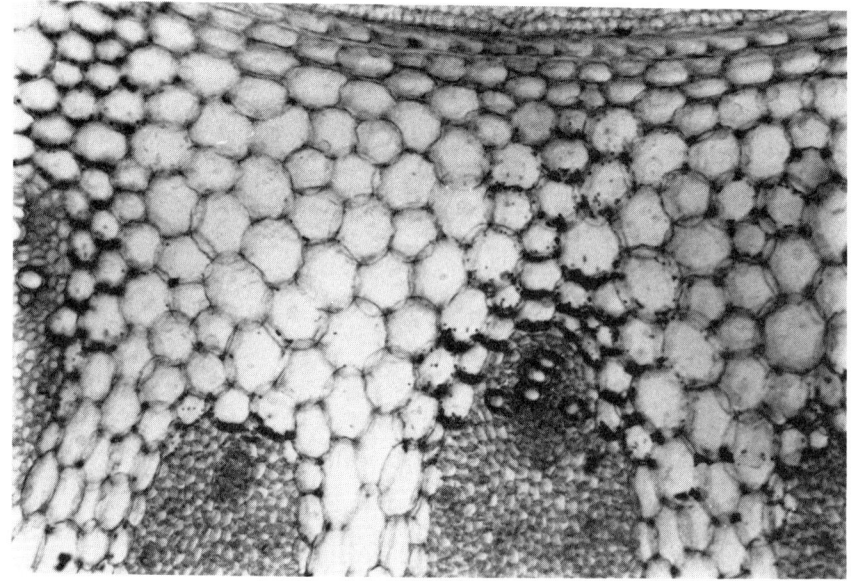

**FIGURE 19.5**

Light micrograph of cross section through oat shoot leaf-sheath pulvinus illustrating starch gravireceptors in bases of cells next to each vascular bundle. (×330.) (From Brock, T. G. and Kaufman, P. B., *The Pulvinus: Motor Organ for Leaf Movement*, 1990, © American Soc. of Plant Physiologists. With permission.)

image formation. For crystalline specimens, transmitted electrons can also produce channeling patterns, diffraction patterns, and Kikuchi patterns. Some materials produce fluorescent light when struck by the electron beam. This light can be detected by a photomultiplier tube and used for image formation.

The versatility of the SEM is further increased because the electron beam causes X-ray photons to be emitted from atoms in the specimen. These photons have energies and wavelengths that are determined by the electron energy levels of the atoms. Measurement of either their energies or wavelengths makes it possible to identify the kinds of atoms present in the specimen (qualitative chemical analysis). Measurement of the rate at which they are emitted under carefully controlled conditions, and comparison with standards of known compositions, permits the concentrations of the different kinds of atoms to be determined (quantitative analysis).

These measurements can be made with a solid state X-ray detector in combination with a multichannel analyzer. The solid state detector is a special silicon diode that produces an electrical pulse each time it receives an X-ray photon. The amplitudes of these pulses are linearly proportional to the energies of the photons. The multichannel analyzer receives the pulses from the detector, after proper amplification, measures their amplitudes, and records their receipt in memory positions (channels) that are assigned numbers proportional to the photon energies. The counts recorded in each channel can be displayed on a CRT, read out numerically, or printed out with a teletypewriter, giving an X-ray spectrum of the specimen. In this way it is possible to determine the major elements present in a specimen and to gain a rough idea of their concentrations (semiquantitative) in a few minutes. Detection of elements present in minor amounts, and accurate concentration determinations, both require more complicated and time-consuming procedures. For this purpose it is often preferable to use a crystal spectrometer, which gives considerably better sensitivity and selectivity than the solid state detector, which requires more time and effort because only one element can be detected

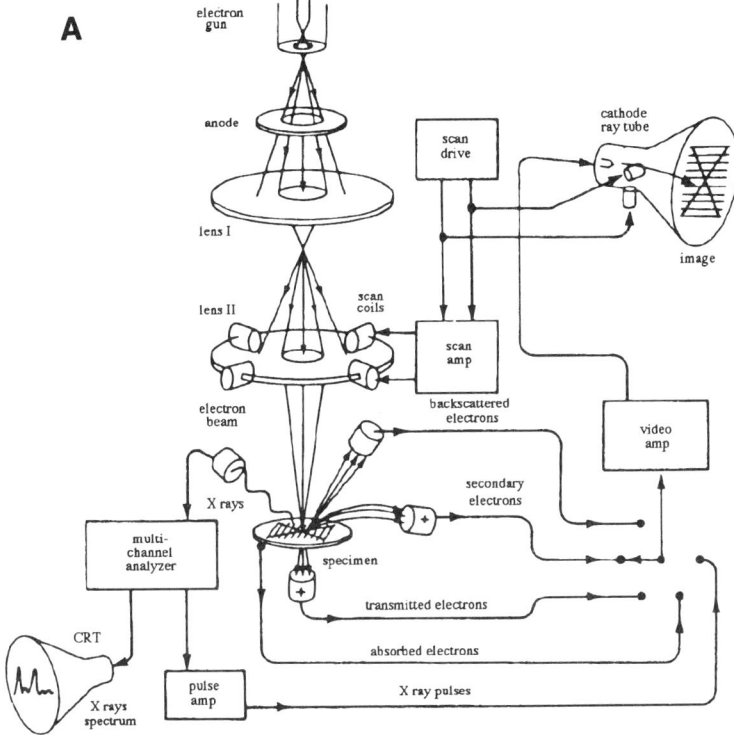

**FIGURE 19.6A**

The basic components of the electron optical system of a scanning electron microscope. (Reprinted with permission from Kaufman, P. B., Labavitch, J., Anderson-Prouty, A., and Ghosheh, N. S., *Laboratory Experiments in Plant Physiology,* Macmillan College Publishing Co, New York, Copyright © 1975.)

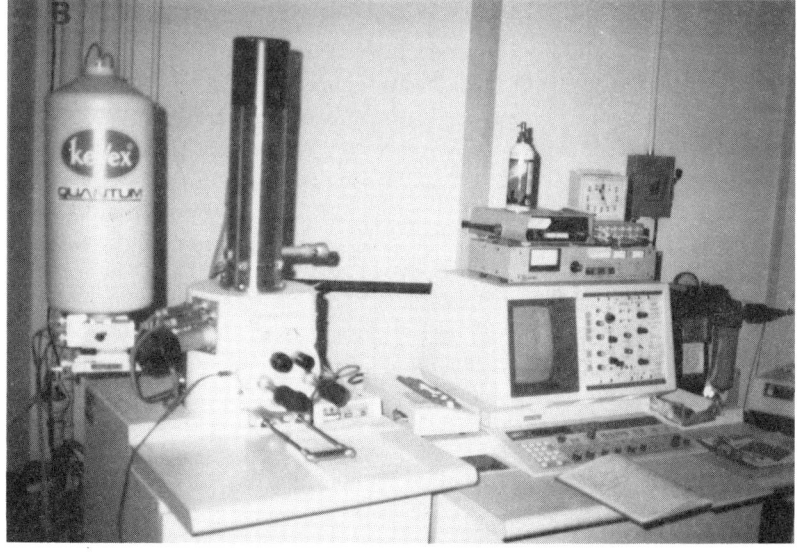

**FIGURE 19.6B**

Illustration of a Hitachi scanning electron microscope. (Photo taken by Najati S. Ghosheh.)

at a time. The Electron Microprobe Analyzer is an instrument that has an electron optical system similar to that of the SEM but is otherwise designed especially for carrying out accurate quantitative analyses with crystal spectrometers. If the multichannel analyzer of the crystal spectrometer is set to feed pulses from only a single element to the video amplifier, each pulse will produce a single dot of light on the CRT. As the electron beam scans over portions of the specimen where the concentration of the selected element is high, the pulses will be received more rapidly and the number of dots per unit area on the CRT will increase. In this way an image can be produced showing the overall distribution of the element in the specimen. When compared with secondary electron or sample current images of the same area of a specimen, such characteristic X-ray images are very useful for showing the overall compositional variations in the specimen and their relationships to various structural features. X-ray images usually do not reveal small changes in the concentrations of major elements, and are otherwise difficult to interpret quantitatively.

The theoretical resolution of the SEM is determined basically by the diameter of the electron beam as it strikes the specimen. The beam diameter is limited by the spherical aberration of the second lens, the electron beam accelerating voltage and current, and the brightness and size of the beam at the electron gun. In practice a number of other factors such as the mechanical stability of the specimen stage, the quality of the detectors and amplifiers, and the cleanliness of the electron optical system also affect the resolution. Many commercial installments are available that can give resolutions in the range from 10 to 20 nm routinely, whereas resolution below 1 nm has been achieved with specially designed experimental instruments.

The resolution is also dependent on the nature of the interaction between the electron beam and the specimen that generated the image-forming signal. In general, resolution with backscattered electrons and sample current is considerably greater than the diameter of the incident electron beam because the high-energy electrons may be backscattered from a depth of several hundred nanometers beneath the surface of the specimen, depending on the electron accelerating voltage used and the density of the specimen. In traveling this far into the specimen the electrons may be scattered laterally by collisions with the atoms of the specimens before finally being backscattered. Best resolution is usually obtained with secondary electrons, because these are produced only in the outer 1 to 5 nm of the specimen before there has been appreciable lateral scattering of the electron beam.

The lateral scattering of the electron beam within the specimen, and its penetration in depth, permits X-ray photons to be emitted from a teardrop-shaped volume of the specimen whose size depends on the density ($\rho$, $[g] \cdot [cm^3]^{-1}$) of the specimen and the accelerating voltage (kV). The depth of this volume ($\mu m$) is roughly as for different types of materials.

| Material type | Accelerating voltage (kV) | | | | |
|---|---|---|---|---|---|
| | 5 | 10 | 15 | 20 | 25 |
| Biological ($\rho = 1$) | 1 | 3 | 6 | 10 | 15 |
| Ceramic ($\rho = 3$) | 0.3 | 1 | 2 | 3 | 5 |
| Metallic ($\rho = 7$) | 0.1 | 0.5 | 0.8 | 1 | 2 |

It will be evident that the resolution of X-ray images is severely limited by this phenomenon. For thick specimens, resolution is essentially independent of electron beam diameter, provided the beam diameter is less than about 0.1 $\mu m$. Resolution can be better for thin specimens, where the beam penetrates through the specimen before it has a chance to spread laterally to any great extent; however, the X-ray intensity falls off sharply because of the smaller volume of specimen being irradiated. This penetration and spreading of the electron beam within the specimen must also be considered when interpreting quantitative data obtained with the electron beam in a fixed spot on the specimen. Perhaps the most remarkable feature of the SEM is the unusually great depth of field it provides. This is a consequence of the characteristics of the electron lenses used to focus the beam. These lenses perform acceptably only when the electron beam has a diameter of less than 500 μm when it passes through the lens. When such a small beam is focused to the finest possible diameter, there is a considerable distance above and below the focus point where it changes diameter very little.

Over this range of distance the beam will give acceptably sharp images. This depth of field varies with the minimum beam diameter, which in turn limits the maximum beam current and magnification, as follows:

| Minimum beam diameter | 10 μm | 200 nm | 10 nm |
|---|---|---|---|
| Depth of field | 0.5 mm | 100/μm | 10 μm |
| Magnification | 20× | 1000× | 20,000× |
| Maximum beam current | $10^{-6}$ Å | $10^{-9}$ Å | $10^{-12}$ Å |

The significance of these values for the depth of field (Table 19–1) can be appreciated by comparing them with the following typical values for optical microscopes at comparable magnifications and resolutions:

| Magnification | 20× | 100× | 1000× |
|---|---|---|---|
| Depth of field | 100 μm | 10 μm | 0.05 μm |
| Resolution | 10 μm | 1 μm | 200 nm |

## Table 19–1 Comparison of Important Features of Light and Scanning Electron Beam Microscopes

| Feature | Average light microscope | Average scanning electron microscope |
|---|---|---|
| Radiation | Visible light | High-speed electrons |
| Wavelength* | ~0.0005 mm | ~0.00000001 mm |
| Lens type | Glass | Electromagnetic |
| Lens function | Form magnified image of specimen | Focus electrons on a small spot on specimen |
| Image | By image forming lens | By electron amplification of signals from specimen |
| Magnification | 50× to 500× | 50× to 50,000× |
| Resolution* | 0.004–0.0004 mm | 0.004–0.00001 mm |
| Depth of field* | 0.05–0.002 mm | 1–0.003 mm |
| Working distance* | 40 to 0.2 mm | 12mm |
| Cost | $200 to $10,000 | $25,000 to $100,000 |
| Operating expense | Nil | $35 per hour |

* To convert to inches multiply values given in millimeters by 0.04 (i.e., 0.004 mm = 0.00016 in.).

Prepared by Wilbur C. Bigelow, Department of Materials and Metallurgical Engineering, University of Michigan, Ann Arbor.

Note that at 1000×, the maximum useful magnification for most optical microscopes, the SEM has a depth of field 200 times greater than the optical microscope, whereas at 20,000× its depth of field is about the same as that of an optical microscope at 100×. This means that the SEM is ideally suited for studies of rough and irregular specimens that are too difficult to examine by optical microscopy or conventional transmission microscopy. Typical examples are microfossils, textile fibers and fabrics, catalyst particles, pollen grains, fracture surfaces, wood fibers, and the external surfaces of plants and insects. Procedures for preparation of specimens for examination in a SEM vary considerably, depending on the type of specimen material and the results desired. Metallic specimens are good conductors of electricity and can usually be mounted for examination without any special preparation, although it may be desirable to treat them in an ultrasonic cleaner to remove traces of oil and adhering dust particles.

Specimens of ceramics and minerals, fossils, teeth and bone, fibers, and self-supporting or "hard" biological tissues, and so on, which are nonconductors, must be coated with a thin layer of conducting material to prevent accumulation of static electric charges that arise from the electron beam and cause image distortion and spurious variations in image intensity. In some cases satisfactory coating can be obtained by evaporating 10 to 50 nm of gold or some other metal in the surface of the specimen in a conventional vacuum evaporator. This technique is generally unsatisfactory for specimens that are very rough, porous, or fiberous because the metal will not cover the undersides of fibers and the walls of pores and asperities. The use of devices that rotate and rock the specimen during evaporation will give better results with such specimens. However, a glow-discharge coater is perhaps simpler and faster and gives best overall results.

The preparation of soft biological tissues presents many difficult problems because they become severely distorted during handling and upon exposure to the vacuum and electron beam inside the SEM. Various procedures, including chemical fixation, freeze-drying, and critical-point drying, have been successfully used to stabilize such tissues for examination. The details of these procedures, which vary with the type of tissue and the objectives of the investigation, can be found in the technical literature. When X-ray methods are to be used, the conductive coating should be as thin as possible and a low-atomic number material such as aluminum or carbon is preferred to minimize attenuation of the X-rays by absorption in the coating layer. If accurate quantitative analyses are to be performed, the specimen surface must be smooth, flat, and undistorted to allow accurate control of experimental variables and accurate knowledge of the various physical parameters involved in computation of results.

# B.     Preparation of Biological Tissues for SEM

In contrast to the techniques for the TEM, tissue preparation procedures for work with the SEM are much easier. Very often, no fixation of the tissue is necessary. Peels or sections made by hand or with a freezing microtome may be examined in the SEM directly or lyophilized (freeze-dried) and then examined. The basic procedures are as follows:

## *Fresh Biological Material*

1.  Take biological materials to the SEM lab and prepare small pieces 1 to 2 mm in diameter. Pollen grains and spores may be mounted directly.

2.  Mount specimens on double-stick Scotch™ tape or directly with electrically conductive silver paint (in butyl acetate solvent) or carbon paint (TV tube-coat) onto 1-cm aluminum or carbon stubs. Carbon stubs are used if you intend to do elemental analysis with the multichannel analyzer

attached to the SEM; use aluminum stubs otherwise (prepared from ringstand rods). Air dry with a hair dryer at room temperature.

3.  If a critical-point drying apparatus is available, use it; your tissue will be much better preserved for scanning with the SEM.

4.  Coat your specimens with gold (Au) or chromium (Cr). If this is not done, you get very bright areas on your tissue appearing on the SEM TV screen or oscilloscope screen as a result of charge build-up.

5.  Place in the SEM to scan. Learn how to adjust the vertical, horizontal, and circular drives that move the specimen. A demonstration of focusing and changing magnification should be given. When you are more adept at using the SEM, learn how to use the MCA (multichannel analyzer) to do elemental analysis in plant tissues.

## Fixed Biological Materials

1.  Fix the sample in 2% formaldehyde/1% glutaraldehyde for 30 min to 1 h.

2.  Wear gloves when dealing with these solutions. This treatment basically serves to kill and fix the cells.

3.  Rinse the solution off the tissue with 0.1 $M$ phosphate buffer for 10 min.

4.  Repeat the rinse in phosphate buffer.

5.  Start dehydrating the tissue in 50% ETOH (ethyl alcohol) for 10 min.

6.  Start dehydrating the tissue in 70% ETOH for 15 min.

7.  Start dehydrating the tissue in 95% ETOH for 15 min.

8.  Repeat 95% ETOH rinse for another 15 min.

9.  Dehydrate the tissue in 100% ETOH for 10 min and repeat twice (total time 30 min).

10. Place in HMDS (hexamethyldisilazane) for 10 min. Repeat. This replaces all water in tissue, fixes tissue, and stops shrinking.

11. Place the sample in an empty vial in a desiccator and store overnight under full vacuum. Place under vacuum for 20 min before mounting. Mount on a stub with carbon-paint adhesive. Carbon adhesive prevents the buildup of charge.

12. Mount on stub then dry with a hair dryer. Pre-grind the aluminum stub with emery paper before mounting.

13. After drying, place the stub with sample in the film cartridge with Drierite™ (absorbs moisture).

## C.     Operating Instructions for the Hitachi S-570 SEM

### Turning on the SEM

*   Turn on (push up) the POWER/DISPLAY switch under column. POWER/EVAC is usually left on
*   Record filament time, etc. on Log. Be certain to record actual use time as well as filament time.

### Specimen Loading

1.  Using gloves, fasten specimen in specimen holder.

2.  Set stage height control (Z) to EX-EX (exchange).

3.  Vent column by pushing and releasing AIR/EVAC next to column.

4.   When hissing sound stops, gently pull forward on stage knobs to open door.

5.   Using gloves, gently insert specimen and holder into stage while holding stage height control (so stage will not be pushed down).

*Note:   If using BSE imaging check clearance between specimen and detector.*

6.   Gently slide door close, press in AIR/EVAC, and hold stage closed until pumping starts.

## Changing Menu Parameters

•   CALL displays the menu. Hit key number, change value, then ENTER value. ENTER again hides menu — back to scanning.

•   Example: CALL, push 0 to select KV (accelerating voltage), type 20, then hit ENTER (Accelerating Voltage, Working Distance (WD), and Magnification Preset are often changed).

•   To set actual working distance: put in a WD; then after beam is on, focus on specimen using the height (Z) control until image is sharp (do not use Focus knobs — they change the WD set in the menu).

*Note:   Do not raise the specimen too high or you may hit the BSE detector or objective lens.*

## Turning on the Beam

1.   Wait until the green "Ready" light (sufficient vacuum) above the keypad comes on before doing anything else (actually it is best to allow about two minutes after the "Ready" light comes on).

2.   Set COND. LENS knobs to 3 (upper knob) and 5 (lower knob).

3.   Set VIDEO SIGNAL SELECTOR (above keypad): Video 1 and 2 switches to SE positions and the last switch to MANUAL.

4.   Turn on ACC. VOLTAGE.

5.   Set scan speed to RAPID SCAN Mode; turn on the WAVE FORM/SCANNING MODE (button next to the viewing screen).

6.   Set the CONTRAST and BRIGHTNESS knobs to about 2:00 and 10:00 respectively.

*Note:   We will saturate using the height (not amplitude) of the waveform as a measure of image brightness. As the FILAMENT CURRENT is increased the brightness will initially go up, increasing until the peak, and then down and finally up to the plateau. We want to be just on the edge of the plateau shown by the X in Figure 19.7 and no further (oversaturation greatly reduces filament life). BRIGHTNESS and CON-TRAST also affect the waveform; if the waveform goes off the screen adjust either or both BRIGHTNESS and CONTRAST to keep it on the screen.*

7.   Increase the FILAMENT CURRENT by rotating the knob clockwise slowly (you must see a vertical deflection of the waveform by the time the knob nears the max marker; if not, something is wrong; turn down the filament and get help). Then slowly increase the current to peak a valley and then up to the plateau. Make sure you stop when the brightness no longer shows a steady increase with current increase. Finally back off slightly so you are right on the edge of the plateau.

8.   Return to the NORMAL/SCANNING MODE; LOW MAG and RAPID SCAN.

*Note:   If you have any difficulties doing this get help from lab staff.*

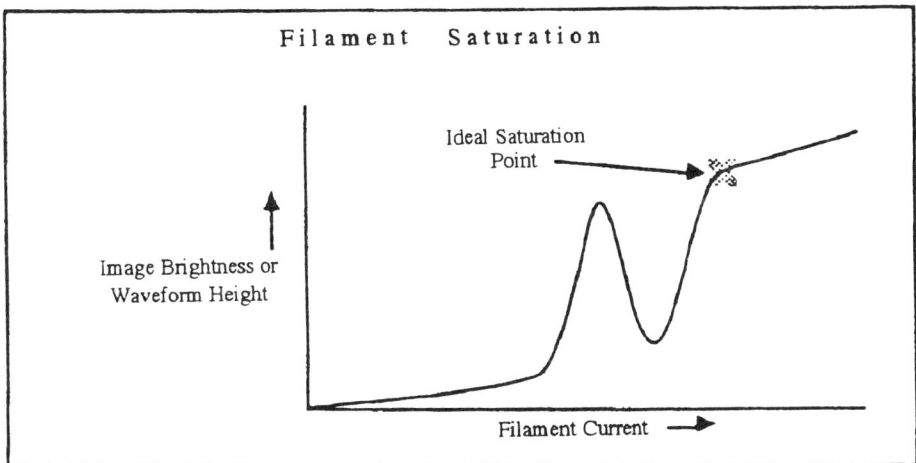

**FIGURE 19.7**
Filament saturation.

## Focusing Four Ways

- Manually using COARSE and FINE FOCUS.
- Using SEARCH sharpen and increase amplitude of waveform; SEARCH also increases the effect of COARSE FOCUS.
- AUTO FOCUS will automatically focus if the focus is not too far off; otherwise, use Search first.
- Focus using the stage height control (Z) if you want to maintain a WD previously set in the menu.

## Setting Brightness and Contrast

- Automatically: Set the VIDEO SIGNAL SELECTOR to AUTO and push the AUTO/MONITOR button.
- Manually: Adjust the BRIGHTNESS and CONTRAST knobs with VIDEO SIGNAL SELECTOR in the MAN position.

## Aperture Alignment

For best resolution the aperture will have to be aligned whenever the condenser or voltage is changed or a different aperture is used.

1. Go to the RAPID SCANNING SPEED.
2. Locate a region with good high-magnification detail using STAGE X and Y controls.
3. Go to fairly high magnification (about 2KX to start).
4. Focus
5. Turn on APERTURE ALIGN (far right side of console).
6. Use knobs on arm extending from the column to steady the image. Two knobs are concentric to each other; use the smaller one, along with the knob which is perpendicular to the arm to reduce the wobbling of the image (the larger of the two concentric knobs changes the aperture).
7. Increase magnification to about 10KX focus and align again. This is usually only necessary if you are working at very high magnification.
8. Turn off APERTURE ALIGN.

## Stigmation

- Generally only necessary at very high magnifications (5KX or greater).
- At high magnifications use X and Y STIGMATION knobs to optimize image sharpness.

## Condenser Setting

- For improved high-magnification resolution use higher number COND. LENS upper and lower knob settings.
- For lower magnifications and applications which require higher beam currents, such as BSE or X-ray analysis, use lower number COND. LENS settings.
- Generally one must realign the aperture after changing condenser settings.

## Miscellaneous Imaging Options

- Display Magnification, KV, Picture Number, and Micron Marker using DATA DISPLAY/ON (far right from keypad).
- Enter text onto image using DATA DISPLAY/KEY IN and the keypad or keyboard.
- Image Shift (far right) can be used to move the image small amounts at high magnifications.
- Rotate image without moving specimen using the RASTER ROTATION (to the right of the CRT). Note that this may cause image distortion at low magnifications.
- Split imaging can be done to slow different magnifications or imaging modes on the same picture using the SPLIT/SCANNING MODE and MODE/DUAL MAG control (next to viewing screen) or VIDEO SIGNAL SELECTOR.
- GAMMA control can be used to reduce the contrast when there are abnormally bright or charging regions; higher gamma numbers are used for high-contrast areas.
- For very high magnifications and small working distances an upper detector is used — get lab staff help for this.

## Taking Pictures

1. Focus at a higher magnification (at least two times the magnification of your desired image).
2. Setting exposure (three ways):
   a. Push AUTO/MONITOR while in the AUTO mode of VIDEO SIGNAL SELECTOR.
   b. If you previously set photo start condition to 2 or 4 (on startup menu), the exposure is set automatically when you take a picture.
   c. You can set the exposure manually using contrast and BRIGHTNESS knobs and WAVE FORM/SCANNING MODE (next to the viewing screen).
3. Put the camera handle in L (Load) position, slide the Polaroid™ PN-55 film all the way in with "This side towards lens" down.
4. Expose film by pulling out film cover until it stops.
5. Push the PHOTO button (to right of keypad).
6. When the photo light (red) goes out, and a "beep" sounds, slide the film cover back all the way in, put the camera handle in P (Process) position and then smoothly pull the whole film packet all the way out.

7. Neatly process negative and print.

*Notes:* *(1) If you have problems setting exposure it may be necessary to do it manually — ask lab staff for help. (2) Be careful when turning on the room light not to push EVAC instead of ROOM LIGHT.*

## Sample Change

1. Go to LOW MAG.
2. Turn off ACC. VOLT (it is not necessary to turn down filament current first).
3. Press AIR/EVAC to vent column.
4. Change the specimen.
5. Press AIR/EVAC to pump down column.
6. Wait two minutes after green "ready" lights come on and turn on ACC. VOLT.

## Shut Down Procedure

1. Turn down FILAMENT CURRENT (Counterclockwise).
2. Turn off ACC. VOLTAGE.

*Note:* *Retract BSE detector if you were using it.*

3. Remove the specimen.
   a. Adjust the stage height knob (Z) to EX-EX position.

*Note:* *If in a high-stage position be careful to rotate the proper way (counterclockwise) to lower stage.*

   b. Vent the column using AIR/EVAC; when the hissing stops, open the door and remove the specimen using gloves (hold stage or Z while removing specimen).
4. Evacuate the column: close door, press in AIR/EVAC, hold door closed until pumping starts.

- **Courtesies to help the next user**
   a. Turn off DATA DISPLAY/ON and KEY IN.
   b. Go to LOW MAG.
   c. Set VIDEO SIGNAL SELECTOR to MAN and SE switch positions. Do not leave in X-RAY or EXTERNAL positions.
   d. Make sure any unusual imaging modes are turned off.
   e. Set COND. LENS back to 3 and 5 upper and lower knob positions.
   f. Normal Aperture setting is Aperture 3.
   g. Turn POWER/DISPLAY off (under column).
   h. Sign out (2 places) and process negatives and prints.
   i. Turn off KEVEX display (push-button switch on back of monitor); **do not** turn off the computer.

Advanced Techniques

Kevex EDS System Notes

**Hardware notes**

- The detector should be always used and left at the 6-cm position. If for some reason you need to change this use extreme caution. The EDS detector is very delicate and expensive. **Do not** position the detector or your specimen where there is **any** danger of contact.

- The Kevex software is set up for a detector distance of 6 cm, a working distance (WD) of 25 mm, and a tilt of 15°. With these settings the quantification program will work properly.

- It may be necessary to increase the beam current in order to obtain a good count rate (1000 to 2000 cps) at a 20 to 30% deadtime on your sample. Use the condenser lens controls and/ or change apertures. Realign as necessary.

- The Kevex console is usually left on, except for the display CRT. To turn this on, press in the button at the rear of the display.

**Software notes**

- Only necessary for printing, quantitative analysis, or advanced imaging.

- Starting up Kevex software:

  a. Boot the computer by pressing the [Program] key typing 5-[return]-5-[return]. Wait for a menu to appear.

  b. When a menu appears, hit return (or "Q") to run Quantex. After the program loads and starts execution (about a minute), you should see a "Quantex>" prompt.

- To plot a spectrum:

  a. Make sure the Okidata printer is on, then press CONTROL-P to print.

  b. Be careful to push CONTROL-P only once.

- For a more complete description of the QUANTEX program and for information on quantitative analysis, see the Kevex manual.

- If the system locks up, (will not respond to keyboard input) or types strange characters, you will need to "warm start" the Kevex by pressing the [Help] and [Clear] keys simultaneously. You may or may not need to reboot the computer (see above) following this action. If the unit still acts strangely, try a "cold start" by pressing the [Help] and [Clear] key twice in quick succession.

SEM applications — Figure 19.8, 19.9, and 19.10 illustrate application of the SEM to studies of stomata in *Taxus canadensis,* Canadian yew (Figure 19.8), gravitropic response and starch-filled chloroplasts (gravisensors) in oat *(Avena sativa)* shoots (Figure 19.9), and stem and leaf surface structure in oat shoots (Figure 19.10).

# IV.   Transmission Electron Microscopy

## A.   The Conventional Transmission Electron Microscope

The basic components of the electron optical system of the conventional transmission electron microscope (CTEM) are shown schematically in Figure 19.11. The electron beam is generated by drawing electrons away from a heated tungsten filament by means of an electron accelerating voltage, which is usually in the range from 50,000 to 10,000 V, but may extend to 1 MV in ultra-high-voltage models. The electron beam is initially focused to a small diameter by the electron gun. It then passes through the anode into the condenser lens system,

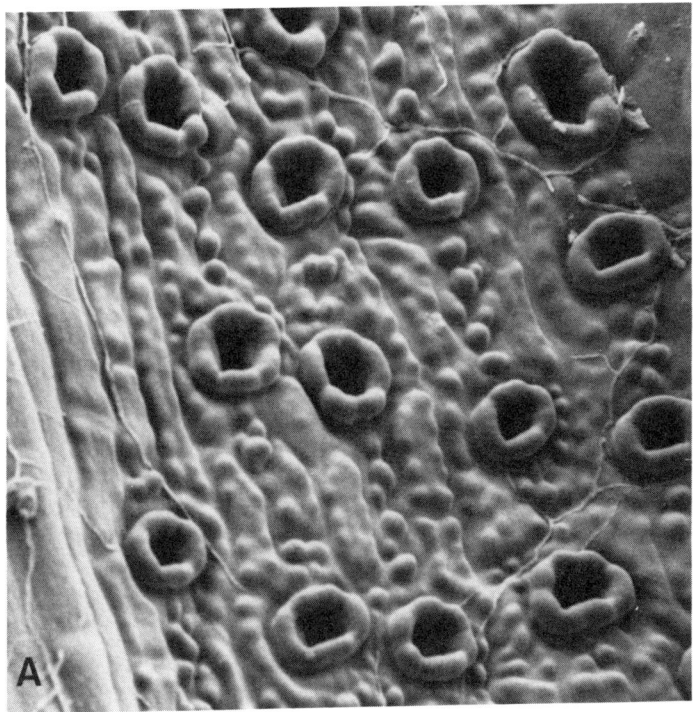

**FIGURE 19.8A**

A: SEM view of the surface of a *Taxus canadensis* (Canadian yew) needle showing "craters", which are loci of stomata (×300); B: SEM view of a single stomatal aperture of Canadian yew (×1500); C: SEM view of the cross section of a Canadian yew stomatal apparatus (×1500). (SEM photos provided by Leland Cseke.)

which serves the primary function of controlling the intensity of illumination on the specimen to give levels of brightness appropriate for viewing and photographing the electron image. This is done by changing the focal length of the lenses to vary the diameter of the beam as it encounters the specimen. Because the beam current (electrons per second) is maintained very constant by the electron gun, the intensity at the specimen (electrons per second per square micrometer) will increase as the diameter of the beam is decreased. The condenser lens system also contributes fundamentally to the size, angle of incidence, and coherence of the beam as it strikes the specimen, and so has an important influence on image contrast, quality, and stability.

Specimens for conventional transmission electron microscopy must be thin enough to transmit substantially all of the electrons in the electron beam. Optimum thickness for instruments operating at 80 kV is 30 to 50 nm. Thicker specimens lose contrast and detail, and are more susceptible to beam damage. Suitably thin specimens of most biological issues can be prepared by ultramicrotomy after appropriate treatments to stabilize them and enhance contrast.

Contrast in biological tissues arises by two basic mechanisms. For structures in the range of 1 nm and larger, contrast arises primarily because some electrons lose small amounts of energy in the specimen and are scattered away from the axis of the objective lens enough to be trapped by the lens aperture. Because the number of electrons that undergo this type of scattering increases with an increase in the atomic number of the atoms in the specimen, contrast can be enhanced by treating the specimen with chemicals such as $KMnO_4$, $OSO_4$, and

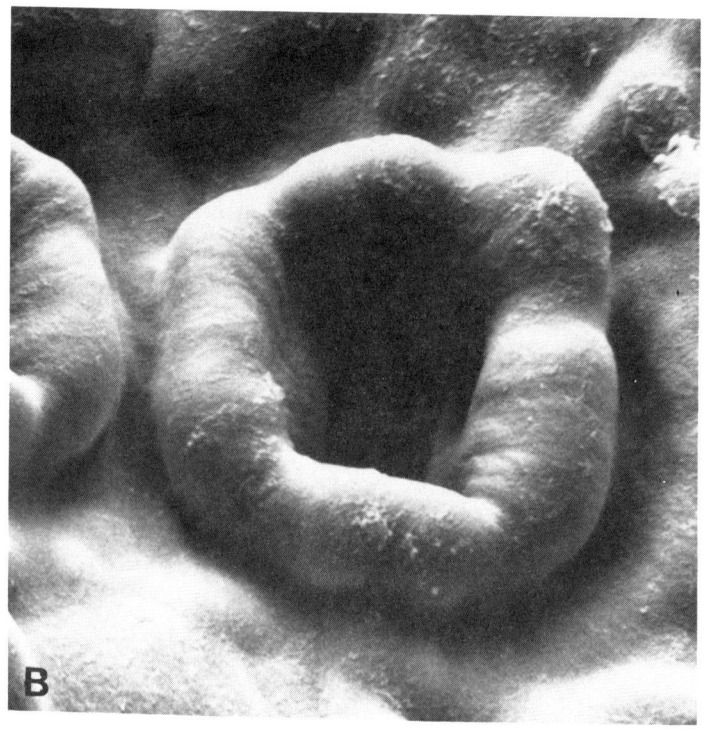

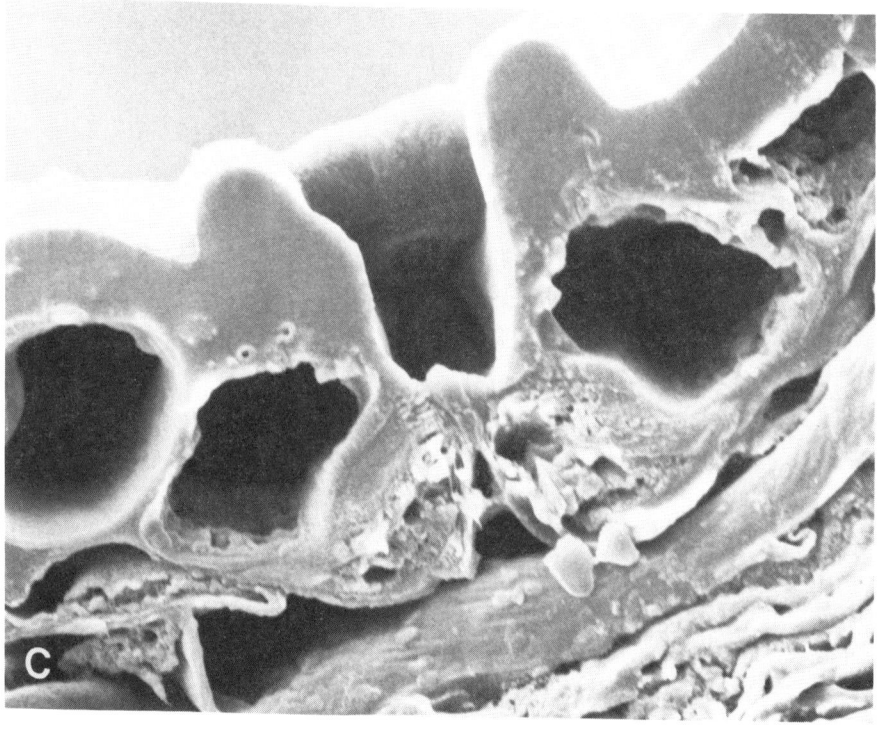

FIGURE 19.8B and C

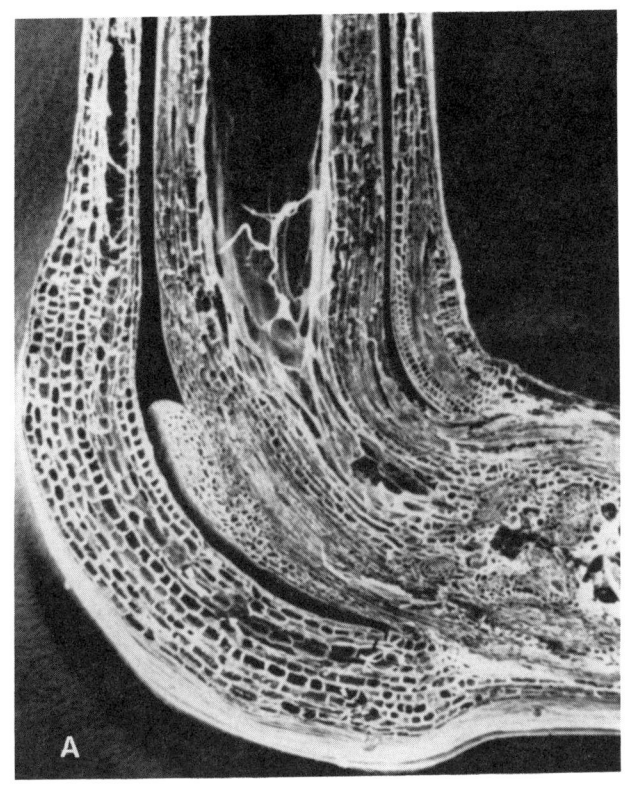

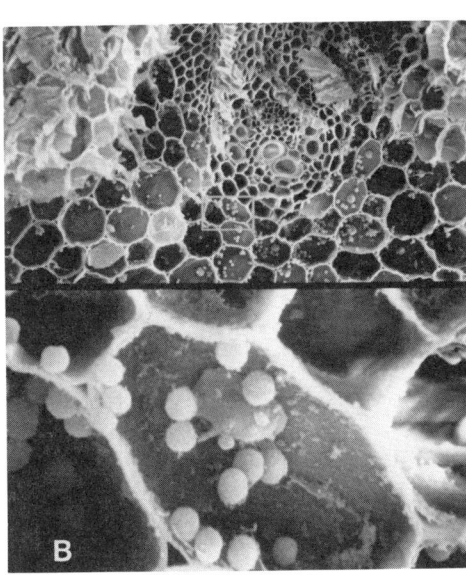

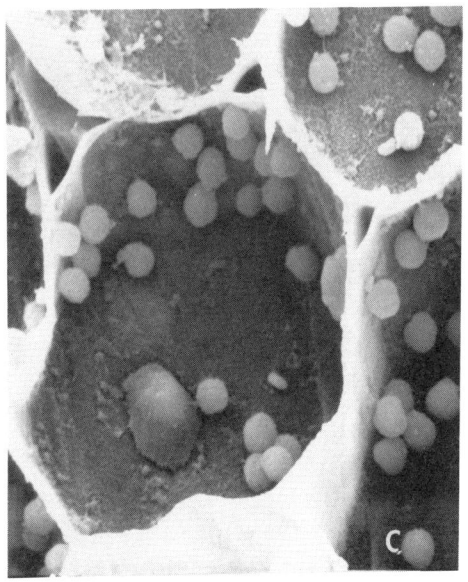

**FIGURE 19.9**

SEM views of graviresponse in pulvinus oat shoot (A) and starch-filled chloroplast gravisensors in leaf-sheath pulvinus cells of oat shoot (B and C). A, ×40; B, ×220; C, ×3000. (SEM photos provided by Peter Kaufman.) (Figure 9A from Brock, T. G. and Kaufman, P. B., *The Pulvinus: Motor Organ for Leaf Movement,* 1990, © American Society of Plant Physiologists. With permission.)

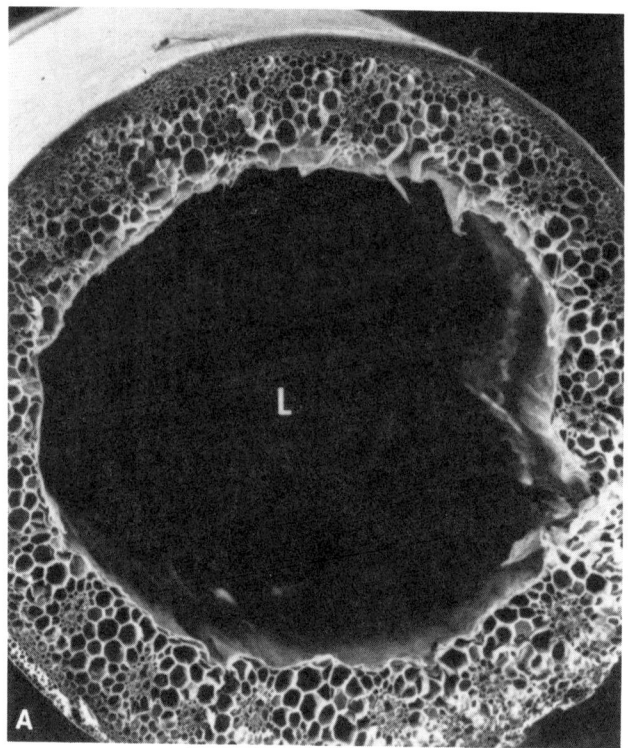

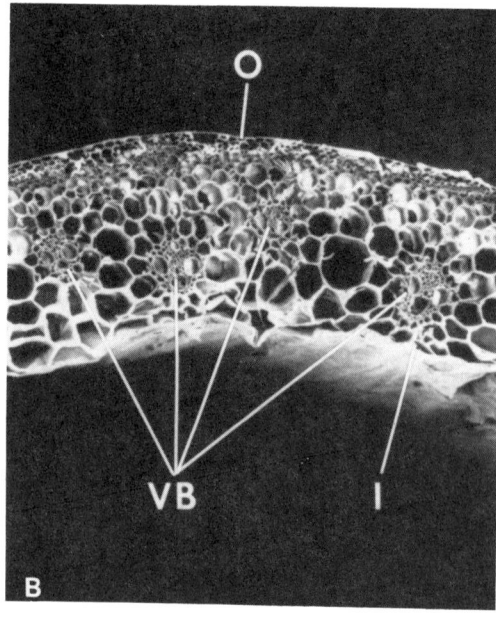

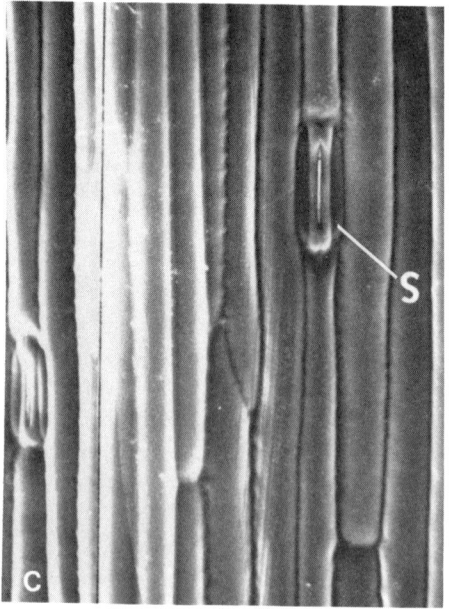

**FIGURE 19.10**

SEM views of three cross sections of an oat *(Avena sativa)* shoot stem (A and B) and surface of an oat plant leaf illustrating three stomata (C). A, ×30; B, ×50; C, × 300. L, lacuna; VB, vascular bundles; O, outer epidermis; I, inner epidermis; S, stomatal apparatus. (From Kaufman, P. B. and Brock, T. G., *Oat Sci. Tech. Agro. Mon.* 33, 1992, © Oat Science and Technology. With permission.)

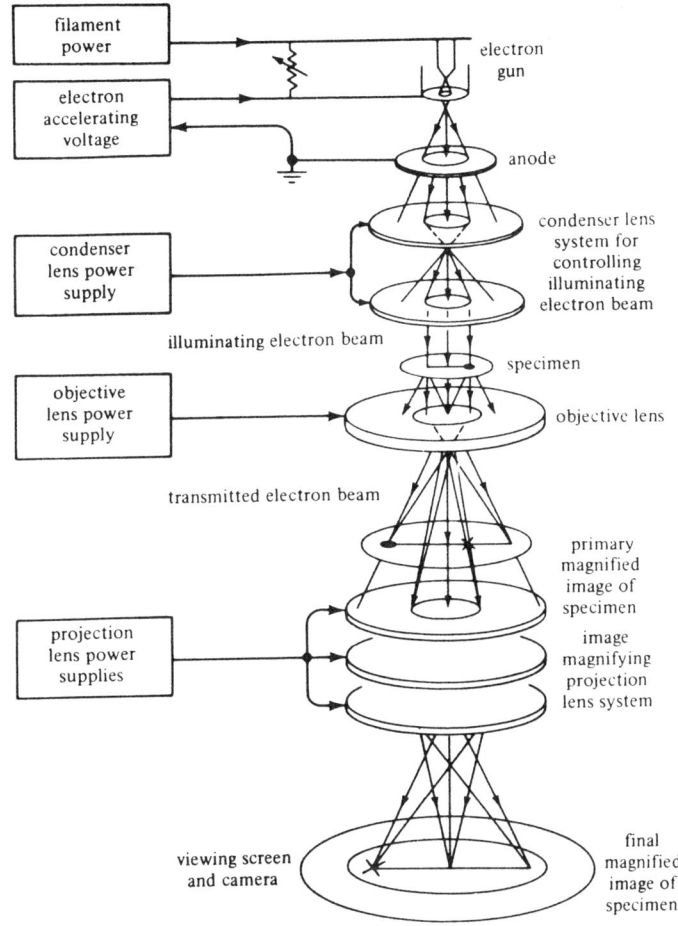

**FIGURE 19.11**
Diagram illustrating inner working components of a transmission electron microscope. (Reprinted with permission from Kaufman, P. B., Labavitch, J., Anderson-Prouty, A., and Ghosheh, N. S., *Laboratory Experiments in Plant Physiology,* Macmillan College Publishing Co., New York, Copyright © 1975.)

PbCitrate, which contain metallic elements with high atomic numbers that attach preferentially to certain types of organelles. Electrons that are not scattered may experience a phase shift at highly localized sites, which results in interference effects that show up as phase-contrast in structures in the range below 1 nm.

After passing through the specimen, the electrons enter the most important component of the electron optical system, the objective lens. This lens serves to produce the primary magnified image of the specimen; hence, its optical quality determines the basic quality of the image produced by the instrument. The highly refined objective lenses of most current electron microscopes can provide resolutions of 0.2 nm or better. These lenses are designed with extremely short focal lengths to minimize aberrations. For routine biological specimens, which do not require such extreme resolution, lenses with somewhat longer focal length may be advantageous because they provide somewhat higher contrast.

The primary image formed by the objective lens usually has a magnification of about 100×. This image is then further enlarged by a series of two or three projector lenses to give the final image, which is displayed on a fluorescent viewing screen or recorded photographically. These lenses, as well as the objective and condenser lenses, of most current CTEMs are electromagnetic lenses. They consist of coils of wire wound on highly refined bobbins of soft iron, which contain magnetic gaps that shape the magnetic fields along the axes of the lenses so that they produce high-quality electron images. By varying the electric currents in the coil windings, the strengths of the magnetic fields can be changed; this, in turn, changes the focal lengths of the lenses making it possible to vary the magnification of the final image continuously from less than 500× to as much as 500,000×.

The fact that the focal length of the lenses can be changed continuously over such wide ranges and with such great ease gives the CTEM greater flexibility than any optical instrument; thus, it is possible to use the TEM to perform a variety of functions in conjunction with basic image formation. These include several special diffraction techniques, operation in scanning mode, and X-ray spectral analysis. A variety of accessories are also available for special types of specimens and for treating specimens in special ways during examination.

## B.      Preparation of Tissues for TEM

Fixation, dehydration, and sectioning tissues for TEM requires about a week. The basic steps include the following:

1. Tissue preparation for fixation
2. Fixation of tissue with $KMnO_4$ (potassium permanganate), glutaraldehyde, and $OSO_4$ (osmium tetroxide)
3. Dehydration steps
4. Embedding in plastic and plastic polymerization
5. Preparing glass knives and use of diamond knives for sectioning
6. Sectioning with the ultramicrotome
7. Mounting sections on grids
8. Staining ultrathin sections
9. Placing sections in the TEM to scan
10. Basic steps in the operation of the TEM

**TEM applications** — Figures 19.12 and 19.13 illustrate two examples of TEM applications. The first is an example of immunolocalization of cell wall-bound invertase in phloem tissue of dwarf pea stems (Figure 19.12). The second is a transmission electron micrograph of three starch-filled chloroplasts (gravisensors in cereal grass shoots) in oat shoot leaf-sheath pulvini (Figure 19.13).

## V.      Confocal Microscopy

Confocal microscopy allows one to image thick biological tissues in three dimensions (3-D) by means of "optical sectioning". This contrasts with conventional microscopy when specimens (smears, peels, sections, or imprints) are observed in two-dimensional projection. It allows one to do quantitative image analysis by obtaining 3-D images from a succession of

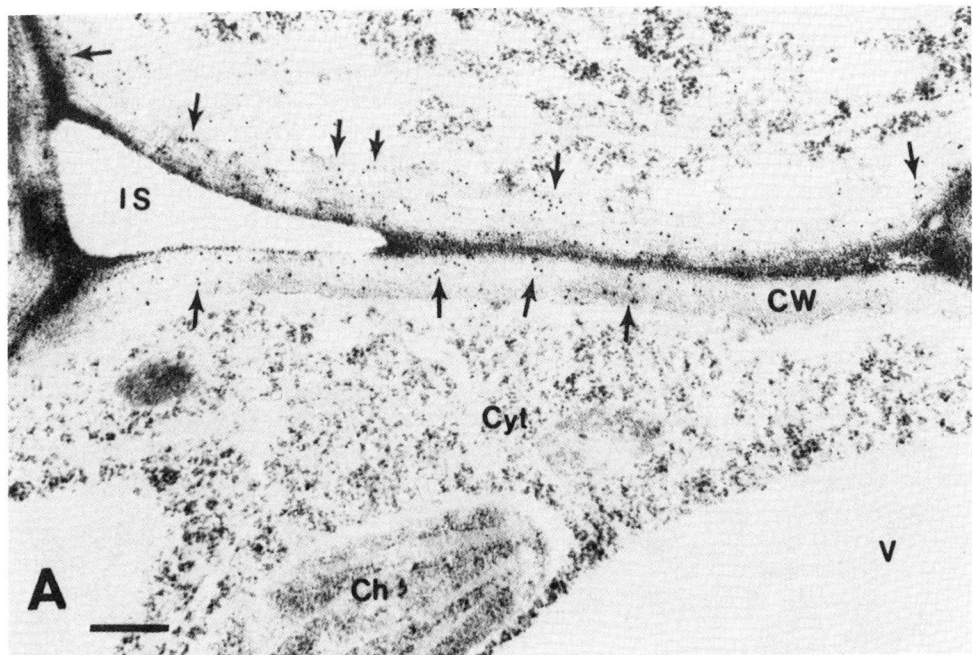

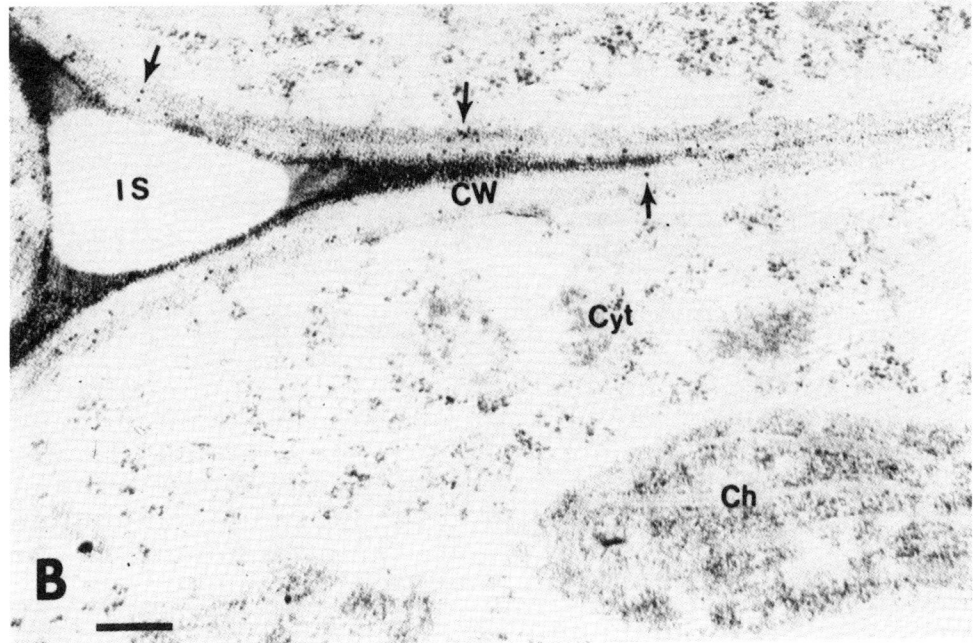

**FIGURE 19.12**

Immunolocalization of the cell wall-bound invertase in the phloem tissue of dwarf pea stems. A, $GA_3$-treated tissue; B, untreated tissue. CW, cell wall; Cyt, cytoplasm; IS, intercellular space; V, vacuole. Bars on electron micrographs ($\times$48,000) equal 0.2 $\mu$m. Some immunogold particles are pointed by arrows. The density of the particle number was statistically analyzed by $x \pm SD \cdot cm^{-2}$ of the micrographs. (From Wu, L-L., Mitchell, J. P., Cohn, N. S., and Kaufman, P. G., *Intl. J. Plant Sci.* 154(2), 280, 1994, © University of Chicago Press. With permission.)

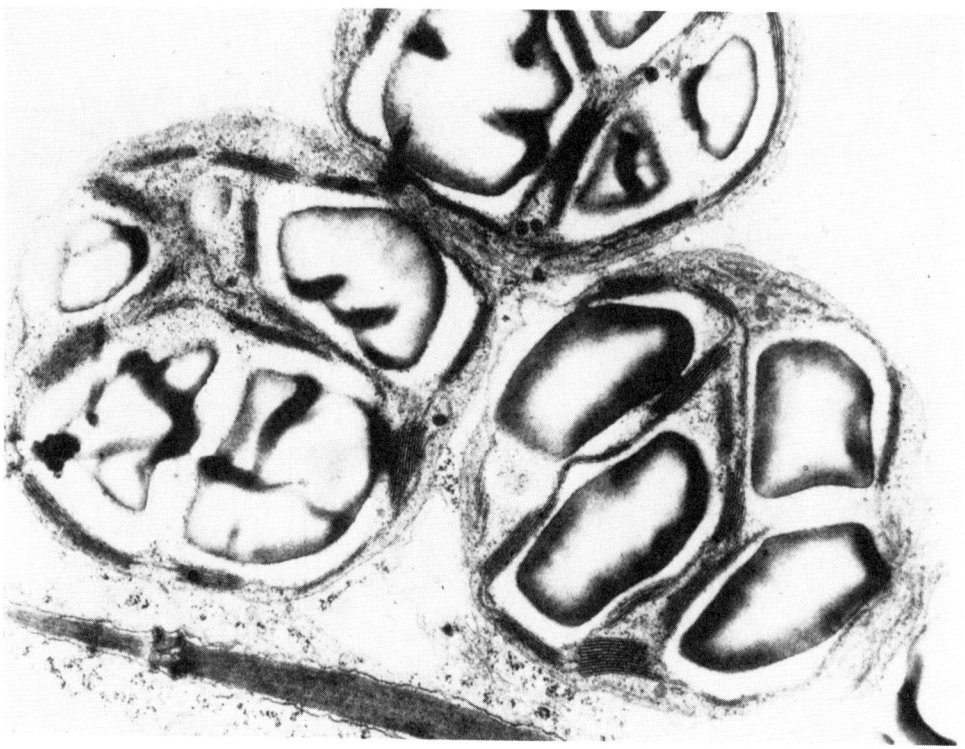

**FIGURE 19.13**

TEM view of three starch-filled chloroplasts of an oat *(Avena sativa)* plant. These plastids, filled with starch, act as the gravisensors in graviresponding cereal grass shoots. (TEM photo provided by Casey Lu.)

thin, independent "optical sections" that are obtained at various levels of unsectioned cell or tissue specimens. These images are obtained by laser beam light source scanning in instruments called confocal scanning laser microscopes (CSLM). Coupled with the CSLM is a computer system that allows one to store successive optical images obtained for a given specimen.

One of the most common applications of the CSLM is to select particular fluorochromes (stains) in order to examine intracellular pH, flavoprotein (FAD), $Ca^{2+}$ distribution and concentrations, and DNA. Markers for DNA include paranosamlin, quinacrine, propidium iodide, acridine orange, and ethidium bromide. Hoescht 33, 342, and DAPI stains require UV excitation, a major problem with CSLM instruments. Indo-1 used for $Ca^{2+}$ imaging and quantitation also has the same problem. Table 19–2 indicates some of the common fluorochromes used for specific applications.

## Table 19–2 Fluorochromes and Their Application

| Use | Fluorochrome | Excitation peak (nm) | Emission peak (nm) |
|---|---|---|---|
| Covalent labeling | Fluorescein-isothiocyanate (FITC) | 490 — 494 | 520 — 521 |
| | Eosin-ITC | 524 — 525 | 548 — 550 |
| | Erythrosin-ITC | 535 — 540 | 558 — 560 |
| | Tetramethylrhodamine-ITC (TRITC) | 541 — 554 | 572 — 573 |
| | Rhodamine X-ITC (XRITC) | 578 — 582 | 601 — 604 |
| | Texas Red sulfonyl chloride | 596 | 615 — 620 |
| | Lissamine rhodamine B sulfonyl chloride | 567 — 570 | 584 — 590 |
| | Cascade Blue | 375 — 378 | 423 — 425 |
| | Cascade Blue | 398 — 399 | 423 — 425 |
| | Phycoerythrin-R | 480 — 565 | 578 |
| | Allophycocyanin | 650 | 660 |
| | Coumarin-phalloidin | 387 | 470 |
| | Bodipy phallicidin | 505 | 512 |
| | Nitrobenzoxadiazole (NBD) | 478 | 520 — 550 |
| | Dansyl hydrazine | 336 | 531 |
| | Indotrimethinecyanines (CY3) | 530 — 550 | 575 |
| | Indopentamethinecyanines (CY5) | 630 — 650 | 670 |
| Nucleic acid labeling | Hoechst 33342 | 340 — 343 | 450 — 483 (DNA) |
| | DAPI | 345 — 350 | 455 — 470 (DNA) |
| | Propidium iodide | 493 — 536 | 623 — 630 (DNA) |
| | Ethidium bromide | 482 — 510 | 595 — 616 (DNA) |
| | ACMA | 430 | 474 (DNA) |
| | Chromomycin A3 | 420 — 445 | 474 — 580 (DNA) |
| | Mithramycin | 420 — 445 | 575 — 580 (DNA) |
| | 7-Aminoactinomycin-D | 523 | 647 (DNA) |
| | Acridine Orange | 480 — 487 | 510 — 520 (DNA) |
| | Acridine Orange | 440 — 470 | 650 (RNA) |
| | Thiazole Orange | 453 — 501 | 480 — 547 (RNA) |
| | Pyronin Y | 497 | 563 (RNA) |
| Thiol conjugates | Lucifer Yellow | 426 — 490 | 525 — 540 |
| | Cascade Blue | 375 — 378 | 423 — 425 |
| | Cascade Blue | 398 — 399 | 423 — 425 |
| Membrane potential | Thiazole Orange | 453 | 480 |
| | $DiOC_n(3)$ | 482 — 485 | 500 — 510 |
| | Tetramethyl-rhodamine ethylester | 549 | 574 |
| Mitochondria | Rhodamine 123 | 505 — 511 | 534 |
| | 2-di$_1$ASP(DASPMI) | 461 | 518 |
| Lipid content | Nile Red | 485 | 525 |
| | NBD phosphatidyl-ethanolamine | 450 | 530 |
| | Diphenylhexatriene (DPH) | 330, 351, 370 | 430 |
| pH | BCECF | 505 — 508 | 530 — 531 (basic pH) |
| | SNAFL- | 479, 508 | 543 (acid pH) |
| | SNAFL- | 537 | 623 (basic pH) |

## Table 19–2 (continued) Flurochromes and Their Application

| Use | Fluorochrome | Excitation peak (nm) | Emission peak (nm) |
|---|---|---|---|
| | SNAFL- | 518 — 548 | 587 (basic pH) |
| | SNAFL- | 574 — 579 | 636 — 640 (basic pH) |
| | DCDHB | 340 — 360 | 420 — 440 (acid pH) |
| | DCDHB | 340 — 360 | 500 — 580 (basic pH) |
| | FD | 490 | 515 |
| Ca²⁺ | Fura-2 | 335 — 340 | 512 — 518 (low Ca) |
| | Fura-2 | 360 | 505 — 510 (high Ca) |
| | Fura-3 | 368 — 373 | 462 — 494 (low Ca) |
| | Fura-3 | 341 | 459 (high Ca) |
| | Indo-1 | 350 | 482 — 485 (low Ca) |
| | Indo-1 | 330 | 390 — 410 (high Ca) |
| | Quin-2 | 332 | 492 (high Ca) |
| | Quin-2 | 352 | 492 (low Ca) |
| | Rhod-2 | 550 | 492 (low Ca) |
| | Fluo-3 | 464 | 525 (low Ca) |

Modified from Table 2 in Häder.[2]

# References

1. **Cherry, R. J., Ed.,** *New Techniques of Optical Microscopy and Microspectroscopy. Topics in Molecular and Structural Biology,* CRC Press, Boca Raton, FL, 1991.

2. **Häder, D-P.,** *Image Analysis in Biology,* CRC Press, Boca Raton, FL, 1992.

3. **Hayat, M. A.,** *Correlative Microscopy in Biology. Instrumentation and Methods,* Academic Press, New York, 1987.

4. **Hayat, M. A.,** *Principles and Techniques of Electron Microscopy,* 3rd ed. CRC Press, Boca Raton, FL, 1989.

5. **Matsumoto, B., Ed.,** *Cell Biological Applications of Confocal Microscopy, Volume 38, Methods in Cell Biology,* Academic Press, New York, 1993.

6. **Monl G.,** *Hybridization Techniques for Electron Microscopy,* CRC Press, Boca Raton, FL, 1993.

7. **Ogawa, K. and Barka, T.,** *Electron Microscopic Cytochemistry and Immunocytochemistry in Biomedicine,* CRC Press, Boca Raton, FL, 1993.

8. **Pawley, J. B.,** *Handbook of Biological Confocal Microscopy,* Revised ed., Plenum Press, New York, 1990.

9. **Shotton D.,** *Electronic Light Microscopy. Techniques in Modern Biomedical Microscopy,* John Wiley & Sons, New York, 1993.

10. **Smith, Robert F.,** *Photomicroscopy and Photomicrography. A Working Manual,* 2nd ed., CRC Press, Boca Raton, FL, 1993.

# Chapter 20

# Bioseparation Techniques and Their Applications

## Contents

This chapter focuses on the primary methods used to extract, separate, and identify biologically important molecules.[1,4,5,7,10-12,14,17,18] We have already covered methods used to separate proteins by 1-d and 2-d gel electrophoresis (SDS-PAGE) and to visualize them using silver and Coomassie blue stains in Chapter 3. In Chapters 1 and 2, we discussed how RNA and DNA are separated by agarose and polyacrylamide gel electrophoresis. In this chapter, we examine: (1) extraction methods for primary and secondary organic metabolites derived from microbial, plant, and animal cells and tissues; (2) purification methods for crude extracts prior to chromatographic separation; (3) chromatographic separation of organic molecules by adsorption chromatography, partition chromatography, and gel filtration or permeation chromatography; and (4) use of mass spectrometry (MS) to identify biologically important molecules. We also include the very recent bioseparation techniques in use today. These include capillary zone electrophoresis and high-performance immobilized metal-ion affinity chromatography.

# I.   An Overview of Chromatographic Separation Techniques

Chromatography is a process of separating gases, liquids, or solids in a mixture or solution by adsorption. For example, separation can be accomplished by selective adsorption on particle clay, silica gel, fine powdered sugar, alumina, or on paper. As an extraction mixture flows through the adsorbent medium or phase, often in a column, each substance in the sample is separated on the basis of differences in hydrophobicity and/or ionic charges. Consequently, each substance appears in the adsorbent medium eluting from the column at a different time. This time difference is called the retention time (Rt); it represents the relative position of a compound(s) with respect to the origin (top) and the bottom of a column.

Primary methods of chromatography in use today include paper chromatography (PC), thin layer chromatography (TLC), liquid column chromatography (LC), gas chromatography (GC), high-performance liquid chromatography (HPLC), fast protein liquid chromatography (FPLC), immobilized metal-ion affinity chromatography, and antibody affinity chromatography. Colored products or bands can be obtained, which can be measured with a spectrophotometer, as with an enzyme assay[18] or with a radioimmunoassay (RIA). Compounds eluting off at particular times during HPLC can also be read spectrophotometrically at wavelengths equivalent to the maximum absorption peaks for such compounds. Compounds appearing at a particular Rt value during TLC can be quantified densitometrically or spectrophotometrically after scraping the compound(s) off the adsorbent phase. Highly volatile compounds (e.g., ethylene and ethane) that exist primarily in the gas phase can be subjected to gas chromatography using a flame ionization detector (FID). GC's also have other types of detectors for selective ion monitoring such as a thermal conductivity detector or even a mass spectrometer.

# II.   General Extraction Procedures for Biologically Important Organic Compounds

The primary ways for extraction of organic molecules of interest to biologists and medical investigators involve breaking open the cells of the organism under investigation. Cell rupturing is accomplished in a variety of ways. The method used depends on the type of organism being dealt with and the type of tissue used. For bacterial cells, one usually uses a French press so as to break open the cell walls. This involves the use of a heavy cylinder with high pressure applied to a piston that compresses the cells into a successively smaller volume within the free cylinder. As the cells leave the cylinder, the rapid drop in pressure causes the cells to lyse. Such a procedure can also be used for plant cells grown in suspension culture or for plant callus tissue. A sonicator can also be used for this purpose. In this case, repeated high-frequency pulses of ultrasonic vibrations rupture the cell membranes. Animal cells and plant cells grown in culture can be ruptured with a glass tissue homogenizer. Highly lignified or silicified plant tissues within organs such as leaves, stems, roots, seeds, or fruits, are usually frozen and pulverized using liquid nitrogen in a mortar and a pestle. Softer plant tissues can be ground in a small volume of buffer in a mortar, using white washed sand and a pestle to rupture the cells.

Once the cells have been ruptured, the actual extraction is performed using techniques that depend on the chemical properties of the compound(s) of interest. Water-soluble compounds and proteins are extracted in water or buffers. Organically soluble compounds are extracted with organic solvents. For example, since taxanes are miscible in methanol, this solvent is often used as the extraction reagent. This is by no means the only solvent that will work. Some compounds, such as cell wall constituents, have no need to be solubilized for extraction because they can be obtained in pellet form by filtration or centrifugation followed by washing with buffer solutions. It is worth noting that integral-membrane proteins often require the use of strong detergents, such as Triton X 100, to be extracted from the membranes.

Two steps are usually critical for good extraction. First, ruptured cells should be ground or homogenized in the extraction solvent, depending on the cell-rupturing technique chosen. For example, taxanes (terpenoid compounds derived from plants) are extracted by grinding the plant tissue in organic solvent (methanol) in the same mortar and pestle that is used along with liquid nitrogen to rupture the cells. Waxes, on the other hand, can be removed from the aqueous phase coming from a French press by homogenizing and partitioning with chloroform and methanol. Second, once ground or homogenized, the extraction mixture should be allowed to stand undisturbed for 0.5 to 24 h at a temperature that will not allow degradation of the compound(s) or protein(s) of interest to occur (e.g., 4°C). This is done simply to allow time for the extraction solvent to penetrate all parts of the ruptured cells.

After these two critical steps, the resulting slurry is filtered to obtain a filtrate free of particulates, or it is centrifuged in order to obtain a cell-wall and/or membrane-pellet fraction and a cell cytosol-containing supernatant fraction. Such crude fractions can be used directly for enzyme reaction assays, or they can be subjected to further purification and clean-up procedures in order to separate and identify the compounds of interest. For example, C-18 Sep Pak™ columns can be used to remove chlorophylls, which interfere with subsequent analysis of taxane extractions.

For the extraction of natural products of potential medicinal or other value in plant samples, as used in the authors' laboratory, the following specific protocols have been employed:

Protocol 1:   Hot Water Extraction of Water-Soluble Medicinal Compounds from Plants

Protocol 2:   Organic Solvent Extraction of Organic Solvent-Soluble Medicinal Compounds from Plants

Protocol 3:   Extraction of Taxanes (diterpenoid compounds) from Yew (*Taxus* spp.) Needles and Stems

Protocol 4:   Extraction of Cuticular Wax from Needles of Yew (*Taxus* spp.) Plants

## A.   Hot Water and Organic Solvent Extraction of Water-Soluble and Organic Solvent-Soluble Medicinal Compounds from Plants

There are so many different ailments and diseases that it is a wonder we ever are healthy. Today, the development of drugs and medicines is a growing industry. U.S. companies are losing the race to be the leading developer of medicines. What are other countries doing that the U.S. is not? Most other countries have made the trend back to nature. They started looking at plants for natural medicines. The U.S. has just joined this trend, but has not been able to produce as many drugs as some other countries, such as Japan. Unfortunately, it takes a lot of money to develop any new drug, and the risk of the drug failing is great.

Many of the plants that have known medicinal values have been used for centuries by different cultures. Natural medicines have been used for centuries in many parts of Asia, such as China and India. In particular, Native Americans have had a profound influence on the natural medicines of today. The Indians of North America have made the largest contribution to this field.

A project conducted in the authors' laboratory at The University of Michigan by DaRhon Conner and Nina Jain involved the preparation of plant extracts from plants that contain natural products of known medicinal value. These extracts are prepared to be screened by Warner-Lambert Pharmaceutical Co. (Ann Arbor, MI). The plants come from a variety of sources ranging from botanical gardens to grocery stores. The plants are cut and stored in a deep-freeze unit at −80°C. These extracts are tested in an 80-sample screen for medicinal compounds in order to determine which are effective against viruses, bacteria, mycoplasms, and fungi that are pathogenic to humans. Many of the plants are naturally toxic and harmful to humans. However, if they are administered in small amounts, the effects on humans are beneficial.

There are two different types of extracts used in this project. One is obtained using hot water and is called an *aqueous extract.* The other uses organic solvents, such as methylene chloride, and is called an *organic extract.* There are two very specific procedures that have been developed for each type of extraction. These procedures are as follows.

### Protocol: Hot Water Extraction of Water-Soluble Medicinal Compounds from Plants

First, we weigh out a 0.5-g sample of a given plant. This is then ground to a fine powder using liquid nitrogen. Next, we place the powder into a test tube full of hot (80°C) water and place the tube into a hot water bath for 10 minutes. The tube is centrifuged at $3000 \times g$ for 10 min. By centrifuging it, all of the particulate plant materials from the grinding get pelleted at the bottom of the tube, leaving a relatively clear liquid containing the water-soluble compounds of interest. This liquid is filtered to make sure that no plant particulates remain in the filtrate. Next, the filtrate is placed in a Petri dish and frozen by placing it on dry ice. After the sample is completely frozen, it is placed in a freeze-drying apparatus and lyophilized. The sample could take anywhere from 2 to 12 h to lyophilize. Freeze-drying is done in order to remove all moisture. This yields a powdered residue, and it is this powder that we send to Warner-Lambert. Figure 20.1 outlines this procedure.

### Protocol: Organic Solvent Extraction of Organic Solvent-Soluble Medicinal Compounds from Plants.

As mentioned above, this procedure uses an organic solvent to extract the compounds of interest. At least one very toxic solvent (methylene chloride) is used and therefore needs to be monitored carefully. The temperature of this extraction must also be carefully watched due to the special equipment that is used. We make use of a Soxhlet extractor, which is basically a specialized glass refluxing unit that is used for organic solvent extractions. The temperature must be maintained at 80°C for 18 h in order to obtain complete extraction of a given sample. If the temperature falls below this, extraction will be slow. If the temperature goes above this, the risk of degrading the compounds of interest becomes great. Figure 20.2 outlines this procedure.

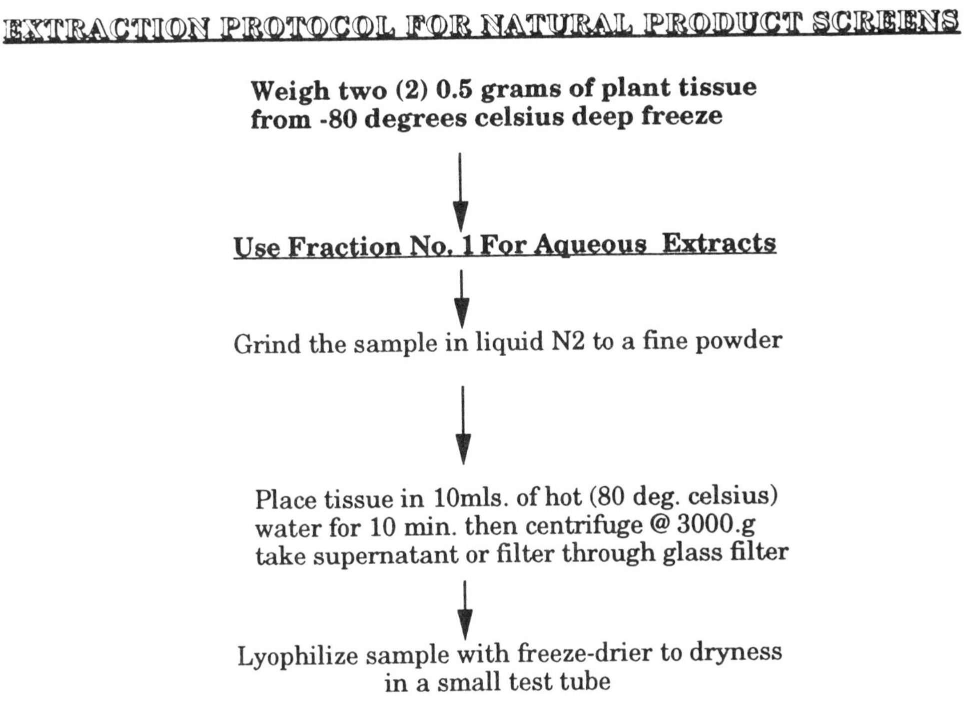

## EXTRACTION PROTOCOL FOR NATURAL PRODUCT SCREENS

**Weigh two (2) 0.5 grams of plant tissue
from -80 degrees celsius deep freeze**

↓

**Use Fraction No. 1 For Aqueous Extracts**

↓

Grind the sample in liquid N2 to a fine powder

↓

Place tissue in 10mls. of hot (80 deg. celsius)
water for 10 min. then centrifuge @ 3000.g
take supernatant or filter through glass filter

↓

Lyophilize sample with freeze-drier to dryness
in a small test tube

**FIGURE 20.1**
Hot-water extraction protocol for water-soluble plant natural product compounds that have potential medicinal value.

## III.  Purification of Crude Extracts Prior to Chromatographic Separation — Taxol and Cuticular Wax Extractions from *Taxus* spp. (Yew) Plants.

For bioseparation, clean-up procedures are usually necessary before samples are analyzed by HPLC or by any of the other chromatographic techniques mentioned above. One application that we have employed is for the analysis of taxol. Taxol is a unique taxane diterpene amide that possesses antitumor and antileukemic properties. Kilograms of this cancer chemotherapeutic agent are needed for clinical treatment of patients having breast cancer; however, taxol exists in only minute quantities — 0.01% of the inner bark and needles of the yew *(Taxus)* species. Until recently, taxol could not be synthesized, and even now, the most economical source of taxol is still from the Pacific yew *(Taxus brevifolia)*. For this reason, large areas of Pacific yew forests in the Pacific Northwest of the United States were destroyed in order to obtain this anticancer drug. The taxol extraction methods used by researchers commonly involve complicated partitioning methods in which the plant is first extracted with methanol and $H_2O$ and then partitioned using methylene chloride to remove chlorophyll and other unwanted compounds. In the process, taxol molecules move from the aqueous methanol to the more hydrophobic methylene chloride. Unfortunately, methylene chloride is a suspected carcinogen, and it seems counterproductive, in our opinion, to extract a cancer chemotherapeutic with a substance that could cause cancer! Hence, it has been our motive in studying taxol to find a cheap, efficient, and easy way to separate taxol from the thousands of other

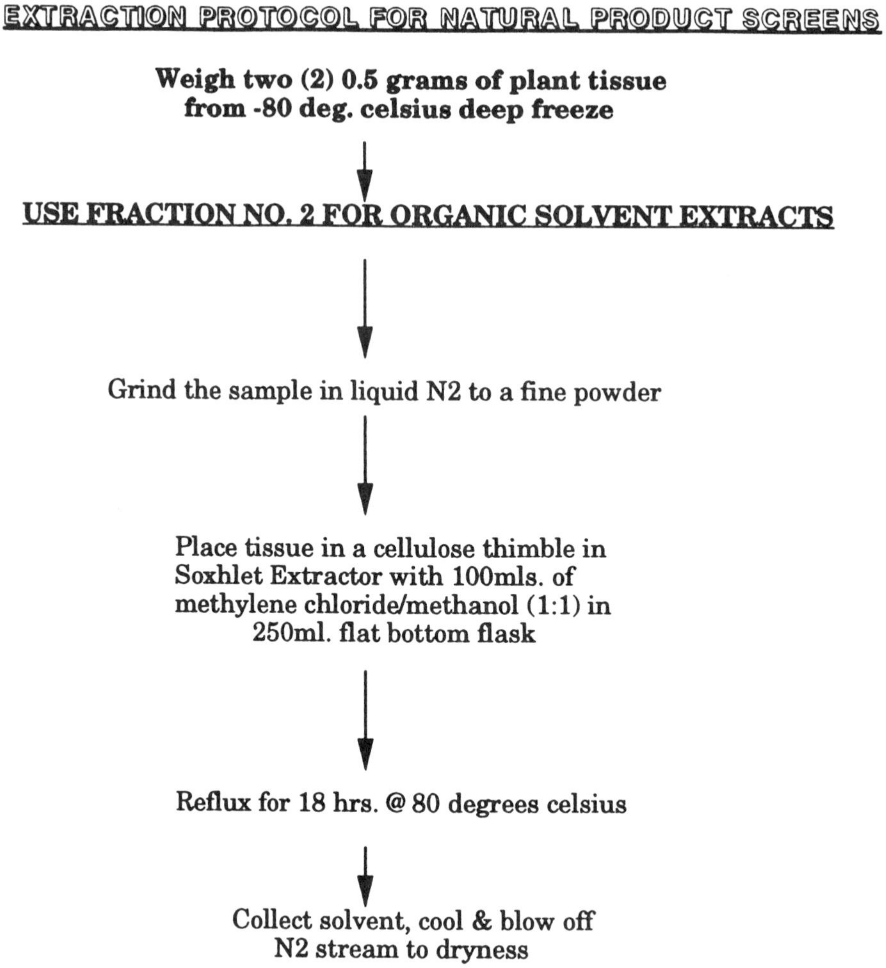

### EXTRACTION PROTOCOL FOR NATURAL PRODUCT SCREENS

**Weigh two (2) 0.5 grams of plant tissue
from -80 deg. celsius deep freeze**

↓

### USE FRACTION NO. 2 FOR ORGANIC SOLVENT EXTRACTS

↓

**Grind the sample in liquid N2 to a fine powder**

↓

**Place tissue in a cellulose thimble in
Soxhlet Extractor with 100mls. of
methylene chloride/methanol (1:1) in
250ml. flat bottom flask**

↓

**Reflux for 18 hrs. @ 80 degrees celsius**

↓

**Collect solvent, cool & blow off
N2 stream to dryness**

**FIGURE 20.2**

Organic solvent extraction protocol for organic solvent-soluble plant natural product compounds that have potential medicinal value.

organic compounds in yew tissues. One of the primary difficulties, in this case, is the removal of chlorophylls from methanolic extracts. Chlorophylls absorb at the same wavelength as taxol and often occur in such large quantities that their resulting peaks interfere with taxane peaks. For this purpose, we use a C-18 reverse phase Sep-Pak™ column. The protocol for doing this follows below. The main point here is that preparatory columns are also useful to remove many unwanted compounds before the actual chromatographic analysis is performed. Figure 20.3 shows the small size of the C-18 Sep-Pak setup as used in our laboratory. HPLC results using C-18 Sep-Pak cleaning vs. no cleaning are shown in Figure 20.4a and b, respectively.

A new, simple, and rapid method that successfully works for extraction and HPLC separation of taxol from crude extracts of *Taxus cuspidata* (Japanese yew) needles and stems has recently been developed in our laboratory and tested by ten groups independently with repeated success. It requires 2 h to perform steps 1 through 21 and 70 min per sample to run them in an automated Shimadzu HPLC apparatus. This long a run time is used to separate the multitude of peaks that result at 228 nm spectrophotometric monitoring when **not** using the

**FIGURE 20.3**
Leland Cseke is shown standing beside the C-18 Sep-Pak™ column apparatus (lower center) used in our laboratory to partially purify yew (*Taxus* spp.) plant extracts in preparation for HPLC analysis of taxane-type diterpenes found in these plants.

C-18 Sep-Pak clean-up method. The advantage of not using clean-up procedures is that the researcher saves time in preparing extracts. In this case, the researcher could go home and sleep while the Shimadzu Sil-6a auto-injector does the work. Figure 20.5 illustrates the Shimadzu automated HPLC apparatus that is used in this procedure.

In some analytical situations, or for cases where publication is desired, it may be necessary to clean the crude extract with a C-18 Sep-Pak™ column. The clean-up procedure is given following the crude extraction protocol below. But this is by no means the only way extracts can be cleaned up.[6] Any form of chromatography can be used to clean up the extracts (see discussion of chromatographic methods). The basic trick is to find the fraction eluted from the clean-up column that contains the compound of interest. In the case of taxol, this was accomplished experimentally, using purified taxol standards dissolved in 10% methanol. At this concentration of methanol, it was found that taxol binds to the C-18 packing material in the column. Then, the concentration of methanol was arbitrarily increased in several steps up to 100% methanol. A fraction of eluted mobile phase (2 to 3 ml) was collected at each concentration increase, and HPLC was performed on each fraction to determine where taxol had eluted off the column. After repeated experiments, the specific concentration of methanol that elutes taxol was identified.

In some cases, chromatography is not necessary for clean-up of the sample. Addition of adsorptive particles such as activated charcoal to the crude extracts followed by filtration may be all that is necessary to remove unwanted compounds.

### Protocol A: Taxol Extraction from Fresh Taxus spp. (Yew) Tissue Resulting in Crude Extract Samples That Can Be Used for HPLC

1. Obtain *Taxus* spp. plant specimens.
2. For each sample, weigh out 0.5 g of needles and 0.5 g of stems.

## Extract Cleaning

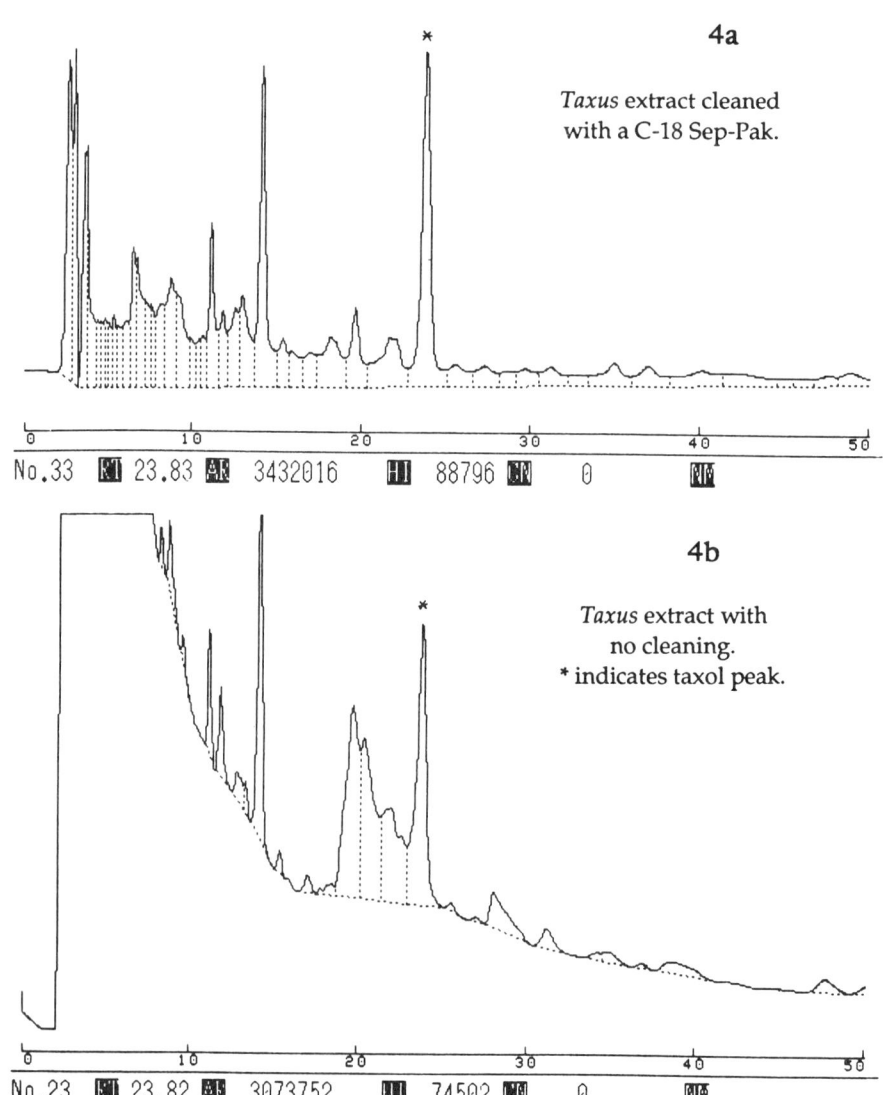

**4a**

*Taxus* extract cleaned
with a C-18 Sep-Pak.

No.33 ▨ 23.83 ▨ 3432016 ▨ 88796 ▨ 0 ▨

**4b**

*Taxus* extract with
no cleaning.
* indicates taxol peak.

No.23 ▨ 23.82 ▨ 3073752 ▨ 74502 ▨ 0 ▨

**FIGURE 20.4**
HPCL taxane analysis results showing chart recordings resulting from C-18 Sep-Pak cleaning (a) vs. no cleaning (b) of extracts from yew (*Taxus* spp.) needles and stems.

3. Grind to a fine powder using liquid $N_2$ in a mortar and pestle.
4. Let the powder warm to room temperature.
5. Add 1 ml of 100% MeOH and grind vigorously.
6. Transfer this slurry to a 15-ml Corex centrifuge tube.
7. Add 2 ml of 100% MeOH and grind the remaining plant material in the mortar.
8. Add this to the tube and label.
9. Add another 1 ml of 100% MeOH to the mortar and rinse again.

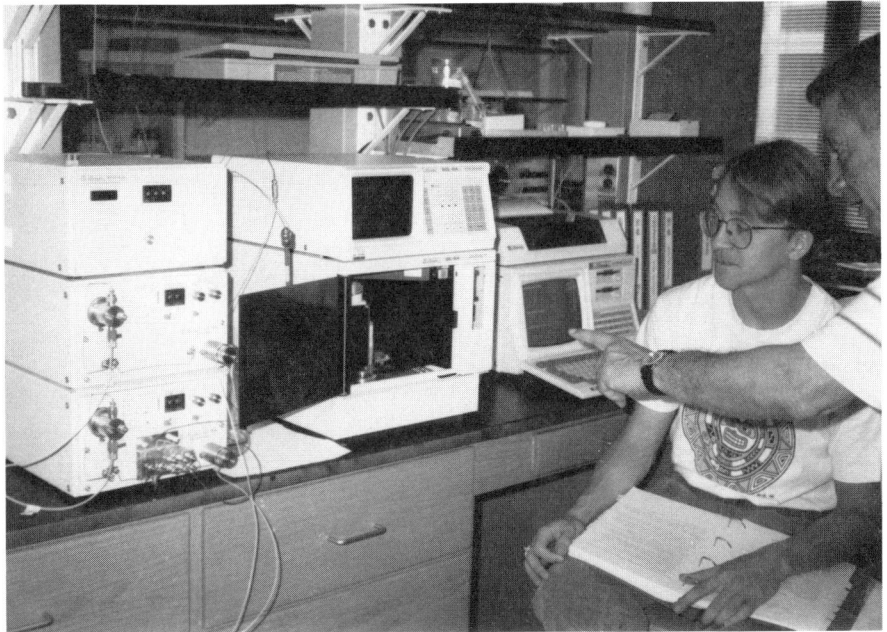

**FIGURE 20.5**

Illustration of the Shimadzu automated HPLC apparatus used in our laboratory. Sitting at the lab bench is Leland Cseke and standing, pointing to the auto-injector, is Najati Ghosheh.

10. Add this to the tube and label.

11. Place Parafilm™ over the tube and vortex or shake for 10 min.

12. Cool at −80°C for 10 min.

13. Balance the tubes by adding more 100% MeOH to the lighter tube.

14. Centrifuge at 12,000 rpm for 15 min.

15. Take off the supernatant and place in a clean, acid-washed tube using a Pasteur pipette; then label.

16. Put this tube into a water bath at 55°C.

17. Blow $N_2$ gas gently into tubes until the samples are completely dry.

18. Add 0.5 ml of ice-cold HPLC-grade 100% MeOH and vortex for several minutes with periodic placement in an ice bath. This allows one to later quantify the amount of taxol extracted because a known volume of sample has been created.

19. When the MeOH is very cool, place the 0.5 ml into a syringe and filter through a 0.2-μm filter into an HPLC bottle.

20. Add an aluminum septum to the top of the bottle and cap.

21. Cover well with Parafilm™ to prevent evaporation and label.

22. Inject the sample into the HPLC apparatus and run through the Curosil G™ column (from Phenomenex Corp.), collecting data at 228 nm. Run each sample for 70 min at 1 ml · min⁻¹ in 36.5% acetonitrile and 63.5% 10 m$M$ ammonium acetate at pH 4.0.

23. Use purified taxol to make several known concentrations of taxol. Run these standards on HPLC as in step 22 to make a standard curve. Figure 20.6 shows an example of a typical standard curve for taxol.

24. Compare your data to the standard curve for taxol and calculate the percent of taxol per unit (g) fresh weight of tissue.

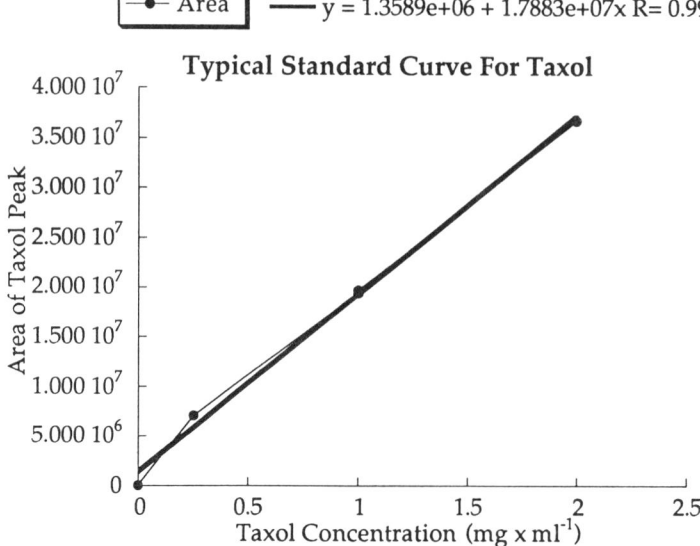

**FIGURE 20.6**
A typical standard curve for taxol as determined by HPLC analysis of taxol standards at a wavelength of 228 nm.

## Protocol B: Taxol Extraction from Fresh Taxus spp. (Yew) Tissue Resulting in Cleaned Extracts That Can Be Used for HPLC–Use of the C-18 Sep-Pak Column

1.  Obtain *Taxus* spp. plant specimens.
2.  Weigh out 0.5 g of needles and 0.5 g of stems.
3.  Grind to a fine powder using liquid $N_2$ in a mortar and pestle.
4.  Let the powder warm to room temperature.
5.  Add 1 ml of 100% MeOH and grind vigorously.
6.  Transfer this slurry to a 15-ml Corex centrifuge tube.
7.  Add 2 ml of 100% MeOH and grind the remaining plant material in the mortar.
8.  Add this to the tube and label.
9.  Add another 1 ml of 100% MeOH to the mortar and rinse again.
10. Add this to the tube and label.
11. Place Parafilm™ over the tube and vortex or shake for 10 min.
12. Cool at –80°C for 10 min.
13. Balance the tubes by adding more 100% MeOH to the lighter tube.
14. Centrifuge at 12,000 rpm for 15 min.
15. Take off the supernatant and place in a clean, acid-washed tube using a Pasteur pipette; then label.
16. Dilute the volume of MeOH supernatant 10-fold with dd.$H_2O$ to produce a 10% MeOH sample.
17. Run all of this diluted sample through a C-18 Sep-Pak column (activated as per instructions from the supplier). Taxol-type molecules will bind to the column at this concentration of MeOH.
18. In the case of taxol itself, wash the column first with 3 ml of 35% MeOH, then with 3 ml of 55% MeOH. This washes compounds that have less affinity for the C-18 adsorbent than taxol out of the column, while the taxol remains bound.

19. Now wash the C-18 column with 2 ml of 65% MeOH and collect the sample. All taxane species of compounds (including taxol, 7-epi-taxol, 7-epi-10 deacetyl taxol, and cephalomanine) are released from the column during this elevation in MeOH concentration, leaving almost all of the problematic chlorophyll still bound to the column-adsorbent phase.

20. The 2 ml of taxol-containing sample is completely dried under a stream of $N_2$ gas in a 55°C water bath while the C-18 column can be cleaned with 100% MeOH and reused.

*Note:*     *No detectable amount of taxol is lost during this procedure.*

21. Add 0.5 ml of ice-cold HPLC-grade 100% MeOH and vortex for several minutes with periodic placement in an ice bath.

22. When the MeOH is very cool, place the 0.5 ml into a syringe and filter through a 0.2-μm filter into an HPLC bottle.

23. Add an aluminum septum to the top of the bottle and cap.

24. Cover well with Parafilm™ to prevent evaporation; then label.

25. Inject the sample into the HPLC apparatus and run through a Curosil G column (from Phenomenex Corp.), collecting data at 228 nm. Run each sample for 30 min at 1 ml · $min^{-1}$ in 40% acetonitrile and 60% 10 m$M$ ammonium acetate at pH 4.0.

26. Use purified taxol to make several known concentrations of taxol. Run these standards on HPLC as in step 25 to make a standard curve (Figure 20.6).

27. Compare your data to the standard curve for taxol and calculate the percent of taxol per unit (g) fresh weight of tissue.

*Note:*     *The run time can be reduced to as little as 10 min by increasing the percentage of acetonitrile. The taxol peak is still clearly resolved in this case.*

## Protocol C: Extraction of Cuticular Wax from Needles of Yew (Taxus spp.) Plants

Waxes are lipids synthesized by plants and animals and serve functionally to keep out infectious organisms (e.g., ear wax in animals) and to prevent desiccation and serve as a barrier against fungal and bacterial pathogens in plants. In yew plants (*Taxus* spp.), waxes are synthesized in increased amounts in response to stresses, such as hyper-g. The protocol we have successfully used to quantitatively analyze cuticular waxes from *Taxus* needles is presented below. This procedure should be adaptable for any species of plant or animal.

1. Weigh out 1.0 g of *Taxus* needles.

2. Dice needles with a razor blade and place in a mortar and pestle.

3. Add a 2:1 mixture of methanol:chloroform to the mortar, and grind thoroughly.

4. Transfer the resulting slurry to a centrifuge tube and centrifuge at 12,000 rpm for 15 min.

5. Remove the supernatant and place in a separate collection flask. This supernatant contains the lipids from the *Taxus* needles.

6. Place the pellet back into a mortar and pestle.

7. Add a 2:1:0.8 mixture of methanol:chloroform:water, and grind thoroughly.

8. Centrifuge this slurry at 12,000 rpm for 15 min.

9. Transfer the supernatant to the above collection flask.

10. Regrind the pellet from the centrifuge tube using the original 2:1 mixture of methanol:chloroform, and repeat steps 8 and 9.

11. While collecting the rinse liquid, vacuum filter the pellet, rinsing with 100% chloroform.

12.  Transfer this rinse liquid to the collection flask.

13.  Transfer all of the collected liquid to a separatory funnel, and add a 1:1 mixture of chloroform:water. This will produce two phases. The lower phase will contain the dissolved lipid.

14.  Empty the bottom layer of chloroform from the separatory funnel and dry it over solid sodium acetate. This absorbs water from the sample.

15.  Weigh a clean rotary evaporator flask (round bottom flask), and record the weight.

16.  Pipette the chloroform solution into the rotary evaporator flask.

17.  Vacuum filter the sodium acetate solid, rinsing with chloroform. Add the collected rinse liquid to the rotary evaporator flask.

18.  Evaporate the chloroform using a rotary evaporator until the sample is completely dry.

19.  Weigh the rotary evaporator flask with the sample, and subtract the recorded initial weight for the empty flask. This will give the amount of cuticular wax from the original sample of *Taxus* needles.

# IV.    Chromatographic Separation of Organic Molecules

Chromatography refers to the separation of chemical compounds by partitioning them between two phases, one of which is stationary and the other is in motion. In this process, the compounds to be separated are distributed between the stationary and mobile phases. This technique is very powerful because it allows one to separate very similar compounds, within a given extract, which may be isomers of each other. It is also fast, easy, and economical.

There are three types of chromatography. These include: (1) adsorption chromatography, (2) partition chromatography, and (3) gel permeation chromatography. Thin-layer chromatography (TLC), gas-liquid chromatography (GLC), affinity chromatography (AC), high-performance liquid chromatography (HPLC), and gel filtration chromatography are adaptations of these three basic types of chromatography, as will be explained in the subsequent sections.

## A.    Adsorption Chromatography

In adsorption chromatography, the compounds of interest are separated by allowing them to adsorb (bind to) to the surface of a solid phase, such as DEAE cellulose (diethylaminoethyl cellulose) or charcoal. The compounds are desorbed (removed) from the solid phase by an eluting solvent such as NaCl of varying concentrations (e.g., a linear gradient of increasing NaCl concentrations from 0 to 1.0 $M$) or by shifting the pH of the mobile phase to lower or higher pHs. The solid phase is poured as a slurry into a column fitted with sintered glass at its base. The column typically has a length-to-width ratio of 10:1 (of the actual poured solid phase). A filter paper disk is added to the top of the buffer-covered (about 1 cm of buffer) solid phase, over which the compounds of interest are gently layered without disturbing the solid phase. Elution is accomplished with the help of gravity and a gradient maker. A fraction collector can be used to collect 1-, 2-, 5-, or 10-ml fractions for later analysis of enzyme activity (using enzyme assays), salt composition and concentrations (using a conductivity meter or a spectrophotometer), and protein composition and concentrations (using Bradford or Lowry or other protein assays in conjunction with a spectrophotometer).

One specialized type of adsorption chromatography is termed **affinity chromatography.**[12] In this case, a receptor, such as an antibody, is linked by covalent bonds to an inert solid support phase. The receptor has a high binding affinity for one of the compounds (its ligand) in the mixture of compounds from the prepared extract. Such binding is both specific and reversible. Inert solid support phases include cross-linked dextrans, cross-linked polyacrylamide, cellulose, and agarose. Due to its great selectivity, affinity chromatography offers a very powerful means of achieving excellent separation and purification of biological molecules in as little as one step. Applications include purification of proteins (including antibodies and enzymes), nucleic acids, or any compound that acts as a ligand for a given bindable receptor. For example, some drugs, such as taxol, can be purified by using their monoclonal antibody as the bound receptor. After eluting unwanted compounds from the column, the bound ligand is then easily eluted from the column by shifting the mobile phase to a low (or in some cases a high) pH. This procedure can be done by immobilizing the ligand or by immobilizing the ligand's receptor to purify the ligand.

One hybrid modification of affinity chromatography and ion-exchange chromatography is **high-performance immobilized metal-ion affinity chromatography** (HPIMAC) used to separate peptides and other organic molecules.[15] It utilizes several types of stationary phases: synthetic polymers, silica, or cross-linked agarose. Basically, chromatographic separation involves a metal-ion chelator bound to the solid support phase. This in turn binds positively charged metal ions. The metal ions frequently used are $Zn^{2+}$, $Mg^{2+}$, and $Cu^{2+}$. Negatively charged side groups on a given protein (or other negatively charged species such as polysaccharides, nucleotides, or nucleic acids) bind to the positively charged metal ions via ionic interactions. Nonbound compounds simply pass through the column and elute off whereas the bound compounds stick to the solid phase until they are eluted off by use of pH shifts or a selective change in ionic strength.

Another variation of adsorption chromatography is **ion-exchange chromatography.**[11] Here, the solid adsorbent phase has charged groups that are linked chemically to an inert solid matrix. What happens during the chromatography is that ions become electrostatically bound to the charged groups of the solid adsorbent. These ions may then be *exchanged* for ions in the mobile aqueous phase. This is accomplished by changing the ionic strength, or pH of the eluting solvent. Two types of ion exchangers are used in ion exchange chromatography: (1) cation exchangers, which are exchangers with chemically bound negative charges and (2) anion exchangers, which are exchangers with chemically bound positive charges. On the exchangers, the charges are balanced by counterions. For this purpose, chloride ions ($Cl^-$) are used for anion exchangers and positively charged metal ions are used for cation exchangers. To elute the molecule of interest from such ion-exchange columns, one can use (1) changes in pH of the eluting buffer, (2) increasing ionic strength of salt (e.g., [NaCl] or [KCl]) in solution, and (3) affinity selection, which depends on both charge (opposite to that of the bound macromolecule), and specific affinity for the bound macromolecule.

## B.    Partition Chromatography

Partition chromatography, often called liquid–liquid partition chromatography, involves two liquid mobile phases. The substances to be analyzed are separated based on their different solubilities in the two liquid phases. An inert support is used in this type of chromatography. Examples of such inert supports include sheets of paper (cellulose) as used in paper chromatography, or a thin layer of silica gel ($SiO_2 \cdot nH_2O$) or powdered alumina on a glass plate, as used in thin-layer chromatography (TLC).

**Paper chromatography** is usually carried out in a large glass tank or cabinet and involves either ascending or descending flow of the mobile phase solvents. Descending paper chromatography is faster due to gravity facilitating the flow of solvents. Large sheets of Whatman No. 1 or No. 2 filter paper (the latter is thicker) are cut into long strips (e.g., 22 × 56 cm) for use in descending paper chromatography or a wide strip of paper (e.g., 25 cm wide) of variable height is used for ascending paper chromatography.

For descending liquid–paper chromatography, substances to be separated are applied as spots (e.g., 25 mm apart) along a horizontal pencil line placed down from the V-trough folded top of the paper. The V-trough folded paper is placed in a glass trough, held down by a glass rod, and when the tank has been equilibrated (vapor-saturated) with "running solvents" (mobile phase), the same solvent is added to the trough via a hole in the lid covering the chromatography tank. The lid is sealed onto the chamber with stopcock grease in order to make the chamber airtight. After the mobile phase trails to the base of and off the paper sheet, the paper is hung up to dry in a fume hood where it can then be sprayed with reagents (e.g., Ninhydrin reagent for amino acids) that give color to the separated compounds of interest in white or UV light. Some compounds of interest have their own distinctive colors, such as chlorophylls, and hence can be purified using this technique. In other cases, the dyes used to stain the location of the compound or protein of interest causes irreversible covalent changes to the compound. In these cases, purification is not possible.

In ascending paper chromatography, the same basic set up and principles apply with the exception that the mobile phase is placed at the bottom of the tank. Separation is achieved when the mobile phase travels up the paper via capillary action.

Another type of liquid–liquid partition chromatography is **thin-layer chromatography** (TLC), which has several advantages over paper chromatography: (1) greater resolving power, (2) faster speed of separation, and (3) availability of a diverse array of adsorbents. The first two of these advantages is attributed to the fine particle size of the solid support adsorbent (less than 0.1-mm diameter particles), which allows more contact of this solid support with the compounds of interest as they travel up the plate. The adsorbents (e.g., silica gel, alumina, cellulose, and derivatives of cellulose) are available commercially on glass plates of various sizes. TLC plates are used in glass tanks, using ascending, or in some cases, descending chromatography. For the former, 1- to 10-μl samples of interest are spotted at 2- to 3-cm intervals across a line 15 to 20 mm from the base of the plate. The spots are allowed to dry; then the plate is placed in the glass chromatographic tank with the solvent previously placed in the bottom of the tank to a depth of 10 mm. Often it is necessary to equilibrate the vapor in the tank by placing filter paper around the sides of the tank. Next, a lid is sealed to the top of the tank with stopcock grease and the solvent is allowed to rise by capillary flow to the top of the plate. Once the mobile phase reaches the top of the plate, the plate is removed and allowed to dry. Then the spots are developed with appropriate reagents for the types of compounds being separated and assessed. However, this procedure, as in paper chromatography, may result in sample destruction.

TLC can be run in two dimensions, using different solvent systems, as can paper chromatography, in order to allow for better separation of compounds. This procedure is very similar to 2-D electrophoresis, which is discussed in Chapter 3. TLC is widely used to separate lipids, fructans, sugars, and hormones.

**Gas-liquid chromatography** (GLC) is another type of partition chromatography where a high-boiling point liquid is the stationary phase and an inert gas is the mobile phase. There is also an inert solid packing used in columns where these two phases are separated. Separation of compounds of interest is achieved due to differentiated solubility of the compounds in the mobile and stationary phases. Thus, as the carrier gas passes through the column, the compounds in the sample come off the column at different times (Rt). A GLC apparatus basically consists of a tank of carrier gas (e.g., helium or nitrogen), an oven containing a coiled metal or glass chromatography column, a sample-injection port, a detector (e.g., a flame ionization detector (FID) or thermal conductivity detector), and a recorder.

With FIDs, hydrogen gas is used to provide fuel for the flame; this is coupled to a flow of air to the detector to provide oxygen that allows the hydrogen to burn. A wire loop is positioned above the flame to detect compounds that pass from the column to the flame; this in turn is connected to the recorder. FIDs are very sensitive to most organic compounds, but not to water, carbon monoxide, carbon dioxide, or the inert gases. Obviously, samples are destroyed when using this type of detector. Thus, GLC is usually not used for purification of compounds. Thermal conductivity detectors, on the other hand, are less sensitive than FIDs. However, they are nondestructive to the samples, and this allows one to completely recover a sample.

One successful application of GLC to the analysis of natural products in plants is the use of capillary GLC to analyze the monoterpene, menthol, in peppermint glandular hairs from the leaves.[8]

**High-performance liquid chromatography** (HPLC) can also be grouped into this section on partition chromatography because it uses the same principles discussed above. It has two main advantages. First, it uses a pump to force the mobile phase through a given type of column at high pressure. The column can be made (or usually purchased) using any of the solid absorbent phases discussed above. Hence, HPLC is commonly used to shorten the running times of any of the above types of chromatography, which are usually time-restricted by gravity or capillary action. Second, because high pressure is used, much smaller adsorbent solid support particles can be used in the columns in conjunction with much smaller column volumes. Together, these two factors allow much better resolution of the compounds passing through the column due to the increased contact of the compound with the solid support adsorbent. **Fast protein liquid chromatography** (FPLC) works in the same manner as HPLC, but it makes use of specialized columns for use in protein purifications. The main disadvantage of HPLC or FPLC is the high cost of the specialized equipment. There is also the disadvantage of the adsorbent only being able to detect substituents on the surface of the compounds or proteins of interest, but this is a disadvantage in all forms of chromatography.

## C.    Gel Filtration or Permeation Chromatography

This type of chromatography is often called **gel filtration chromatography**.[10] It involves the use of porous gel molecules of agarose, cross-linked dextran, or polymers of acrylamide, allowing the separation of compounds based on their molecular sizes/weights. One commercial series called Sephadex™ (Pharmacia Fine Chemicals, Inc.) is used for this purpose. These types of column packings must be hydrated before they are functional as a separation medium. The hydration process causes the pores in the Sephadex to swell to the appropriate size for the given Sephadex type. For example, G-10 Sephadex, during hydration, gains 1 ml of water per gram of dry gel; G-200 Sephadex gains 20 ml of water per gram of dry gel. Bio-gels™ from Bio-Rad Laboratories consist of long polymers of acrylamide that are cross-linked to $N,N'$-ethylene-*bis*-acrylamide. These gels have a larger range of pore sizes than the Sephadex G series. Still another porous gel with an even wider pore size is agarose. It is made up of the neutral polysaccharide fraction from agar. Agarose and polyacrylamide are used to separate viruses, ribosomes, nucleic acids, and proteins. Sephadex is widely used in purification of proteins and in determining their molecular weights.

The general rationale for separation is as follows: (1) a gel having an appropriate pore size is chosen in relation to the size of the molecule of interest; (2) samples are added to the top of the gel column and are washed through using an appropriate mobile phase that is based on the solubility of the molecule of interest; (3) molecules that are too big to fit in the pores of the solid support will travel around the gel particles, and hence, elute first from the column; (4) molecules that fit in the pores will elute at different times according to their mobility through the gel pores.

Very fast, efficient separation of macromolecules is now possible by a technique termed **capillary zone electrophoresis**.[2] This is not a form of chromatography, but it deserves recognition as an extremely powerful technique for separating compounds of interest via electrophoresis. Basically, capillary electrophoresis utilizes small-bore open capillary tubes (e.g., 200 µm internal diameter) in a system equipped with a grounded high-voltage power supply, solvent reservoir in a Plexiglas™ box connected to the capillary tube, a detector, a solvent reservoir after the detector, and a power supply for current flow to ground. Once a very small sample (e.g., nanogram quantity) is loaded into the capillary tube at one end, negatively charged species, such as the negatively charged side groups of proteins, are separated by the same mechanism as ordinary electrophoresis. The major advantage here is that the capillary tube has a large surface-to-volume ratio allowing rapid dissipation of the heat produced by the electric current. Consequently, much higher voltages can be used in capillary zone electrophoresis than can be used in normal electrophoresis. High voltages in normal electrophoresis tend to cause heat convection within the gel. This results in distortions and blurring of the separation bands. High voltages in capillary zone electrophoresis allow much better resolution of related species of compounds as well as much faster running times. In addition, the capillary tube's inner surface is negatively charged, and thus, it attracts positively charged species of molecules. As the buffer travels through the capillary tube via the electrical current, there is an electroosmotic flow produced that carries these positively charged species of molecules in the same direction as the negatively charged species. Hence, both negatively and positively charged species can be separated and analyzed at the same time. Capillary zone electrophoresis utilizes several types of detectors including spectrophotometers, MS, electrochemical detectors, and radiometric detectors instead of the cumbersome stains used in ordinary electrophoresis.

## D.    Use of Mass Spectrometry to Identify Biologically Important Molecules

In order to analyze the compound(s) or protein(s) of interest, one must be able to unambiguously identify their positions or retention times (Rt) on or through the adsorbent phase used in any given type of chromatography. For example, HPLC chart recordings produced by a chart recorder monitoring a spectrophotometer during a sample run will produce a number of peaks whose identities are not necessarily known. In order to determine which peak is the compound or protein of interest, known standards are run on HPLC, prior to running the collected sample. This will determine the Rt of the compound or protein of interest. The Rt is the time in minutes on the chart recording where a given peak of interest occurs. Unknown compounds whose Rt are similar to those of the standards can be tentatively identified using multiple forms of chromatography. However, a mass spectrometer (MS) or nuclear magnetic resonance (NMR) analysis of the collected unknown peak must be performed in order to unambiguously identify the compound of interest. A cogent application is the characterization of taxol by MS performed by McClure et al.[9] and by NMR performed by Falzone et al.[3]

The amount of the compound of interest under a given peak on a chromatogram is determined by measuring the area under the peak for this compound. Several methods can be used:

1. Measure the half-peak height times its width at the base and compare with the areas of standard peaks analyzed in the same manner to produce a standard curve (see Figure 20.6). The advantage of this technique is that it is very easy. However, only very crude estimations of the area, and hence, concentration, are obtainable using this method.

2. Cut out the curve on the chart paper and weigh it in comparison to the weight of a known standard peak. This procedure is slightly more accurate than procedure 1 — but only if the paper is of uniform density.

**FIGURE 20.7**
View of the Shimadzu C-R4A Chromatopac electronic integrator used to analyze and record areas under
successive peaks obtained from running samples on an HPLC apparatus.

3.  Integrate the area under the peak using an electronic integrator/scanner (see Figure 20.7 for a view
    of the Shimadzu C-R4A Chromatopac electronic integrator used in our lab). This is by far the most
    accurate of these procedures because computer-controlled integrators can implement very
    complicated mathematical algorithms for integration, giving very accurate results. This procedure
    also has the advantage of requiring no mathematical manipulations by the researcher.

It is very difficult to say exactly what precise compound is contained in a given peak from
a purified extract that shows up from HPLC. Mass spectrometry (MS) is very useful to clear
up such ambiguities. MS functions by bombarding a compound with high-energy electrons (or
other particles), causing the loss of an electron from each molecule, yielding *molecular ions.*
If the energy of the electron beam is high enough, the molecular ions will have enough excess
vibrational and electronic energy to break the molecules into various positive and negative ion
fragments. These ion fragments are then passed through a very strong magnetic field which,
depending on the charge of the fragment, deflects the fragments at an angle into a detector.
This results in a spectrum of the compound's ion fragments. Every compound has its own
unique mass spectrum. A good example is the mass spectrum for taxol, which has been
determined by fast atom bombardment (FAB) MS.[9] In that study, three ion series were
observed: (1) the M-series, which are characteristic of the intact taxol molecule; (2) the T-
series whose ion fragments are derived from the taxane ring; and (3) the S-series that
represents the C-13 side chain of taxol (see Figure 20.8).

Gas chromatography (GC) can be combined with MS to first separate the compounds of
interest in an extract according to their boiling points. Then, as each compound comes through
the MS, it is broken up by the particle beam and produces a new mass spectrum. The mass
spectrum for each compound is then compared to the known library of mass spectra that can
be easily accessed from a computer to tell you exactly what your compound is with no
ambiguity. This technique has been successfully used by Pichersky and Raguso et al. to
characterize monoterpene scent compound production in flowers.[13,16]

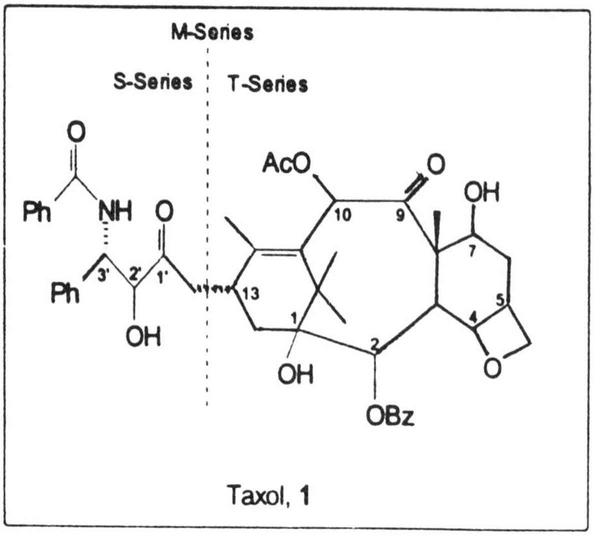

**FIGURE 20.8**

Illustration of the taxol molecule and the 3-ion series observed following fast atom bombardment (FAB) mass spectrometry of taxol. (From McClure, T. D., Schram, K. H., and Reiner, M. L. J., *J. Am. Mass Spectrom.*, 3, 672, 1992. With permission from Elsevier Science Inc. © American Society for Mass Spectrometry.)

# References

1. **Bruno, T. J.,** *Chromatography and Electrophoretic Methods,* Prentice-Hall, Englewood Cliffs, NJ, 1991.

2. **Ewing, A. G., Wallingford, R. A., and Olefirowicz, T. M.,** Capillary electrophoresis, *Anal. Chem.,* 61(4), 271, 1989.

3. **Falzone, C. J., Benesi, A. J., and Lacomte, J. T. J.,** Characterization of taxol in methylene chloride by NMR spectroscopy, *Tetrahedron Lett.,* 33(9), 1169, 1992.

4. **Heftmann, E.,** *Chromatography,* 5th ed., part B: Applications, Journal of Chromatography Library, Vol. 51 B, Elsevier, New York, 1992.

5. **Heftmann, E.,** *Chromatography,* 5th ed., part A: Fundamentals and Techniques, Journal of Chromatography Library, Vol. 51 A, Elsevier, New York, 1992.

6. **Ketchum, R. E. B. and Gibson, D. M.,** Rapid isocratic reversed phase HPLC of taxanes on new columns developed specifically for taxol analysis, *J. Liquid Chromatogr.,* 16 (12), 2519, 1993.

7. **Lehman, J. W.,** *Operational Organic Chemistry: A Laboratory Course.* 2nd ed., Allyn and Bacon Inc., Needham Heights, MA, 1988.

8. **McCaskill, D., Gershenzon, J., and Croteau, R.,** Morphology and monoterpene biosynthetic capabilities of secretory cell clusters isolated from glandular trichomes of peppermint (Mentha piperita L.), *Planta,* 187, 445, 1992.

9. **McClure, T. D., Schram, K. H., and Reiner, M. L. J.,** The mass spectrometry of taxol, *J. Am. Soc. Mass Spectrom.,* 3, 672, 1992.

10. **Pharmacia Fine Chemicals,** *Gel Filtration Theory and Practice,* Pharmacia Fine Chemicals AB, Uppsala, Sweden, 1980.

11. **Pharmacia Fine Chemicals,** *Ion Exchange Chromatography Principles and Methods,* Pharmacia Fine Chemicals AB, Uppsala, Sweden, 1980.

12. **Pharmacia Fine Chemicals,** *Affinity Chromatography Principles and Methods,* Pharmacia Fine Chemicals AB, Uppsala, Sweden, 1979.

13. **Pichersky, E. Raguso, R. A., Lewinsohn, E., and Croteau, R.,** Floral scent production in *Clarkia breweri.* I. Localization and developmental modulation of monoterpene emision and linalool synthase activity, *Plant Physiol.,* in press.

14. **Poole, C. F. and Poole, S. K.,** *Chromatography Today,* Elsevier, New York, 1991.

15. **Porath, J.,** High-performance immobilized-metal-ion affinity chromatography of peptides and proteins, *J. Chromatog.,* 443, 3, 1988.

16. **Raguso, R. A. and Pichersky, E.,** Floral volatiles from *Clarkia breweri* and *C. concinna* (Onagraceae): recent evolution of floral scent and moth pollination, *Plant Systematics and Evolution,* in press.

17. **Robyt, J. F. and White, B. J.,** *Biochemical Techniques. Theory and Practice,* Brooks/Cole Publishing Co., Monterey, CA, 1987.

18. **Segal, I.,** *Enzyme Kinetics,* Wiley Interscience, New York, 1975.

Chapter **21**

# Preparation of Monoclonal and Polyclonal Antibodies Against Specific Protein(s)

## Contents

# I.    Introduction

Antibodies are proteins in nature and have long been used as molecular probes in several aspects:[1] (1) detection, measurement, and purification of biological molecules of interest; (2) *in situ* immunocytochemical localization of specific protein/enzyme in cells and tissues; (3) immunoscreening of expressional cDNA libraries to identify cDNAs of interest; and (4) treatment of some human diseases. The preparation of antibodies is based on the fact that the immune system of an animal has specific immune responses to and produces antibodies against foreign substances (antigens) such as carbohydrates, nucleic acids, and proteins. The antibodies produced are then secreted into the serum by lymphoid cells in several organs including bone marrow, spleen, and lymph nodes.[2] The serum can be obtained by bleeding an immunized animal. There are two types of antibodies:[1-5] (1) monoclonal antibodies raised in mice, and (2) polyclonal antibodies that are usually raised in rabbits. Monoclonal antibodies contain a single antibody specificity, a single affinity, and a single immunoglobulin isotype. However, a preparation of polyclonal antibodies has a variety of antibody molecules directed against the antigen as well as antibodies that do not react with the antigen of interest. These differences account for the fact that monoclonal antibodies are more specific as compared with polyclonal antibodies. This chapter describes in detail the preparation of both monoclonal and polyclonal antibodies against protein/enzyme of interest.

# II.    Production of Monoclonal Antibodies

Monoclonal antibodies are produced by a monoclonal population of cells that are derived from a single cloned cell, so that all the molecules are identical to each other. The general procedures include: (1) purification of the protein of interest as an antigen;[1,2] (2) *in vivo* immunization of mice by injecting the antigen; (3) removal of plasma cells in the spleen from the immunized mice and hybridization of plasma cells with a myeloma cell line; (4) cloning of cell colonies; (5) screening and selection of a monoclonal population of cells; and (6) harvesting and purification of the monoclonal antibodies secreted into the medium (Figure 21.1).

## Protocol A: Purification of Antigen for Immunization

The general procedures are described in Chapter 3. It is recommended that the protein of interest as an antigen be as pure as possible. Pure antigens can make the procedure of screening and selection of monoclonal colonies of cells much easier. For glycoproteins, deglycosylation should be carried out because the animal can raise antibodies against the sugar residues, which will reduce the specificity of the monoclonal antibodies and may result in potential cross-reactions between the antibodies and a variety of proteins.

## Protocol B: In Vivo Immunization of Mice with the Purified Antigen

1.    Resuspend the purified antigen in PBS in a sterile microcentrifuge tube to a concentration of 1 to 5 μg/μl, and use 50 to 500 μg per injection depending on the particular antigen to be used.

2.    Combine 100 μl (100 to 150 μg) of the antigen sample with an equal volume of complete Freund's adjuvant to a final volume of 200 μl for the first injection per mouse. Mix thoroughly to obtain an emulsion using a syringe or a pipette. Slowly take up the emulsified mixture with a 1-ml disposable syringe equipped with a 18- or 20-gauge needle and remove air bubbles.

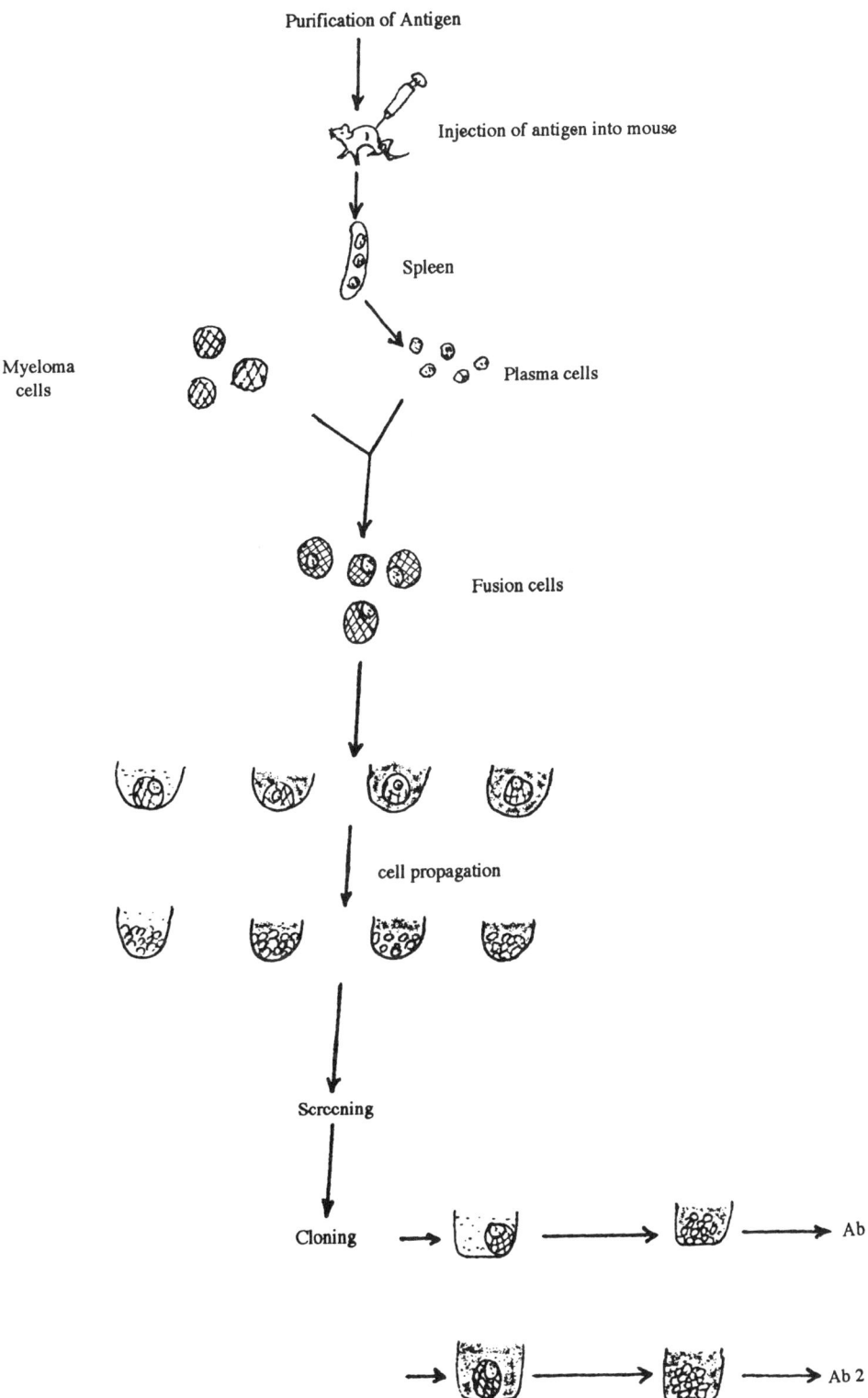

**FIGURE 21.1**
Schematic representation of procedures used to produce monoclonal antibodies.

3.  Take a blood sample (0.5 to 1.0 ml) as a preimmune serum from the vein behind the eye of a mouse before the first injection. Allow the blood to clot by placing the sample at 4°C for approximately 30 min, and centrifuge at $7000 \times g$ for 10 min. Transfer the serum to a fresh tube, dilute it at 1:100–500 with PBS and store at –20°C until use.

4.  Carry out immunization of two to four mice (BALB/c strain, 6 to 8 weeks old) by injecting the antigen mixture (200 µl per mouse) into the mice. It is recommended that one should have a helper to hold the mouse for injection or use an appropriate restraining device if no helper is available. Briefly clean the area for injection with a cotton swab wetted with 70 to 75% ethanol and inject the mixture into the mouse using one of the following methods:

    a.  Intravenous injection into the tail vein; this is the best method for stimulation of the spleen

    b.  Intraperitoneal injection (do not push the needle too deep so as to avoid injecting the needle into stomach)

    c.  Subcutaneous or intramuscular injection into the thigh muscle

5.  Place the mice back in their cage(s). If the mice were injected with the same antigen, they can be caged together. If different mice were immunized by different antigens, they must be labeled properly and kept in different cages.

6.  Mix 100 µl (100 to 150 µg) of the antigen with 100 µl of incomplete Freud's adjuvant to obtain an emulsion and then make the first booster injection 4 weeks after the first injection (primary injection).

7.  Repeat booster injections two to three times at 4-week intervals as in step 6.

8.  Take a blood sample from the same source of mice as immune serum in order to check the antibody titre by the use of enzyme-linked immunosorbent assay (ELISA) using the preimmune serum as a negative control, and purified mouse immunoglobulin standard as a positive control.

9.  Boost the mice with another 100 µl (100 µg) of antigen mixed with 100 µl of Freund's incomplete adjuvant 3 to 4 days prior to doing the cell fusion.

## Reagents Needed

### Phosphate-Buffered Saline (PBS)

$KH_2PO_4$ (anhydrous) (0.92 g)
$K_2HPO_4 \cdot 3H_2O$ (6.39 g)
NaCl (28.6 g)
Add $H_2O$ to a final volume of 3.5 l. The pH should be 7.2 to 7.4.
Autoclave.

### Complete Freund's Adjuvant

Sigma Chemicals

### Incomplete Freund's Adjuvant

Sigma Chemicals

## Protocol C: Determination of Mouse Immunoglobulin Concentration by ELISA

1.  Design a 96-well ELISA plate such that the following should be included for each aliquot of immunoglobulin (Ig) supernatant to be tested:

| Components | Ig supernatant | Control 1 (negative) | Control 2 (positive) | Control 3 (positive) | Blank |
|---|---|---|---|---|---|
| Antigen (Ag) | + | + | + | + | – |
| First antibody (1° Ab) (immune serum or Ig supernatant) | + | – | – | – | – |
| Mouse Ig standard | – | – | – | + | – |
| Preimmune serum | – | + | – | – | – |
| Second antibody (2° Ab) | + | + | – | + | – |
| Substrate (S) | + | + | + | + | – |

*Notes:* *(1) Five concentrations (0.001 to 1 μg/μl) of mouse immunoglobulin standard should be used in the same plate or another plate under the same conditions in order to generate a standard curve. (2) For each of the above controls including the Ig supernatant to be assayed, negative controls, blank control, and five different concentrations of mouse Ig standard, two to four wells of replicates should be set up in order to obtain an average assay value. A 96-well ELISA plate has 12 columns named in the order from 1 to 12, and 8 rows from A to H. All 96 wells can be used for multiple samples, or 48 wells for a few samples, leaving the other 48 wells used as blanks.*

2.  Dilute the antigen sample to 50 μg/ml in PBS (pH 7.2). According the above design, coat appropriate wells of a disposable ELISA plate by adding the antigen to the wells (50 μl per well). Cover the plate and incubate at room temperature for 2 h or at 4°C overnight.

3.  Decant the antigen solution from wells by dumping the plate upside down on a paper towel two to three times. Fill each well (100 to 200 μl per well) with 1 to 3% (w/v) bovine serum albumin (BSA) in PBS with 0.05% Tween-20 to block any remaining binding sites in the wells. Incubate at room temperature for 30 to 60 min.

4.  Dump out the blocking solution as in step 3 and add 50 μl of the first antibody solution including preimmune serum, immune serum or Ig supernatant, and different concentrations of mouse Ig standard into appropriate wells according to the design in step 1. Allow to incubate at room temperature for 60 min.

5.  Carefully remove the first antibody solution from appropriate wells and immediately wash the plate by filling up each well with washing buffer followed by dumping out the buffer after 2 min. Repeat washing twice.

6.  According to the design in step 1, add 50 μl per well of diluted (1:2000–5000 in PBS with 1% BSA) second antibody to the appropriate wells. The second antibody is goat antimouse immunoglobulin-peroxidase-conjugate or goat antimouse immunoglobulin-alkaline phosphatase conjugate (Sigma Chemicals). Incubate the plate at room temperature for 60 min.

7.  Wash the plate as in step 5.

8.  As quickly as possible, add 50 μl per well of freshly prepared substrate solution to the appropriate wells according to the design in step 1. Allow the enzyme reaction to proceed for 15 min at room temperature in the dark. A color change will occur as the reaction takes place between the enzyme and its substrate. The color developed indicates a positive reaction.

9.  Stop the reaction by adding 50 μl per well of 12.5% (v/v) $H_2SO_4$.

10.  Place the plate in a cassette in an autoreader (e.g., Bio-Tek, E1310) and read the absorbance or the optical density at 410 or 490 nm according to the manufacturer's instructions. Draw a standard curve on a graph paper using the concentrations as the ordinate and the optical density or

absorbance as the abscissa. To determine the concentration of antibodies in the sample, find its density and draw a horizontal line crossing to the standard curve. Starting from the cross-point on the standard curve, draw a vertical line crossing to the ordinate at which point the concentration of mouse Ig standard is equal to the concentration of antibodies in the sample.

## Reagents Needed

### Phosphate-Buffered Saline (PBS)

$KH_2PO_4$ (anhydrous) (0.92 g)
$K_2HPO_4 \cdot 3H_2O$ (6.39 g)
NaCl (28.6 g)
Add $H_2O$ to a final volume of 3.5 l. The pH should be 7.2 to 7.4.
Autoclave.

### Washing Buffer or Blotto

5% (w/v) Nonfat powdered milk in PBS or 1.5% BSA in PBS with 0.05% (v/v) Tween-20 detergent

### Substrate Solution for Goat Antimouse Ig-Peroxidase-Conjugate

0.2% (w/v) o-Phenylenediamine (OPD)
0.03% (v/v) Hydrogen peroxide
In a buffer (pH 6.0) containing 17 mM citric acid and 65 mM sodium phosphate (dibasic)

### Substrate Solution for Goat Antimouse Ig-Alkaline Phosphatase-Conjugate

0.1% (w/v) p-Nitrophenyl phosphate (NPP)
10 mM $MgCl_2$
In 50 mM sodium carbonate buffer (pH 9.8 adjusted with sodium bicarbonate)

### Protocol D: Preparation of Peritoneal Exudate Cells (PECs) under Sterile Conditions

Peritoneal exudate cells (PECs) or endothelial cell growth supplement (ECGS) derived from bovine hypothalamus have been demonstrated to be necessary as the feeder cells for culture of hybridoma cells.

1. Thoroughly clean a laminar flow cabinet as follows:
   a. Turn on the air flow.
   b. Thoroughly clean the cabinet using 95% ethanol.
   c. Thoroughly clean the cabinet 10 min after step b using 70% ethanol.
   d. Thoroughly clean the cabinet 10 min after step c using 70% ethanol.
2. Sacrifice the nonimmunized mouse (the same strain as the immunized mouse) two days before cell fusion by cervical dislocation or $CO_2$ asphyxiation (by placing the mouse in a closed container with a few pieces of dry ice). Soak the dead mouse briefly in a beaker of 95% ethanol prior to placing it in the laminar flow cabinet.
3. Make a small incision in the skin covering the abdominal region and split the skin to expose the peritoneal cavity using sterile scissors and forceps.

4.  Inject 3 to 5 ml of serum-free medium (DMEM) into the peritoneal cavity using a disposable syringe with a 20- or 23-gauge needle and gently squeeze the abdomen for 2 to 3 min using the fingers.

5.  Slowly withdraw the mixed fluid containing the peritoneal exudate cells and place in a disposable Petri dish and dilute the cells to 10 ml with serum-free DMEM.

6.  Count the cells as follows:

    a.  Dilute 0.1 ml of the cell suspension to 1 ml in a microcentrifuge tube containing 0.9 ml of staining solution [0.1 ml of 2 to 10% (w/v) trypan blue in distilled water is mixed with 0.8 ml of serum-free medium (DMEM)].

    b.  Place a 0.4-mm dry coverslip on a dry hemaocytometer slide and place a drop of the diluted cell suspension on the groove at the edge of the slide. The drop should be drawn by capillary action onto the slide and it should form an even film to the edges without overflowing. This indicates the correct volume that is used for counting.

    c.  Focus on the lines under a microscope and the cells should be visible using a 10× objective lens.

    d.  Count the number of cells within the four corner areas.

    e.  Calculate the cell numbers using the following equation:

$$\text{Cells per milliliter} = \frac{\text{total cells counted}}{\text{number of corners counted}} \times 10^4 \times 10 \text{ (dilution factor)}$$

For example,

| Corner | Cells counted |
|--------|---------------|
| 1 | 80 |
| 2 | 100 |
| 3 | 120 |
| 4 | 100 |
| **Total** | **400** |

Cells per milliliter = 400/4 × $10^4$ × 10 (1:10 dilution) = 1 × $10^7$

Total cells of original cell suspension = 1 × $10^7$ × total volume. Usually one mouse can yield 1 × $10^6$ to 1 × $10^7$ cells.

7.  Dilute the cells to 4 × $10^5$ cells/ml with serum-free DMEM and add one drop per well (approximately 50 µl per well, 2 × $10^3$ to 2 × $10^4$ per well) of the cell suspension into a 96-well plate using a sterile 5-ml pipette and cover the plate.

8.  Place the plate containing the PECs in a cell culture incubator at 37°C with 5% $CO_2$. Allow cells to incubate for 2 days and check for possible contamination prior to their being used for hybridoma culture.

## Reagents Needed

### Serum-Free Medium (DMEM)

Liquid media are commercially available. For commercial powder, slowly empty one bottle of Dulbecco's modified Eagle medium (DMEM) powder (Sigma Chemicals) into a clean beaker containing 800 ml distilled, deionized water (dd.$H_2O$) and a stirring bar. After completely dissolving, add 3.7 g

$NaHCO_3$. Adjust the pH to 7.2 with 1 $N$ HCl. Add 10 ml of antibiotic-antimycotic (ABAM, usually contains penicillin or streptomycin) stored at $-20°C$ to the medium and add dd.$H_2O$ to a final volume of 1000 ml. In a sterile laminar flow cabinet, immediately filter the sterilized the medium into a sterile bottle using a disposable Nalgene filtration unit with 0.22-$\mu$m filter under vacuum. Cover the bottle tightly and store at 4°C until use. Prewarm the medium to approximately 37°C and thoroughly clean the outside surface with 70% ethanol before opening.

## Protocol E: Preparation of Spleen Cells and Myeloma Cells for Fusion

1.  Thoroughly clean the laminar flow hood as follows:
    a.  Turn on the air flow.
    b.  Thoroughly clean the cabinet using 95% ethanol.
    c.  Thoroughly clean the cabinet 10 min after step b using 70% ethanol.
    d.  Thoroughly clean the cabinet 10 min after step c using 70% ethanol.

2.  Sacrifice the immunized mouse 2 days before cell fusion by cervical dislocation or $CO_2$ asphyxiation (by placing the mouse in a closed container with a few pieces of dry ice). Soak the dead mouse briefly in a beaker of 95% ethanol prior to placing it on the sterile bench in the laminar flow cabinet.

3.  Lift a little skin of the abdominal region with one hand using sterile forceps and make a small incision in the skin with the other hand with sterile scissors. Tear the skin to expose the abdominal wall and make a small incision in the abdominal wall. Carefully remove the spleen (maroon color) using forceps and place it in a disposable Petri dish containing 10 ml of serum-free medium (DMEM). Carefully remove the connective tissue such as fatty tissue from the spleen by use of a bent needle and a pair of forceps.

4.  Inject 2 to 5 ml of serum-free medium (DMEM) into the spleen with one hand using a disposable syringe fitted with a 20- or 23-gauge needle and hold the spleen with the other hand using a pair of forceps. This causes the spleen to swell.

5.  Gently tease the tissue apart using sterile forceps and a disposable syringe needle. The spleen cells will be released into the dish.

6.  Remove the spleen cell clumps as much as possible and tilt the Petri dish so as to cause the cells to flow to one side of the dish. Carefully transfer the cells to a sterile 15 to 30-ml centrifuge tube with a V-shaped bottom (avoid transfer of cell clumps to the tube).

7.  Centrifuge at 600 rpm for 5 to 10 min in a IEC Centra-4B centrifuge or its equivalent at room temperature. While the cells are being spun down, prepare working Geys' hemolytic medium that can reduce any damage to the lympoid cells. The working Geys' hemolytic medium consists of: 8 ml of Geys' solution A, 29 ml dd.$H_2O$, and 2 ml of Geys' solution B. The pH is adjusted to 7.2 using pH indicator and sterile-filtered into a sterile container. The medium should be used within 30 min (20 ml per spleen).

8.  Carefully remove the supernatant from the cell pellet as much as possible and with the fingers, flick the bottom of the tube in order to loosen the cell pellet. Add 4 ml of freshly prepared Geys' hemolytic medium to the cells and gently suspend the cells using a sterile Pasteur pipette followed by adding 16 ml of the Geys' hemolytic medium. Mix well and allow to stand for exactly 5 min at room temperature.

9.  While the spleen cells are being hemolyzed, collect myeloma cells, grown to mid-log phase in T-flasks (mouse myeloma cell lines are commercially available), into a sterile centrifuge tube with a V-shaped bottom by scraping the cells with a sterile rubber policeman.

10. Centrifuge both spleen cells and myeloma cells as in step 7 and resuspend the cells in 10 ml of serum-free DMEM at room temperature. Repeat centrifugation for myeloma cells and resuspend in 10 ml of serum-free DMEM.

## Reagents Needed

### Serum-Free Medium (DMEM) or Incomplete DMEM

Liquid media are commercially available. For commercial powder preparation, slowly empty one bottle of Dulbecco's modified Eagle medium (DMEM) powder (Sigma Chemicals) into a clean beaker containing 800 ml distilled, deionized water (dd.$H_2O$) and a stirring bar. After completely dissolving, add 3.7 g $NaHCO_3$. Adjust the pH to 7.2 with 1 $N$ HCl. Add 10 ml of antibiotic-antimycotic (ABAM, usually contains penicillin or streptomycin) stored at $-20°C$ to the medium and add dd.$H_2O$ to a final volume of 1000 ml. In a sterile laminar flow cabinet, immediately filter-sterilize the medium into a sterile bottle using a disposable Nalgene filtration unit with 0.22-$\mu$m filter under vacuum. Cover the bottle tightly and store at 4°C until use. Prewarm the medium to approximately 37°C and thoroughly clean the outside surface with 70% ethanol before opening.

### Complete Medium (DMEM)

Add 5 to 20% (v/v) sterile, commercial fetal calf serum (FCS) to incomplete DMEM depending on the percentage of FCS required for the particular medium. Thoroughly clean the outside of the FCS bottle prior to opening with 70% ethanol in a laminar flow cabinet under sterile conditions. Warm the complete DMEM to 37°C before use.

### Geys' Solution A (500 ml)

$NH_4Cl$ (17.5 g)
KCl (0.93 g)
$Na_2HPO_4 \cdot 12H_2O$ (0.75 g)
  (disodium hydrogen orthophosphate)
$KH_2PO_4$ (0.06 g)
  (potassium dihydrogen orthophosphate)
D-Glucose (2.5 g)
Phenol red (0.03 g)
Gelatine (12.5 g)

Dissolve well after each addition in 400 ml of dd.$H_2O$ and add dd.$H_2O$ to a final volume of 500 ml. Aliquot into 20 to 50 ml per vial and autoclave at 15 psi for 15 min. Store at room temperature for up to 1 year.

### Geys' Solution B (200 ml)

$MgCl_2 \cdot 6H_2O$ (8.4 g)
$MgSO_4 \cdot 7H_2O$ (2.8 g)
$CaCl_2$ (6.8 g)

Dissolve well after each addition in 150 ml dd.$H_2O$ and add dd.$H_2O$ to a final volume of 200 ml. Aliquot 10 ml per vial and autoclave at 10 psi for 10 min. Store at room temperature for up to 1 year.

## Protocol F: Fusion of Spleen Cells and Myeloma Cells under Sterile Conditions

1. Count the cells as described in Protocol D. The viability should be >80% and only a few red cells will be visible.

2. Using all of the spleen cells, make a mixture of spleen and myeloma cells (the ratio of spleen cells to myeloma cells is 5:1 or 10:1) in a sterile centrifuge tube with a 'V' bottom.

3. Prepare aliquots of 50% (w/v) polyethylene glycol (PEG, 6000) in serum-free DMEM medium and place in a 37°C water bath.

4. Centrifuge the spleen/myeloma cell mixture at 1000 rpm at room temperature for 5 min and completely remove the supernatant.

5. Tap the bottom of the tube to loosen the cell pellet and slowly add 1 ml of 50% (w/v) PEG in serum-free medium to cells at a density of $10^7$ to $10^8$ cells/ml. The addition should be made dropwise (1 drop) with constant swirling. Rapidly mix the cells in the viscous solution by pipetting up and down. After 1 to 2 min, start the dilution in the next step.

6. Add 3 ml (three times the volume of PEG solution) of the warmed, serum-free DMEM to the mixture dropwise (1 drop/1.5 s) over a 1.5-min period with gentle swirling in a 37°C water bath.

7. Add 12 ml of warmed serum-free DMEM as in step 6 over 3 min.

8. Centrifuge as in step 4, decant the supernatant, and resuspend the "fusion" cells in 10 ml of prewarmed complete DMEM containing 20% (v/v) FCS. Count the cells and dilute them to a final density of $10^5$ to $10^6$ cells/ml in complete DMEM.

9. Add 50 μl per well of the diluted "fusion" cell suspension into disposable 96-well plate(s). Add one drop per well (approximately 50 μl per well) of the feeder cell suspension (PECs) prepared in Procedure D in 96-well plate(s). Cover the plate(s) and incubate in an incubator (e.g., Queue Cell Culture Incubator) at 37°C with 5% $CO_2$ for 10 to 24 h prior to the next procedure.

## Protocol G: Selection and Propagation of Hybridoma Cells

The spleen cells contain genes coding for antibodies against the antigen of interest. The myeloma cells contain genes that are able to multiply indefinitely but cannot code for their own immunoglobulin (Ig). However, they can secrete Ig coded by spleen cell genes after fusion. The myeloma cell line is usually hypoxanthine, aminopterin, and thymidine (HAT) selective, in which aminopterin blocks the main biosynthetic pathway for nucleic acids. The myeloma cells are also usually mutants that do not contain the gene coding for hypoxanthine guanine phosphoribosyl transferase (HGRPT) that is required for the salvage pathway selection. The basic principle of generating hybridized cells is that, after fusion, the mutant parent myeloma cells can proliferate in the absence of aminopterin, but do not survive in HAT medium. The normal parent spleen cells can survive in the HAT medium via the salvage pathway, but they die out due to their sensitivity to ouabain or amphotericin B methylester in the HAT medium. Only the fusion products or hybrid cells, which contain the HGPRT gene from the parent spleen cells and the ouabin resistance gene from the parent myeloma cells, can multiply indefinitely and secrete antibodies against the antigen of interest.

1. Add 50 μl per well of prewarmed (37°C) 2X HAT in complete DMEM containing 20% FCS to the growing cells in a sterile laminar flow cabinet and return the plate to the incubator.

2. Feed the cells by adding 50 μl per well prewarmed 1X HAT in complete DMEM containing 20% (v/v) FCS once every 3 days.

3. Check for possible contamination and monitor the formation of colonies once every 2 to 4 days under a microscope set near the incubator, and quickly return the plate to the incubator.

4. When the hybridoma colonies (usually a heterogeneous mixture of hybridomas) are visible, start isolation and propagation of monoclonal hybridomas by the limiting dilution method.

    a. Carefully transfer the cells from the well to a sterile tube and add 4 ml of prewarmed 1X HAT in complete DMEM containing 20% (v/v) FCS supplemented with 5% ECGS.

    b. Count the cells as described previously.

    c. Dilute the cells to 8, 4, 2, and 1 cell(s) in 50 μl in a series of tubes using prewarmed 1X HAT in complete DMEM containing 20% (v/v) FCS supplemented with 10% PECs or 10% ECGS.

    d. Add 50 μl per well containing approximately 1 cell to the 96-well plate and retain in the incubator for 10 to 14 days or until colonies become visible.

## Reagents Needed

### 100X HAT Medium (100 ml)

> Hypoxanthine (136 mg)
> Aminopterin (1.9 mg)
> Thymidine (38.8 mg)
> Dissolve well after each addition in 100 ml distilled water and autoclave. Aliquot and store at −20°C.

### 2X HAT Medium

> Dilute 100X HAT medium in complete DMEM containing 20% (v/v) FCS and 10% ECGS.

### 100X HT Medium

> The same as 100X HAT medium except no aminopterin is included.

## Protocol H: Screening of Hybridoma Supernatant by ELISA and Harvesting of Monoclonal Antibodies[5]

1. When the supernatant in the well containing healthy and uncontaminated hybridomas turns from bright red to orange or yellow, it indicates that the supernatant is acid due to the respiration of the hybridoma cells. The supernatant should be carefully removed from the well and prewarmed, fresh 1X HAT in complete DMEM containing 20% FCS and 10% PECs should be added to the well, which is then returned to the incubator.

2. Carry out ELISA to identify positive supernatants as described in Protocol C. A positive supernatant from a specific well contains antibodies against the antigen that was used for the immunization of the mouse.

3. Reclone the hybridoma cells secreting positive supernatant using the limiting dilution method, as the first single-limiting dilution usually does not ensure monoclonality. Normally, by two or three successive clonings, the monoclonal antibody-producing cells can be obtained. The medium should now be changed to HT in complete DMEM containing 10% (v/v) FCS for the subclonings.

4.  Carry out ELISA to verify the positive supernatant as in step 2.

5.  The healthy and uncontaminated cell lines that produce monoclonal antibodies should be transferred to 24-well plates or T-25 or T-75 disposable culture flasks containing 30 to 60 ml of complete DMEM with 10% FCS for large-quantity production of monoclonal antibodies.

*Note:*    *If one uses T-flasks, the flasks should not be overfilled with medium, and they should be capped tightly.*

6.  When the supernatant changes from red to yellow, harvest the supernatant by transferring the culture to a centrifuge tube and centrifuging at 1000 rpm for 10 min at room temperature. In a sterile laminar flow hood, transfer the supernatant to a fresh tube and adjust the pH to 7.2. After adding sodium azide to a final concentration of 0.01 m$M$, the supernatant can be stored at –20°C until use. The cell pellet can be resuspended in the complete DMEM for further culture or for being frozen.

## Protocol I: Isotyping of Monoclonal Antibodies

Antibodies or immunoglobulin (Ig) contain an enzymelike specificity for certain structural epitopes of the antigen. Antibodies can be classified into five subclasses: IgA, IgD, IgE, IgG, and IgM. IgG can be divided into IgG1, IgG2a, IgG2b, and IgG3. An antibody molecule contains two heavy chains and two light chains, which are held together by disulfide bounds. The moloclonal antibodies obtained can be isotyped as follows:

1.  Dilute the antigen sample to 50 μg/ml in PBS (pH 7.2). According the above design, coat appropriate wells of a disposable 96-well polystyrene ELISA plate by adding the antigen to the wells (50 μl per well). Cover the plate and incubate at room temperature for 2 h or at 4°C overnight.

2.  Decant the antigen solution from wells by dumping the plate upside-down on a paper towel two to three times. Fill each well (100 to 200 μl per well) with 1 to 3% (w/v) bovine serum albumin (BSA) in PBS with 0.05% Tween-20 to block any remaining binding sites in the wells. Incubate at room temperature for 30 to 60 min.

3.  Dump out the blocking solution as in step 2 and add 50 μl per well of monoclonal antibody supernatant as with the first antibodies. Allow to incubate at room temperature for 60 min.

4.  Carefully remove the first antibody solution from appropriate wells and immediately wash the plate by filling up each well with washing buffer followed by dumping out the buffer after 2 min. Repeat the washing twice.

5.  Add 50 μl per well of appropriate subclass-specific rabbit anti-mouse Ig such as IgG, IgG2a, IgG2b, IgG3, IgA, IgD, and IgM to the appropriate wells and incubate at 37°C for 60 min.

6.  Wash the plate as in step 4.

7.  Add 50 μl per well of horseradish peroxidase-labeled, affinity-purified goat anti-rabbit antibody diluted at 1:50 with washing buffer to each well. Allow to incubate at 37°C for 60 min.

8.  Wash the plate as step 4.

9.  Add 100 μl of freshly prepared peroxidase substrate solution to each well. The solution contains 2,2-azino-di(3-ethylbenzthiazoline sulfonic acid) (ABTS) that is diluted at 1:50 in 100 m$M$ citrate phosphate buffer (pH 4.2) and 0.03% (v/v) hydrogen peroxide. Incubate at room temperature for 30 min. A positive reaction should be bluish green in color.

10. Place the plate in a cassette in an autoreader (e.g., Bio-Tek, E1310) for quantitative measurement and read the absorbance or the optical density at wavelengths of 405 to 415 nm according to the manufacturer's instructions.

## Protocol J: Verification of Antibodies by Western Blot Analysis

In addition to the use of ELISA to test for positive reactions of the antibodies obtained, the specificity of the antibody produced should also be verified by Western blotting using a protein mixture containing the antigen for immunization. Putative monoclonal antibodies should detect the expected protein of known size in the protein mixture. The detailed procedures are described under the Western blotting protocol in Chapter 5.

## Protocol K: Purification of Monoclonal Antibodies Using Affinity Chromatography[4]

1.  Prepare an appropriate size of column by equilibrating Sephadex or equivalent beds conjugated with goat anti-mouse IgG or IgM antibody with a buffer (pH 7.2) containing 10 m$M$ sodium phosphate and 100 to 200 m$M$ NaCl. Allow the column to stand in a cold room for 1 h and drain away the liquid.

2.  Load the antibody supernatant onto the column, collect the eluate, and reload onto the column two or three times in order to obtain optimal binding between the column adsorbant matrix and antibody.

3.  Wash the column twice with 10 volumes of the total column volume using the equilibrating buffer.

4.  Elute the bound antibodies with elution buffer (pH 2.3) containing 100 m$M$ glycine-HCl and 150 m$M$ NaCl.

5.  Neutralize the acidic eluate with 1 $M$ Tris/150 m$M$ NaCl to pH 7.2 to 7.4. The purified antibody can be directly stored at –20°C until use. If it is too dilute, the antibody can be precipitated with 50% (w/v) solid ammonium sulfate, centrifugation, resuspension, and dialysis. Alternatively, the eluate can be concentrated by inserting an appropriate concentration tube and centrifuging at 2000 to 3000 × $g$ for an appropriate period of time.

## Protocol L: Freezing and Thawing Cell Lines under Sterile Conditions

Once the monoclonal antibody-producing cell lines are obtained, it is necessary to freeze some of the cells for long-term storage. The frozen cells can be thawed for future culture in order to prepare fresh antibody.

### 1.   Freezing Cell Lines

1.  Transfer healthy, uncontaminated, and mid-log grown cells to a centrifuge tube with a V-shaped bottom by scraping the cells with a sterile rubber policeman.

2.  Centrifuge at 1000 rpm for 5 min at room temperature and remove the supernatant. Tap the bottom of the tube to loosen the cell pellet and resuspend the cells in 1 ml of HB 101-DMEM (serum-free).

3.  Count the cells as described previously and dilute the cells to $10^6$ cells/ml with HB 101-DMEM.

4.  Add an equal volume of cold (4°C) 20% (v/v) dimethyl sulfoxide (DMSO) in HB 101-DMEM to the cell suspension with gentle swirling. Both HB 101-DMEM and 10% DMSO will permit the cells to survive during freezing.

5.  Aliquot the cell suspension (1 to 1.5 ml per vial) into 2-ml cryogenic vials set in an abucket of ice in a sterile laminar flow cabinet. Tightly cap the vials, allow the vials to set on ice for 10 to 15 min, and place the vials inside a styrofoam-insulated cooler. Place the cooler at –80°C for 2 to 3 days to slowly freeze the cells.

*Notes:*   *The vials should be clearly labeled with the cell line name. The monoclonal antibody-producing cell line can be named progressively according to the order of subcloning procedure for specific cell lines. For example, if a primary hybridoma is cultured in the A4-well of the 96-well plate, then transferred to the B2-well that produces a positive supernatant, then sucloned into the B4-well of a 24-well plate and further transferred into T-flask no. 5, which produces monoclonal antibodies, the name for this cell line can be XA4B2B4T5, where X refers to the antigen or protein name. Alternatively, the cell line can be named according to antigen name and be given a number such as X2 or X101.*

6.   Transfer the frozen cells to a liquid nitrogen tank for extended storage. Check and keep the liquid nitrogen above the safety level.

## 2.        Thawing Frozen Cells

1.   Remove the cell line of interest from the liquid nitrogen tank and quickly immerse the vial in a 37°C water bath. Gently swirl the vial and stop thawing when the last ice crystals disappear. The cell suspension should still be cold.

2.   Clean the outside of the vial with 75% ethanol prior to opening and then transfer the cell suspension into a 50-ml centrifuge tube containing 20 ml of complete DMEM with 20% FCS. Mix well.

3.   Centrifuge at 1000 rpm for 10 min at room temperature and remove the supernatant. Tap the bottom of the tube to loosen the cell pellet and resuspend the cells in 1 to 2 ml complete DMEM containing 20% FCS.

4.   Transfer the cell suspension to one or two wells of a culture plate and culture in an incubator at 37°C with 5% $CO_2$ as described previously.

# III.    Preparation of Polyclonal Antibodies

## *Procedures*

1:   Purify the antigen for immunization.

The general procedures are described in Chapter 3. It is recommended that the protein of interest as an antigen be as pure as possible. Pure antigen can make the procedures for screening and selection of monoclonal colonies of cells much easier. For glycoproteins, deglycosylation should be carried out because animals can raise antibodies against the sugar residues, which will reduce the specificity of the monoclonal antibody and may result in potential cross-reactions between the antibody and a variety of proteins.

2.   Carry out *in vivo* immunization of female rabbits with the purified antigen.

   a.   Resuspend the purified antigen in PBS in a sterile microcentrifuge tube to a concentration of 1 to 5 µg/µl, and use 50 to 200 µg per injection depending on the particular antigen to be used.

   b.   Combine 100 µl (100 to 150 µg) of the antigen sample with an equal volume of complete Freund's adjuvant to a final volume of 200 µl for the first injection per rabbit. Mix thoroughly to obtain an emulsion using a sterile syringe or a pipette. Slowly take up the emulsified mixture in a 1-ml disposable syringe fitted with an 18- or 20-gauge needle and remove any air bubbles.

   c.   Take a blood sample (0.5 to 1.0 ml) as preimmune serum from the ear vein before the first injection. Allow the blood to clot by placing the sample at 4°C for approximately 30 min, and centrifuge at 7000 × g for 10 min. Transfer the serum to a fresh tube, dilute it at 1:100–500 with PBS and store at −20°C until use.

    d. Place the rabbit (New Zealand White) in a cage, and briefly clean the areas on the back or the butt area of the rabbit with 70% ethanol. Slowly inject the antigen mixture into the muscle in two to four areas. Return the rabbit to the cage.

    e. Boost the rabbit by injecting the same amount of antigen as for the first injection. The antigen is mixed with incomplete Freund's adjuvant to a final volume of 200 µl two to four times at 21-day intervals.

3.   Collect blood from the immunized rabbit.

Two or four days after the third or fourth injection, take blood from the ear vein up to 20 to 30 ml per rabbit. This can be done by placing the immunized rabbit in a cage, removing the hair from the ear vein area, and cleaning the area with 95% ethanol. To achieve a good flow of blood, spray the area with some xylene and make a cut in the ear vein using a disposable knife. Lower the ear and collect the blood in a sterile tube. Stop the bleeding with a dry cotton swab by pressing it on the cut area using the fingers until the bleeding stops. Return the rabbit to the cage.

4.   Harvest the serum containing polyclonal antibodies against the antigen.

Place the blood sample at 4°C for 30 to 90 min and centrifuge the clotted sample at $8000 \times g$ for 10 min at 4°C. Carefully transfer the serum containing polyclonal antibodies to a fresh tube, aliquot, and store at –20°C until use.

5.   Purify the polyclonal antibodies.[3]

The serum may be purified with an appropriate Sephadex 4B or agarose column conjugated with purified antigen. Load the serum onto the column, collect the eluate, and reload the column several times. The bound antigen and antibody complex can be eluted by using the appropriate buffer. Store the purified antibodies at –20°C until use.

6.   Check the quality of the antibodies by ELISA as described previously.

7.   Verify the quality of the antibodies by Western blot analysis as described previously.

# References

1. **Harlow, E. and Lane, D.,** *Antibodies: A Laboratory Manual,* Cold Spring Harbor Press, New York, 1988.

2. **Zola, H.,** *Monoclonal Antibodies: A Manual of Techniques,* CRC Press, Inc., Boca Raton, Florida, 1984.

3. **Diano, M., Le Bivic, A., and Hirn, M.,** A method for the production of highly specific polyclonal antibodies, *Anal. Biochem.,* 166, 224, 1987.

4. **Olmstead, J. B.,** Affinity purification of antibodies from diazotized paper blots of heterogenous protein samples, *J. Biol. Chem.,* 256, 11955, 1981.

5. **Gaastra, W.,** Enzyme-linked immunosorbent assay (ELISA), in *Methods in Molecular Biology, Vol. 1,* Walker, J. M., Ed., Humana Press, Clifton, NJ, 1984, chap. 38.

# Index

## C

# T